Control Systems: Advances in Technology

Control Systems: Advances in Technology

Editor: Scott Crosby

NY RESEARCH PRESS
New York

Published by NY Research Press
118-35 Queens Blvd., Suite 400,
Forest Hills, NY 11375, USA
www.nyresearchpress.com

Control Systems: Advances in Technology
Edited by Scott Crosby

International Standard Book Number: 978-1-63238-593-2 (Hardback)

Cataloging-in-Publication Data

Control systems : advances in technology / edited by Scott Crosby.
p. cm.
Includes bibliographical references and index.
ISBN 978-1-63238-593-2
1. Automatic control. 2. Control theory. 3. Technology. I. Crosby, Scott.
TJ213 .C66 2018
629.8--dc23

Contents

Preface

The behavior of varied devices is managed and regulated by control systems. Control systems are of two types, open loop control systems and closed loop control systems. Programmable logic controllers, PID controllers, microcontrollers are some common types of control systems. While understanding the long-term perspectives of the topics, the book makes an effort in highlighting their impact as a modern tool for the growth of the discipline. The book studies, analyses and upholds the pillars of control systems and its utmost significance in modern times. Students, researchers, experts and all associated with control systems will benefit alike from this book.

This book unites the global concepts and researches in an organized manner for a comprehensive understanding of the subject. It is a ripe text for all researchers, students, scientists or anyone else who is interested in acquiring a better knowledge of this dynamic field.

I extend my sincere thanks to the contributors for such eloquent research chapters. Finally, I thank my family for being a source of support and help.

Editor

1

A polynomial algorithm for diagnosability of fair discrete event systems

Pradeep Kumar Biswal and Santosh Biswas*

Department of Computer Science and Engineering, Indian Institute of Technology, Guwahati, India

The discrete event system (DES) has been used for failure detection and diagnosis (FDD) of a wide range of systems. The major reason for resorting to the DES framework is the simplicity in modelling and the low complexity of the FDD algorithms. Pure DES models cannot directly capture systems with continuous dynamics. However, the DES paradigm surmounts the problem by partitioning the continuous state space and capturing each subspace as a discrete state. Conventional methods for FDD consist in constructing a diagnoser, the complexity of which is exponential in the number of system states. In the case of non-diagnosability, the diagnoser needs to be reconstructed after taking suitable measures such as increase in measurements, etc. The conventional schemes have two issues, namely, exponential complexity of diagnoser and erroneous diagnosability conclusions for fair systems. Several works address the first issue where checking diagnosability does not involve a diagnoser and has polynomial time complexity. Once a fault is diagnosable, a diagnoser is constructed for concurrent system monitoring. Regarding the second issue, the abstraction employed in DES modelling may obliterate the fairness property for systems having continuous dynamics, leading to erroneous inferences. Works addressing this issue have augmented fairness to the model and the diagnoser, and new diagnosability conditions have been proposed for fair systems. However, all the Fair DES diagnosability frameworks are based on diagnoser and hence have exponential complexity. In this paper, we purpose a new DES diagnosability framework that suffices for fair systems but at the same time has polynomial complexity.

Keywords: fair discrete event systems; failure diagnosis; diagnosability; diagnoser

1. Introduction

Diagnosis of failures in large, complex systems is a crucial and involved task. The increasingly stringent requirements on performance and reliability of complex systems have necessitated the development of systematic methods for timely and accurate diagnosis of system failures. As a result, the failure diagnosis problem has been an area of recent research interest in the contexts of centralized systems (Bavishi and Chong, 1994; Lafortune, Teneketzis, Sampath, Sengupta, and Sinnamohideen, 2001; Lin and Wonham, 1988), decentralized systems (Debouk, Lafortune, and Teneketzis, 1998), timed systems (Lamperti and Zanella, 1999; Zad, Kwong, and Wonham, 2005) and un-timed systems (Sampath, Sengupta, Lafortune, Sinnamohideen, and Teneketzis, 1995, 1996; Thorsley and Teneketzis, 2005; Zad, Kwong, and Wonham, 2003). In these works, the systems have been abstracted using discrete event system (DES) models. A DES model is characterized by a discrete state space and event-driven dynamics. A state-based DES model that captures both normal and failure conditions of a system comprises two categories of states, namely, the normal states and the failure states. Any failure state differs from a normal state in the value of a state variable (status variable) which cannot be measured.

A *DES diagnosability analysis* procedure takes as input any DES model with the measurable subset of the state variables and determines whether it is diagnosable (Sampath et al., 1995, 1996; Zad et al., 2003; Thorsley and Teneketzis, 2005). Typically, such a procedure consists in constructing a *diagnoser*, which is a kind of state estimator of the model. Then, it is checked whether the diagnoser can ascertain that the model is in a failure state within finite time of the concurrence of the failure. Some of such methods are discussed at length in Sampath et al. (1995) and Zad et al. (2003). The DES diagnosability analysis procedures, referred to above, suffer from two drawbacks.

First, in the worst case, the number of diagnoser states may be exponential in the number of system states. Hence, if a failure turns out to be non-diagnosable, then the diagnoser needs to be reconstructed after taking proper steps such as increasing the measurements. Thus, it is preferable that diagnosability be analysed using structures that are computationally simpler than a diagnoser. Construction of the diagnoser can be taken up only when the system is found to be diagnosable. Secondly, these procedures turn out to be inadequate for many systems which are fair. The transitions in the DES models corresponding to many real-life systems, such as those with continuous dynamics for

*Corresponding author. Email: santoshbiswas402@yahoo.com

example, are *fair*. A transition is fair if it occurs infinitely often in all traces that visit the state from where the transition emanates, infinitely often. Many such systems, in which occurrences of a failure manifest themselves in finite time, are adjudged to be non-diagnosable by the diagnosability procedures mentioned above because fairness of transitions is not considered. In this paper, an example illustrates this fact.

To handle these two issues, several works have been reported in the literature. To address the first issue, procedures for diagnosability analysis have been reported in the literature for centralized systems (Jiang, Huang, Chandra, and Kumar, 2001; Tae-Yoo and Lafortune, 2002), decentralized systems (Moreira, Jesus, and Basilio, 2011) and fuzzy systems (Liu, 2014), which do not require the construction of a diagnoser. Instead, they construct a two-dimensional product model and check for a suitable condition over that model. The procedures accordingly have complexity polynomial in the number of system states.

The second issue has been handled in Thorsley and Teneketzis (2005) and Biswas, Sarkar, Mukhopadhyay, and Patra (2010), where the basic DES model was augmented with the concept of fairness. The classical DES frameworks declare a fault as non-diagnosable if there is a cycle through failure states which cannot be distinguished from a similar cycle through normal states. Thorsley and Teneketzis (2005) showed that the mere presence of such a cycle through faulty states to declare the failure non-diagnosable does not hold for many system, for example, once having continuous dynamics. Thorsley et al. proposed a new DES paradigm where the classical DES model was augmented with probabilities of transitions. In the stochastic framework, failure is considered diagnosed when it is found that probability of the system traversing though failure states is higher than a threshold. Latter, Biswas et al. (2010) have proposed another DES paradigm to handle similar systems, where fairness was augmented to the classical DES model. The claim was, the abstraction employed in obtaining DES models from many systems, for example, those having continuous dynamics often obliterates the fairness property. The diagnosability condition in this case checks if there exists equivalent strongly connected components (SCCs) involving failure states and normal states. Comparing these works, it may be said that a transition having positive probability can be assumed to have fairness. In fact, the equivalence of the stochastic DES scheme (Thorsley and Teneketzis, 2005) and fair DES (FDES) scheme has been established in Biswas (2013). However, both these schemes which address fairness are based on diagnoser and hence exponential in the number of system states. To the best of our knowledge, no work has been reported for failure detection in the DES framework that suffices for fair systems and at the same time has polynomial time complexity.

The present paper aims at devising a polynomial time diagnosability testing procedure for FDES models. The major contributions of this paper are as follows:

- Identifying the issues in failure detection and diagnosis (FDD)-based DES frameworks having polynomial time complexity (Jiang et al., 2001; Tae-Yoo and Lafortune, 2002) when applied for fair systems. A brief discussion regarding these schemes and complexity analysis has been given. Following that, how wrong diagnosability inferences can be given if these techniques are applied on *fair systems* has been identified and demonstrated using a simple example.
- FDD schemes in DES frameworks have been enhanced to handle fair systems (Biswas et al., 2010; Thorsley and Teneketzis, 2005). A brief discussion is given in this paper regarding these schemes and how they handle fairness is highlighted. However, it is also demonstrated that these schemes involve exponential complexity with respect to the system states.
- To address the above situation, we developed an FDD framework for FDES models but involving only polynomial time complexity.
  - An algorithm has been proposed that takes in an FDES model and generates a strict composition with itself after eliminating the unmeasurable transition sequences.
  - A diagnosability condition has been proposed in the model obtained after composition, which ascertains if an FDES model is diagnosable under a given measurement limitation.
  - The correctness and completeness of the condition have been proved formally.
  - The complexity of the scheme, that is, generating the composition and checking the diagnosability condition have been analysed and shown to be polynomial with respect to the number of system states.
  - The entire theory has been illustrated using an example.

The paper is organized as follows. Section 2 presents the DES model, failure modelling, a formal definition of (DES) diagnosability and the polynomial time diagnosability analysis algorithm reported in Jiang et al. (2001) and Tae-Yoo and Lafortune (2002). In Section 3, an example has been used to illustrate that the DES diagnosability condition checked by this algorithm fails for many practical systems with continuous dynamics. In this section, we also present in brief the traditional diagnoser-based algorithm (Biswas et al., 2010; Thorsley and Teneketzis, 2005) for handling FDD of FDES models. We also discuss the computational complexity of the algorithm and highlight that it is exponential in number of system states.

FDES models are introduced next in Section 4. A new condition for FDES diagnosability analysis is introduced in the same section incorporating the property of fairness and the necessity and sufficiency of the condition are formally established. Then, we provide the computational complexity of checking this FDES diagnosability condition. The paper is concluded in Section 5.

2. DES models

A *DES model* G is defined as

$$G = \langle V, X, \Im, X_0 \rangle, \tag{1}$$

where $V = \{v_1, v_2, \ldots, v_n\}$ is a finite set of discrete variables assuming values from some finite sets, called the domains of the variables, X is a finite set of states, $\Im$ is a finite set of transitions and $X_0 \subseteq X$ is the set of initial states. A state x is a mapping of each variable to one of the elements of the domain of the variable. A *transition* $\tau \in \Im$ from a state x to another state x^+ is an ordered pair $\langle x, x^+ \rangle$, where x is denoted as initial(τ) and x^+ is denoted as final(τ). We assume that any state of G is reachable from some initial state.

A *trace* of a DES model G is an infinite sequence of transitions of G and denoted as $s \triangleq \langle \tau_1, \tau_2, \ldots, \rangle$, where initial($\tau_1$) is an initial state in X_0 and the consecution property holds, that is, initial(τ_{i+1}) = final(τ_i), for $i \geq 1$. A finite prefix of a trace is refereed to as a 'finite trace'. Henceforth, we assume the consecution property for any 'sequence of transitions'. For any trace $s = \langle \tau_1, \tau_2, \ldots, \rangle$, initial($s$) = initial($\tau_1$) and for a finite prefix $s = \langle \tau_1, \tau_2, \ldots, \tau_f \rangle$, final($\tau_f$) = final($s$). The set of all traces of G and their finite prefixes are the language of G, denoted as $L(G)$. The set $L_f(G)$ denotes the subset of $L(G)$ comprising the finite prefixes of the members of $L(G)$. Naturally, $L(G) - L_f(G)$ is a subset of $\Im^w$, where $\Im^w$ is the set of all infinite sequences of $\Im$; $L_f(G)$ is a subset of $\Im^*$, the Kleene closure of $\Im$. The post language of G after a finite prefix s of a trace, denoted as $L(G)/s$, is defined as

$$\frac{L(G)}{s} = \left\{ t \in \Im^w | \text{ st } \in L(G) \right\}. \tag{2}$$

$L_f(G)/s \subset L(G)/s$ comprises finite prefixes of the traces of $L(G)/s$.

2.1. Models with measurement limitations

The set of variables are partitioned into two disjoint subsets, V_m and V_u, of *measurable* and *unmeasurable* variables, respectively. Given such a partition, the transitions are partitioned into two sets, $\Im_m$ and $\Im_u$, of *measurable* and *unmeasurable* transitions, respectively, as follows.

DEFINITION 1 *Measurable Transitions: A transition $\tau = \langle x, x^+ \rangle$ is said to be measurable if $x|_{V_m} \neq x^+|_{V_m}$, where $x|_{V_m}$ is the restriction of the function (defined by the state) x to V_m. A transition which is not measurable is an unmeasurable transition.*

DEFINITION 2 *Two states x and y are said to be (measurement) equivalent, denoted as xEy, if $x|_{V_m} = y|_{V_m}$.*

DEFINITION 3 *Two measurable transitions $\tau_1 = \langle x_1, x_1^+ \rangle$ and $\tau_2 = \langle x_2, x_2^+ \rangle$ are equivalent, denoted as $\tau_1 E \tau_2$, if $x_1 E x_2$ and $x_1^+ E x_2^+$.*

DEFINITION 4 *A projection operator $P : \Im^* \to \Im_m^*$ can now be defined in the following manner:*

$$\begin{aligned} P(\epsilon) &= \epsilon, \quad \text{null string,} \\ P(\tau) &= \tau \text{ if } \tau \in \Im_m, \\ P(\tau) &= \epsilon \text{ if } \tau \in \Im_u, \\ P(s\tau) &= P(s)P(\tau) \quad \text{where } s \in L_f(G), \tau \in \Im. \end{aligned} \tag{3}$$

The function P erases the unmeasurable transitions from the argument finite trace. $P(s)$ is termed as the *measurable finite trace* corresponding to the finite trace s.

DEFINITION 5 *Two finite traces s and s' are measurement equivalent if $P(s) = \langle \tau_1, \tau_2, \ldots, \tau_n \rangle$, $P(s') = \langle \tau_1', \tau_2', \ldots, \tau_n' \rangle$ and $\tau_i E \tau_i'$, $1 \leq i \leq n$.*

We use the symbol E to denote measurement equivalence of finite traces as well as that of transitions, with slight abuse of notation. The equivalence of finite traces s and s' implies that if measurable transitions are extracted from s and s' by the use of the operator P, then all the transitions are measurement equivalent.

The inverse projection operator $P^{-1} : \Im_m^* \to 2^{\Im^*}$ is defined as

$$P^{-1}(s) = \{s' \in L_f(G) | sEs'\}. \tag{4}$$

Thus, $P^{-1}(s)$ includes all possible sequences of transitions that are equivalent to the finite trace s. The projection function P, the inverse function P^{-1} and the measurement equivalence E of finite traces can be extended to traces $\in \Im^w$, in a natural way.

2.2. Failure diagnosis

Failure modelling: Each state x is assigned a failure label by an unmeasurable status variable $C \in V$ with its domain $= \{N\} \cup 2^{\{F_1, F_2, \ldots, F_p\}}$, where F_i, $1 \leq i \leq p$, denotes *permanent* failure status and N denotes normal status. For a normal state x_N, $x_N(C) = \{N\}$. The set of all normal states is denoted as X_N. Similarly, for a failure state (i.e. an F_i-state) x_{F_i}, $F_i \in x_{F_i}(C)$ and for a state $x_{F_i F_j}$ having simultaneous failures F_i and F_j, $F_i, F_j \in x_{F_i F_j}(C)$. The set of all states x s.t. $F_i \in x(C)$ is denoted as X_{F_i}. In the sequel, a G-transition $\langle x, x^+\rangle$ is called a *normal* (F_i) *G-transition* if $x, x^+ \in X_N(X_{F_i})$. A transition $\langle x, x^+\rangle$, where $x(C) \neq x^+(C)$, is called a *failure* transition indicating the first occurrence of some failure in $x^+(C) - x(C)$.

Since failures are assumed to be *permanent*, there is no transition from any x_{F_i} to any x_N or any $x_{F_iF_j}$ to any x_{F_i}. Also, for a transition $\langle x, x^+\rangle$, $x(C) \neq \{N\} \Rightarrow x(C) \subseteq x^+(C)$.

The following definition, proposed in Sampath et al. (1995), formalizes the notion of DES diagnosability of the failure F_i. Let $\Psi(X_{F_i}) = \{s \in L_f(G)|$ the last transition of s is measurable and final$(s) \in X_{F_i}\}$.

DEFINITION 6 *F_i-Diagnosability: A DES model G is said to be F_i-diagnosable for the failure F_i under a measurement limitation if the following holds $\exists n_{F_i} \in \mathbb{N}$ s.t. $[\forall s \in \Psi(X_{F_i})\{\forall t \in L_f(G)/s)(|t| \geq n_{F_i} \Rightarrow D)\}]$, where the condition D is $\forall u \in P^{-1}[P(st)]$, final$(u) \in X_{F_i}$.*

The above definition means the following. Let s be any finite prefix of a trace of G that ends in an F_i-state and let t be any sufficiently long continuation of s. Condition D then requires that every sequence of transitions, measurement equivalent with st (i.e. belonging to $P^{-1}(P(st))$), shall end into an F_i-state. This implies that, along every continuation t of s, one can detect the occurrence of failure corresponding to F_i within a finite delay, specifically in at most n_{F_i} transitions of the system after s.

In the next subsection, we present in brief the polynomial time algorithm for diagnosability analysis of DES models as given in Jiang et al. (2001). The diagnosability result obtained by the scheme proposed in Tae-Yoo and Lafortune (2002) is equivalent to the one obtained by the technique proposed in Jiang et al. (2001). Both the algorithms are polynomial time in the number of system states and works on the DES framework but uses slightly different diagnosability analysis procedures. In this paper, we compare our technique with the algorithm proposed in Jiang et al. (2001).

The algorithm presented in Jiang et al. (2001) is based on an *event-based DES framework* first proposed in Sampath et al. (1995). In contrast, the algorithm proposed in our present work is based on the *state-based framework* similar to the one used in Zad et al. (2003). We present the algorithm proposed in Jiang et al. (2001) after adapting it for the state-based framework. The state-based DES framework has two advantages compared to event-based one, as identified in Zad et al. (2003) and discussed in brief as follows.

In the event-based framework, the problem of fault diagnosis is to use observations to detect if a failure event (unobservable) had occurred in the system since the start of diagnosis. In state-based frameworks, there is an additional assumption that the state set of the system can be partitioned according to the condition (failure status) of the system. The assumption has two benefits. First, instead of detecting failure events, for the purpose of fault diagnosis, the diagnoser determines the system condition. This is particularly useful in cases where the failure might have occurred before the start of diagnosis. Another benefit of the aforementioned assumption is that it simplifies the transition function of the diagnoser. Since the system condition is a function of state, after receiving a new observation (i.e. output symbol), updating the estimate of the systems state is enough to update the condition in the diagnoser. In this way, label propagation as done in event-based models can be avoided.

Therefore, in our present work, we use the state-based framework. As the works reported on polynomial time failure diagnosis schemes (Jiang et al., 2001; Tae-Yoo and Lafortune, 2002) are based on event-based frameworks, we recast them in terms of state-based framework.

2.3. *A polynomial time algorithm for testing diagnosability of DES models*

The algorithm for polynomial time diagnosability analysis (Jiang et al., 2001) is as follows.

ALGORITHM 1 *Input: A DES Model G with failures F_i, $1 \leq i \leq p$.*
Output: Diagnosability of G for a fault F_i for some $i \in [1, p]$

(1) Obtain a new DES $G_n = \langle V, X_n, \Im_n, X_0\rangle$ from $G = \langle V, X, \Im, X_0\rangle$ as follows:
  (a) $X_n = \{x | x \in X_0$ or $x \in X$ and there exists a measurable transition $\tau \in \Im$ such that final$(\tau) = x\}$,
  (b) $\Im_n = \{\langle x, x^+\rangle | x, x^+ \in X_n$ and there exists a sequence s of G-transitions of the form $s = \langle \tau_1, \tau_2, \ldots, \tau_k, \tau\rangle$, where initial$(\tau_1) = x$, final$(\tau) = x^+$ and $P(s) = \langle \tau\rangle\}$, i.e. τ_1 through τ_k are unmeasurable transitions.

  The language generated by G_n is $L(G_n)$. It may be noted that G_n can also be considered as a DES model but without any unmeasurable transition. Hence, all the definitions for DES model G also holds for G_n.

(2) Compute $G_d = (G_n || G_n)$, the strict composition of G_n with itself as the ordered tuple $G_d = \langle V, X_{d0}, \Im_d, X_d\rangle$, where
  (a) X_{d0} is the set of initial states of G_d defined as $X_{d0} = \{(x_1, x_2) | x_1 E x_2$ and $x_1, x_2 \in X_0\}$,
  (b) $\Im_d$ is the set of transitions of G_d defined by the following rules:
    (i) All unordered pairs of G_n-transitions of the form (τ_1, τ_2) are members of $\Im_d$, where $\tau_1 = \langle x_1, x_1^+\rangle$, $\tau_2 = \langle x_2, x_2^+\rangle$, $(x_1, x_2) \in X_{d0}$ and $x_1^+ E x_2^+$.
    (ii) All unordered pairs of G_n-transitions of the form (τ_1, τ_2) are members of $\Im_d$, where

$\tau_1 = \langle x_1, x_1^+ \rangle$, $\tau_2 = \langle x_2, x_2^+ \rangle$, $x_1^+ E x_2^+$ and $\exists(\tau_{O1}, \tau_{O2}) \in \Im_d$ such that final$(\tau_{O1}) = x_1$ *and* final$(\tau_{O2}) = x_2$.

(c) *X_d is the set of states of G_d defined by the following rules:*
   (i) *All members of X_{d0} are members of X_d.*
   (ii) *All unordered pairs of G_n-states of the form (x_1, x_2) such that for some $(\tau_1, \tau_2) \in \Im_d$, $x_1 =$ final(τ_1), $x_2 =$ final(τ_2) and* (initial(τ_1), initial$(\tau_2)) \in X_d$].

Thus, a state of G_d comprises two measurement equivalent G_n-states x_1 and x_2, which are either initial G_n-states or are reachable from some pair of initial G_n-states through two measurement equivalent finite sequences of transitions of G_n. A G_d-state (x_1, x_2), where $x_1 = x_2$, is also possible.

(3) *The F_i-diagnosability condition: There does not exist any cycle c_d of the form $\langle x_{d_1}, x_{d_2}, \ldots, x_{d_k}, x_{d_1} \rangle$, where, for any j, $1 \le j \le k$, $x_{d_j} = (x_1, x_2)$, $F_i \in x_1(C)$ and $F_i \notin x_2(C)$. Such a cycle is called an 'offending cycle' for failure F_i.*

Henceforth, in this paper the states (transitions) of G, G_n and G_d would be termed as G-states (transitions), G_n-states (transitions) and G_d-states (transitions), respectively.

DEFINITION 7 *N-G_d-state: A G_d-state x_d of the form (x_1, x_2), where $x_1(C) = x_2(C) = N$ is called an N-G_d-state. F_i-G_d-state: A G_d-state of the form (x_1, x_2), where $F_i \in x_1(C)$ or $F_i \in x_2(C)$ is called an F_i-G_d-state.*

DEFINITION 8 *F_i-certain-G_d-state: A G_d-state of the form (x_1, x_2) where $F_i \in x_1(C)$ and $F_i \in x_2(C)$ is called an F_i-certain-G_d-state.*

DEFINITION 9 *F_i-uncertain-G_d-state: A G_d-state of the form (x_1, x_2) where $F_i \in x_1(C)$ and $F_i \notin x_2(C)$ is called an F_i-uncertain-G_d-state.*

It may be noted that an offending cycle for failure F_i is a cycle in G_d with only F_i-uncertain-G_d-states.

PROPERTY 1 *If s_n is any trace in $L(G_n)$, then there exists a trace s in $L(G)$ such that $s = P^{-1}(s_n)$. Roughly speaking, if only the measurable transitions of any trace s of G are considered in juxtaposition, then we can find a corresponding trace s_n of G_n.*

PROPERTY 2 *For any trace $s_n = \langle x_{n_0}, x_{n_1}, \ldots, x_{n_k}, \ldots, x_{n_l}, x_{n_k} \rangle$ of G_n, where $x_{n_0} \in X_0$ and $x_{n_i} \in X_n, 1 \le i \le l$, the following holds:*

(1) *$x_{n_k}(C) = x_{n_{(k+1)}}(C) \cdots = x_{n_l}(C)$, as the states form a cycle and any failure is permanent.*
(2) *There exists a trace s of $L(G)$ of the form $s't$ such that $P(s') = \langle \langle x_{n_0}, x_{n_1} \rangle, \langle x_{n_1}, x_{n_2} \rangle, \ldots, \langle x_{n_{(k-1)}}, x_{n_k} \rangle \rangle$ and $P(t) = \langle \langle x_{n_k}, x_{n_{(k+1)}} \rangle, \ldots, \langle x_{n_{(l-1)}}, x_{n_l} \rangle, \langle x_{n_l}, x_{n_k} \rangle \rangle$.*

PROPERTY 3 *For any path $s_d = \langle x_{d_0}, x_{d_1}, \ldots, x_{d_k}, \ldots, x_{d_l}, x_{d_k} \rangle$ of G_d, where $x_{d_0} \in X_{d0}$ is of the form $(x^1_{n_0}, x^2_{n_0})$ and for $1 \le i \le l$, $x_{d_i} \in X_d$ is of the form $(x^1_{n_i}, x^2_{n_i})$, there are two measurement equivalent traces of $L(G_n)$ corresponding to s_d as follows:*

(1) $s^1_n = \langle x^1_{n_0}, x^1_{n_1}, \ldots, x^1_{n_k}, \ldots, x^1_{n_l}, x^1_{n_k} \rangle$,
(2) $s^2_n = \langle x^2_{n_0}, x^2_{n_1}, \ldots, x^2_{n_k}, \ldots, x^2_{n_l}, x^2_{n_k} \rangle$.

Two traces s^1_n and s^2_n may be the same G_n-trace also. If the G_d-cycle $\langle x_{d_k}, \ldots, x_{d_l}, x_{d_k} \rangle = c_d$ is an offending cycle for any failure F_i, then each of the G_d-state of c_d comprises one F_i-G_n-state and one non-F_i-G_n-state. In that case s^1_n and s^2_n are two different, but measurement equivalent, G_n-traces.

THEOREM 1 *A DES model is F_i-diagnosable iff there is no offending cycle for failure F_i in G_d.*

An intuitive proof is presented for completeness; for details the reader may refer to Jiang et al. (2001). From Property 3, the existence of an offending cycle for failure F_i in G_d implies that there are at least two different, but measurement equivalent, *syntactic* cycles in G_n (corresponding to the offending cycle for F_i). By Property 2, there are two G-cycles, one comprising only non-F_i-states and the other comprising F_i-states. These cycles can be reached by measurement equivalent G traces (from a state in X_0). This renders the DES model F_i-non-diagnosable because, at each point in the cycle, there exists uncertainty regarding the occurrence of F_i and the system may not exit from such a cycle. Therefore, F_i may not be diagnosed in finite time. For detailed proof and the complexity analysis, the reader is referred to Jiang et al. (2001).

Since the modified diagnosability analysis algorithm (for FDES models) presented in this paper also needs the computation of G_n and G_d, we now determine the complexity of these two steps of Algorithm 1.

Complexity of Algorithm 1: Let the number of G-states be n; the number of G-transitions t is $O(n^2)$.

Step 1 Construction of X_0 in G_n involves searching all the G-states which is $O(n)$. Construction of $X_n - X_0$ involves $O(n^2 \cdot t)$ steps because there are n^2 state pairs and studying reachability of each of them involves $O(t)$ steps. Hence, generation of G_n requires $O(n + n^2 \cdot t) = O(n^2 \cdot t) = O(n^2 \cdot n^2) = O(n^4)$ steps.

Step 2 Generating G_d from G_n requires

(i) Construction of X_{d0} by rule 2(a);
(ii) Construction of G_d-transitions emanating from each member $\langle x_1, x_2 \rangle$ of X_{d0} by rule 2(b)(i) and adding the successor pair (x_1^+, x_2^+) as non-initial G_d-state to X_d.

(iii) Construction of G_d-transitions emanating from each member $\langle x_1, x_2 \rangle$ of $X_d - X_{d0}$ (non-initial G_d-states) by rule 2(b)(i) and adding the successor pair (x_1^+, x_2^+) as non-initial G_d-state to X_d.

Step (i) is $O(n^2)$ and steps (ii) and (iii) together are $O(t^2) = O(n^4)$ because no transition pair of G_n needs to be checked more than once. The non-initial G_d-states are generated hand-in-hand with the G_d-transitions as discussed above and thus do not require additional complexity. Hence, generation of G_d from G_n requires $O(n^4)$ steps.

Step 3 The complexity of checking the presence of any offending cycle for a failure F_i is of the $O(n^4)$, explained as follows (Jiang et al., 2001). First, only G_d states of type $x_{d_j} = (x_1, x_2)$, such that $F_i \in x_1(C)$ and $F_i \notin x_2(C)$ are retained. In other words, G_d states $x_{d_l} = (x_3, x_4)$, such that $F_i \in x_3(C)$ and $F_i \in x_4(C)$ are deleted. Also, transitions that emanate from or lead to G_d states which are deleted are also eliminated. In the resultant G_d, if there is a cycle then it is an offending cycle for the failure F_i. The complexity of checking the presence of a cycle in a directed graph can be done using depth first search and involves complexity of the $O(|V| + |E|)$, where $|V|$ is the number of states and $|E|$ is the number of edges in the graph. As the number of states and transitions in G_d are n^2 and n^4, respectively, in the worst case checking of offending cycle involves complexity of the $O(n^2 + n^4) \equiv O(n^4)$.

The complexity of Algorithm 1 is $O(n^4 + n^4 + n^4) = O(n^4)$, that is, polynomial in the number of system states.

The algorithm above works with DES models where fairness of transitions is not ensured. In a system with all transitions fair, if there is an offending cycle with some transitions leading out of the cycle, then that cycle cannot execute infinitely at a stretch. Thus, the above theorem does not hold for such systems. In the next section, we depict a situation involving a failure which is diagnosable because of fairness of transitions but the corresponding G_d contains an offending cycle for the failure F_i. Then, we preset in brief the traditional diagnoser-based scheme (Thorsley and Teneketzis, 2005) that solves the failure diagnosis problem in FDES models.[1] There we highlight how the scheme addresses the fairness issue and how the complexity is exponential in a number of system states.

3. DES diagnosability for a fair system

3.1. *DES diagnosability for the chemical reaction chamber*

Let us consider a chemical reaction chamber where temperature (T) is maintained between 50°C and 70°C by switching a heater ON and OFF and pressure (P) is maintained between 80 and 90 cms of Hg by opening and closing a valve. A DES model having states x_1 through x_8 is shown in Figure 1(a). We consider only a heater stuck-off failure, designated as F_1. States x_1 to x_4 are normal and x_5 to x_8 are F_1-states. When the temperature falls below 50°C, a temperature controller changes C_H from 0 to 1 to put the heater ON. This is captured by transitions 1 and 5 in the normal condition and transitions 9 and 12 in the case of F_1. Similarly, when temperature reaches 70°C or more, the temperature controller changes C_H from 1 to 0 to put the heater OFF. This is captured by transitions 2 and 6 in the normal condition. It may be noted that in the case of failure F_1, the temperature does not rise and hence the corresponding transitions to put the heater OFF are absent. Similar control logic holds for the (normal) pressure control loop. The controller commands C_H (for the heater), C_V (for the valve) and the signs of the rates d_T of T and d_P of P are measurable discrete variables. The following symbols are used to designate the states of the heater and the valve at various G states: H_F: heater off, H_{SF}: heater stuck-off, H_N: heater on, V_O: valve open, and V_C: valve closed.

The model G_n for the system model G is shown in Figure 1(b). In G_n, the two unmeasurable transitions of G, labelled as *fault* in Figure 1(a), are replaced by transitions numbered 15, 16, 17 and 18. The model G_d is illustrated in Figure 1(c). Consider the cycle marked $c_d = \langle (x_1, x_5), (x_4, x_8) \rangle$ in Figure 1(c). $x_1(C) = N \neq x_5(C) = \{F_1\}$. Similarly, $x_4(C) = N \neq x_8(C) = \{F_1\}$. Hence, c_d is an offending cycle for failure F_1. Hence, by the DES diagnosability condition reported in Jiang et al. (2001) and presented as Theorem 1 in the previous section, the failure is non-diagnosable.

The failure, however, is manifested in finite time. It may be noted that in the cycle $\langle x_5, x_8 \rangle$, the temperature keeps decreasing up to 50°C whereupon the cycle is exited. Consequently, the system reaches state x_6 or x_7 in finite time where failure can be detected because the corresponding normal states x_2 and x_3 have a different value for the measurable variable d_T. More specifically, if the system is normal and the temperature falls below 50°C, the system reaches state x_2 (or x_3), where $C_H = 1$, $C_V = 0$, $d_P = -1$ and $d_T = 1$ ($C_H = 1$, $C_V = 1$, $d_P = 1$ and $d_T = 1$). In contrast, if the failure F_1 has occurred, then the system reaches state x_6 (or x_7) where $C_H = 1$, $C_V = 0$, $d_P = -1$ but $d_T = -1$ ($C_H = 1$, $C_V = 1$, $d_P = 1$ but $d_T = -1$). It may be noted that the difference is due to the fact that in the failure condition, even when $C_H = 1$, the temperature keeps falling whereas it should be rising as in the normal condition. In short, the fact that the cycle $\langle x_5, x_8 \rangle$ cannot execute infinitely long results in manifestation of F_1 in finite time. This happens because the transition 9 from x_5 and the transition 12 from x_8 are fair.

This property is not apparent in the DES model because of the following reason.

The chemical reaction chamber is basically a hybrid system because it has, in addition to discrete variables C_H, C_V, etc., continuous dynamics corresponding to temperature and pressure. The model illustrated in Figure 1(a) is a DES model where the enabling conditions

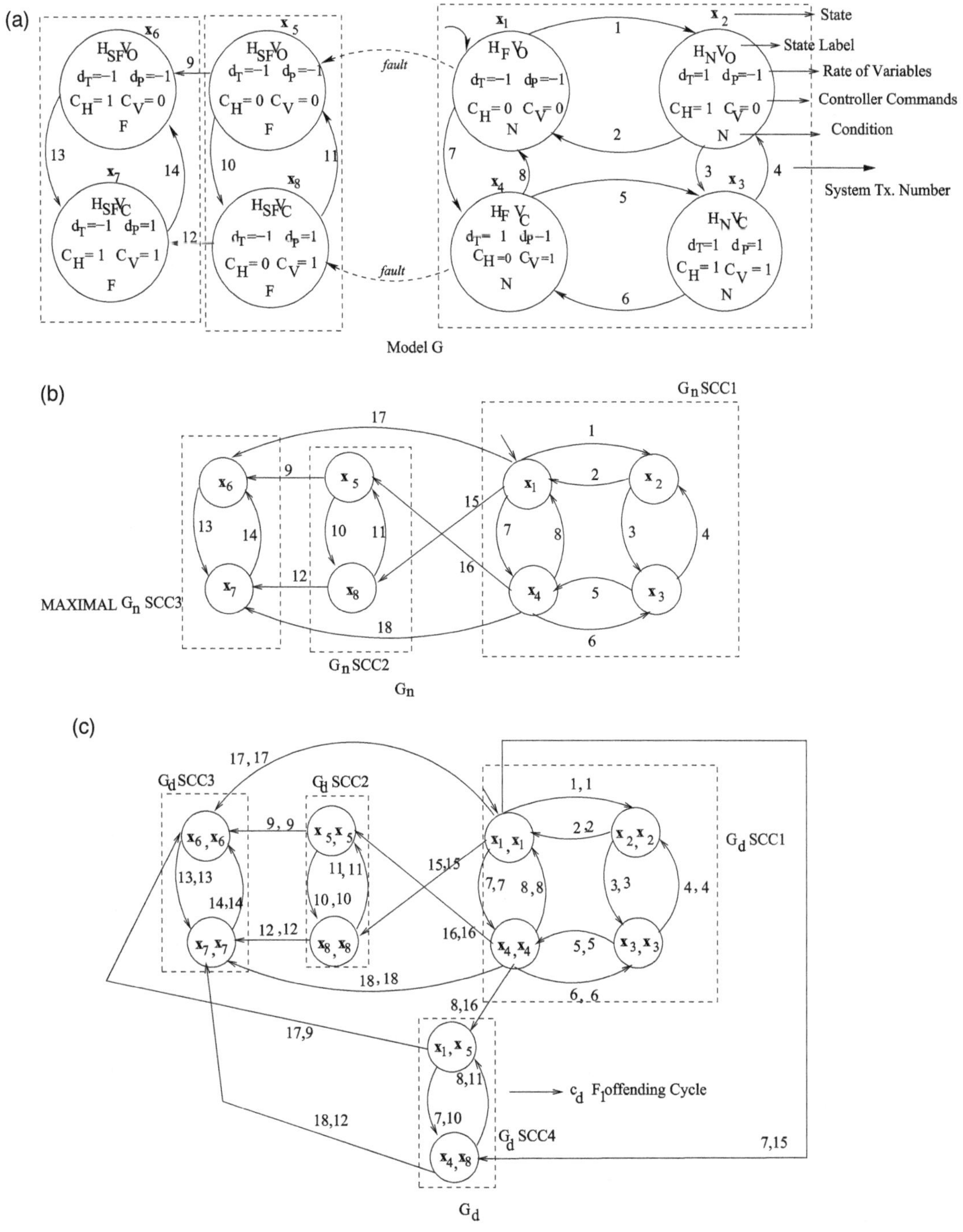

Figure 1. A chemical reaction chamber – its model, G_n and G_d.

of the transitions and the continuous dynamics in the model states are abstracted out. It may be noted that with each model transition there is an implicit enabling condition that is a linear inequality over either of the continuous variables; for example, if the continuous variable T (for temperature) were used explicitly, then the transitions 1, 5, 9, and 12 would have $T \leq 50°C$ as the enabling condition. The enabling conditions are abstracted out in the DES model as we do not model (or measure) the continuous variables T and P. Furthermore, in the DES model we retain *only the signs* of the rates of the continuous variables, e.g. $d_T = -1$ or $+1$ and not the accurate continuous dynamics of these (continuous) variables. Moreover, there is no special semantics associated with d_T or d_P unlike the rates used in the hybrid system model where the (absolute) rates are interpreted in terms of a 'clock tick' (i.e. say the rate of T is $+5°$C per unit time). As 'clock tick' is not modelled in DES, we cannot ascribe any semantics to the signs of the rates of the continuous variables to capture the accurate (continuous) dynamics. This abstraction of the continuous dynamics, in turn, abstracts out the fairness of transitions as indicated below.

Any system trace with the infinite suffix $\langle x_5, x_8, x_5, x_8, \ldots, \rangle$, which implies infinitely long execution of the cycle $\langle x_5, x_8 \rangle$, cannot happen in the system because the temperature keeps on decreasing without bound in such a suffix trace. The transitions $\langle x_5, x_6 \rangle$ and $\langle x_8, x_7 \rangle$ are unfair in the trace $\langle x_1, \ldots, x_5, x_8, x_5, x_8, \ldots, \rangle$ as the trace visits the states x_5 and x_8 infinitely many times without taking these transitions infinitely often. We rule out such traces from the language generated by the DES models of the systems with (implicit) continuous dynamics (where all transitions are fair). In other words, fairness of transitions is explicitly imposed as an assumption of the DES models and we designate such models as FDES models. The following definitions are in order.

A transition $\tau \triangleq \langle x, x^+ \rangle$ is said to be *enabled* in a trace s if x is in s and is said to be *taken* in s if both x and x^+ are in s.

DEFINITION 10 (Fair transitions:) *A transition $\tau = \langle x, x^+ \rangle$ is fair if there is no trace that visits x indefinitely often without taking τ in it indefinitely often.*

DEFINITION 11 (FDES Models) *A DES model in which all transitions are fair is called an FDES model.*

The fairness assumption restricts the traces actually generated by G. Any sequence of transitions, constructed syntactically satisfying the consecution property, is not necessarily a member of $L(G)$. The assumption also implies that any cycle in G having at least one transition out of the cycle can only be traversed a finite number of times consecutively.

3.2. Diagnoser-based FDD scheme for FDES models (Thorsley and Teneketzis, 2005)

In this subsection we discuss in brief the traditional diagnoser-based FDD scheme proposed in Thorsley and Teneketzis (2005). The diagnosability condition to be checked for ascertaining if a fault is diagnosable in a fair system is called the A-diagnosability condition and comprises the following steps:

- Construct a logical diagnoser.
  The diagnoser is a directed graph $D = \langle Z, A \rangle$, where Z is the set of diagnoser states and A is the set of diagnoser transitions. Each diagnoser state $z \in Z$ is an ordered set comprising a subset of equivalent G-states representing the uncertainty about the actual G-state and each diagnoser transition $a \in A$ is a set of equivalent G-transitions representing the uncertainty about the actual transition that occurs.
- Associate probability matrix Φ_a with each diagnoser transition $a = \langle z, z^+ \rangle$.
  Φ_a represents probability of transitions from G-states in the diagnoser state z to G states in the diagnoser state z^+. If z has i G-states and z^+ has j G-states, Φ_a is of dimension $i \times j$ (i rows and j columns). For a G-state $x_l \in z$ and another G-state $x_m \in z^+$, the probability of the transition $\tau \in a$ (from x_l to x_m in the model G) is represented by the l, m element of Φ_a and denoted as $\Phi_a[l, m]$.
- Find the recurrent and the non-recurrent G-states in the diagnoser states.
  A Markov matrix is generated from the diagnoser with the help of the probability matrices. Given each diagnoser state, it is determined if the G-states in the diagnoser state are recurrent or non-recurrent. A recurrent (non-recurrent) G-state if visited once, then the probability of visiting it again in finite number of transitions is 1 (0) and this condition holds indefinitely often.
- Diagnosability condition (termed as A-diagnosability condition) check.
  If any F_i-uncertain diagnoser state contains a recurrent F_i-G-state, then fault is non-diagnosable, else diagnosable.The basic idea behind the condition is, if there is a recurrent G-state in any diagnoser state then that diagnoser state will be visited again and again with probability 1. If the diagnoser state under question is F_i-uncertain and the recurrent G-state is F_i-G-state, then F_i-certain diagnoser states will not be reached; this renders F_i to be non-diagnosable. On the other hand, if all F_i-uncertain diagnoser states have non-recurrent F_i-G-states then F_i-certain diagnoser states will be reached in finite time after occurrence of failure F_i with probability 1; this renders F_i to be diagnosable.
  This condition can adequately handle FDES models. If there is a cycle of F_i-G-states with an outward transition then these F_i-G-states are non-recurrent. Therefore, even if these F_i-G-states are embedded in an F_i-uncertain diagnoser cycle, the fault is diagnosable because the F_i-uncertain diagnoser states (involved in the F_i-uncertain diagnoser cycle) do not have recurrent F_i-G-states.

Now we discuss the computational complexity of the A-diagnosability checking scheme.

Computational complexity

- If n is the number of G-states then the number of diagnoser states can be $O(2^n) = n_d$. As each

diagnoser state comprises a subset of (measurement equivalent) G-states, so all possible diagnoser states is power set of G-states. Therefore, the complexity of constructing a diagnoser is $O(2^n)$. The number of G-transitions can be $O(n^2)$ denoted as t and diagnoser transitions can be $O(n_d^2)$ denoted as t_d.
- The probability transition matrices have dimensions of the $O(n \times n)$ because each diagnoser state can have $O(n)$ G-states. The value of any l, m element of such a matrix is the probability of the corresponding G-transition. The complexity of generating one matrix is $O(n^2)$ because it involves considering probability values of *all* G-transitions in the worst case. Complexity of generating these matrices for all diagnoser transitions is $O(n^2 \times t_d)$.
- From the Markov matrix, recurrent and non-recurrent G-states in the diagnoser states can be determined using the scheme proposed in Xie and Beerel (1998a) and involves a complexity $O((n_d.n)^2)$.
- A-diagnosability condition checking would involve verifying each F_i-uncertain diagnoser state and containment of any recurrent F_i-G-state; that is the worst case would involve a complexity $O(n_d \times n)$.

Hence, the total complexity of A-diagnosability is $O(2^n + (n^2 \times t_d) + ((n_d \times n)^2) + (n_d \times n))$. Substituting t_d with n_d^2 we obtain the complexity of A-diagnosability as $O(2^n + (n^2 \times n_d^2) + ((n_d \times n)^2) + (n_d \times n))$. As $n_d = O(2^n)$, the complexity can be written as $O(2^n + (n^2 \times 2^{2.n}) + (2^{2.n} \times n^2) + (2^n \times n))$. Therefore, it can be concluded that the traditional diagnoser-based A-diagnosability scheme is exponential (i.e., $O(2^n)$) in number of G-states.

Note: We have represented the steps of A-diagnosability analysis and their computational complexity in an over simplified manner, just required to highlight the fairness handling capability and the exponential complexity. For a more complete analysis, the readers are referred to Thorsley and Teneketzis (2005) and Biswas et al. (2010).

4. A new polynomial time condition for FDES diagnosability

In this section, we propose a new condition for FDES diagnosability taking fairness of transitions into account. Henceforth, we use G to represent FDES models; whenever G is used for DES models, it would be denoted explicitly.

Before we formalize, the new condition certain definitions are introduced.

DEFINITION 12 *Recurrent States: A state (node) x in a directed graph G is said to be recurrent if any (infinite) trace s of G that visits x once, visits x infinitely often.*

DEFINITION 13 *SCC of a Directed Graph: A maximal connected sub-graph of a directed graph G is called a connected component of G. A connected component of a directed graph G is said to be an SCC , if for every pair $\langle x_1, x_2\rangle$ of states in the component (subgraph), there is a path from x_1 to x_2 and vice-versa.*

The set of SCCs of G is denoted as C. Any $c \in C$ comprises a subset of G-states and such G-states are termed as 'G-states in the SCC c'. Similarly, 'G-transitions in c' is given by the set $\{\tau \mid \tau \in \Im$ and $\text{initial}(\tau), \text{final}(\tau) \in c\}$.

DEFINITION 14 *Maximal SCCs: Let a partial order $\preceq \subseteq (C \times C)$ be defined as: $c_1 \preceq c_2$, if some state in c_2 is reachable from some state in c_1.*

The set of maximal elements of the poset $\langle C, \preceq\rangle$ is denoted as $C_{\max}$. Clearly, from any $c_{\max} \in C_{\max}$, there is no path to any other SCC. The SCCs in $C_{\max}$ are referred to as maximal SCCs.

It may be observed that in a directed graph corresponding to an FDES model, the states in any $c \notin C_{\max}$ are the non-recurrent states. This is because, if some state in a $c_{\max} \in C_{\max}$ is reached, then only the states in $c_{\max}$ can be visited infinitely often and a trace taken to reach $c_{\max}$ passes through the intermediate states that are never visited again after it reaches $c_{\max}$. The details on the relationship of recurrence of states with SCCs can be found in Nuutila and Soisalon-Soininen (1994) and Xie and Beerel (1998b).

DEFINITION 15 *F_i-Diagnosability for an FDES model: An FDES model G is said to be F_i-diagnosable for a failure F_i under a measurement limitation if the following holds $\exists n \in \mathbb{N}[\forall s \in \Psi(X_{F_i})\{\forall t \in L_f(G)/s)(|t| \geq n \Rightarrow D)\}]$, where D is $\forall u \in P^{-1}[P(st)], \text{final}(u) \in X_{F_i}$.*

It may be noted that the definition for F_i-Diagnosability for DES models (Definition 6) and that for F_i-Diagnosability for FDES models is the same except for the language $L(G)$.

4.1. A condition for FDES diagnosability

As shown in Section 3, a failure may be diagnosable in spite of the presence of an offending cycle for the failure (F_1) in G_d (Figure 1). If there is an outward transition from the F_i-G_n-cycle, then this cycle will not move infinitely long at a stretch. Furthermore, corresponding to this outward F_i-G_n-transition (from the F_i-G_n-cycle), if there is no measurement equivalent non-F_i-G_n-transition (from the non-F_i-G_n-cycle), then the failure is diagnosed in finite time. It may be noted that in the example of the chemical reaction chamber (shown in Figure 1(c)), corresponding to the F_1-offending cycle $\langle(x_1, x_5), (x_4, x_8)\rangle$ the F_1-G_n-cycle is $\langle x_5, x_8\rangle$ and the non-F_1-G_n-cycle is $\langle x_1, x_4\rangle$. There are outward F_1-G_n-transitions from the F_1-G_n-cycle $\langle x_5, x_8\rangle$, namely, 9 and 12; there is no measurement equivalent non-F_1-G_n-transition corresponding to 9 and 12 emanating from the non-F_1-G_n-cycle $\langle x_1, x_4\rangle$. Hence, the presence of this F_1-offending cycle does not hamper F_1-diagnosability. In contrast, the traces of an FDES model (e.g. G_n) may remain infinitely long only in the maximal SCCs (of G_n). Hence, to analyse failure diagnosability of F_i, we need to find SCCs in G_d with F_i-uncertain states that has an embedded maximal SCC of G_n. This motivates the following definition.

DEFINITION 16 *Terminal offending F_i-SCCs: An SCC c_d of G_d is said to be a terminal offending F_i-SCC, if it comprises F_i-uncertain G_d-states $(x^1_{n_1}, x^2_{n_1}), (x^1_{n_2}, x^2_{n_2}), \ldots, (x^1_{n_k}, x^2_{n_k})$ such that there exists a maximal G_n-SCC $c_{n(\max)} = \{x^1_{n_1}, x^1_{n_2} \cdots x^1_{n_k}\}$, where $x^1_{n_1}, \ldots, x^1_{n_k}$ are F_i-uncertain G_d-states.*

The intuitive logic behind the equivalence of F_i-diagnosability of an FDES model with the absence of terminal offending F_i-SCC is as follows. A terminal offending F_i-SCC c_d contains a (maximal) F_i-G_n-SCC having all transitions emanating from states of that F_i-G_n-SCC in some G_d-transitions of c_d. The states in c_d are F_i-uncertain. Hence, corresponding to the F_i-G_n-SCC in question, there is another G_n-SCC of non-F_i states and the transitions involved in the traces generated by traversing both these SCCs are measurement equivalent. Thus, as long as the system (after failure F_i) moves in the F_i-G_n-SCC the failure cannot be diagnosed. As the F_i-G_n-SCC is maximal, the system after failure may move in that F_i-G_n-SCC under question infinitely long (even in an FDES framework); so the failure will not be diagnosed in finite time.

The structure G_d for the chemical reaction chamber example shown in Figure 1(c) has an F_i-uncertain SCC c_d, which is not a terminal offending F_1-SCC because the F_1-G_n-states x_5 and x_8 contained in c_d do not belong to any $c_{n(\max)} \in C_{n(\max)} = \{\{x_6, x_7\}\}$. Hence, the failure is FDES-diagnosable in keeping with the intuitive discussion included in the last section.

Before we prove formally the equivalence of F_i-diagnosability of FDES models and the absence of a terminal offending F_i-SCC in G_d, we introduce the following properties that follow from the construction of G_n and G_d.

PROPERTY 4 *For any G_d-SCC $c_d = \{x_{d_1}, \ldots, x_{d_k}\}$ there are two SCCs in G_n, not necessarily distinct, given by*

(1) $c^1_n = \{x^1_{n_1}, \ldots, x^1_{n_k}\} \cup X^1_n$, where $x^1_{n_i} \in x_{d_i}, 1 \le i \le k$, (i.e. these are states of c^1_n that are in c_d) and X^1_n is the subset of states of c^1_n that are not in c_d,
(2) $c^2_n = \{x^2_{n_1}, \ldots, x^2_{n_k}\} \cup X^2_n$, where $x^2_{n_i} \in x_{d_k}, 1 \le i \le k$, (i.e. these are states of c^2_n that are in c_d) and X^2_n is the subset of states of c^2_n that are not in c_d, such that $x^1_{n_i} E x^2_{n_i} 1 \le i \le k$.

Proof The fact that the SCCs c^1_n and c^2_n may contain G_n-states which do not pair up as c_d states follows from the following observation. There may be transitions emanating from a state in $\{x^1_{n_1}, \ldots, x^1_{n_k}\}$ and terminating in a state in X^1_n for which there is *no measurement equivalent transition* emanating from a state in $\{x^2_{n_1}, \ldots, x^2_{n_k}\}$ and terminating in a state in X^2_n, or vice-versa.

This may be observed in the example of the chemical reaction shown in Figure 1(c). In the G_d-SCC c_d of the example, G_n-SCC2 is embedded totally in c_d; however, only two states (namely, x_1 and x_4) of the G_n-SCC1 are contained in c_d while other two states (namely, x_2 and x_3) are not. So, $X^1_n = \phi$ and $X^2_n = \{x_2, x_3\}$.

PROPERTY 5 *Let s be a trace of G such that $P(s) = \langle \tau_1, \ldots, \tau_l\rangle$. There is a path in G_d corresponding to s, namely $s_d = \langle \tau_{d_1}, \ldots, \tau_{d_l}\rangle$, such that $\tau_i \in \tau_{d_i}$, $1 \le i \le l$. It implies that each transition of s_d contains a transition from $P(s)$ as one of its two members. If s_d also corresponds to another sequence of G-transitions, s' say, then s' starts from an* initial *state of G and is measurement equivalent with s.*

The proof follows from the definition of G_d-transitions.

PROPERTY 6 *In any G_d-SCC c_d, F_i-ceratin G_d-states cannot coexist with F_i-uncertain G_d-states or N-G_d-states.*

The proof follows from the permanence of failures.

THEOREM 2 *An FDES model is F_i-diagnosable iff there is no terminal offending F_i-SCC in G_d.*

Proof ($\Rightarrow$:) Let G be F_i-diagnosable. Let there be a terminal offending F_i-SCC in G_d, namely $c_d = \{x_{d_1}, x_{d_2}, \ldots, x_{d_k}\}$. By Property 4, there are two SCCs of G_n in c_d namely, c^1_n and c^2_n. From the definition of terminal offending F_i-SCC, as $F_i \in x^1_{n_j}(C)$ and $F_i \notin x^2_{n_j}(C)$ for $1 \le j \le k$, c^1_n and c^2_n are not the same. Also, $c^1_n \in C_{n(\max)}$ and has no transition that is contained in a G_d-transition that leads out of c_d. In other words, c^1_n is completely embedded in c_d and hence $X^1_n = \phi$. Furthermore, since all the model states are reachable, there is a G-trace that leads to a state in c^1_n from an initial G-state. Obviously, $\text{final}(s_1) = X_{F_i}$. All the extensions of this G-trace remain confined in c^1_n because it is a maximal SCC; also, $X^1_n = \phi$. Hence, $\forall n_{F_i} \ge 1$, an extension of s_1 can be constructed as $s_1 t_1$ which does not leave c^1_n. By Property 5, there is also a trace u say, (which is measurement equivalent to $s_1 t_1$, i.e., $u \in P^{-1}(P(s_1 t_1)))$ for c^2_n that comprises non-F_i-states and hence $\text{final}(u) \notin X_{F_i}$. So G is F_i-non-diagnosable. (Contradiction)

($\Leftarrow$:) Let there be no terminal offending F_i-SCC in G_d. Let G be F_i-nondiagnosable. Then by negating the diagnosability Definition (15),

$$\forall n_{F_i} \in \mathbb{N}, \exists s \in \Psi(X_{F_i}), \exists t \in L(G)/$$
$$s[|t| > n \wedge \exists u \in P^{-1}(P(st)) \wedge \text{final}(u) \notin X_{F_i}], \quad (5)$$

$\text{final}(st) = \text{final}(t) \in X_{F_i}$ because $\text{final}(s) \in X_{F_i}$ and F_i is permanent. By Property 5, there is a path $s_d = \langle \tau_1, \ldots, \tau_l \rangle$ in G_d corresponding to st. Since $u \in P^{-1}(P(st))$, (i.e. u is measurement equivalent to st), s_d also corresponds to u. As s_d corresponds to st, where $\text{final}(st) \in X_{F_i}$, and to u, where $\text{final}(u) \notin X_{F_i}$, s_d ends in an F_i-uncertain G_d state, x_d say. Since (Equation (5)) holds for any $n_{F_i} \in \mathbb{N}$, t can be taken to be arbitrarily long so that the length of the G_n-path t is greater than the number of states in G_n. Hence, t is of the form t_1t_2 where t_2 is confined in a maximal G_n-SCC c_n, say. Furthermore, st and u are measurement equivalent traces; let u lead to u_2 such that u_2 is measurement equivalent to t_2. Thus, the G_n-states of t_2 are embedded entirely in some G_d-SCC, c_d say, and the G_n-states of u_2 are also in c_d. As t_2 comprises F_i-G_n-states and u_2 comprises non-F_i-G_n-states, c_d is an F_i-uncertain G_d-SCC. As there is a maximal F_i-G_n-SCC c_n that is embedded in an F_i-uncertain G_d-SCC, from the definition of 'terminal offending F_i-SCC' (Definition 16), c_d is a terminal offending F_i-SCC in G_d. (Contradiction)

The diagnosability analysis algorithm that follows from Theorem 2 for the F_i-diagnosability analysis of FDES models (for a given failure F_i) is given below:

ALGORITHM 2
Input: An FDES Model G.
Output: Diagnosability of G for a fault F_i.

(1) *Obtain DES $G_n = \langle V, X_n, \Im_n, X_0 \rangle$ from $G = \langle V, X, \Im_n, X_0 \rangle$ (same as Step-1 of Algorithm 1).*
(2) *Compute G_d (same as Step-2 of Algorithm 1).*
(3) *Compute C_d, the set of SCCs of G_d.*
(4) *Check if any $c_d \in C_d$ is a terminal offending F_i-SCC. If there is one such SCC in G_d, then the system in non-diagnosable for failure F_i, otherwise, it is diagnosable for failure F_i.*

4.2. *Complexity analysis*

Complexity of Algorithm 2

Both *Step 1* and *Step 2* are $O(n^4)$ as discussed in Section 2.3.

Step 3 This step requires $O(n^2 + t^2) = O(n^4)$, i.e. the time for finding the G_d-SCCs (Nuutila and Soisalon-Soininen, 1994).

Step 4 For each $c_d \in C_d$ the following sub-steps are required to determine if c_d is a terminal offending F_i-G_n-SCC.

- Step 4(i) Check that one of the component G_n-states in any state of c_d is an F_i-state and the other is a non-F_i-state.
- Step 4(ii) Generate G_n-SCCs and mark the maximal SCCs, i.e. construct $C_{n(\max)}$.
- Step 4(iii) Locate a maximal SCC $c_{n(\max)} \in C_{n(\max)}$ such that all the F_i-states of G_n that are in any state of c_d are from the maximal SCC $c_{n(\max)}$.
- Step 4(iv): Check that none of the transitions of $c_{n(\max)}$ is in an outgoing transition from c_d.
- Step 4(i) through step 4(iv) are repeated for all G_d-SCCs.

Step 4(i) is $O(n^2)$ as in the worst case, each G_d-SCC may have n^2 states and this step involves vising each state in c_d and checking the failure labels of the two component G_n-states. Step 4(ii) requires $O((n+t)+t)$ steps as follows. Generating G_n-SCCs is $O(n+t)$ and finding $C_{n(\max)}$ involve construction of the relational matrix $R_{\preceq}$ of $\langle C_n, \preceq \rangle$ and identification of the rows of $R_{\preceq}$ having all elements as zeros except the diagonal elements. This can be done in $O(t)$ time by examining the G_n-transitions. This is because, in the relational matrix of any partial order, the row corresponding to any maximal element has all zeros except in the diagonal element. Step 4(iii) requires checking the F_i-G_n-states in c_d vis-a-vis all the G_n-states in any $c_{n(\max)}$; this involves $((n^2) + (n \cdot n \cdot n)) = O(n^3)$ steps. This is because finding all the F_i-G_n-states that are in c_d requires checking the F_i-G_n-states in each of the states of c_d. Furthermore, checking if all the G_n-states in c_d are in $c_{n(\max)}$ requires checking the G_n-states in c_d vis-a-vis the G_n-states in $c_{n(\max)}$ and there can be $O(n)$ maximal G_n-SCCs (i.e. $|C_{n(\max)}| = n$). Step 4(iv) requires checking each G_n-transition of $c_{n(\max)}$ vis-a-vis the outgoing transitions of c_d, which in the worst case is $O(t \cdot t^2)$. Complexity of step 4(i) through step-4(iv) is $O((n^2) + ((n+t)+t) + (n^3) + (t \cdot t^2)) = O(n^2 + n + 2 \cdot t + n^3 + t^3) = O(t^3) = O(n^6)$. As Step 4(i) through Step 4(iv) are to be repeated for all G_d-SCCs, detecting terminal offending F_i-SCC is $O(n^2 \cdot n^6) = O(n^8)$ as in the worst case there may be $O(n^2)$ G_d-SCCs (i.e. when each state of G_d is an SCC).

Thus, the overall complexity of Algorithm 2 is $O(n^4 + n^4 + n^4 + n^8) = O(n^8)$. Without loss of generality, if we assume that all failures involve the same number of system states and transitions, then it may be shown that F_i-diagnosability analysis for FDES models is $O((n^8)/p)$. Therefore, when we repeat for all failures, the complexity figure is not increased. The detailed calculation is avoided for simplicity. Thus, the overall complexity of the scheme for FDES diagnosability analysis is polynomial in the number of system states.

As already discussed, the complexly of classical polynomial time algorithm (Algorithm 1) is $O(n^4)$. Therefore, we can see that complexity of FDES polynomial time diagnosability analysis is higher than that of classical polynomial time diagnosability analysis, however, the order is same. Therefore, we can conclude that incorporating fairness in the polynomial time diagnosability framework increases the complexity but maintains the order.

5. Conclusion

The DES framework has been found to be applicable for FDD of a wide range of systems, even including those having continuous dynamics. For such hybrid systems, DES paradigm partitions the continuous state space and captures each subspace as a discrete state. The DES framework facilitates simplistic modelling and failure analysis algorithms. Furthermore, it has been seen that most of the failures can be detected even if their effect is abstracted in terms of discrete state space. The first series of FDD algorithms were based on a diagnoser which is a state estimator of the system. It basically performs the following two functions–(i) diagnosability analysis, where certain conditions on the diagnoser are checked off line to determine if failure can be diagnosed within finite time of its occurrence and (ii) online fault diagnosis. The classical scheme suffers from two drawbacks, namely, (i) exponential complexity and (ii) improper diagnosis of fair systems having continuous dynamics. These two problems have been solved *independently* in the literature. In this paper, we have addressed both these problems together in an unified approach.

We have discussed in this paper how the DES framework has been enhanced to handle fair systems (Thorsley and Teneketzis, 2005; Biswas et al., 2010) by using the concept of recurrent and non-recurrent states. The major idea is to check the impact of *only* the recurrent states in diagnosability analysis. No system trace passes over the non-recurrent states after a finite time and hence they need not be used in diagnosability checking. Then, the computational complexity of these schemes has been derived and shown to be exponential with respect to the system states. We have also presented in this paper the schemes that have handled the complexity issue (Jiang et al., 2001; Tae-Yoo and Lafortune, 2002). These frameworks use a product automata, instead of diagnoser to derive the diagnosability results. As the use of diagnoser is avoided, these schemes have polynomial time complexity. However, the product automata cannot be used for online diagnosis. The motivation of still using these polynomial schemes is to first check diagnosability using a simple scheme. Only when diagnosability is assured a diagnoser can be constructed else, appropriate steps, such as increasing measurement points, improving sensor quality, etc. need to be taken. However, how wrong diagnosability inferences can be given if these polynomial schemes are applied on *fair systems* have been identified and demonstrated using a simple example in this paper.

To address the above situation we developed an FDD framework for FDES models but involving only polynomial time complexity. Given an FDES model, first unmeasurable transitions are eliminated. Then, an automaton is generated by strict composition of the resultant FDES model with itself. In that automaton, we check for 'terminal offending SCCs'. Such SCCs are composed of F_i-uncertain states and may not exit in finite time because the corresponding system states are recurrent. Hence, the presence of such SCCs render a failure nondiagnosable. If no such SCCs are present, the fault is diagnosable. The working of the theory has been illustrated using the example of a simple chemical reaction chamber. Also, the scheme has been formally shown to be working properly by proving the correctness and completeness of the 'terminal offending SCCs' based condition.

The scheme proposed in this paper works for FDES where all transitions are fair. In general, however, a system may have some transitions fair and some unfair (Biswas et al., 2010). If the unfair and fair transitions are identified then there exits an algorithm (Biswas et al., 2006) that can transform the DES model with unfair and fair transitions to an equivalent DES model with only fair transitions (i.e. FDES model). This generalizes the given diagnosability algorithm for FDES models to general DES models with fair and unfair transitions *provided unfair transitions are identified explicitly*.

It is desirable that given any DES model, the unfair transitions be detected algorithmically. This work can be accomplished using symbolic model checking of the corresponding hybrid system models. Several computational tree logic model checkers have been reported in the literature for hybrid systems, namely, HyTech (Alur, Courcoubetis, Henzinger, and Ho, 1993), CheckMate (Chutinan and Krogh, 1999) and PHAVer (Frehse, 2005). To verify unfairness of a transition $\tau = \langle x, x^+ \rangle$, we need to verify symbolically that (i) there is a path where x is visited infinitely often, and (ii) in that path, e_τ is not satisfied in future forever. All these techniques call for specifying the initial states/values of the continuous variables. However, it may be noted that this technique is incomplete for generalized classes of hybrid automata and partially decidable for linear hybrid systems (LHS) (Alur et al., 1993; Alur, Henzinger, and Ho, 1996). The details of symbolic model checking algorithms for LHS are discussed in Alur et al. (1993) and Alur et al. (1996).

Disclosure statement

No potential conflict of interest was reported by the authors.

Note

1. Two approaches (Thorsley and Teneketzis, 2005; Biswas et al., 2010) have solved the same problem in a slightly different manner. In this paper, we discuss the scheme proposed in Thorsley and Teneketzis (2005); equivalence between these schemes has been shown (Biswas, 2013)

References

Alur, R., Courcoubetis, C., Henzinger, T. A., & Ho, P.-H. (1993). Hybrid automata: An algorithmic approach to the specification and verification of hybrid systems. In R. L. Grossman, A. Nerode, A. P. Ravn, & H. Rischel (Eds.), *Hybrid Systems, LNCS 736* (pp. 209–229). Heidelberg: Springer.

Alur, R., Henzinger, T. A., & Ho, P.-H. (1996). Automatic symbolic verification of embedded systems. *IEEE Transactions on Software Engineering*, *22*, 181–201.

Bavishi, S., & Chong, E. (1994). Automated fault diagnosis using a discrete event systems framework. *IEEE International Symposium Intelligent Control*, 213–218.

Biswas, S. (2013). *Equivalence of fair diagnosability and stochastic diagnosability of discrete event systems.* IEEE SMC, Manchester, pp. 378–383.

Biswas, S., Karfa, C., Kanwar, H., Sarkar, D., Mukhopadhyay, S., & Patra, A. (2006). *Fairness of transitions in diagnosability analysis of hybrid systems.* American Control Conference 2006, Minneapolis, MN, pp. 2664–2669.

Biswas, S., Sarkar, D., Mukhopadhyay, S., & Patra, A. (2010). Fairness of transitions in diagnosability of discrete event systems. *Journal of Discrete Event Dynamic Systems: Theory and Applications*, *20*(3), 349–376.

Chutinan, A., & Krogh, B. (1999). Verification of polyhedral-invariant hybrid automata using polygonal flow pipe approximations. *International Workshop on Hybrid Systems: Computation and Control*, 76–90.

Debouk, R., Lafortune, S., & Teneketzis, D. (1998). *Coordinated decentralized protocols for failure diagnosis of discrete event systems.* Proceeding of the 37th IEEE Conference on Decision and Control, Tampa, FL, pp. 3763–3768.

Frehse, G. (2005). Phaver: Algorithmic verification of hybrid systems past hytech. *International Workshop on Hybrid Systems: Computation and Control*, 258–273.

Jiang, S., Huang, Z., Chandra, V., & Kumar, R. (2001). A polynomial algorithm for testing diagnosability of discrete-event systems. *IEEE Transactions on Automatic Control*, *46*, 1318–1321.

Lafortune, S., Teneketzis, D., Sampath, M., Sengupta, R., & Sinnamohideen, K. (2001). *Failure diagnosis of dynamic systems: An approach based on discrete event systems.* American Control Conference, Arlington, VA, pp. 2058–2071.

Lamperti, G., & Zanella, M. (1999). *Diagnosis of discrete event systems integrating synchronous and asynchronous behavior.* 9th International workshop on principles of diagnosis (pp. 129–139).

Lin, F., & Wonham., W. M. (1988). On observability of discrete event system. *Information Sciences*, *44*, 173–198.

Liu, F. (2014). Polynomial-time verification of diagnosability of fuzzy discrete event systems. *Science China Information Sciences*, *57*(6), 1–10.

Moreira, M. V., Jesus, T. C., & Basilio, J. C. (2011). Polynomial time verification of decentralized diagnosability of discrete event systems. *IEEE Transaction on Automation Control*, *56*(7), 1679–1684.

Nuutila, E., & Soisalon-Soininen, E. (1994). On finding the strongly connected components in a directed graph. *Information Processing Letters*, *49*, 9–14.

Sampath, M., Sengupta, R., Lafortune, S., Sinnamohideen, K., & Teneketzis, D. (1995). Diagnosability of discrete-event systems. *IEEE Transaction on Automatic Control*, *40*(9), 1555–1575.

Sampath, M., Sengupta, R., Lafortune, S., Sinnamohideen, K., & Teneketzis, D. C. (1996). Failure diagnosis using discrete-event models. *IEEE Transaction on Control Systems Technology*, *4*(2), 105–124.

Thorsley, D., & Teneketzis, D. (2005). Diagnosability of stochastic discrete-event systems. *IEEE Transaction on Automatic Control*, *50*(4), 476–492.

Xie, A., & Beerel, P. A. (1998a). Efficient state classification of finite-state Markov chains. *IEEE Transaction on CAD Circuits System*, *17*(12), 1334–1339.

Xie, A., & Beerel, P. A. (1998b). Efficient state classification of finite-state Markov chains. *IEEE Transaction on CAD of Circuits and Systems*, *17*(12), 1334–1339.

Yoo, T.-S., & Lafortune, S. (2002). Polynomial time verification of diagnosability of partially observed discrete event systems. *IEEE T Automatic Control*, 129–139.

Zad, S. H., Kwong, R. H., & Wonham, W. M. (2003). Fault diagnosis in discrete-event systems: Framework and model reduction. *IEEE Transaction on Automatic Control*, *48*(7), 1199–1212.

Zad, S. H., Kwong, R. H., & Wonham, W. M. (2005). Fault diagnosis in discrete-event systems: Incorporating timing information. *IEEE Transaction on Automatic Control*, *50*(7), 1010–1015.

2

Robust stability on averaging behaviour of linear time-varying uncertain systems

Ping-Min Hsu[a] and Chun-Liang Lin[b*]

[a]*Automotive Research & Testing Center, Lukang, Changhua 50544, Taiwan;* [b]*Department of Electrical Engineering, National Chung Hsing University, Taichung 402, Taiwan*

This research proposes a new type of robust stability – mean robust asymptotic stability for linear time-varying (LTV) uncertain systems and its extension in robust asymptotic stability analysis. The system will feature this kind of stability if its state mean varies towards an equilibrium point. Robust input–output finite-time stability of the LTV uncertain system over a bounded time interval is first analysed by proposing a criterion to characterize the stability condition. The possible cases of steady-state error and state oscillation are then considered and tackled. Finally, some case studies are presented to demonstrate superiority of the proposed work.

Keywords: asymptotic stability; linear systems; robust stability; stability analysis; stability robustness

1. Introduction

Finite-time stability (FTS) was introduced based on the definition that a system with bounded initial states would be finite-time stable if the system states do not exceed a threshold over a bounded time interval. An extension of FTS, named input–output FTS (IO-FTS), has recently been investigated for the systems with norm-bounded input over a bounded time interval (Ma & Jia, 2011). It is said to be IO-FTS if its output does not exceed a predefined threshold over the time interval. Sufficient conditions of IO-FTS for stochastic Markovian jump systems and linear singular systems were proposed by Juan, June, Lin, and Yu (2011) and Ma and Jia (2011). In Ma and Jia (2011), the IO finite-time mean square stability of the stochastic Markovian jump system was studied. In Juan et al. (2011), the IO-FTS for the time-varying singular system was solved.

Amato, Carannante, De Tommasi, and Pironti (2012) summarized the sufficient and necessary conditions of IO-FTS for the linear systems; in particular, Theorem 3 in this work proposed a sufficient condition of stability for the linear time-varying (LTV) systems. Although it works well for the demonstrative case in Amato et al. (2012), this theorem would not be valid for systems having a singular system matrix at certain time instant or intervals. Moreover, robust stability analysis of the LTV system remains to be studied due to the existence of unknown noise input or modelling uncertainty. Robust IO-FTS was introduced in Amato, Cosentino, and De Tommasi (2011) for the linear system, which was treated as robust IO-FTS if it is IO-FTS for admissible uncertainties/disturbances. Nevertheless, the robust IO-FTS system may still suffer state oscillation, which should not be ignored in practical applications. Robust asymptotic stability analysis (Bilman, 2004; Cheng, Su, & Chien, 2012; Corradini, Cristofaro, & Orlando, 2010; Kelly, 1996; Kerrigan & Maciejowski, 2004; Li, Gao, & Lu, 2011; Li, Wang, & Lu, 2012; Lu & Chen, 2009; Lu & Chen, 2010; Xu & Lam, 2005) is an appropriate method to remedy the above drawback. Although this analysis has been extensively discussed in the manner of linear matrix inequalities (LMI), quantized H_∞ approach (Li et al., 2011), time-varying sliding surface (Corradini et al., 2010), new type of stage cost (Kerrigan & Maciejowski, 2004), and Lyapunov approach (Cheng et al., 2012; Li et al., 2012), the issue related to robust asymptotic stability study from the viewpoint of the state's averaging behaviour is still absent. Especially, such discussions are useful for the network systems, since for that kind of systems, only its averaging behaviour can be technically characterized.

Motivated by the problems depicted above, we propose in this research a novel IO-FTS theorem focusing on the improvement of the result in Amato et al. (2012). For the IO-FTS LTV uncertain system, the issue of stability robustness is discussed, in the sense that it is robustly asymptotically stable with respect to its mean if the state mean values approach the equilibrium state. This contributes the benefit of analysing robust asymptotic stability by identifying the state oscillation and brings the lunching point of the idea.

*Corresponding author. Email: chunlin@dragon.nchu.edu.tw

2. Problem statement

In the remaining parts, O represents a zero matrix with appropriate dimensions, 0 denotes a zero vector, and $\lambda_i(M(t))$ denotes the i-th eigenvalue of the matrix $M(t)$ at t. Moreover, under the assumption that the ergodic hypothesis holds, the expected value of the random vector $d(t) = [v_{dij}(t)]_{m_1\times 1}$ is defined as $d_m(t) = E[d(t)] \triangleq \left[\int_{t_0}^{t} v_{dij}(\tau)d\tau/(t-t_0)\right]_{m_1\times 1}$.

2.1. System description

Consider the LTV uncertain system described by

$$\dot{x}(t) = A(t)x(t) + B_1(t)u(t) + B_2(t)w(t), \quad (1)$$
$$y(t) = C(t)x(t),$$

where $x(t) \in \mathbb{R}^n$, the modelling uncertainty $w(x(t), u(t), d(t)) = [v_{wij}(t)]_{m\times 1}$, the control input $u(t) \in \mathbb{R}^{n_u}$, $B_1(t) \in \mathbb{R}^{n\times n_u}$, $B_2(t) = [v_{bij}(t)]_{n\times m}$, the system output $y(t) \in \mathbb{R}^q$, $A(t) = [v_{aij}(t)]_{n\times n}$, and $C(t) \in \mathbb{R}^{q\times n}$. Referring to Amato et al. (2012), we suppose that (i) $u(t) \in U_2(\Omega, R(\cdot)) \equiv \{u(\cdot) \in L_2(\Omega) : \|u\|_{2,R} \le 1\}$ where $\Omega \triangleq [t_0, T]$, $L_2(\Omega) = \left\{l(\cdot) \middle| \left(\int_\Omega \|l(t)\|^2 dt\right)^{1/2} < \infty\right\}$, $\|l\|$ denotes the Euclidian norm of the vector l with appropriate dimensions, $\|u\|_{2,R} \triangleq \left(\int_\Omega u^T(t)R(t)u(t)dt\right)^{1/2}$, and $R(\cdot)$ is the symmetric norm-bounded continuous positive definite matrix-valued function; and (ii) $w(t) \in U_2(\Omega, R(\cdot))$. If $u(t) = K(t)x(t)$ where $K(\cdot)$ is the piecewise continuous matrix-valued function, the closed-loop system of (1) is

$$\dot{x}(t) = A_{cl}(t)x(t) + B_2(t)w(t), \quad (2)$$
$$y(t) = C(t)x(t),$$

where $A_{cl}(t) = A(t) + B_1(t)K(t) = [v_{aij}(t)]_{n\times n}$.

2.2. Introduction of robust IO-FTS

The idea of robust IO-FTS (2) is defined below:

DEFINITION 1 (Amato et al., 2011): *Given U_2, Ω, and a continuous positive definite matrix-valued function $Q(\cdot)$, the system (2) is said to be robust IO-FTS corresponding to $(U_2, Q(\cdot), \Omega)$ if $y^T(t)Q(t)y(t) < 1$, $\forall t \in \Omega$ for any admissible $w(t)$.*

Theorem 3 in Amato et al. (2012) can be applied to guarantee robust IO-FTS of (2) based on Definition 1. It is stated as

THEOREM 1 (Amato et al., 2012): *Consider the system (2). Given $U_2(\Omega, R(\cdot))$, $Q(\cdot)$, and Ω. The system is robust IO-FTS with respect to $(U_2, Q(\cdot), \Omega)$ if there is $(K(\cdot), P(\cdot) > 0)$ satisfying*

$$\begin{pmatrix} \dot{P}(t) + A_{cl}^T(t)P(t) + P(t)A_{cl}(t) & P(t)B_2(t) \\ B_2^T(t)P(t) & -R(t) \end{pmatrix} < 0, \quad (3)$$

$$P(t) > C^T(t)Q(t)C(t), \quad (4)$$

over Ω.

In Amato et al. (2012), the LMI (3) has been equalized to

$$\dot{P}(t) + P(t)A_{cl}(t) + A_{cl}^T(t)P(t) + P(t)B_2(t)R^{-1}(t)B_2^T(t)P(t) + Q_p(t) = 0, \quad (5)$$

by using Schur complements, where $Q_p(t) > 0$, $\forall t \in \Omega$.

2.3. Problem statement

Theorem 1, requiring $P(t)$ satisfying (4) and (5) over Ω, is invalid for (2) because of violation of (5) over Ω if it features the singular $A_{cl}(t)$ at some $t \in \Omega$ (such as a linearized system of variable stiffness joint dynamics (Choi, Hong, Lee, Kang, & Kim, 2011) around its equilibrium point). More precisely, (5) is a Lyapunov matrix differential equation if $B_2(t) = O$; it does not uniquely admit

$$P(t) = \Phi_A^{-T}(t, t_0)P(t_0)\Phi_A^{-1}(t, t_0) - \int_{t_0}^{t} \Phi_A^T(t, \tau)Q_p(\tau)\Phi_A(t, \tau)d\tau, \quad (6)$$

proposed by Abou-Kandil, Freiling, Ionesco, and Jank (2003) and Davis, Gravagne, Marks, and Ramos (2010) at these instances with the state transition matrix $\Phi_A(t, t_0)$ of (2) since $\Phi_A^{-1}(t, t_0)$ does not exist at the same times.

To illustrate, without loss of generality, let $A_{cl}(t) \equiv A_\alpha(t) - A_\beta(t)$ where $A_\alpha(t)$ is non-singular at $t \in U_t \equiv \{t \in \Omega | A_{cl}(t) \text{ is singular at } t\}$ and $A_\alpha(t) \ne A_\beta(t)$, $\forall t \in U_t$. By contradiction, we suppose that $\Phi_A^{-1}(t, t_0)$ exists at all $t \in U_t$. Furthermore,

$$\det(\dot{\Phi}_A(t, t_0)) = \det(A_{cl}(t))\det(\Phi_A(t, t_0)) = 0, \quad \forall t \in U_t, \quad (7)$$

where $\det(A_{cl}(t)) = 0$ over U_t. Substituting $A_{cl}(t) = A_\alpha(t) - A_\beta(t)$ into (7) gives

$$\det(A_\alpha(t)\Phi_A(t, t_0) - A_\beta(t)\Phi_A(t, t_0)) = \det(A_\alpha(t))\det(\Phi_A(t, t_0))\det(I - A_\alpha^{-1}(t)A_\beta(t)) = 0, \quad \forall t \in U_t, \quad (8)$$

where $\det(I - A_\alpha^{-1}(t)A_\beta(t)) \ne 0$ and $\det(A_\alpha(t)) \ne 0$, $\forall t \in U_t$. However, (8) implies $\det(\Phi_A(t, t_0)) = 0, \forall t \in U_t$. Hence, $\Phi_A^{-1}(t, t_0)$ does not exist at all $t \in U_t$. Nevertheless, the considered system may still be IO-FTS with respect to $(U_2, Q(\cdot), \Omega)$.

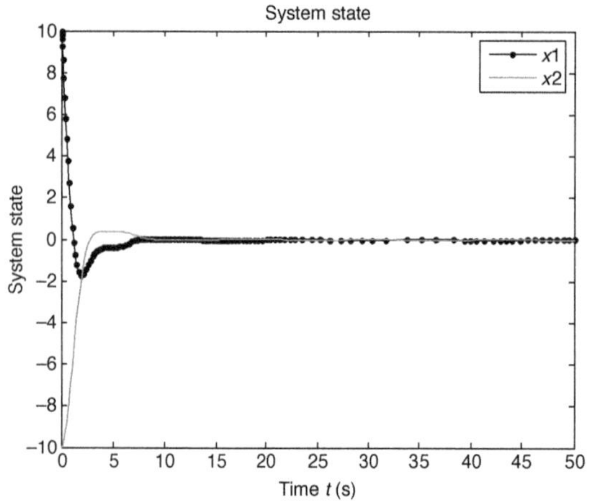

Figure 1. Deficiency of Theorem 1.

To demonstrate, let us consider

$$\dot{x}(t) = A_{\text{cl}}(t)x(t),$$
$$y(t) = x(t), \tag{9}$$

where $A_{\text{cl}}(t) = \begin{bmatrix} -1 & \sin(t) \\ -1 & -1 \end{bmatrix}$ is singular at $t = 3\pi k/2$ (s), $k \in \mathbb{N}$ over $\Omega = [0, 50]$(s). Theorem 1 is inapplicable to (9) since $P(t)$ guaranteeing (5) does not exist at $t = 3\pi k/2$ (s), $k \in \mathbb{N}$ over Ω but (9) is IO-FTS with respect to $(U_2, Q(\cdot), \Omega)$ at $x(0) = \begin{bmatrix} 10 & -10 \end{bmatrix}^{\text{T}}$ with $Q = I/120$, as illustrated in Figure 1. This gives rise to a demand to amend Theorem 1.

Robust IO-FTS of (2) with respect to the unknown $w(t) \in U_2(\Omega, R(\cdot))$ is studied first in the following sections. However, it may encounter steady-state error or oscillation. This gives rises to another concern: is it possible to ensure asymptotic stability around at the equilibrium point of (2) by the control law $u(t)$? This will be addressed as well.

3. Main results

This section deals with three issues: (i) robust IO-FTS by improving Theorem 1; (ii) robust asymptotic stability with respect to its mean; and (iii) robust asymptotic stability. The first issue requires a redefinition of U_2 with $U_{2b}(T) \equiv \{u(\cdot) \in L_2(\Omega) : \|u\|_{2,\Omega} \le \varepsilon_{b0}\}$ where $\|u\|_{2,\Omega} \equiv \left(\int_\Omega u^{\text{T}}(t)u(t)\text{d}t\right)^{1/2}$. $U_2(\Omega, R(\cdot)) \subseteq U_{2b}(T)$ under $\varepsilon_{b0} \equiv 1/\gamma_0^{1/2}$ since $\gamma_0^{1/2}\|u\|_{2,\Omega} \le \|u\|_{2,R} \le 1$ where $u^{\text{T}}(t)R(t)u(t) \equiv \gamma(t)u^{\text{T}}(t)u(t)$, $0 < \gamma(t) < \infty$, $\forall t \in \Omega$, and $\gamma_0 \le \min_{t\in\Omega}\gamma(t)$. The next discussion is fulfilled by assuming $w(t) \in U_{2b}(T)$.

3.1. IO-FTS analysis of the LTV system

THEOREM 2 *Consider the system (2). Suppose that $Q_1(t) \ge Q_2(t)$, and $Q_3(t) > 0$, $\forall t \in \Omega$. The system is robust IO-FTS with respect to $(U_{2b}(T), \tilde{Q}(\cdot), \Omega)$ where $\tilde{Q}(t) = Q_3(t)/[\varepsilon_{b0}^2 + x^{\text{T}}(t_0)\tilde{P}(t_0)x(t_0)]$ if there is $(K(\cdot), A_1(\cdot), A_2(\cdot), \tilde{P}(\cdot))$ ensuring*

$$A(t) + B_2(t)K(t) = A_1(t) - A_2(t), \tag{10}$$

$$A_1^{\text{T}}(\tau)\tilde{P}(\tau) + \tilde{P}(\tau)A_1(\tau) = -Q_1(\tau), \quad \tau \in [t_0, t], \tag{11}$$

$$\dot{\tilde{P}}(\tau) = A_2^{\text{T}}(\tau)\tilde{P}(\tau) + \tilde{P}(\tau)A_2(\tau) - \tilde{P}(\tau) B_2(\tau)B_2^{\text{T}}(\tau)\tilde{P}(\tau) + Q_2(\tau), \quad \tau \in [t_0, t], \tag{12}$$

$$\tilde{P}(t) \ge C^{\text{T}}(t)Q_3(t)C(t), \tag{13}$$

where $\tilde{P}(\cdot) > 0$, $\forall t \in \Omega$.

Proof Set $V(\tau) = x^{\text{T}}(\tau)\tilde{P}(\tau)x(\tau)$ then

$$\dot{V}(\tau) = x^{\text{T}}(\tau)[A_{\text{cl}}^{\text{T}}(\tau)\tilde{P}(\tau) + \tilde{P}(\tau)A_{\text{cl}}(\tau) + \dot{\tilde{P}}(\tau)] \cdot x(\tau) + w^{\text{T}}(\tau) B_2^{\text{T}}(\tau)\tilde{P}(\tau)x(\tau) + x^{\text{T}}(\tau)\tilde{P}(\tau)B_2(\tau)w(\tau), \tag{14}$$

where $A_{\text{cl}}(\tau) = A_1(\tau) - A_2(\tau)$ due to (10). From (11)–(12), it follows

$$\dot{V}(\tau) = -x^{\text{T}}(\tau)[Q_1(\tau) - Q_2(\tau)]x(\tau) + w^{\text{T}}(\tau)w(\tau) - a^{\text{T}}(\tau)a(\tau) < w^{\text{T}}(\tau)w(\tau), \tag{15}$$

over $t \in [t_0, t]$ where $Q_1(t) - Q_2(t) \ge O$ and $a(\tau) = w(\tau) - B_2^{\text{T}}(\tau)\tilde{P}(\tau)x(\tau)$. Integrating (15) over $[t_0, t]$ yields

$$x^{\text{T}}(t)\tilde{P}(t)x(t) - x^{\text{T}}(t_0)\tilde{P}(t_0)x(t_0) < \int_{t_0}^{t} w^{\text{T}}(\tau)w(\tau)\text{d}\tau \le \int_\Omega w^{\text{T}}(\tau)w(\tau)\text{d}\tau \le \varepsilon_{b0}^2, \tag{16}$$

since $w^{\text{T}}(t)w(t) \ge 0$ and $w(t) \in U_{2b}(T)$. That is, $x^{\text{T}}(t)\tilde{P}(t) x(t) < \varepsilon_{b0}^2 + x^{\text{T}}(t_0)\tilde{P}(t_0)x(t_0)$. From (13), $y^{\text{T}}(t)Q_3(t)y(t) < \varepsilon_{b0}^2 + x^{\text{T}}(t_0)\tilde{P}(t_0)x(t_0)$. That is $y^{\text{T}}(t)\tilde{Q}(t)y(t) < 1$ where $\tilde{Q}(t) = Q_3(t)/[\varepsilon_{b0}^2 + x^{\text{T}}(t_0)\tilde{P}(t_0)x(t_0)]$. According to Definition 1, the system (2) is thus possessing robust IO-FTS with respect to $(U_{2b}(T), \tilde{Q}(\cdot), \Omega)$.

Remark 1 The pair $(K(\cdot), A_1(\cdot), A_2(\cdot), \tilde{P}(\cdot))$ can be decided via five steps: (i) determine $A_1(t)$ as a stable upper triangular matrix over Ω so that (11) must admit $\tilde{P}(\cdot)$; (ii) evaluate $A_2(\cdot)$ by substituting $\tilde{P}(\cdot)$ into (12); (iii) compute $K(\cdot)$ via (10); (iv) verify (13) with the obtained $\tilde{P}(\cdot)$; and (v) repeat Step (i) by setting another $A_1(\cdot)$ such that $\tilde{P}(\cdot)$ has larger $\text{Re}[\lambda_i(\tilde{P}(t))]$ for all i if (13) does not work.

Comparing Theorem 1 with Theorem 2 when $R(t) = I$, the matrix inequality (3) can be derived by combining (11) and (12) while (13) remains invariant as (4). To illustrate, combining (11) and (12) yields

$$\dot{\tilde{P}}(t) + A_{\rm cl}^{\rm T}(t)\tilde{P}(t) + \tilde{P}(t)A_{\rm cl}(t) + \tilde{P}(t)B_2(t)B_2^{\rm T}(t)\tilde{P}(t) = -[Q_1(t) - Q_2(t)], \quad (17)$$

where $Q_1(t) - Q_2(t) > O$ is determined for the system (2) with the non-singular $A_{\rm cl}(t)$ over Ω. Equation (17) is consistent with (5); as a result, the sufficient condition in Theorem 1 becomes a special case to Theorem 2 when $R(t) = I$.

3.2. *Robust asymptotic stability of IO-FTS systems with respect to its mean*

This section focuses on the steady-state error rejection of (2) with $\bar{B}_2(t) = \left[\int_{t_0}^{t} v_{bij}(\tau)d\tau/(t - t_0)\right]_{n\times m} \neq O$ and $w(t) \in \tilde{U}_2(T) \equiv \{u(t) : \|u(t)\|_2 \leq \varepsilon_0, t \in [t_0, T]\}$.

DEFINITION 2 *The system (2) is said to be robustly asymptotically stable with respect to its mean if* $\lim_{t\to\infty} E[x(t)] = 0$.

The major difference between robustly asymptotic stability and that in Definition 2 is that the former investigates actual state response while the proposed stability specifies the averaging state behaviour. The definition is useful for stability characterization of the local network system (Misra, Gong, & Towsley, 2000), whose averaging dynamic behaviour can be illustrated by the following fluid-flow model (Hollot, Misra, & Towsley, 2002):

$$\dot{\bar{w}}(t) = \frac{3}{4}\frac{[1 - p(t - R_a)]}{R_a(t - R_a)} - \frac{7}{20}\frac{\bar{w}(t)\bar{w}(t - R_a)p(t - R_a)}{R_a(t - R_a)},$$
$$\dot{\bar{q}}(t) = \bar{w}(t)n_f/R_a(t) - C_p, \quad (18)$$

where $R_a(t) = a + \bar{q}/C_p$ denotes the averaging round-trip time, a refers to the propagation delay, $\bar{q}/\bar{w}$ represents the averaging queue length/window size, C_p is the capacity, n_f is the number of Transmission Control Protocol flows, and p is the dropping probability. Performing linearization with respect to (18) around its equilibrium point (w_0, q_0, p_0) yields the linearized model of the following form:

$$\Delta\dot{\bar{w}} = -\frac{n_f}{R_0^2C_p}\Delta\bar{w} - \frac{n_f}{R_0^2C_p}\Delta\bar{w}(t - R_a) - \frac{R_0C_p^2}{2n_f^2}\Delta p(t - R_a), - \frac{\Delta\bar{q}}{R_0^2C_p} + \frac{\Delta\bar{q}(t - R_a)}{R_0^2C_p}, \quad (19)$$
$$\Delta\dot{\bar{q}} = \frac{n_f\Delta\bar{w}}{R_0} - \frac{\Delta\bar{q}}{R_0},$$

where $R_0 = a + q_0/C_p$, $\Delta\bar{w} = \bar{w} - w_0$, $\Delta\bar{q} = \bar{q} - q_0$, and $\Delta p = p - p_0$. The linearized system describing the behaviour of (q, w) around (w_0, q_0, p_0) is focused here with $q(t)/w(t)$ being the real queue length/window size. It can be deduced by applying (19) and extended network disturbances (END) (Hsu & Lin, 2012). Using $\Delta p(t) = K(t)x(t)$, the closed-loop system can be formulated in the form of (2) where $A_{\rm cl}(t) = A_x + B_xK(t), B_2(t) = I_2, C(t) = I_2$, $A_x = \begin{bmatrix} -2n_f/R_0^2C_p & 0 \\ n_f/R_0 & -1/R_0 \end{bmatrix}$, $B_x = \begin{bmatrix} -R_0C_p^2/2n_f^2 \\ 0 \end{bmatrix}$, $x(t) = \begin{bmatrix} \Delta w(t) & \Delta q(t) \end{bmatrix}^{\rm T}$, $w_n(t) = -B_xd_{\rm end}(t) - \begin{bmatrix} \dot{d}_{yw}(t) & \dot{d}_{yq}(t) \end{bmatrix}^{\rm T}$, $d_{\rm end}(t)$ denotes END, $\Delta w(t) = w(t) - w_0$, $\Delta q(t) = q(t) - w_0$, $d_{yw}(t) = w(t) - \bar{w}(t)$, and $d_{yq}(t) = q(t) - \bar{q}(t)$. Although (19) can be asymptotically stabilized with $\Delta p = K\begin{bmatrix} \Delta\bar{w} & \Delta\bar{q} \end{bmatrix}^{\rm T}$ (Hollot et al., 2002), this does not mean that (2) simultaneously holds asymptotic stability. For instance, consider Example 2 in Hollot et al. (2002), in which $A_x = \begin{bmatrix} -0.5288 & 0 \\ 243.9024 & -4.065 \end{bmatrix}$, $B_x = \begin{bmatrix} -480.4688 & 0 \end{bmatrix}^{\rm T}$, and $\Delta p = K\begin{bmatrix} \Delta\bar{w} & \Delta\bar{q} \end{bmatrix}^{\rm T}$ with $K = \begin{bmatrix} 0 & 5.8624 \cdot 10^{-5} \end{bmatrix}$. Figure 15(a) in Hollot et al. (2002) shows that the network system implementing this controller leads to queue oscillation. This motivates the thought to express Δp in terms of $K(t)$ so that $\Delta p(t)$ becomes more sensitive to the queue variation as $t \to \infty$. This brings the possibility of reducing state oscillation in (2).

For the steady-state error in (2), it can be analysed from the viewpoint of the state's mean value because its averaging system behaviour is precisely governed by (18).

The robust stability of (2) is studied under the following conditions:

(C1): $\lim_{t\to\infty} E[B_2(t)w(t)] = 0$,
(C2): $\lim_{t\to\infty} E[B_2(t)w(t)] \neq 0$.

The model (2) in (C1) would be robustly asymptotically stable with respect to its mean by Definition 2, if, and only if, the system $E[\dot{x}(t)] = E[A_{\rm cl}(t)x(t)]$ has a norm-bounded state transition matrix approaching O as $t \to \infty$. However, this is invalid in (C2). To proceed with the robust stability of (2) in (C2), we show the following result with $\Omega = [t_0, \infty)$.

LEMMA 1 *For* $w_m(t) = E[w(t)]$ *with bounded* $\|w_m(t_0)\|_2$, $\lim_{t\to\infty} \dot{w}_m(t) = 0$, *if, and only if,* $w(t) \in \tilde{U}_2(\infty)$.

Proof For the sufficient condition, one has $w_m(t) = \left[\int_{t_0}^{t} v_{wij}(\tau)d\tau/(t - t_0)\right]_{m\times 1} \in \tilde{U}_2(\infty)$ since $\|w_m(t)\|_2 \leq \int_{t_0}^{t} \|w(\tau)\|_2 d\tau/(t - t_0) \leq \varepsilon_0, \forall t \in [t_0, \infty)$ where $\|w(t)\|_2 \leq \varepsilon_0$ over Ω and $w(t) \in \tilde{U}_2(\infty)$. Since $\dot{w}_m(t) = [w(t) -$

$w_m(t)]/(t-t_0)$ then

$$\lim_{t\to\infty}\|\dot{w}_m(t)\|_2 \le \frac{\lim_{t\to\infty}2\varepsilon_0}{t-t_0} = 0, \tag{20}$$

where $\|w(\infty)\|_2 \le \varepsilon_0$ and $\|w_m(\infty)\|_2 \le \varepsilon_0$ because $w(t), w_m(t) \in \tilde{U}_2(\infty)$. Clearly, (20) reveals $\lim_{t\to\infty}\|\dot{w}_m(t)\|_2 = 0$. As a result, $\lim_{t\to\infty}\dot{w}_m(t) = 0$.

For the necessary part with $\lim_{t\to\infty}\dot{w}_m(t) = 0$ and $\|\dot{w}_m(t_0)\|_2 = 0$, this reveals that $\|\dot{w}_m(t)\|_2$ is bounded and inversely proportional to t. Hence, without loss of generality, $\|\dot{w}_m(t)\|_2 \triangleq \varepsilon_1(t)(t-t_0)$ and $(\varepsilon_1(t_0), \dot{\varepsilon}_1(t_0)) = (0,0)$. Substituting it into (20) gives $\varepsilon_1(t) = \|w(t)-w_m(t)\|_2$ that implies

$$\varepsilon_1(t) \ge \|w(t)\|_2 - \|w_m(t)\|_2, \tag{21}$$

where $\|w_m(t)\|_2$ is bounded over $[t_0,\infty)$ due to the invariant $\lim_{t\to\infty}\|w_m(t)\|_2$ and $\|w_m(t_0)\|_2 < \infty$. That is, $\|w_m(t)\|_2 = \varepsilon_2(t) < \infty, \forall t \in [t_0,\infty)$ and $\dot{\varepsilon}_2(\infty) = 0$. Substituting $\|w_m(t)\|_2 = \varepsilon_2(t)$ into (21) gives $\|w(t)\|_2 \le \varepsilon_1(t)+\varepsilon_2(t) < \infty$ where there is a finite constant $\varepsilon_{30} \in \mathbb{R}^+$ such that $\|w(t)\|_2 \le \varepsilon_1(t)+\varepsilon_2(t) < \varepsilon_{30}$, $\forall t \in [t_0,\infty)$. Hence $w(t) \in \tilde{U}_2(\infty)$.

In addition to $\tilde{U}_2(T)$, Lemma 1 can be extended to the case of

$$\tilde{U}(T,F,\eta) = \{F(\cdot)\,|\,\|F(t)\|_2 \le \varepsilon_{M0}, t \in [t_0,T]\}, \tag{22}$$

where $\varepsilon_{M0} < \infty$, $F(t) = [v_{fij}(t)]_{n_1\times n_2}$, the induced norm $\|F(t)\|_2 = \sup_{\eta(t)\ne 0}\|F(t)\eta(t)\|_2/\|\eta(t)\|_2$ with $\eta \in \mathbb{R}^{n_2}$.

LEMMA 2 *Given* $\tilde{U}$ *and* $\bar{F}(t) = \left[\int_{t_0}^t v_{fij}(\tau)\mathrm{d}\tau/(t-t_0)\right]_{n_1\times n_2}$ *with bounded* $\|\bar{F}(t_0)\|_2$, $\lim_{t\to\infty}\dot{\bar{F}}(t) = O$, *if, and only if,* $F(t) \in \tilde{U}(\infty,F,\eta)$.

Proof Follow the similar process of the proof of Lemma 1.

We now present the following result with $\bar{F}_1(t) = \left[\int_{t_0}^t v_{f_1ij}(\tau)\mathrm{d}\tau/(t-t_0)\right]_{n_1\times n_2}$, $F_1(t) = [v_{f_1ij}(t)]_{n_1\times n_2}$, $w_{1m}(t) = E[w_1(t)] = [v_{w_1ij}(t)]_{n_2\times 1}$, and the random vector $w_1(t) \in \mathbb{R}^{n_2\times 1}$.

LEMMA 3 *Given* $\lim_{t\to\infty}E[F_1(t)(w_1(t)-w_{1m}(t))] = \alpha_0$, $\lim_{t\to\infty}E[F_1(t)w_1(t)] = \bar{F}_1(\infty)w_{1m}(\infty)+\alpha_0$, *if* $F_1(t) \in \tilde{U}(\infty,F_1,w_1)$ *and* $w_1(t) \in \tilde{U}_2(\infty)$.

Proof Let $F_1(t)w_{1m}(t) = [v_{hij}(t)]_{n_1\times 1}$,

$$\lim_{t\to\infty}\left\{\left[\int_{t_0}^t \frac{v_{hij}(\tau)\mathrm{d}\tau}{t-t_0}\right]_{n_1\times 1} - \bar{F}_1(t)w_{1m}(t)\right\}$$
$$= \lim_{t\to\infty}\{[\bar{v}_1(t)\ \cdots\ \bar{v}_{n_2}(t)]^{\mathrm{T}}\}, \tag{23}$$

where $\bar{v}_i(t) = \sum_{k=1}^{n_2}\int_{t_0}^t v_{f_1ik}(\tau)[v_{w_1k1}(\tau)-v_{w_1k1}(\infty)]\,\mathrm{d}\tau/(t-t_0)$ and $w_{1m}(\infty) = [v_{w_1ij}(\infty)]_{n_2\times 1}$ is constant from Lemma 1 due to $w_1(t) \in \tilde{U}_2(\infty)$. This reveals $v_{f_1ik}(t)[v_{w_1k1}(t)-v_{w_1k1}(\infty)] \to 0$ in (23) as $t\to\infty$ for $i = 1 \sim n_2$ since $|v_{f_1ik}(t)| < \infty$ over $[t_0,\infty)$ upon $F_1(t) \in \tilde{U}(\infty,F_1,w_1)$. In other words, $\lim_{t\to\infty}\bar{v}_i(t) = 0$ because $\lim_{t\to\infty}\int_{t_0}^t v_{f_1ik}(\tau)[v_{w_1k1}(\tau)-v_{w_1k1}(\infty)]\mathrm{d}\tau/(t-t_0) = 0$ if $\int_{t_0}^\infty v_{f_1ik}(\tau)[v_{w_1k1}(\tau)-v_{w_1k1}(\infty)]\mathrm{d}\tau < \infty$; otherwise,

$$\lim_{t\to\infty}\int_{t_0}^t \frac{v_{f_1ik}(\tau)[v_{w_1k1}(\tau)-v_{w_1k1}(\infty)]\,\mathrm{d}\tau}{t-t_0}$$
$$= \lim_{t\to\infty} v_{f_1ik}(t)[v_{w_1k1}(t)-v_{w_1k1}(\infty)] = 0.$$

Therefore, $\lim_{t\to\infty}\left[\int_{t_0}^t v_{hij}(\tau)\mathrm{d}\tau/(t-t_0)\right]_{n_1\times 1} = \bar{F}_1(\infty)w_{1m}(\infty)$ from (23). As a result,

$$\lim_{t\to\infty}E[F_1(t)w_1(t)] = \lim_{t\to\infty}\left[\int_{t_0}^t \frac{v_{hij}(\tau)\mathrm{d}\tau}{t-t_0}\right]_{n_1\times 1}$$
$$+\lim_{t\to\infty}E[F_1(t)(w_1(t)-w_{1m}(t))]$$
$$= \bar{F}_1(\infty)w_{1m}(\infty)+\alpha_0, \tag{24}$$

where from Lemma 1, α_0 is constant since $F_1(t)[w_1(t)-w_{1m}(t)] \in \tilde{U}_2(\infty)$, in which $w_{1m}(t) \in \tilde{U}_2(\infty)$ by Lemma 1 when $w_1(t) \in \tilde{U}_2(\infty)$.

Lemma 3 shows that $E[F_1(t)w_1(t)] - \bar{F}_1(t)w_{1m}(t) - \alpha_0$ is proportional to $\dot{\bar{F}}_1(t)$ and $\dot{w}_{1m}(t)$; consequently, let

$$E[F_1(t)w_1(t)] = \bar{F}_1(t)w_{1m}(t) + \dot{\bar{F}}_1(t)\Sigma_1(t)$$
$$+\Sigma_2(t)\dot{w}_{1m}(t)+\alpha_0, \tag{25}$$

where $\Sigma_1(t) \in \tilde{U}_2(\infty)$ and $\Sigma_2(t) \in \tilde{U}(\infty,\Sigma_2,\dot{w}_{1m})$ with appropriate dimensions. Taking differentiation of (25):

$$\frac{\mathrm{d}E[F_1(t)w_1(t)]}{\mathrm{d}t} = \frac{F_1(t)w_1(t)-E[F_1(t)w_1(t)]}{t-t_0}, \tag{26}$$

gives

$$\lim_{t\to\infty}\frac{\mathrm{d}E[F_1(t)w_1(t)]}{\mathrm{d}t}$$
$$= \lim_{t\to\infty}\frac{F_1(t)w_1(t)-\bar{F}_1(t)w_{1m}(t)-\dot{\bar{F}}_1(t)\Sigma_1(t)-\Sigma_2(t)\dot{w}_{1m}(t)}{t-t_0}$$
$$= \lim_{t\to\infty}\left[F_1(t)-\frac{\Sigma_2(t)}{t-t_0}\right]\dot{w}_{1m}(t)$$
$$+\lim_{t\to\infty}\dot{\bar{F}}_1\left[w_{1m}(t)-\frac{\Sigma_1(t)}{t-t_0}\right] \tag{27}$$
$$= F_1(\infty)\dot{w}_{1m}(\infty)+\dot{\bar{F}}_1(\infty)w_{1m}(\infty).$$

where $F_1(t) = \dot{\bar{F}}(t)(t-t_0)+\bar{F}_1(t)$ and $w_1(t) = \dot{w}_{1m}(t)(t-t_0)+w_{1m}(t)$.

THEOREM 3 *Consider the system (2) which satisfies (C2). Suppose* $\alpha_1 > 0$, $B_2(t) \in \tilde{U}(\infty, B_2, w)$, $\bar{A}_{cl}(t) = \left[\int_{t_0}^{t} v_{aij}(\tau)d\tau/(t-t_0)\right]_{n\times n}$, $\tilde{A}_T(t) = \begin{bmatrix} O & I \\ \dot{\bar{A}}_{cl}(t) & \bar{A}_{cl}(t) \end{bmatrix}$, *and* $Q_{m1}(t) = Q_{m1}^T(t) \geq Q_{m2}(t) = Q_{m2}^T(t)$, $\forall t \in [t_1, \infty)$. *This system is robustly asymptotically stable with respect to its mean, if there are* $K(\cdot), \tilde{A}_a(\cdot), \tilde{A}_b(\cdot)$, and $\tilde{P}_1(\cdot) = \tilde{P}_1^T(\cdot) > 0$ *satisfying*

$$\tilde{A}_T(K(t)) = \tilde{A}_a(t) - \tilde{A}_b(t), \tag{28}$$

$$\tilde{A}_a^T(t)\tilde{P}_1(t) + \tilde{P}_1(t)\tilde{A}_a(t) = -Q_{m1}(t), \tag{29}$$

$$\dot{\tilde{P}}_1(t) = \tilde{A}_b^T(t)\tilde{P}_1(t) + \tilde{P}_1(t)\tilde{A}_b(t) - \alpha_1\tilde{P}_1^2(t) + Q_{m2}(t), \tag{30}$$

$\forall t \in [t_1, \infty)$ and $\tilde{A}_T(\infty)$ is full rank.

Proof Consider the associated model

$$E[\dot{x}(t)] = E[A_{cl}(t)x(t) + B_2(t)w(t)]. \tag{31}$$

It is desirable to introduce a transformation to convert it into the form without uncertainties by Lemmas 1–3. Expanding (31) at $x(t_0)$ gives

$$\begin{aligned} E[\dot{x}(t)] &= \frac{[x(t) - x_m(t)]}{t - t_0} + \frac{[x_m(t) - x(t_0)]}{t - t_0} \\ &= \dot{x}_m(t) + \frac{[x_m(t) - x(t_0)]}{t - t_0} \\ &= E[A_{cl}(t)x(t)] + E[B_2(t)w(t)]. \end{aligned}$$

Then,

$$\dot{x}_m(t) = E[A_{cl}(t)x(t)] + E[B_2(t)w(t)] - \frac{x_m(t)}{t - t_0} + \frac{x(t_0)}{t - t_0}. \tag{32}$$

The system would approach to a limit system (Lee, Liaw, & Chen, 2001) as $t \to \infty$. First, since $B_2(t) \in \tilde{U}(\infty, B_2, w)$ and $w(t) \in \tilde{U}_2(\infty)$, expanding $\lim_{t\to\infty} E[B_2(t)w_n(t)]$ in (32) and using Lemma 3 gives

$$\lim_{t\to\infty} E[B_2(t)w(t)] = \bar{B}_2(\infty)w_m(\infty) + \phi_0, \tag{33}$$

where ϕ_0 is a constant. Substituting (33) into (32) and performing differentiation gives

$$\begin{aligned} \lim_{t\to\infty} \ddot{x}_m(t) &= \lim_{t\to\infty} \left\{ \frac{dE[A_{cl}(t)x(t)]}{dt} + \dot{\bar{B}}_2(t)w_m(t) \right\} \\ &+ \lim_{t\to\infty} \left\{ \bar{B}_2(t)\dot{w}_m(t) - \frac{\dot{x}_m(t)}{t - t_0} + \frac{x_m(t)}{(t - t_0)^2} \right\} \\ &= \lim_{t\to\infty} \left\{ \frac{dE[A_{cl}(t)x(t)]}{dt} - \frac{\dot{x}_m(t)}{t - t_0} + \frac{x_m(t)}{(t - t_0)^2} \right\}, \end{aligned} \tag{34}$$

where $\dot{w}_m(\infty) = 0$ and $\dot{\bar{B}}_2(\infty) = O$ (from Lemmas 1 and 2), because $w(t) \in \tilde{U}_2(\infty)$ and $B_2(t) \in \tilde{U}(\infty, B_2, w)$. Second, $dE[A_{cl}(t)x(t)]/dt$ in (34) is expanded as (27) where $(F_1(t), w_1(t)) \triangleq (A_{cl}(t), x(t))$. Substituting this into (34) gives $\lim_{t\to\infty}\dot{\tilde{x}}_m(t) = \lim_{t\to\infty}\hat{A}_T(t)\tilde{x}_m(t)$ where $\hat{A}_T(t) = \begin{bmatrix} O & I \\ \dot{\bar{A}}_{cl}(t) + I/(t-t_0)^2 & \bar{A}_{cl}(t) - I/(t-t_0) \end{bmatrix} \to \tilde{A}_T(t)$ as $t \to \infty$, $\tilde{x}_m(t) = \begin{bmatrix} x_m(t) & \dot{x}_m(t) \end{bmatrix}^T$, and $\tilde{A}_T(\infty)$ is full rank. This reveals that (31) reaches a limit system defined by

$$\dot{\tilde{x}}_m(t) = \tilde{A}_T(t)\tilde{x}_m(t), \tag{35}$$

which has the equilibrium state $\tilde{x}_m = 0$.

Next, consider a Lyapunov function candidate given by $V_m(t) = \tilde{x}_m^T(t)\tilde{P}_1(t)\tilde{x}_m(t)$. Then

$$\dot{V}_m(t) = \tilde{x}_m^T(t)[\tilde{A}_T^T(t)\tilde{P}_1(t) + \tilde{P}_1(t)\tilde{A}_T(t) + \dot{\tilde{P}}_1(t)]\tilde{x}_m(t), \tag{36}$$

where $\tilde{A}_T(t) = \tilde{A}_a(t) - \tilde{A}_b(t)$ from (28). Substituting (29) and (30) into (36) gives

$$\begin{aligned} \dot{V}_m(t) &= -\tilde{x}_m^T(t)[Q_{m1}(t) - Q_{m2}(t)]\tilde{x}_m(t) - \alpha_1\tilde{x}_m^T(t)\tilde{P}_1^2(t)\tilde{x}_m(t) \\ &\leq -\alpha_1\tilde{x}_m^T(t)\tilde{P}_1^2(t)\tilde{x}_m(t) < 0, \end{aligned} \tag{37}$$

for $t \in [t_1, \infty)$ where $Q_{m1}(t) \geq Q_{m2}(t)$. As a result, (2) converges to (35). That is, the system is robustly asymptotically stable with respect to its mean.

The quaternate set $(K(\cdot), \tilde{A}_a(\cdot), \tilde{A}_b(\cdot), \tilde{P}_1(\cdot))$ can be designed by following the steps depicted in Remark 1 where $\tilde{A}_a(\cdot), \tilde{A}_b(\cdot)$, and $\tilde{P}_1(\cdot)$ correspond, respectively, to $A_1(\cdot), A_2(\cdot)$, and $\tilde{P}(\cdot)$ and (10)–(12) are substituted by (28)–(30), respectively.

Theorem 3 presents a function for identifying the existence of the steady-state error. It is suitable for theoretical analysis of the results in Example 2 (Hollot et al., 2002). This work did not provide explanations why the system was not suffering from steady-state error under the treatment $(\bar{w}, \bar{q}) = (w, q)$ for realization of the controller in terms of $(\bar{w}, \bar{q})$. Theorem 3 can be used to enhance the required analysis. The network work system (2), equipped with the congestion controller (Hollot et al., 2002) in terms of (w, q) instead, in Section 3.2 will not reveal steady-state error if it satisfies Theorem 3. This reveals advantage of Theorem 3.

The system (2), meeting Definition 2, might work without steady-state errors; however, there might exist oscillation in state responses. The problem will be addressed next.

3.3. *Robust asymptotic stability analysis via behaviour identification*

This section deals with oscillation behaviour of the system (2) where the system has been robustly asymptotically stabilized with respect to its mean by determining $K(\cdot)$ according to Theorem 3. Without loss of generality, its solution is denoted by $x(t) \equiv r(t) + x_m(t)$ where $x_m(\infty) = 0$ and $E[r(t)] = 0$ over $[t_0, \infty)$.

THEOREM 4 *Consider the robustly asymptotically stabilized system (2) with respect to its mean. Suppose* $\alpha_2 > 0$, $\tilde{A}_r(t) = \begin{bmatrix} O & I \\ \dot{A}_{\mathrm{cl}}(t) + \dot{\bar{A}}_{\mathrm{cl}}(t) & A_{\mathrm{cl}}(t) \end{bmatrix}$ *with* $\tilde{A}_r(\infty)$ *being full rank,* $\dot{A}_{\mathrm{cl}}(t) \in \tilde{U}(\infty, \dot{A}_{\mathrm{cl}}, x_m)$, $\dot{\bar{A}}_{\mathrm{cl}}(t) \in \tilde{U}(\infty, \dot{\bar{A}}_{\mathrm{cl}}, x_m)$, *and* $Q_{r1}(t) = Q_{r1}^{\mathrm{T}}(t) \geq Q_{r2} = Q_{r2}^{\mathrm{T}}(t)$ *for all* $t \in [t_1, \infty)$. *This system is robustly asymptotically stable, if there are* $\tilde{A}_{r1}(\cdot)$ *and* $\tilde{P}_2(\cdot) = \tilde{P}_2^{\mathrm{T}}(\cdot) > O$ *satisfying*

$$[\tilde{A}_r(t) + \tilde{A}_{r1}(t)]^{\mathrm{T}}\tilde{P}_2(t) + \tilde{P}_2(t)[\tilde{A}_r(t) + \tilde{A}_{r1}(t)] = -Q_{r1}(t), \quad \forall t \in [t_1, \infty), \tag{38}$$

$$\dot{\tilde{P}}_2(t) = \tilde{A}_{r1}^{\mathrm{T}}(t)\tilde{P}_2(t) + \tilde{P}_2(t)\tilde{A}_{r1}(t) - \alpha_2\tilde{P}_2^2(t) + Q_{r2}(t), \quad \forall t \in [t_1, \infty). \tag{39}$$

Proof Substituting $x(t) = r(t) + x_m(t)$ into (2) yields

$$\begin{aligned} \dot{r}(t) &= A_{\mathrm{cl}}(t)r(t) + A_{\mathrm{cl}}(t)x_m(t) + \dot{x}(t) - A_{\mathrm{cl}}x(t) - \dot{x}_m(t) \\ &= A_{\mathrm{cl}}(t)r(t) - A_{\mathrm{cl}}(t)\dot{x}_m(t)(t - t_0) + \ddot{x}_m(t)(t - t_0) \\ &\quad + \dot{x}_m(t), \end{aligned} \tag{40}$$

where $x_m(t) = x(t) - \dot{x}_m(t)(t - t_0)$. Then

$$\begin{aligned} \lim_{t\to\infty} \dot{r}(t) &= \lim_{t\to\infty} [A_{\mathrm{cl}}(t)r(t) - A_{\mathrm{cl}}(t)\dot{x}_m(t)(t - t_0) \\ &\quad + \ddot{x}_m(t)(t - t_0) + \dot{x}_m(t)] \\ &= \lim_{t\to\infty} \{A_{\mathrm{cl}}(t)r(t) + [A_{\mathrm{cl}}(t) - \bar{A}_{\mathrm{cl}}(t)]x_m(t)\}, \end{aligned} \tag{41}$$

where $\lim_{t\to\infty}\ddot{x}_m(t) = \dot{\bar{A}}_{\mathrm{cl}}(\infty)x_m(\infty) + A_{\mathrm{cl}}(\infty)\dot{x}_m(\infty)$ from (35), $\dot{x}_m(\infty) = 0$, and $\bar{A}_{\mathrm{cl}}(t) = [A_{\mathrm{cl}}(t) - \bar{A}_{\mathrm{cl}}(t)]/(t - t_0)$. In other words,

$$\begin{aligned} \lim_{t\to\infty} \ddot{r}(t) &= \lim_{t\to\infty} \{\dot{A}_{\mathrm{cl}}(t)r(t) + A_{\mathrm{cl}}(t)\dot{r}(t) \\ &\quad + [\dot{A}_{\mathrm{cl}}(t) - \dot{\bar{A}}_{\mathrm{cl}}(t)]x_m(t) \\ &\quad + [A_{\mathrm{cl}}(t) - \bar{A}_{\mathrm{cl}}(t)]\dot{x}_m(t)\} \\ &= \lim_{t\to\infty} [\dot{A}_{\mathrm{cl}}(t)r(t) + A_{\mathrm{cl}}(t)\dot{r}(t) + \dot{\bar{A}}_{\mathrm{cl}}(t)r(t)], \end{aligned} \tag{42}$$

where $\dot{x}_m(t) = r(t)/(t - t_0)$, $\dot{\bar{A}}_{\mathrm{cl}}(t) = [A_{\mathrm{cl}}(t) - \bar{A}_{\mathrm{cl}}(t)]/(t - t_0)$, and $[\dot{A}_{\mathrm{cl}}(\infty) - \dot{\bar{A}}_{\mathrm{cl}}(\infty)]x_m(\infty) = 0$ since $x_m(\infty) = 0$ and $\dot{A}_{\mathrm{cl}}(t) - \dot{\bar{A}}_{\mathrm{cl}}(t) \in \tilde{U}(\infty, \dot{A}_{\mathrm{cl}} - \dot{\bar{A}}_{\mathrm{cl}}, x_m)$ with $\dot{A}_{\mathrm{cl}}(t) \in \tilde{U}(\infty, \dot{A}_{\mathrm{cl}}, x_m)$ and $\dot{\bar{A}}_{\mathrm{cl}}(t) \in \tilde{U}(\infty, \dot{\bar{A}}_{\mathrm{cl}}, x_m)$. Since $\tilde{A}_r(\infty)$ is full rank, the limit system $\dot{x}_r(t) = \tilde{A}_r(t)x_r(t)$ has an equilibrium point $x_r = 0$ where $x_r(t) = \begin{bmatrix} r^{\mathrm{T}}(t) & \dot{r}^{\mathrm{T}}(t) \end{bmatrix}^{\mathrm{T}}$.

Given the equilibrium point, the Lyapunov function candidate $V_r(t) = x_r^{\mathrm{T}}(t)\tilde{P}_2(t)x_r(t)$ is defined where $\tilde{P}_2(t) > O$, $\forall t \in [t_1, \infty)$. From which one has

$$\begin{aligned} \dot{V}_r(t) &= x_r^{\mathrm{T}}(t)[\tilde{A}_r^{\mathrm{T}}(t)\tilde{P}_2(t) + \tilde{P}_2(t)\tilde{A}_r(t) + \dot{\tilde{P}}_2(t)]x_r(t) \\ &= -x_r^{\mathrm{T}}(t)[Q_{r1}(t) - Q_{r2}(t)]x_r(t) - \alpha_2 x_r^{\mathrm{T}}(t)\tilde{P}_2^2(t)x_r(t) \\ &\leq -\alpha_2 x_r^{\mathrm{T}}(t)\tilde{P}_2^2(t)x_r(t) < 0, \end{aligned} \tag{43}$$

because of (38) and (39) where $Q_{r1}(t) \geq Q_{r2}$, $\forall t \in [t_1, \infty)$. The dynamic equation (2) thus admits the solution $x(t) = r(t) + x_m(t)$ where $x_m(\infty) = 0$ and $r(\infty) = 0$ imply robust asymptotic stability.

As mentioned in Section 3.2, although (19) can be asymptotically stabilized with $\Delta p = K\begin{bmatrix}\Delta\bar{w} & \Delta\bar{q}\end{bmatrix}^{\mathrm{T}}$ (Hollot et al., 2002), this does not imply that (2) possesses asymptotic stability. Asymptotic stability of (2) requires suppression of the steady-state error and queue oscillation. The former can be fulfilled using Theorem 3 while the latter could be ensured by Theorem 4, which is capable of identifying existence of the oscillation nature.

4. Case study

4.1. *Verification of IO-FTS LTV system*

As mentioned in Section 2.3, Theorem 1 is invalid for (9). Theorem 2 was used here to study its robust IO-FTS over $\Omega = [0, 50]$ with $A(t) = A_1(t) - A_2(t)$, $Q_2 = I$, $Q_1(t) = [Q_1(i,j)]_{2\times 2}$, and $Q_3(t) = [Q_3(i,j)]_{2\times 2}$ as shown in Figure 2(a) and 2(b), respectively. Based on the setting, solving (10)–(13) yields $A_1(t) = \begin{bmatrix} -3 & \sin(t) \\ 0 & -2 \end{bmatrix}$, $A_2(t) =$

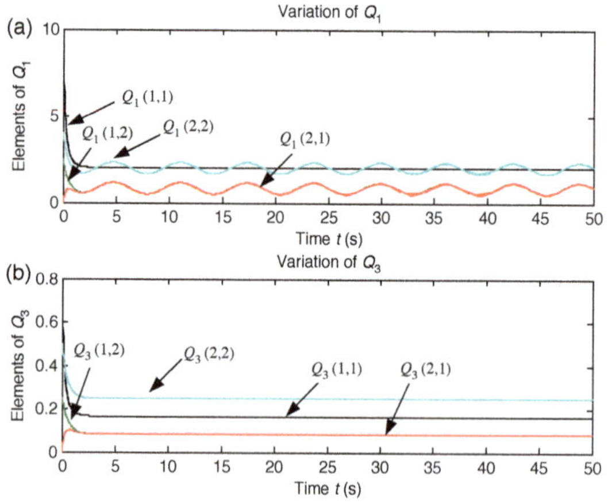

Figure 2. Responses of (a) $Q_1(t)$ and (b) $Q_3(t)$.

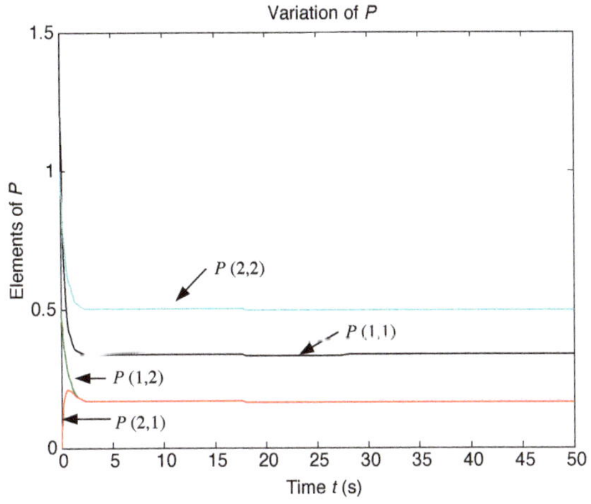

Figure 3. Response of $\tilde{P}(t)$.

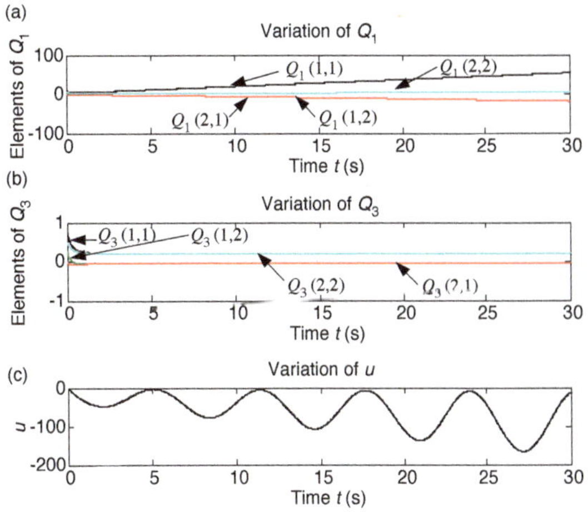

Figure 4. Responses of $Q_1(t)$, $Q_3(t)$, and u, (a) and (b) for $Q_1(t)$ and $Q_3(t)$, and (c) for the control input.

$\begin{bmatrix} -2 & 0 \\ 1 & -1 \end{bmatrix}$, and $\tilde{P}(t)$ approaches $\begin{bmatrix} 0.3333 & 0.1667 \\ 0.1667 & 0.5 \end{bmatrix}$ (see Figure 3) as $t \to 50$. Consequently, from Theorem 2, the system (9) is IO-FTS with respect to $(\tilde{U}_2(T), \tilde{Q}(\cdot), \Omega)$.

4.2. Case study of robust stability

Consider the system (1) with

$$A(t) = \begin{bmatrix} t & 2 \\ 3t+2 & -t+1 \end{bmatrix}, \; B_1 = B_2 = \begin{bmatrix} 1 \\ 1 \end{bmatrix}, \; C = I_2,$$

$$w(t) = r_w(t) = 10 \, \sin(t) + 10.$$

To demonstrate the advantage of Theorems 3 and 4, Theorem 2 was first applied to guarantee its robust IO-FTS by considering $u(t) = K(t)x(t)$ with $Q_2 = I$ (see Figure 4(a)), and Q_3 (see Figure 4(b)). Note that $Q_1(2,1)$ approaches $Q_1(1,2)$ gradually. In spite of determining the diverging $Q_1(t)$ in Theorem 2 to ensure $Q_1(t) \geq Q_2$ over $\Omega \in [0, 30]$, conducting the steps depicted in Remark 1 yields $K(t) = \begin{bmatrix} -3t-2 & t-3 \end{bmatrix}$, $A_1(t) = \begin{bmatrix} -2t-3 & t-1 \\ 0 & -3 \end{bmatrix}$, $A_2(t) = \begin{bmatrix} -1 & 0 \\ 0 & -1 \end{bmatrix}$, and $\tilde{P}(t) > 0$ over $\Omega \in [0, 30]$ (see Figure 5(a)) with $\tilde{P}(30) = \begin{bmatrix} 0.433 & -0.067 \\ -0.067 & 0.433 \end{bmatrix}$. The system (2) is thus possessing robust IO-FTS with respect to $(\tilde{U}_2(T), \tilde{Q}(\cdot), \Omega)$. However, using the control input in Figure 4(c), the result of Figure 5(b) exhibits steady-state error caused by the nonzero $w_m(t)$ and state oscillation induced by $w(t)$. This should thus be improved by redesigning $K(t)$.

To remedy, Theorem 3 was adopted to design the control law $u(t) = K(t)x(t)$. Under the settings of $Q_{m2} = 2I$, $\alpha_1 = 0.1$, $t_0 = 1$, and $Q_{m1}(t) = \begin{bmatrix} Q_a(t) & Q_b(t) \\ Q_c(t) & Q_d(t) \end{bmatrix}$ with $Q_a = [Q_a(i,j)]_{2\times2} \to \begin{bmatrix} 1.999 & 0 \\ 0 & 1.999 \end{bmatrix}$,

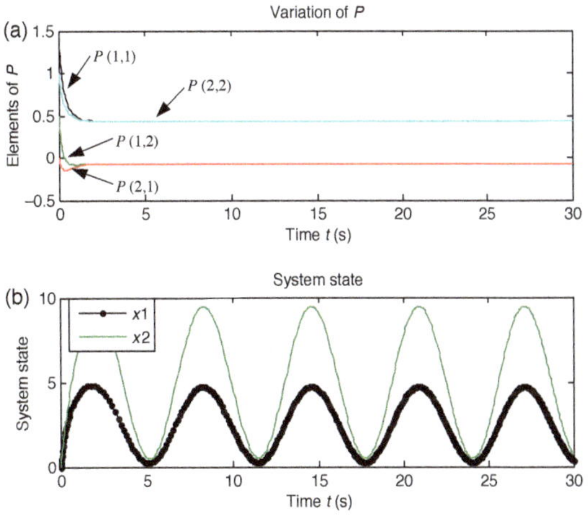

Figure 5. (a) Response of $\tilde{P}(t)$ and (b) state response.

$Q_b = [Q_b(i,j)]_{2\times2} \to \begin{bmatrix} 0 & 0 \\ 0 & -0.05 \end{bmatrix}$, $Q_c = [Q_c(i,j)]_{2\times2} \to \begin{bmatrix} 0 & 0 \\ 0 & -0.05 \end{bmatrix}$, and $Q_d = [Q_d(i,j)]_{2\times2}$ with $Q_d(1,2) = Q_d(2,1) \to -0.2999$ as $t \to 40$ in Figure 6, solving (28)–(30) upon Remark 1 gives $\tilde{A}_b(t) = -10I$, $\tilde{A}_a(t) = \begin{bmatrix} -10 & 0 & 1 & 0 \\ 0 & -10 & 0 & 1 \\ -1 & 0 & -2t-10 & 1 \\ 0 & -0.5 & 2 & -t-10 \end{bmatrix}$, $K(t) = \begin{bmatrix} -3t & -1 \end{bmatrix}$, and $\tilde{P}_1(t) = \begin{bmatrix} P_a(t) & P_b(t) \\ P_c(t) & P_d(t) \end{bmatrix} > 0$ for $t \geq 10$ with $P_{a,b,c,d} = [P_{a,b,c,d}(i,j)]_{2\times2}$. See Figure 7, it shows $\tilde{P}_1(40) = 0.1I$.

The system is thus robustly asymptotically stable with respect to its mean in spite of setting the diverging $Q_{m1}(t)$ in Theorem 3 to meet $Q_{m1}(t) \geq Q_{m2}$ for $t \geq 10$.

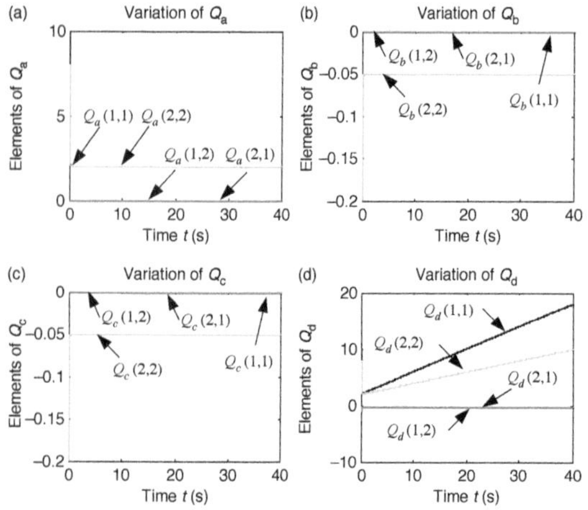

Figure 6. Response of $Q_{m1}(t)$, (a)–(d) for Q_a, Q_b, Q_c, and Q_d, respectively.

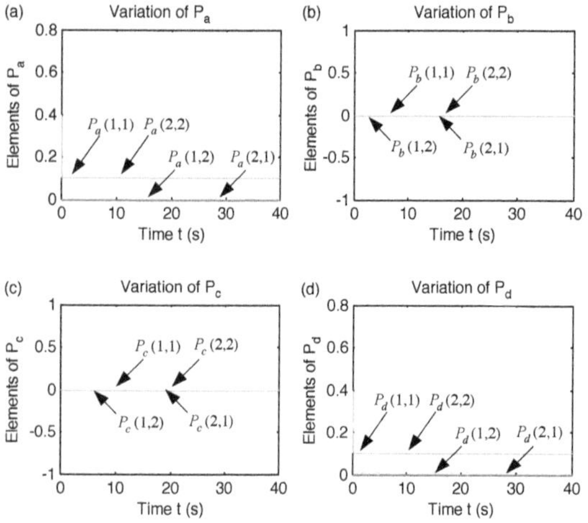

Figure 7. Response of $\tilde{P}_1(t)$, (a) P_a, (b) P_b, (c) P_c, and (d) P_d.

Third, robust asymptotic stability of the system (1) with $u(t) = K(t)x(t) = \begin{bmatrix} -3t & -1 \end{bmatrix} x(t)$ was examined using Theorem 4, in which $\tilde{A}_{r1}(t) = -50I$ and $\tilde{P}_2(t) = \begin{bmatrix} P_a(t) & P_b(t) \\ P_c(t) & P_d(t) \end{bmatrix} > 0$ for $t \in [10, \infty)$ with $\tilde{P}_2(40) = 0.1I$. Response of $\tilde{P}_2(t)$ is shown in Figure 8, when the matrix was evaluated for $Q_{r2}(t) = 10I$ and $\alpha_2 = 0.1$, and $Q_{r1}(t) = \begin{bmatrix} Q_a(t) & Q_b(t) \\ Q_c(t) & Q_d(t) \end{bmatrix}$ was selected to guarantee $Q_{r1}(t) \geq Q_{r2}$ for $t \geq 10$ where $Q_a \to \begin{bmatrix} 9.999 & 0 \\ 0 & 9.999 \end{bmatrix}$, $Q_b \to \begin{bmatrix} 0.2 & 0 \\ 0 & 0.05 \end{bmatrix}$, $Q_c \to \begin{bmatrix} 0.2 & 0 \\ 0 & 0.05 \end{bmatrix}$, and $Q_d(1,2) \to -0.3$. as $t \to 40$ (see Figure 9). By Theorem 4, the system (2) is thus robustly asymptotically stable.

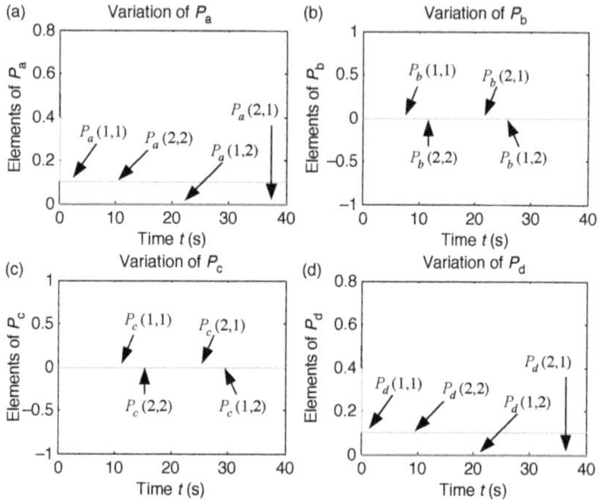

Figure 8. Response of $\tilde{P}_2(t)$, (a)–(d) for P_a, P_b, P_c, and P_d, respectively.

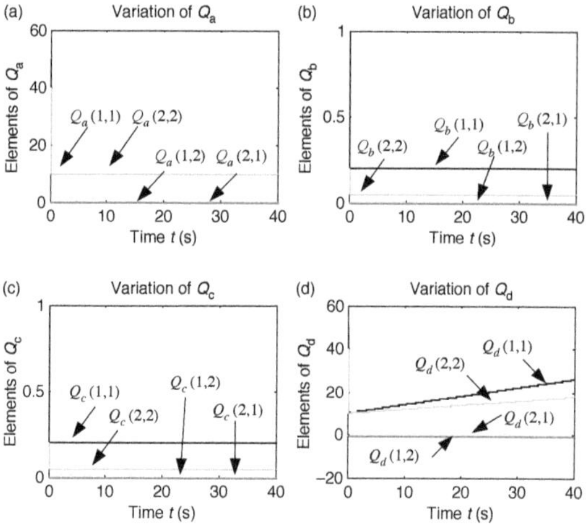

Figure 9. Response of $Q_{r1}(t)$, (a) Q_a, (b) Q_b, (c) Q_c, and (d) Q_d.

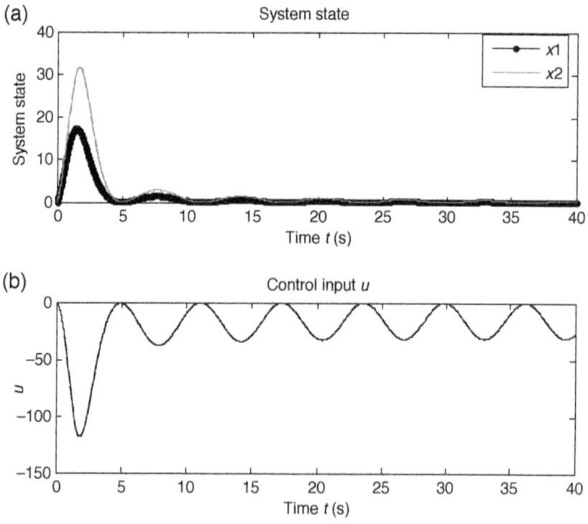

Figure 10. (a) State response and (b) control command.

Finally, the result of Figure 10(a) illustrates that the states response converge to zero as $t \rightarrow \infty$. Figure 10(b) displays the control command.

5. Conclusions

This paper has proposed a novel type of stability robustness–mean robust asymptotic stability, and applies it to a class of LTV uncertain systems. The results are proposed to enhance the insufficiency of robust IO-FTS for certain LTV uncertain system – in particular, when the dynamic equation exists, a singular system matrix at some time over a bounded time interval. For the situation of the systems exhibit steady-state errors or state oscillation, a refinement is further proposed to by identifying the oscillating behaviour. Our main results guarantee robust stability for a class of LTV systems with modelling uncertainty. All the propositions have been demonstrated via case studies.

Disclosure statement

No potential conflict of interest was reported by the authors.

Funding

This research was sponsored by the Ministry of Science and Technology, Taiwan, R.O.C. [grant number NSC 101-2221-E-005-015-MY3], [grant number MOST 103-2218-E-005-005-MY2].

References

Abou-Kandil, H., Freiling, G., Ionesco, V., & Jank, G. (2003). *Matrix Riccati equations in control and systems theory* (pp. 1–20). Birkhäuser verlag: Basel.

Amato, F., Carannante, G., De Tommasi, G., & Pironti, A. (2012). Input-output finite-time stability of linear systems: Necessary and sufficient conditions. *IEEE Transactions on Automatic Control*, *57*(12), 3051–3063.

Amato, F., Cosentino, C., De Tommasi, G. (2011, September). *Sufficient conditions for robust input-output finite-time stability of linear systems in presence of uncertainties*. Proceedings of IFAC World Congress, Milano, 7643–7647.

Bilman, P. A. (2004). A convex approach to robust stability for linear systems with uncertain scalar parameters. *SIAM Journal on Control and Optimization*, *42*(6), 2016–2042.

Cheng, C. C., Su, G. L., & Chien, C. W. (2012). Block backstepping controllers design for a class of perturbed non-linear systems with *m* blocks. *IET Control Theory and Applications*, *6*(13), 2021–2030.

Choi, J., Hong, S., Lee, W., Kang, S., & Kim, M. (2011). A robot joint with variable stiffness using leaf springs. *IEEE Transactions on Robotics*, *27*(2), 229–238.

Corradini, M. L., Cristofaro, A., & Orlando, G. (2010). Robust stabilization of multi input plants with saturating actuators. *IEEE Transactions on Automatic Control*, *55*(2), 419–425.

Davis, J. M., Gravagne, I. A., Marks II, R. J., & Ramos, A. A. (2010). *Algebraic and dynamic Lyapunov equations on time scales*. Proceedings of South Eastern Symposium on System Theory, Tyler, Texas, 329–334.

Hollot, C. V., Misra, V., & Towsley, D. (2002). Analysis and design of controllers for AQM routers supporting TCP flows. *IEEE Transactions on Automatic Control*, *47*(6), 945–959.

Hsu, P. M., & Lin, C. L. (2012, July). *Time-delay compensation for time-varying delayedsystems using extended network disturbance*. Proceedings of IEEE Industrial Electronics and Applications, Singapore, 388–393.

Juan, Y., June, F., Lin, S., & Yu, Z. (2011, July). *Input-output finite-time stability of time-varying linear singular systems*. Proceedings of Chinese Control Conference, Yantai, 62–66.

Kelly, R. (1996). Robust asymptotically stable visual servo of planar robots. *IEEE Transactions on Robotics and Automation*, *12*(5), 759–766.

Kerrigan, E. C., & Maciejowski, J. M. (2004). Feedback min-max model predictive control using a single linear program: Robust stability and the explicit solution. *International Journal of Robust and Nonlinear Control*, *14*, 395–413.

Lee, T. C., Liaw, D. C., & Chen, B. S. (2001). A general invariance principle for nonlinear time-varying systems and its applications. *IEEE Transactions on Automatic Control*, *46*(12), 1989–1993.

Li, C., Wang, J., & Lu, J. (2012). Observer-based robust stabilization of a class of non-linear fractional-order uncertain systems: A linear matrix inequality approach. *IET Control Theory & Applications*, *6*(18), 2757–2764.

Li, H., Gao, H., & Lu, H. (2011). Robust quantised control for active suspension systems. *IET Control Theory & Applications*, *5*, 1955–1969.

Lu, J. G., & Chen, G. (2009). Robust stability and stabilization of fractional-order interval systems: An LMI approach. *IEEE Transactions on Automatic Control*, *54*(6), 1294–1299.

Lu, J. G., & Chen, Y. Q. (2010). Robust stability and stabilization of fractional-order interval systems with the fractional order α: The $0 < \alpha < 1$ case. *IEEE Transactions on Automatic Control*, *55*(1), 152–158.

Ma, H., & Jia, Y. (2011, December). *Input-output finite-time stability and stabilization of stochasticMarkovian jump systems*. Proceedings of IEEE Decision and Control and European Control Conference, Orlando, 8027–8031.

Misra, V., Gong, W. B., & Towsley, D. (2000, August). *Fluid-based analysis of a network of AQM routers supporting TCP flows with an application to RED*. Proceedings of ACM SIGCOMM, Stockholm, 151–160.

Xu, S., & Lam, J. (2005). Improved delay-dependent stability criteria for time-delay systems. *IEEE Transactions on Automatic Control*, *50*(3), 384–387.

3

Robust H_∞ reliable control for nonlinear delay switched systems based on the average dwell time approach

Yanhui Li*, Chao Zhang and Xiujie Zhou

College of Electrical and Information Engineering, Northeast Petroleum University, Daqing, Heilongjiang Province, 163318, People's Republic of China

This paper is concerned with the robust H_∞ reliable control problem for a class of Lipschitz nonlinear delay switched systems. Considering the case of actuator fault, the sufficient condition is proposed to guarantee the global exponential stability with a guaranteed H_∞ performance for the nonlinear delay switched systems by using the piecewise Lyapunov functional and average dwell time (ADT) approach. Then, the corresponding solvability condition for the desired robust H_∞ reliable controllers is established, and the controller design is cast into a convex optimization problem which can be efficiently obtained by numerical software. A numerical example shows validity and feasibility of the proposed design method.

Keywords: nonlinear switched systems; global exponential stability; average dwell time; robust H_∞ reliable control

1. Introduction

As an important abstraction form of hybrid systems, switched systems have received increasing attention within the past few decades. The so-called switched systems consist of a finite number of subsystems and a switching strategy deciding the switching sequence along the system trajectory at a time instant (Lu, Zhang, & Karimi, 2013). In general, the switching strategy of subsystems is considered as completely unknown in advance, and the switching times of subsystems are considered to be either arbitrary or constrained. Switched systems can describe a class of practical engineering. Many real-world processes and systems can be modelled as switched systems, such as the automobile direction-reverse systems, mechanical processes, multiple work points control systems of airplanes. Thus, switched systems have a wide engineering background and deserve a thorough study on theories (Wu & Lam, 2009). Zhai and Hu had investigated the stability of the linear switched systems containing unstable switching models by using the common quadratic Lyapunov function (Zhai, Hu, & Michel, 2001). The average dwell time (ADT) approach was introduced for linear switched systems to study their stability and stabilization (Balochian & Sedigh, 2012; Wu, Qi, & Feng, 2009). In Zhai, He, and Wu (2000), a class of switched systems, which could be described by differential, had been studied for stability analysis. And their sufficient conditions of the asymptotic stability could be obtained based on the Lyapunov function approach. Hu and Michel (2000) analysed the local asymptotic stability of nonlinear switched systems by using the dwell time approach. In Kim, Park, and Ko (2004), the nonlinear switched systems were studied for the robustness and the capacity to improve disturbance attenuation based on the H_∞ optimal control approach. It has achieved lots of certain results about the study of linear switched systems. However, a further study of nonlinear switched systems remains to be investigated.

On the other hand, with the development of science and technology, control systems have become more and more complicated. It is inevitable that actuator failures may exist in some specific operating conditions which will affect the stability and other performance of the systems. Then, the demands for reliability, safety and efficiency of the systems become higher and higher (Du, Lin, & Li, 2013). Therefore, it is of practical importance to design control systems, which can revise malfunction actuators. At present, some achievements have been acquired by the study of reliable control problems. For example, reliable control problems had been studied for linear systems, see Wang, Sun, and Liu (2006) and the references therein. Yang, Zhengrong, Qingwei, and Weili (2006) researched a robust fault-tolerant control problem of delay switched systems, then the state feedback reliable controllers were designed by using the convex combination technique, in order that the closed-loop systems became asymptotically stable. In Lin and Antsaklis (2009), the robust reliable

*Corresponding author. Email: liyanhui@nepu.edu.cn

control problem for linear switched systems with time-varying delay was researched. In Liu (2000), the reliable control problem is investigated for a class of stochastic nonlinear time-delay systems with multiplicative noises. For linear systems, some results of reliable control have been obtained. However, the study of reliable control for nonlinear switched systems has not been fully investigated, which motivates the present study. Therefore, the exploration on robust reliable control problem of nonlinear switched systems is still serious and challenging.

This paper deals with the problem of robust H_∞ reliable control for nonlinear delay switched systems. Choose the continuous gain actuator failure model to describe the actuator failure part in this paper, which is more practical than the discrete actuator failure model. Based on piecewise Lyapunov stability theory and the ADT approach, the robust H_∞ feedback reliable controllers are designed, in order that global exponential stability of the closed-loop systems will be ensured and the H_∞ performance index will be satisfied when the actuator is running normally or malfunctioning. Finally, a numerical simulation shows that the designed control method can meet the control requirements.

2. System description and preliminaries

Consider the following uncertain nonlinear delay switched systems with disturbance:

$$\begin{aligned}\dot{x}(t) &= A_{\sigma(t)}x(t) + A_{d\sigma(t)}x(t-d(t)) + B_{\sigma(t)}u^f_{\sigma(t)}(t)\\&\quad + D_{\sigma(t)}\omega(t) + L_{\sigma(t)}f_{\sigma(t)}(x(t)),\\z(t) &= C_{\sigma(t)}x(t) + G_{\sigma(t)}u^f_{\sigma(t)}(t) + N_{\sigma(t)}\omega(t),\\x(t) &= \phi(t), t \in [-\rho, 0],\end{aligned} \tag{1}$$

where $x(t) \in R^m$ is the state vector, $u^f_{\sigma(t)}(t) \in R^l$ is the control input of the system with partial fault, $\omega(t) \in R^q$ is the disturbance input which belongs to $L_2[0,\infty)$, $z(t) \in R^p$ is the output signal to be controlled. The function $\sigma(t) : [t_0, +\infty) \rightarrow \aleph = \{1, 2, \ldots, n\}$ is the switching signal which is right continuous. $d(t)$ is the time-varying delay of the control systems satisfying $0 \leq d(t) \leq \rho < \infty$, $\dot{d}(t) \leq \mu < 1$, and $\phi(t)$ is a continuous vector-valued initial function. For an arbitrary value $i \in \aleph$, C_i, G_i, N_i are constant matrices. A_i, A_{di}, B_i, D_i, L_i are uncertain real-valued matrices of appropriate dimensions. And the uncertain forms are as follows:

$$\begin{aligned}[A_i\ A_{di}\ B_i\ D_i\ L_i] &= [\bar{A}_i\ \bar{A}_{di}\ \bar{B}_i\ \bar{D}_i\ \bar{L}_i] + H_iF_i(t)\\&\quad [E_{1i}\ E_{di}\ E_{2i}\ E_{3i}\ E_{fi}],\end{aligned} \tag{2}$$

where $\bar{A}_i$,$\bar{A}_{di}$, $\bar{B}_i$, $\bar{D}_i$, $\bar{L}_i$, E_{1i}, E_{di}, E_{2i}, E_{3i}, E_{fi} are known constant matrices of appropriate dimensions. $F_i(t)$ are unknown time-varying matrices which are assumed to be norm-bonded, which satisfy the inequality: $F_i^{\mathrm{T}}(t)F_i(t) \leq I$. $f_i(\cdot) : R^m \rightarrow R^m$ is the unknown nonlinear function and satisfies the global Lipschitz condition:

$$\| f_i(x(t)) \| \leq \| U_ix(t) \| , \tag{3}$$

where U_i is known as the Lipschitz constant matrix.

In actual control systems, there inevitably occur faults in the operation process of actuators. Thus, the control input of actuator fault can be described as

$$u_i^f(t) = M_iu_i(t), \quad i \in \aleph, \tag{4}$$

where $u_i(t)$ represents the normal control inputs, $u_i^f(t)$ represents the abnormal control inputs and M_i is the actuator fault matrix with the following form:

$$\begin{aligned}M_i &= \operatorname{diag}\{m_{i1}, m_{i2}, \ldots, m_{il}\}, \quad 0 \leq m_{ik} \leq 1,\ k\\&= 1, 2, \ldots, l,\end{aligned} \tag{5}$$

Remark 2.1 For the kth actuator control signal of the ith subsystem, there are three cases as follows. When $m_{ik} = 1$, it means the normal operation. When $m_{ik} = 0$, it covers the case of the complete fault. When $0 < m_{ik} < 1$, it corresponds to the case of partial fault. It is noted that m_{ik} $(k = 1, 2, \ldots, l)$ cannot be 0 at the same time.

Construct the state-feedback reliable controllers with the following form:

$$u_i^f(t) = M_iK_ix(t), \tag{6}$$

Then, the closed-loop switched systems (1) can be written as the following form:

$$\begin{aligned}\dot{x}(t) &= \tilde{A}_ix(t) + A_{di}x(t-d(t)) + D_i\omega(t) + L_if_i(x(t)),\\z(t) &= \tilde{C}_ix(t) + N_i\omega(t),\\x(t) &= \phi(t), t \in [-\rho, 0],\end{aligned} \tag{7}$$

where for any $i \in \aleph, \tilde{A}_i = A_i + B_iM_iK_i, \tilde{C}_i = C_i + G_iM_iK_i$.

DEFINITION 2.2 *For the switching law of the systems* (1) *and any* $t_2 > t_1 > 0$, $\Re_{\sigma(t)}(t_1, t_2)$ *is the number of switching between* t_1 *and* t_2. *If the inequality* $\Re_{\sigma(t)}(t_1, t_2) \leq \Re_0 + (t_2 - t_1/\tau_a)$ *holds for any* $\Re_0 \geq 0, \tau_a > 0$, τ_a *is said to be ADT of the switched systems* (1), $\Re_0$ *is vibration amplitude.*

DEFINITION 2.3 *If there exist a switching law* $\sigma(t)$ *and the state trajectory of the systems (7) satisfying*

$$\| x(t) \|_2 \leq \alpha \| x(t_0) \|_2 \, \mathrm{e}^{-\beta(t-t_0)} \tag{8}$$

then, the systems (7) are globally exponentially stable, where $\alpha \geq 0$, $\beta \geq 0$, $t \geq t_0$, *and* β *is exponential damping decrement.*

Robust H_∞ Reliable Control (RHRC) Problem. For a given constant scalar $\gamma > 0$, design the state-feedback controllers in the form of the equality (6) for systems (1) such that

(1) Closed-loop systems are globally exponentially stable under the designed switching strategy;
(2) Under zero-initial condition, the following inequality holds:

$$\| z(t) \|_2^2 \leq \gamma^2 \| \omega(t) \|_2^2, \quad \forall \omega(t) \in L_2[0, \infty). \tag{9}$$

In this case, the equality (6) is said to be robust H_∞ reliable controllers and the systems (7) are globally exponentially stable with a H_∞ disturbance attenuation level γ.

LEMMA 2.4 *(Xie, Wang, & Hao, 2003) Let L, M, W and Z be real matrices of appropriate dimensions, and Z satisfies $Z = Z^{\mathrm{T}}$. Then for all M satisfying $M^{\mathrm{T}}M \leq I$, we have*

$$Z + LMW + W^{\mathrm{T}}M^{\mathrm{T}}L^{\mathrm{T}} < 0 \tag{10}$$

if and only if there exists a scalar $\xi > 0$ such that

$$Z + \xi LL^{\mathrm{T}} + \xi^{-1}W^{\mathrm{T}}W < 0. \tag{11}$$

3. Main results

We shall first investigate the global exponential stability with a guaranteed H_∞ performance of the nonlinear delay switched systems (7). The following theorem presents an existence sufficient condition of state-feedback reliable controllers for the systems (7).

3.1. Switched systems performance analysis

THEOREM 3.1 *Considering the nonlinear delay switched systems (7), for the given scalars $\rho > 0$, $\eta > 1$, $\alpha > 0$, if there exist scalars $\gamma > 0$, $\beta_i > 0$ and positive-definite symmetric matrices P_i, Q_i, such that the following inequalities hold:*

$$\begin{bmatrix} \Sigma_i & P_iA_{di} & P_iD_i + \tilde{C}_i^{\mathrm{T}}N_i \\ * & -(1-\mu)e^{-\alpha\rho}Q_i & 0 \\ * & * & N_i^{\mathrm{T}}N_i - \gamma^2 I \end{bmatrix} < 0, \tag{12}$$

$$P_i \leq \eta P_j, Q_i \leq \eta Q_j \tag{13}$$

and the ADT holds that

$$\tau_a > \tau_a^* = \frac{ln\eta}{\alpha} \tag{14}$$

then the closed-loop systems (7) are globally exponentially stable with an H_∞ performance level γ under the switching law $\sigma(t)$, where

$$\Sigma_i = \tilde{A}_i^{\mathrm{T}}P_i + P_i\tilde{A}_i + Q_i + \alpha P_i + \beta_iP_iL_iL_i^{\mathrm{T}}P_i + \beta_i^{-1}U_i^{\mathrm{T}}U_i + \tilde{C}_i^{\mathrm{T}}\tilde{C}_i, \quad 0 < j < i, \ i,j \in \aleph.$$

Proof For the ith subsystem, we construct the following Lyapunov functional candidate:

$$V_i(t) = x^{\mathrm{T}}(t)P_ix(t) + \int_{t-d(t)}^{t} e^{\alpha(\tau-t)}x^{\mathrm{T}}(\tau)Q_ix(\tau)\,d\tau. \tag{15}$$

Along state trajectory of the systems (7), the time derivative of $V_i(t)$ is given by

$$\begin{aligned} \dot{V}_i(t) \leq{}& x^{\mathrm{T}}(t)(\tilde{A}_i^{\mathrm{T}}P_i + P_i\tilde{A}_i + Q_i)x(t) + 2x^{\mathrm{T}}(t) \\ & P_iA_{di}x(t-d(t)) + 2x^{\mathrm{T}}(t)P_iL_{i}f_i(x(t)) \\ & - (1-\mu)e^{-\alpha\rho}x^{\mathrm{T}}(t-d(t))Q_ix(t-d(t)) \\ & - \alpha\int_{t-d(t)}^{t} e^{\alpha(\tau-t)}x^{\tau}(t)Q_ix(\tau)d\tau + 2x^{\mathrm{T}}(t)P_iD_i\omega(t). \end{aligned}$$

From Lemma 1 and the inequality (3), it is established that

$$\begin{aligned} \dot{V}_i(t) \leq{}& x^{\mathrm{T}}(t)(\tilde{A}_i^{\mathrm{T}}P_i + P_i\tilde{A}_i + Q_i + \beta_iP_iL_iL_i^{\mathrm{T}}P_i \\ & + \beta_i^{-1}U_i^{\mathrm{T}}U_i)x(t) \\ & - (1-\mu)e^{-\alpha\rho}x^{\mathrm{T}}(t-d(t))Q_ix(t-d(t)) \\ & - \alpha\int_{t-d(t)}^{t} e^{\alpha(\tau-t)}x^{\tau}(t)Q_ix(\tau)\,d\tau \\ & + 2x^{\mathrm{T}}(t)P_iD_i\omega(t) + 2x^{\mathrm{T}}(t)P_iA_{di}x(t-d(t)). \end{aligned} \tag{16}$$

Considering $\omega(t) = 0$, it follows that

$$\begin{aligned} \dot{V}_i(t) + \alpha V_i(t) \leq{}& x^{\mathrm{T}}(t)(\tilde{A}_i^{\mathrm{T}}P_i + P_i\tilde{A}_i + Q_i + \alpha P_i \\ & + \beta_i^{-1}U_i^{\mathrm{T}}U_i + \beta_iP_iL_iL_i^{\mathrm{T}}P_i)x(t) \\ & - (1-\mu)e^{-\alpha\rho}x^{\mathrm{T}}(t-d(t))Q_ix(t-d(t)) \\ & + 2x^{\mathrm{T}}(t)P_iD_i\omega(t) + 2x^{\mathrm{T}}(t)P_iA_{di}x(t-d(t)). \end{aligned}$$

According to the inequality (12), it is not hard to know that $\dot{V}_i(t) \leq -\alpha V_i(t)$, then $V_i(t)$ satisfies the following inequality:

$$V_i(t) \leq V_i(t_0^{(i)})e^{-\alpha(t-t_0^{(i)})}, t \geq t_0^{(i)},$$

where $t_0^{(i)}$ is the initial time of the ith activated subsystem.

The switching time is divided into $[t_0, t_1), \ldots, [t_{n-1}, t_n)$. From the inequalities (13), we know that for any $i,j \in \aleph$, $V_i(t) \leq \eta V_j(t)$, where $\eta > 1$, according to Definition 1, the Lyapunov functional candidate satisfies the following inequality:

$$\begin{aligned} V_i(t) &\leq V_{\delta(t_k)}(t_k)e^{-\alpha(t-t_k)} \leq \eta V_{\delta(t_{k-1})}(t_{k-1})e^{-\alpha(t-t_{k-1})} \\ &\leq \cdots \leq \eta^k V_{\delta(t_0)}(t_0)e^{-\alpha(t-t_0)} \\ &\leq V_{\delta(t_0)}(t_0)e^{-(\alpha - ln\eta/\tau_a)(t-t_0)}. \end{aligned}$$

Let $a = \min_{i\in\aleph}\lambda_{\min}(P_i), b = \max_{i\in\aleph}\lambda_{\max}(P_i) + \rho\max_{i\in\aleph}\lambda_{\max}(Q_i)$, we know that $\| x(t) \| \leq \sqrt{b/a} \| x(t_0) \|$

$e^{-1/2(\alpha-\ln\eta/\tau_a)(t-t_0)}$. Thus, it can be obtained that if the ADT satisfies the inequalities (13), the switched systems are globally exponentially stable.

Based on the Schur Complement lemma,

$$
\begin{aligned}
&z^{\mathrm{T}}(t)z(t)-\gamma^2\omega^{\mathrm{T}}(t)\omega(t)+\dot{V}_i(t)+\alpha V_i(t)\\
&\quad\le x^{\mathrm{T}}(t)(\tilde{A}_i^{\mathrm{T}}P_i+P_i\tilde{A}_i+Q_i+\beta_iP_iL_iL_i^{\mathrm{T}}P_i+\beta_i^{-1}U_i^{\mathrm{T}}U_i\\
&\quad+\alpha P_i+\tilde{C}_i^{\mathrm{T}}\tilde{C}_i)x(t)\\
&\qquad-(1-\mu)e^{\alpha\rho}x^{\mathrm{T}}(t-d(t))Q_ix(t-d(t))\\
&\quad+2x^{\mathrm{T}}(t)P_iA_{di}x(t-d(t))\\
&\qquad+\omega^{\mathrm{T}}(t)(N_i^{\mathrm{T}}N_i-\gamma^2I)\omega(t)+2x^{\mathrm{T}}(t)(P_iD_i\\
&\quad+\tilde{C}_i^{\mathrm{T}}N_i)\omega(t)\\
&\quad=\begin{bmatrix} cx(t)\\ x(t-d(t))\\ \omega(t)\end{bmatrix}^{\mathrm{T}}\Psi_i\begin{bmatrix} x(t)\\ x(t-d(t))\\ \omega(t)\end{bmatrix}
\end{aligned}
$$

where

$$
\Psi_i=\begin{bmatrix}\Sigma_i & P_iA_{di} & P_iD_i+\tilde{C}_i^{\mathrm{T}}N_i\\ * & -(1-\mu)e^{-\alpha\rho}Q_i & 0\\ * & * & N_i^{\mathrm{T}}N_i-\gamma^2I\end{bmatrix}.
$$

Obviously, the inequality (12) is equivalent to $\Psi_i<0$, then

$$
z^{\mathrm{T}}(t)z(t)-\gamma^2\omega^{\mathrm{T}}(t)\omega(t)+\dot{V}_i(t)+\alpha V_i(t)<0. \tag{17}
$$

Considering the zero-initial condition,

$$
J=\int_{t_0}^{\infty}(z^{\mathrm{T}}(t)z(t)-\gamma^2\omega^{\mathrm{T}}(t)\omega(t))\,\mathrm{d}t. \tag{18}
$$

Define the piecewise Lyapunov function $V(t)=V_i(t)$, $t\in[t_{v-1},t_v)$, $v\in\aleph$. Combine the inequality (17) and the equality (18),

$$
\begin{aligned}
J&=\int_{t_0}^{\infty}(z^{\mathrm{T}}(t)z(t)-\gamma^2\omega^{\mathrm{T}}(t)\omega(t))\,\mathrm{d}t\\
&\le\int_{t_0}^{\infty}(z^{\mathrm{T}}(t)z(t)-\gamma^2\omega^{\mathrm{T}}(t)\omega(t)+\dot{V}_i(t)+\alpha V_i(t))\,\mathrm{d}t\\
&=\int_{t_0}^{t_1}(z^{\mathrm{T}}(t)z(t)-\gamma^2\omega^{\mathrm{T}}(t)\omega(t)+\dot{V}_{i(0)}(t)+\alpha V_{i(0)}(t))\,\mathrm{d}t\\
&+\int_{t_1}^{t_2}(z^{\mathrm{T}}(t)z(t)-\gamma^2\omega^{\mathrm{T}}(t)\omega(t)+\dot{V}_{i(1)}(t)+\alpha V_{i(1)}(t))\,\mathrm{d}t\\
&+\cdots<0,
\end{aligned}
$$

$$
\int_{t_0}^{\infty}z^{\mathrm{T}}(t)z(t)\,\mathrm{d}t\le\gamma^2\int_{t_0}^{\infty}\omega^{\mathrm{T}}(t)\omega(t)\,\mathrm{d}t. \tag{19}
$$

Therefore, $\| z(t)\|_2^2\le\gamma^2\|\omega(t)\|_2^2$ will always hold for any nonzero external disturbance, $\omega(t)\in L_2[0,\infty)$. The proof is concluded.

Remark 3.2. In Theorem 1, the Lipschitz condition has been applied on linearizing the nonlinear parts. For meeting the exact linearization conditions, we need to leave out some nonlinear characteristics. Then, the omitted parts, which are linearized approximately by Lipschitz functions, continue to exist in the linearized systems in order to improve the control accuracy of the systems.

Remark 3.3. The exponential stability has a fast convergence rate and good dynamic properties in contrast with the asymptotical case. On the other hand, we consider the relation between the parameters α, η and systems stability. When we increase η, it can be seen from Theorem 3.1 that the existence of the multiple Lyapunov functions will be increased. In other words, for a given α, the system stability directly depends on the choice of η.

3.2. *The design method of robust H_∞ reliable controllers*

THEOREM 3.4 *Considering the nonlinear switched systems (7), for the positive scalars $\rho>0$, $\eta>1$, $\alpha>0$, the closed-loop system is globally exponentially stable with a H_∞ performance level γ for existing uncertainties, time-varying delays and actuator faults, if there exist scalars $\gamma>0$, $\varepsilon_i>0$, $\beta_i>0$, positive-definite symmetric matrices X_i, S_i and matrix Y_i, that satisfying the inequalities (13), (14) and the following LMI:*

$$
\begin{bmatrix}
\Lambda_i & \bar{A}_{di}S_i & X_i & \bar{D}_i & \beta_i\bar{L}_i & X_iU_i^{\mathrm{T}} & \Theta_i^{\mathrm{T}} & \varepsilon_iH_i & \Phi_i^{\mathrm{T}}\\
* & -(1-\mu)e^{-\alpha\rho}S_i & 0 & 0 & 0 & 0 & 0 & 0 & S_iE_{di}^{\mathrm{T}}\\
* & * & -S_i & 0 & 0 & 0 & 0 & 0 & 0\\
* & * & * & -\gamma^2I & 0 & 0 & N_i^{\mathrm{T}} & 0 & E_{3i}^{\mathrm{T}}\\
* & * & * & * & -\beta_iI & 0 & 0 & 0 & \beta_iE_{fi}^{\mathrm{T}}\\
* & * & * & * & * & -\beta_iI & 0 & 0 & 0\\
* & * & * & * & * & * & -I & 0 & 0\\
* & * & * & * & * & * & * & -\varepsilon_iI & 0\\
* & * & * & * & * & * & * & * & -\varepsilon_iI
\end{bmatrix}<0, \tag{20}
$$

where $\Lambda_i=(\bar{A}_iX_i+\bar{B}M_iY_i)+(\bar{A}_iX_i+\bar{B}M_iY_i)^{\mathrm{T}}+\alpha X_i$, $\Theta_i=C_iX_i+G_iM_iY_i$, $\Phi_i=E_{1i}X_i+E_{2i}M_iY_i$. Moreover, the parameters of the robust reliable controllers are given by

$$
K_i=Y_iX_i^{-1}. \tag{21}
$$

Proof For the ith subsystem of systems (7), considering the inequality (12), based on the Schur Complement lemma and Lemma 1, it is established that

$$
\begin{bmatrix}
\Pi_i & P_iA_{di} & \Delta_{1i} & P_iL_i & U_i^{\mathrm{T}} & \tilde{C}_i^{\mathrm{T}}\\
* & \Delta_{2i} & 0 & 0 & 0 & 0\\
* & * & \Delta_{3i} & 0 & 0 & 0\\
* & * & * & -\beta_i^{-1}I & 0 & 0\\
* & * & * & * & -\beta_iI & 0\\
* & * & * & * & * & -I
\end{bmatrix}<0,
$$

where $\Pi_i=\tilde{A}_i^{\mathrm{T}}P_i+P_i\tilde{A}_i+Q_i+\alpha P_i$, $\Delta_{1i}=P_iD_i+\tilde{C}_i^{\mathrm{T}}N_i$, $\Delta_{2i}=-(1-\mu)e^{-\alpha\rho}Q_i$, $\Delta_{3i}=N_i^{\mathrm{T}}N_i-\gamma^2I$.

According to the equalities (2), based on Lemma1, through equivalent transformation, it can be obtained that

$$\Upsilon_i + \varepsilon_i \Gamma_i \Gamma_i^{\mathrm{T}} + \varepsilon_i^{-1} \Xi_i^{\mathrm{T}} \Xi_i < 0,$$

where

$$\Upsilon_i = \begin{bmatrix} \Omega_i & P_i\bar{A}_{di} & \Delta_{4i} & P_i\bar{L}_i & U_i^{\mathrm{T}} & \tilde{C}_i^{\mathrm{T}} \\ * & \Delta_{2i} & 0 & 0 & 0 & 0 \\ * & * & \partial_i & 0 & 0 & 0 \\ * & * & * & -\beta_i^{-1}I & 0 & 0 \\ * & * & * & * & -\beta_i I & 0 \\ * & * & * & * & * & -I \end{bmatrix}$$

$$\Xi_i = [E_{1i} + E_{2i}M_iK_i \;\; E_{di} \;\; E_{3i} \;\; E_{fi} \;\; 0 \;\; 0], \quad \Gamma_i = [H_i^{\mathrm{T}}P_i \;\; 0 \;\; 0 \;\; 0 \;\; 0 \;\; 0]^{\mathrm{T}},$$

$$\Omega_i = (\bar{A}_i + \bar{B}_iM_iK_i)^{\mathrm{T}}P_i + P_i(\bar{A}_i + \bar{B}_iM_iK_i) + \alpha P_i + Q_i,$$

$$\Delta_{4i} = P_i\bar{D}_i + \tilde{C}_i^{\mathrm{T}}N_i.$$

Then, based on the Schur Complement lemma, it is converted to

$$\begin{bmatrix} \Omega_i & P_i\bar{A}_{di} & P_i\bar{D}_i & P_i\bar{L}_i & U_i^{\mathrm{T}} & \tilde{C}_i^{\mathrm{T}} & P_iH_i & (E_{1i}+E_{2i}M_iK_i)^{\mathrm{T}} \\ * & \Delta_i & 0 & 0 & 0 & 0 & 0 & E_{di}^{\mathrm{T}} \\ * & * & -\gamma^2 I & 0 & 0 & N_i^{\mathrm{T}} & 0 & E_{3i}^{\mathrm{T}} \\ * & * & * & -\beta_i^{-1}I & 0 & 0 & 0 & E_{fi}^{\mathrm{T}} \\ * & * & * & * & -\beta_i I & 0 & 0 & 0 \\ * & * & * & * & * & -I & 0 & 0 \\ * & * & * & * & * & * & -\varepsilon_i^{-1}I & 0 \\ * & * & * & * & * & * & * & -\varepsilon_i I \end{bmatrix} < 0. \tag{22}$$

Make a congruence transformation for the inequality (22) via $\mathrm{diag}\{P_i^{-1}, Q_i^{-1}, I, \beta_i I, I, I, \varepsilon_i I, I\}$ and the Schur Complement lemma , and denote $X_i = P_i^{-1}$, $S_i = Q_i^{-1}$, $Y_i = K_iP_i^{-1}$. Then, the inequality (22) is equivalent to the inequality (20). Obviously, it is easy to obtain the inequality (21). This proof is completed.

Remark 3.5. According to Theorem 2, if there exists a feasible solution of the inequality (20), the robust H_∞ reliable controllers can be designed in the form of the equality (21). In addition, we can also treat γ^2 as the optimization variable to obtain the optimal disturbance attenuation level, that is,

$$\min \gamma^2 \quad \text{subjected to(13), (14), (20)}$$

by solving the above convex optimization problem, we can obtain the corresponding controller parameters.

4. Illustrative example

In this section, we provide an example to validate the effectiveness of the controller design method proposed in the previous section. Consider the nonlinear delay switched systems (1) with the following state-space matrices:

Subsystem 1:

$$\bar{A}_1 = \begin{bmatrix} -9 & 0.4 \\ 0.9 & -7 \end{bmatrix}, \quad \bar{B}_1 = \begin{bmatrix} 6 & -1 \\ 1 & 0 \end{bmatrix},$$

$$\bar{A}_{d1} = \begin{bmatrix} -0.09 & 0 \\ -0.01 & -0.7 \end{bmatrix}, \quad \bar{D}_1 = \begin{bmatrix} -0.1 & 0 \\ 0.1 & 0.1 \end{bmatrix},$$

$$\bar{L}_1 = \begin{bmatrix} -0.19 & 0 \\ 0 & 0.1 \end{bmatrix}, \quad C_1 = \begin{bmatrix} -0.5 & 0 \\ -0.1 & 0 \end{bmatrix},$$

$$G_1 = \begin{bmatrix} 1 & 0.3 \\ 0.3 & 0.2 \end{bmatrix}, \quad N_1 = \begin{bmatrix} 1 & 0.9 \\ 0.3 & 0.15 \end{bmatrix},$$

$$U_1 = \begin{bmatrix} -0.01 & 0.1 \\ 0.1 & 0 \end{bmatrix}, \quad H_1 = \begin{bmatrix} -0.01 & 0.1 \\ 0 & 0.1 \end{bmatrix},$$

$$E_{11} = \begin{bmatrix} 0.35 & 0 \\ 0.1 & 0 \end{bmatrix}, \quad E_{21} = \begin{bmatrix} 0 & 0 \\ 0 & 0.6 \end{bmatrix},$$

$$E_{d1} = \begin{bmatrix} 0 & -0.3 \\ 0 & 0.6 \end{bmatrix}, \quad E_{31} = \begin{bmatrix} 0 & 0.1 \\ 0 & 0.4 \end{bmatrix},$$

$$E_{f1} = \begin{bmatrix} 0 & -0.2 \\ 0 & 0.8 \end{bmatrix}.$$

Subsystem 2:

$$\bar{A}_2 = \begin{bmatrix} 20 & 0.9 \\ 0.7 & -3 \end{bmatrix}, \quad \bar{B}_2 = \begin{bmatrix} 8 & -1 \\ 1 & -2 \end{bmatrix},$$

$$\bar{A}_{d2} = \begin{bmatrix} -0.01 & -0.02 \\ -0.08 & -0.1 \end{bmatrix}, \quad \bar{D}_2 = \begin{bmatrix} 0 & -0.5 \\ -0.1 & -0.3 \end{bmatrix},$$

$$\bar{L}_2 = \begin{bmatrix} 0 & -0.2 \\ -0.1 & -0.2 \end{bmatrix}, \quad C_2 = \begin{bmatrix} 0 & -0.1 \\ 0 & -0.1 \end{bmatrix},$$

$$G_2 = \begin{bmatrix} -0.1 & 0.1 \\ 0.4 & 0.15 \end{bmatrix}, \quad N_2 = \begin{bmatrix} 1 & 0.1 \\ 0.4 & 0.15 \end{bmatrix},$$

$$U_2 = \begin{bmatrix} 0.1 & 0.1 \\ 0 & -0.01 \end{bmatrix}, \quad H_2 = \begin{bmatrix} 0.01 & 0 \\ 0 & -0.1 \end{bmatrix},$$

$$E_{12} = \begin{bmatrix} 0 & 0.5 \\ 0 & 0.1 \end{bmatrix}, \quad E_{22} = \begin{bmatrix} 0.1 & 0 \\ 0.8 & 0 \end{bmatrix},$$

$$E_{d2} = \begin{bmatrix} 0.1 & 0 \\ 0 & 0.5 \end{bmatrix}, \quad E_{32} = \begin{bmatrix} 0.2 & 0 \\ 0.6 & 0 \end{bmatrix},$$

$$E_{f2} = \begin{bmatrix} 0.3 & 0 \\ 0 & 0.6 \end{bmatrix}.$$

The minimum ADT of the switched systems (1) is $\tau_a^* = \ln \eta/\alpha = \ln 4.6679/0.0800 = 1.925$, so we select the ADT $\tau_a = 2$, and the uncertain time-varying matrices are $F_1(t) = F_2(t) = \mathrm{diag}\{\sin t, \sin t\}$.

Let the initial condition of the systems and the disturbance signal be

$$x(0) = [0 \;\; 0]^{\mathrm{T}}, \quad \omega(t) = [8\sin(-0.1\pi t)\mathrm{e}^{-0.1t} \;\; 10\sin(-0.02\pi t)\mathrm{e}^{-0.1t}]^{\mathrm{T}}.$$

The state response of the nonlinear control system within periodic switching signal, which is shown in Figures 1 and 2, can be acquired. It can be seen that, for

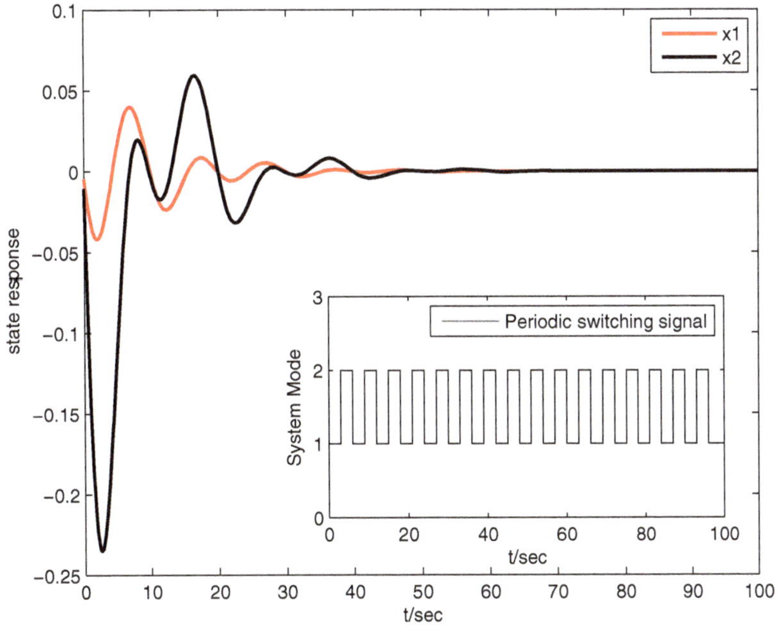

Figure 1. State response for normal operation of the actuators within the periodic switching signal.

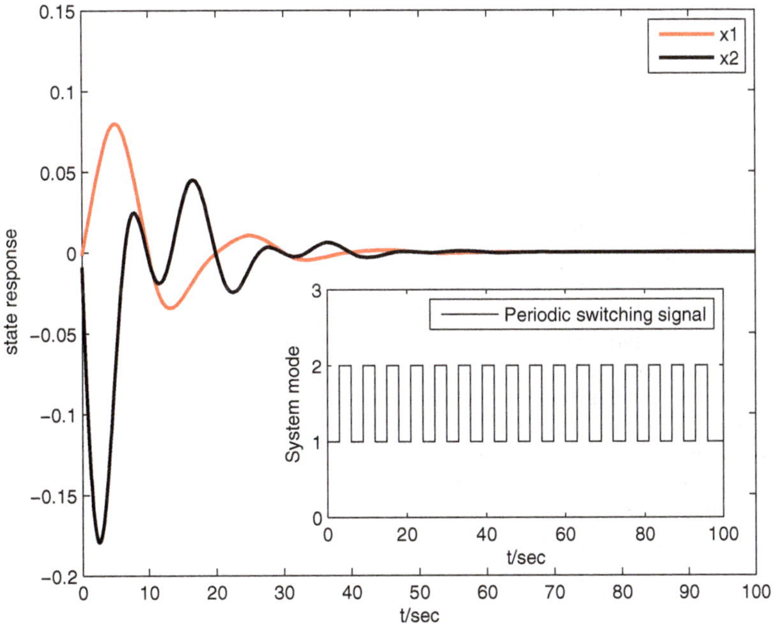

Figure 2. State response for partial fault of the actuators within the periodic switching signal.

two cases that normal operation of the actuators and partial fault of the actuators, the designed robust H_8 reliable controllers can make the nonlinear switched systems globally exponentially stable. Figure 3 gives the state response of the control system within the arbitrary switching signal, which shows that the designed controllers are also effective on robust stabilization problem of nonlinear switched systems within arbitrary switching signal.

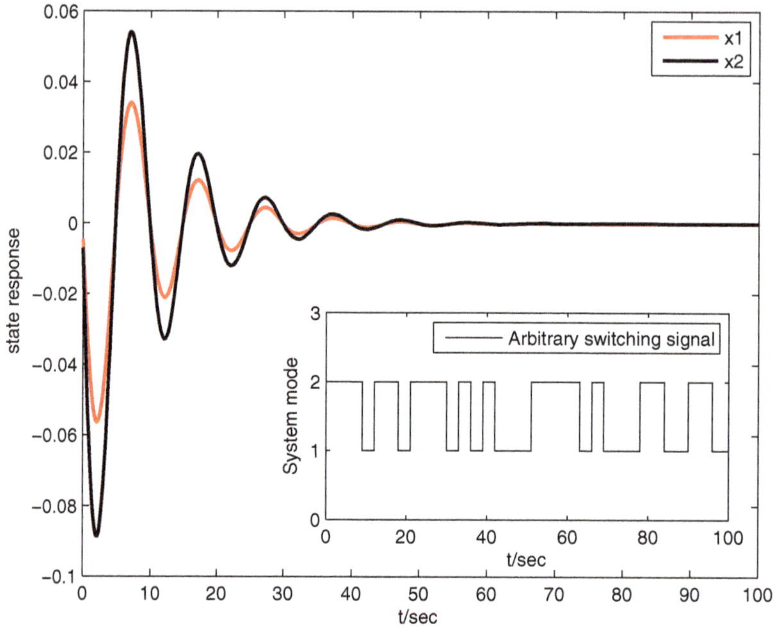

Figure 3. State response for partial fault of the actuators within the arbitrary switching signal.

5. Conclusion

In this paper, the robust H_∞ reliable controllers have been investigated for the nonlinear delay switched systems. Based on the piecewise Lyapunov functional and ADT approach, sufficient conditions have been established for the existence of the reliable controllers that guarantee the nonlinear delay switched systems to be globally exponentially stable with a guaranteed H_∞ performance. However, the nonlinear functions of switched systems in this paper are constrained by Lipschitz conditions. If the Lipschitz conditions are not satisfied any more, our designed control method may not be effective for the closed-loop systems. Thus, the reliable control problems for the switched systems, which have general nonlinear parts, will be good topics. For example, the study for the switched systems, which have stochastic nonlinear parts with multiplicative noise, may be meaningful researches in the future, and the explorations will be big challenges.

Disclosure statement

No potential conflict of interest was reported by the authors.

Funding

This work was supported by the Natural Science Foundation of Heilongjiang Province of China [F201403], the Postdoctoral Science-research Developmental Foundation of Heilongjiang Province of China [LBH-Q13177], and the Fostering Foundation of Northeast Petroleum University (XN2014112), the Northeast Petroleum University Graduate Student Innovation Research Project (YJSCX2014-029NEPU).

References

Balochian, S., & Sedigh, A. K. (2012). Sufficient conditions for stabilization of linear time invariant fractional order switched systems and variable structure control stabilizers. *ISA Transactions*, *51*, 65–73.

Du, H. B., Lin, X. Z., & Li, S. H. (2013). *Finite-time stability and stabilization of switched linear systems.* Joint 48th IEEE conference on decision and control and 28th Chinese Control Conference, Shanghai, pp. 1938–1943.

Hu, B., & Michel, A. N. (2000). Stability analysis of digital feedback control systems with time-varying sampling periods. *Automatica*, *36*, 897–905.

Kim, D. K., Park, P. G., & Ko, J. W. (2004). Output-feedback H_∞ control of systems over communication networks using a deterministic switching system approach. *Automatica*, *40*, 1205–1212.

Lin, H., & Antsaklis P.J. (2009). Stability and stabilizability of switched linear systems: A short survey of recent results. *IEEE Trans on Automatic Control*, *54*, 308–322.

Liu, Y. S. (2000). *Reliable control and filtering for some class of stochastic nonlinear time-delay systems*. Dalian: Dalian University of Technology.

Lu, Q. G., Zhang, L. X., & Karimi, H. R. (2013). H_∞ control for asynchronously switched linear parameter-varying systems with mode-dependent average dwell time. *Control Theory and Applications, IET*, *7*, 673–685.

Wang, L. M., Sun, C. C., & Liu, Y. Z. (2006). Robust fault-tolerant control of uncertain nonlinear delay switched systems. *Journal of Shenyang Normal University*, *24*, 136–139.

Wu, L., & Lam, J. (2009). Weighted H_{∞} filtering of switched systems with time-varying delay: average dwell time approach. *Circuits System Signal Process*, *28*, 1017–1036.

Wu, L., Qi, T., & Feng, Z. (2009). Average dwell time approach to $L_2 - L_{\infty}$ control of switched delay systems via dynamic output feedback. *IET Control Theory and Applications*, *10*, 1425–1436.

Xie, D. M., Wang, L., & Hao, F. (2003). *Robust stability analysis and control synthesis for discrete-time uncertain switched systems*. Proceedings of the 42nd IEEE conference on decision and control, pp. 4812–4817. Retrieved from http://ieeexplore.ieee.org/xpl/login.jsp?tp = &arnumber = 1272350&url = http%3A%2F%2Fieeexplore.ieee.org%2Fxpls%2Fabs_all.jsp%3Farnumber%3D1272350.

Yang, S., Zhengrong, X., Qingwei, C., & Weili, H. (2006). Robust reliable control of switched uncertain systems with time-varying delay. *International Journal of Systems Science*, *37*, 1077–1087.

Zhai, C. L., He, W., & Wu, Z. M. (2000). Stability analysis of switched systems and the design method of controllers. *Information and Control*, *29*, 21–26.

Zhai, G., Hu, B., & Michel, A. N. (2001). Stability analysis of switched systems with stable and unstable subsystems: an Average dwell time approach. *International Journal of Systems Science*, *32*, 1055–1061.

4

Enhanced OCV prediction mechanism for a stand-alone PV-lithium ion renewable energy system

Thomas Stockley*, Kary Thanapalan, Mark Bowkett and Jonathan Williams

Faculty of Computer, Engineering and Sciences, University of South Wales, Pontypridd, Wales

This paper aims to improve the estimation of state of charge (SoC) of the battery component for a small-scale photovoltaic stand-alone system through the use of a simple summing equation, at a set measurement interval. The system uses a pre-defined parameter to accurately predict the open-circuit voltage (OCV) of a cell at a much reduced measurement time of 5 minutes, while maintaining a maximum prediction error of less than 1%SoC. A simulation model has been provided that allows measurement of the cell voltage and current for prediction of the equilibrated OCV. The simulation can be used for single cell, modules and battery packs which use lithium-based technologies. Validation of the model has been performed using experimental data from tests conducted at the Centre for Automotive and Power System Engineering (CAPSE) laboratories, at the University of South Wales. An application has been proposed for this work, which includes a photovoltaic module for energy generation to power an illuminated advertizing sign. The energy is stored in a lithium-based battery model which uses a combination of a battery management system and remote monitoring for real-time data acquisition.

Keywords: PV-lithium; smart monitoring; prediction mechanism; BMS; module simulation

1. Introduction

The abundant sources of power in the form of solar (photovoltaic) and wind energy has been an important tool in the past decade, as efforts are made to reduce global warming while preserving precious dwindling fossil fuel supplies. Popularity of these energy generation tools is still seeing a steady increase, especially in the case of photovoltaic (PV), with reports showing an increase in European capacity by approximately 11 GW in 2013 (European Photovoltaic Industry Association). This is due to advances in the solar technology and government incentives to ensure that growth continues to achieve a target of 20% of European energy being generated by renewable sources by 2020 (European Union. The European Commission, 2007).

A PV array can be operated as a stand-alone or grid-connected system (Salas, Olías, Barrado, & Lázaro, 2006). Mobile stand-alone PV systems are mainly used in the conservation of the environment by using solar energy in locations without access to electricity, and therefore, act as an indispensable electricity source for remote areas. The technology for harvesting the solar energy shows much promise; however, the unreliability of the generation methods mean that they need to be coupled with an energy storage technique for efficient and practical use (Hadjipaschalis, Poullikkas, & Efthimiou, 2008). Due to the emerging advantages of lithium-ion battery technologies, these are being investigated as an alternative energy storage component for renewable energy stations. It is possible to consider lithium batteries for a wider range of applications because of falling manufacturing costs and increasing performance benefits (Nair Nirmal-Kumar & Garimella, 2010).

Although the manufacturing costs of the lithium cell are dropping, the requirement for internal monitoring electronics results in an inflated price when compared to alternative battery technologies. This in turn makes use of lithium technologies in small-scale energy generation systems as a promising next step rather than a common practice, with designers favouring the cheaper more robust lead acid battery.

However, the "expensive" monitoring and management equipment is a necessary addition to the lithium battery pack to stop the battery causing damage to itself, the generation equipment, or a threat of harm to the end user (Daowd, Omar, Bossche, & Mierlo, 2011). This is a wise precaution as was experienced in 2002, where lithium cells were causing damage in high-end electrical appliances (Sima, 2006). The battery management system (BMS) is the equipment charged with keeping each cell within its safe operation limits, and does so by carefully monitoring the cell voltage, temperature and state of charge (SoC). The SoC is the main focal point on a modern BMS, where the knowledge of remaining battery charge is becoming more and more essential.

*Corresponding author. Email: thomas.stockley@southwales.ac.uk

Coulomb counting is the simplest, albeit crudest form of SoC estimation. It allows a very quick to market implementation phase, and requires very little mathematics or background processing (Ng, Moo, Chen, & Hsieh, 2009). Added to these advantages is the fact that the technique can be applied during a charge or discharge state irrelevant of current magnitude and it is easy to see why it is the most common SoC estimation method. The method operates by counting the Amp-hour (Ah) that have been added or removed from a cell during a charge or discharge state and comparing the rolling total to a predefined cell capacity. The cell capacity is usually measured by a low current capacity test. There are, however, disadvantages to using the coulomb counting technique. Unavoidable accumulation errors are common and an unknown initial SoC renders the method useless; however, correction mechanisms are being researched to reduce these errors (Piller, Perrin, & Jossen, 2001). Therefore, the coulomb counting method is seldom used by itself in high-end BMSs where the SoC is requested as a highly accurate estimation.

A second method for the SoC estimation of a lithium cell is the use of an open-circuit voltage (OCV)–SoC lookup table. The OCV–SoC estimation method works by taking a voltage measurement at an equilibrated state (after a 3-hour rest period) and comparing it to a predefined lookup table containing the equivalent SoC of the cell for each voltage value. This method has the disadvantages of being a more complicated technique with a longer implementation time, as the lookup table needs to be generated, and a long open-circuit period is needed to ensure that the cell is at an equilibrated state (Chiang, Sean, & Ke, 2011). These disadvantages mean that the OCV–SoC estimation technique is not one which can be used solely on applications such as PV or wind microstations, where the battery bank is likely to be in a charge or discharge for a majority of the time. The high accuracy recorded by these systems in previous works makes the technique an ideal complement to the more simplistic coulomb counting method, where OCV–SoC can be used to find the initial SoC and then the coulomb counting method can be used to provide an updated running SoC for the battery (Piller et al., 2001; Stockley, Thanapalan, Bowkett, & Williams, 2013).

The need for the OCV measurement to be taken during an equilibrated state is due to hysteresis of the terminal voltage, which produces a false OCV measurement when the cell initially finishes a charge or discharge state. The cell voltage relaxes quickly initially when open-circuit commences, but the rate of the voltage relaxation decays with time as the terminal voltage becomes closer to the OCV. This is an issue that could cause a delay of up to 4 hours as the cell reaches an equilibrated state. This disadvantage is not only common to lithium cells, as alternative chemistries also require the need for a long open-circuit period before an accurate measurement can be made. One solution for predicting the OCV from a 30-minute measurement was proposed by Aylor, Thieme, and Johnso (1992). A simple summing technique was firstly proposed by Aylor et al. (1992), but it was a second technique using asymptotes and a semi-logarithmic scale that was finally implemented. The first method (a simple summing technique) was disregarded due to the fact that lead acid cell properties tend to differ greatly, even between cells in the same battery. However, Stockley et al. (2013) proved that lithium cells are much more uniform because of their chemical construction (the electrodes do not physically break down like a lead acid cell) and a much more automated manufacturing process. This allowed the simple summing method to be used following a charge or discharge rate for both $LiNiMnCoO_2$ and $LiFePO_4$ cells. Stockley et al. (2013) and Stockley, Thanapalan, Bowkett, and Williams (2014a, 2014b) proved that the summing method could be used on lithium cells using a 30-minute measurement, and then reduced this time to 8 minutes to make the prediction mechanism more available to a wider range of applications. The maximum error obtained from the tests conducted in Stockley et al. (2013, 2014a, 2014b) was just 6 mV for the 30-minute rate and 8.3 mV for the 8-minute rate, less than a 1%SoC error.

The second method of OCV estimation using the asymptotes on a logarithmic scale proved a success for Aylor et al. (1992), with a maximum error of just 5%SoC when using a measurement time of 6.6 minutes. Pop, Bergveld, Danilov, Regiten, and Notten (2008) used the asymptote method produced by Aylor to validate a new OCV prediction model and in doing so proved that the asymptote method could be used on lithium-based cells with an error of 0.92%. A combinational model using voltage change over time and temperature was also used to validate the new model, giving a very large error of 20.19%. The model proposed by Houseman (2005) relies on the voltage relaxation curve and achieved an error of just 0.19%. Added to this very low error rate, the model also accounts for the cell temperature, charge/discharge rate and state of health of the cell.

Several other OCV prediction mechanisms are available including: (i) polynomial-based curve estimators as used by Hu, Li, Peng, and Sun (2012), which rely on an extended Kalman filter for prediction. (ii) A sigmoid function approach as used by Weng, Sun, and Peng, (2013), which was compared to several polynomial-based models to produce an error of 2.5 mV, an improvement of 4.8 mV on the most accurate polynomial system. (iii) Linear regression models such as the work by Pei, Wang, Lu, and Zhu (2014), where the diffusion process of a lithium cell was found to have a linear relationship with the OCV, a relationship found through modelling of the cell using equivalent Resistve–Capacitive circuits.

Alternative techniques to estimate the SoC of the lithium cell under load have been proposed (Kutluay, Cadirci, Ozkazanc, & Cadirci, 2005). These methods measure the voltage while the cell is in use and thus the current has a large factor in the cell voltage. For this reason, the loaded voltage technique is more difficult to implement and calls for a much larger voltage to SoC relationship modelling phase. Therefore, the OCV–SoC technique has been chosen as the SoC estimation technique for this scope of work.

Generation and storage of "clean energy" is the focus of reducing the reliance on non-renewable energy sources, but the efficiency of how energy is being stored and consumed also needs to be considered. For this purpose, remote systems or smart metres are now being employed to monitor the power consumption of a specific load (Houseman, 2005). The smart metre was initially prioritized for home installations as part of the Smart Metering Implementation Programme to install metres in every home by 2020 (Great Briton. Department of Energy and Climate Change, 2012), in an aim to reduce the carbon consumption of the UK. Attention is now turning to monitoring energy usage in renewable energy systems both grid connected and stand-alone (Wolfe, 2008; Yu, Zhang, Xiao, & Choudhury, 2011). Wolfe (2008) suggests that by using remote monitoring and management equipment, the renewable energy system could greatly improve the conservation of the generated energy. However, as was discussed by Stefanakos and Thexton (1997) and Zezhong, Hongliang, and Ting (2010), a battery system also benefits from remote monitoring. Operating lithium cells at elevated temperatures, too high charge or discharge rates or in SoC ranges outside of the 80% to 20% boundary results in storage efficiency or cycle life becoming significantly lower (Guena & Leblanc, 2006; Shim, Kostecki, Richardson, Song, & Striebel, 2002). For this reason, the system shown further in this paper will contain a remote energy management and monitoring system, developed at the University of South Wales, to allow instant feedback of incorrect operating parameters.

The remaining sections of this work are organized as follows: Section 2 gives an overview of the OCV–SoC theory, an explanation of previous work and the test setup. Section 3 explains the results of the OCV–SoC prediction tests conducted on the lithium module. Section 4 provides the OCV–SoC model developed in Simulink, while the implemented BMS and design information can be found in Section 5. Lastly, a discussion and conclusion to this paper is provided in Section 6.

2. Research methodology and theory

As mentioned in the Introduction, the OCV–SoC estimation technique usually requires the cell to relax for a period of up to 3 hours after a charge or discharge state. The relaxation time can be reduced by the use of the simple formula provided in the following equation:

$$V_{OC} = V_{tr} \pm K_V, \quad (1)$$

where V_{OC} is the equilibrated OCV, V_{tr} is the voltage at the measurement interval and K_V is a predefined parameter

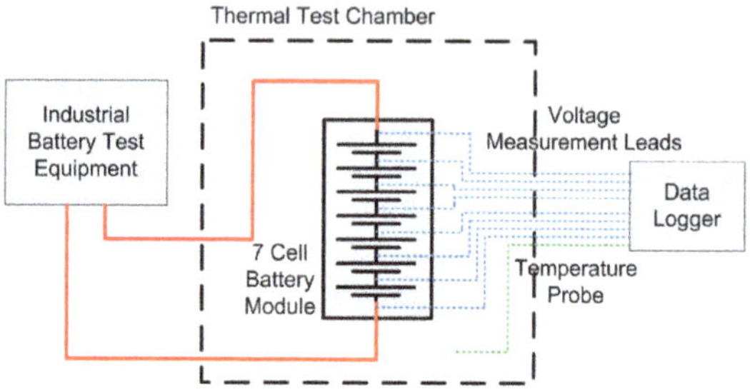

Figure 1. Block diagram of experimental setup for module testing.

found by applying the equation $V_{OC} \pm V_{tr}$ to a set of controlled tests. The voltage at the measurement interval (V_{tr}) was initially set as the voltage after 30 minutes of an open-circuit condition (Stockley et al., 2013, 2014a). However, advances in the research allowed the simple equation to be successfully tested at an improved measurement time of 8 minutes with a maximum SoC error of less than 1% (Stockley et al., 2014b). The work in this paper investigates the further improvement in the rate of prediction with an aim to reduce the time to within 5 minutes.

Testing of the OCV prediction method was conducted at the Centre for Automotive and Power Systems Engineering (CAPSE) labs at the University of South Wales. The tests were carried out using industrial battery test equipment, a thermal chamber to control the boundary conditions and a datalogger to measure the voltage of each of the cells. The setup can be seen in Figure 1. The module under test was made for use in a PV microsystem, and was constructed of seven 40 Ah $LiNiMnCoO_2$ cells as shown in Figure 2. Single cell tests were conducted on a 20-Ah $LiNiMnCoO_2$ cell and although the 40 Ah cells are the focus of this work, the previous work in Stockley et al. (2013, 2014a, 2014b) was conducted on the 20 Ah cells and they are referenced throughout this paper.

The test process included a discharge relaxation test, a charge relaxation test and a mixed state relaxation test. It was expected that by using the OCV prediction mechanism to aid the OCV–SoC technique, a system could be easily implemented to monitor and accurately track the SoC of not only the full battery module, but also each individual cell, ensuring that the module remains correctly balanced.

Each test consisted of a conditioning stage, where the cell was charged to 100%SoC or discharged to 0%SoC depending on the test requirement. The conditioning stage made use of a constant current/constant voltage charge or a constant current discharge. A constant current step was used to alter the cell's SoC by 20%, followed by a four-hour open-circuit measurement period to record the relaxation curve. Initial testing was conducted on the smaller capacity 20 Ah $LiNiMnCoO_2$ cell at 0.33, 1 and 3 C for both the charge and discharge states, resulting in

Figure 2. Seven-cell 40 Ah $LiNiMnCoO_2$ module.

a prediction error of just 6 mV using a 30-minute measurement. This equates to a SoC error of 0.4%SoC, using the fact that previous OCV–SoC tests resulted in a relationship of 9.9 mV is equal to a 1%SoC. Although the prediction error was very low, the application of this mechanism seemed to be very limited with many devices not being able to afford the cells to be out of use for a period of 30 minutes. To resolve this issue, the prediction error was found at several intervals to find the best compromise between measurement time and prediction error. The results can be seen in Figure 3.

As it can be seen in Figure 3, the ideal time for prediction is at 5 minutes where the measurement error is 9.25 mV, slightly less than the 9.9 mV level, where the error would equal 1%SoC. This was used in the new OCV prediction work for the 20 Ah lithium pouch cell, producing a maximum error of just 9.5 mV (0.96%SoC) while reducing the prediction time by 25 minutes from the work conducted in Stockley et al. (2013, 2014a).

3. Module OCV prediction testing

To ensure that the prediction mechanism would transfer from single-cell level to modules and even battery packs, relaxation tests were conducted on the seven-cell $LiNiMnCoO_2$ module. The module voltage was monitored throughout a 0.3 C discharge test, a 0.3 C charge test and a 0.3 C mixed state test, with four-hour open-circuit periods at 80%, 60%, 40% and 20%SoC. Figure 4 provides the relaxation curves for the mixed state test, where each SoC adjust involved both charge and discharge steps. From the shape of the relaxation curves in Figure 4, it can be seen that the open-circuit periods at 80% and 40%SoC followed a discharge state, whereas 60% and 20%SoC relaxed from a charge state. The duty cycle used to achieve these results can be seen in Figure 5.

The value of the parameter K_V for the module was derived by previous testing as explained in Stockley et al. (2014a) as 0.055 and 0.035 following a discharge and charge state, respectively. This allowed a fast yet accurate OCV prediction when using Equation (1). Table 1 contains the mixed state prediction results as an example, where a maximum error of just 45 mV was observed. The module charge and discharge relaxation tests also provided promising results with a maximum error of just 15 and 5 mV for the discharge and charge test, respectively. With an

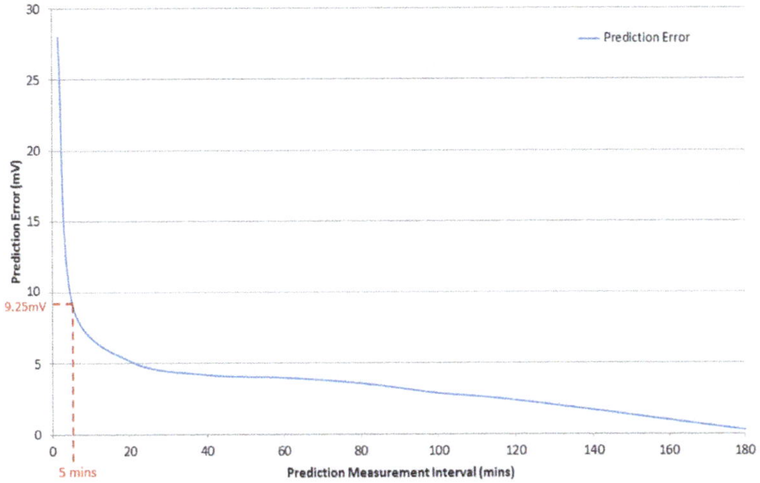

Figure 3. Error vs prediction time for lithium 20 Ah pouch cell.

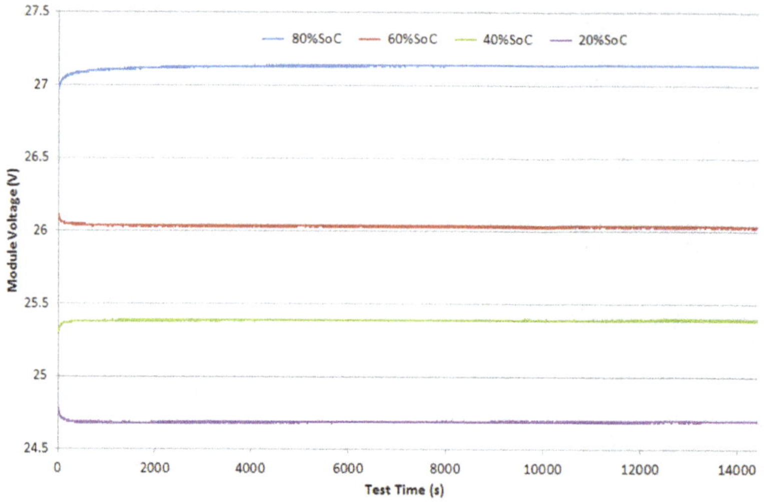

Figure 4. Relaxation curves for module mixed state test.

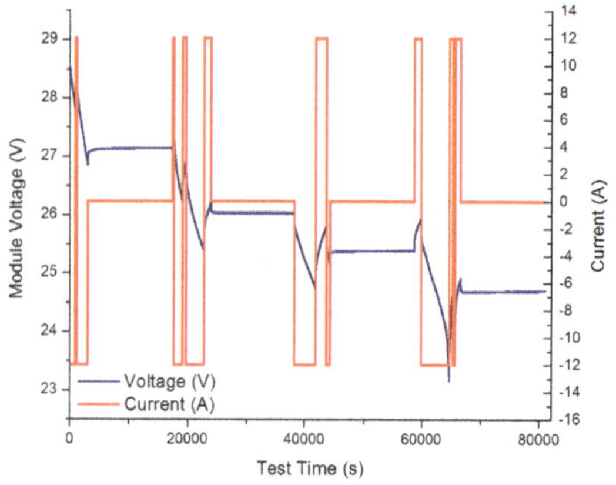

Figure 5. Mixed relaxation test current and voltage duty cycle.

OCV–SoC relationship of approximately 45.5 mV equals 1%SoC, a maximum error of 45 mV is proof that the prediction mechanism can be successfully adapted to lithium modules.

To ensure that the values of parameter K_V for a charge and discharge relaxation curve were chosen to provide the minimum error possible, the OCV prediction model was adapted for design optimization. The functions used for the optimization of parameter K_V can be seen in the following equations:

Table 2. Test parameters for K_V optimization.

Parameter	Value	Unit
Temperature	25	°C
C-rate	0.3	C
State of charge	20–80%	%SoC

$$\min_{(kv)} = f, \tag{2}$$

$$f = \sum_{t=0}^{t=ts} (V_{predict} - V_{measured})^2, \tag{3}$$

where f is the objective function, $V_{predict}$ is the predicted OCV using the parameters of 0.55 and 0.35 for a charge and discharge, respectively, and $V_{measured}$ is the measured equilibrated OCV. The boundary model parameters for the optimization can be seen in Table 2.

To optimize the K_V values for a charge and discharge relaxation curve, two optimization techniques were used. The simplex search method algorithm (SSM) was chosen

Table 1. Comparison of recorded data and OCV prediction for 7-cell module mixed state 0.3 C test.

	Open-circuit voltage (OCV) (V)						
Cell SoC (%)	0 minute	5 minute	8 minute	30 minute	180 minute real	180 minute calc.	Error (mV)
80	26.97	27.08	27.09	27.12	27.14	27.135	5
60	26.11	26.04	26.05	26.04	26.04	26.005	35
40	25.3	25.38	25.38	25.38	25.39	24.435	45
20	24.78	24.69	24.68	24.68	24.7	24.655	45

Table 3. Optimized values of K_V from SSM and PSM algorithms.

Cell condition	Derived K_V	SSM-optimized K_V	PSM-optimized K_V
Discharge relaxation	0.55	0.56	0.572
Charge relaxation	0.35	0.36	0.41

due to its high optimization speed which as Thanapalan, Wang, Williams, Liu, and Rees (2008) explain is due to the fact that the SSM optimization is a hill climbing technique. The method basically works by altering the value of the optimization parameter (n) for each iteration and then measuring the response of the model error. If the error produced by the new parameter (n) is less than the previous parameter value ($n - 1$), then the new value (n) is adopted. If the error of n is greater than the value resulting from the use of $n - 1$, then the new value is rejected.

The second optimization parameter used was a genetic algorithm by use of the pattern search method (PSM). The GAs are computational programmes based on the natural interactions of the genetics seen in nature. As Thanapalan et al. (2008) explain, the method incorporates a survival of the fittest where any parameter values outside of the normal scope is disregarded as the system parameters move closer to the optimized value.

The optimized parameters for K_V can be seen in Table 3 where they are compared to the calculated value of K_V from Table 1.

From the results in Table 3, it can be seen that the optimized results are very close to the initial results produced as in Stockley et al. (2014a, 2014b) and previous in this paper. The use of the optimized values from the SSM algorithm increased the error by 1 mV when compared to the derived K_V parameter values. The PSM-optimized parameter values also resulted in a higher error for the 60%, 40% and 20%SoC tests. The 80%SoC test resulted in a lower error value, however, a reduction in the error of 2.3 mV is negligible when considering that the OCV–SoC relationship of 45 mV is equal to 1%SoC. The results of the optimization parameter values in OCV prediction can be seen in Table 4.

The results in Table 4 show that the derived values of K_V from previous works calculated from the average curves provide the best error value. This means that optimization does not need to be carried out on further cell testing which makes the prediction mechanism simpler and quicker to transfer to other cell types.

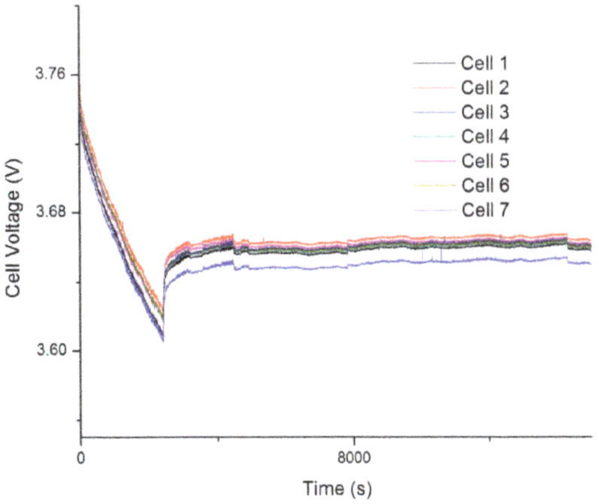

Figure 6. Single cell relaxation curved following 0.3 C discharge to 40%SoC.

The use of the derived values of the K_V parameter also has the advantage of simpler implementation when moving between single cells and modules. This can be seen by the work in Section 4.

4. OCV prediction testing of module's cells

As the module contains seven $LiNiMnCoO_2$ cells, each cell's relaxation voltage had to be monitored through the use of a datalogger, which allowed results such as the curves in Figure 6 to be created. The relaxation curves in Figure 6 follow a 0.3 C discharge to 80%SoC. It should be noted at this point that there is a slight discrepancy between some of the cells in the module. A capacity test and impedance test on each of the cells suggest that the module was not fully balanced prior to the testing phase, rather than some of the cells being damaged or aged. The slight voltage imbalance in the module can be seen in Figure 6, with the maximum equilibrated voltage of 3.667 V held by Cell 2 and the lowest equilibrated voltage of 3.653 V by Cell 3. However, the difference in cell voltage was negligible for the purpose of this research and had little effect on the further results.

The parameter (K_V) for each individual cell was calculated from the parameter value for the seven-cell module.

Table 4. 5-minute OCV prediction results using derived, SSM and PSM K_V parameter values.

Cell SoC (%)	Measured OCV (V)	Derived K_V calc. (V)	Derived K_V error (mV)	SSM K_V calc. (V)	SSM K_V error (mV)	PSM K_V calc. (V)	PSM K_V error (mV)
80	27.14	27.135	5	27.136	4	27.1373	2.7
60	26.04	26.005	35	26.004	36	25.999	41
40	25.39	25.435	45	25.436	46	25.437	47
20	24.7	24.655	45	24.654	46	24.649	51

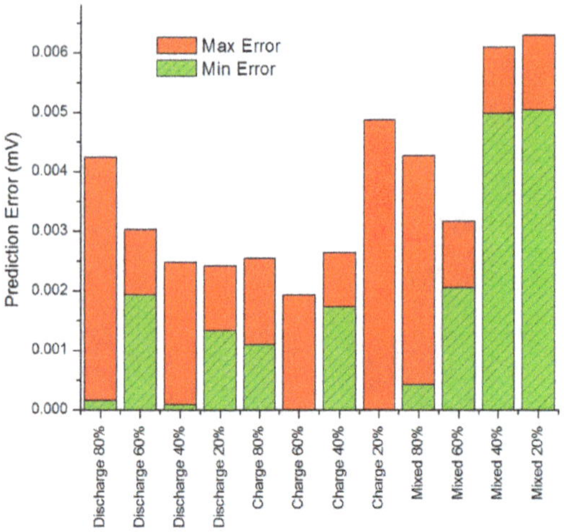

Figure 7. Comparison of minimum and maximum error for each of the single-cell tests.

As the module voltage is a simple summation of the single cells, it stands to reason that the relaxation voltage of a module is equal to the summation of the relaxation voltage of the single cells. Therefore, the module parameter of 0.055 for a discharge and 0.035 for a charge relaxation curve were divided by the number of cells to give a single-cell parameter of 0.0078 and 0.005 for the single-cell discharge and charge curves. As the amount of collected data is large, Figure 7 has been provided as a summary of the tests performed. Figure 7 shows the maximum and minimum errors from each 0.33 C relaxation test noted at the start of Section 3.

The error is greatest for the mixed relaxation tests at 40% and 20%SoC, giving the only prediction errors above 5 mV. From OCV–SoC relationship tests conducted on the 40 Ah $LiNiMnCoO_2$ pouch cell, a relationship of 6.5 mV that equals 1%SoC was found. This results in a maximum SoC error of 0.8% for a measurement time of just 5 minutes.

5. Modelling of OCV prediction mechanism

The OCV prediction mechanism was modelled in Matlab-Simulink which will allow it to be used for simulations and implemented into a real-world application. The developed prediction mechanism consists of several subsystems and is displayed in Figure 8. The top branch of the model is used to control the model, whereas the bottom branch is used to carry out the required action.

The State_Determination subsystem is used to determine whether the battery module is recovering from a charge or discharge state based on the current drawn from or applied to the battery module. This is important as the relaxation curve following a discharge state rises, whereas it falls after a charge state. The KV_Assignment subsystem block is used to assign the correct K_V value to the KvSelect storage variable. A second function of the State_Determination subsystem is to ensure that the prediction mechanism is only used in an open-circuit state. An open-circuit state is identified by a current measurement of less than 0.5 A, a value chosen to account for any measurement errors or system noise. The Timing_System uses the result of the State_Determination subsystem as a trigger to start an internal clock when an open-circuit period is entered. The Timing_System also recognizes when the measurement interval is reached and triggers the system to measure the OCV value. The Process_System monitors the module voltage and predicts the equilibrated 3-hour OCV voltage. The prediction is made based on the OCV value at the measurement interval and the correct K_V value provided by the KV_Assignment subsystem block. The

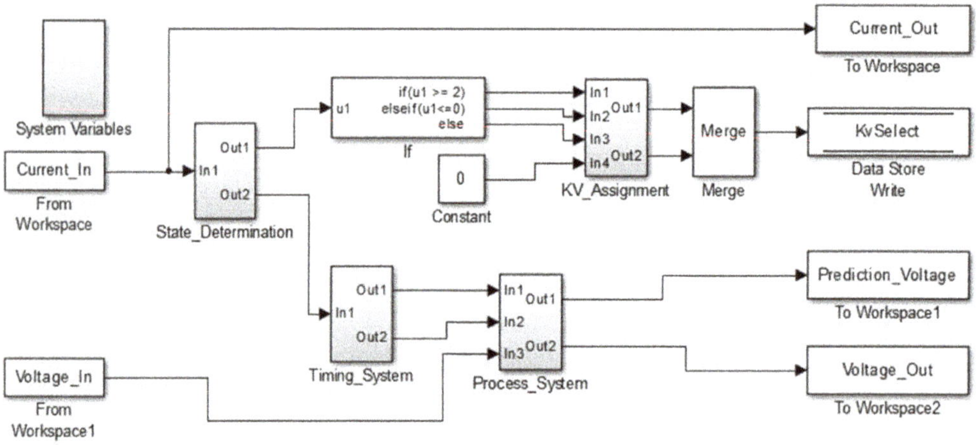

Figure 8. Simulink OCV prediction mechanism model.

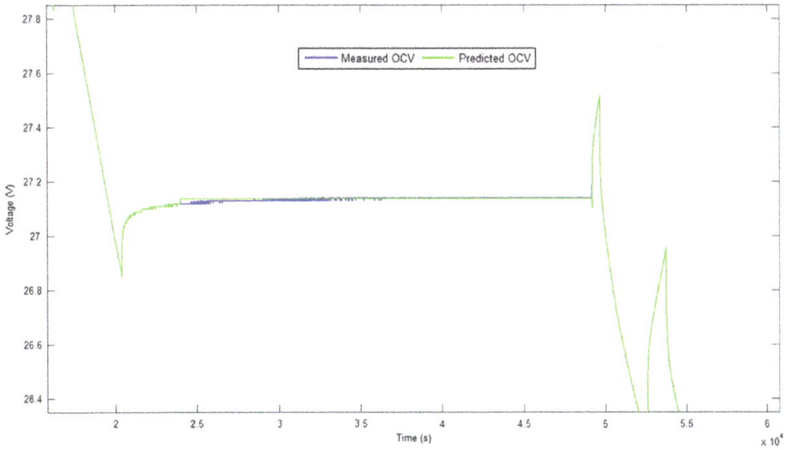

Figure 9. Comparison of measured and simulated relaxation curve at 80%SoC.

results of the model are exported to the Matlab workspace using the components Current_Out, Voltage_Out and Prediction_Voltage.

Figure 9 shows the relaxation curve of the 80%SoC discharge state relaxation test on the seven-cell module. As it can be seen in Figure 9 and noted in Section 3, the relaxation curve follows a discharge and therefore is rising. It is clear to see that the voltage of the module is monitored closely by the prediction model. When the open-circuit period reaches the 5-minute measurement interval, the prediction voltage is calculated and shown as the small jump in the green curve. At this point, the measured OCV can be seen to rise gently towards the predicted OCV measurement value. At the 3-hour mark, the measured and predicted voltages are re-aligned, proving that the model successfully predicted the OCV from the 5-minute measurement.

Although the model in Figure 8 is being used to predict the OCV of the full seven-cell module, it can be easily adapted for single cell or pack prediction. To use the model to predict a single cell, the parameter values of K_V (stored in the KV_Assignment subsystem) are changed from the module parameters in Section 3 to the single-cell K_V values in Section 4. For example, the parameter value would change from 0.055 to 0.0078 for a relaxation curve following a discharge state. As the module K_V value is proven to equal the single-cell K_V value multiplied by the number of cells in the module, it is recognized that a module model could also be accurately represented by summing the output of seven single-cell models. This method allows prediction of each individual cell and the full module.

This mechanism can also be applied to a larger energy storage system if required. As modules can be installed in series to increase the voltage of the battery system, the parameter K_V of the module can be multiplied by the amount of modules to find the K_V value for the battery pack. Likewise with the single-cell-based model, a set of module models could be used to represent a battery pack.

Figure 10. Prototype BMS with embedded OCV prediction mechanism.

6. OCV prediction mechanism in a real-world application

To prove that the prediction mechanism works in a real-world system, the model has been integrated into a BMS currently being developed in the University of South Wales CAPSE labs. The BMS can be seen in Figure 10.

The BMS is a low cost design which monitors the cell and module voltage, the module current and cell temperature. As the voltage and current are already being monitored by the BMS, the prediction mechanism can be monitored without the addition of any hardware. As mentioned in the Introduction, the BMS uses the prediction mechanism coupled to an OCV–SoC lookup table for the initial high accuracy estimation, and the coulomb counting method as a running SoC estimation. Figure 11 shows the flow chart for the prediction mechanism of the BMS. The

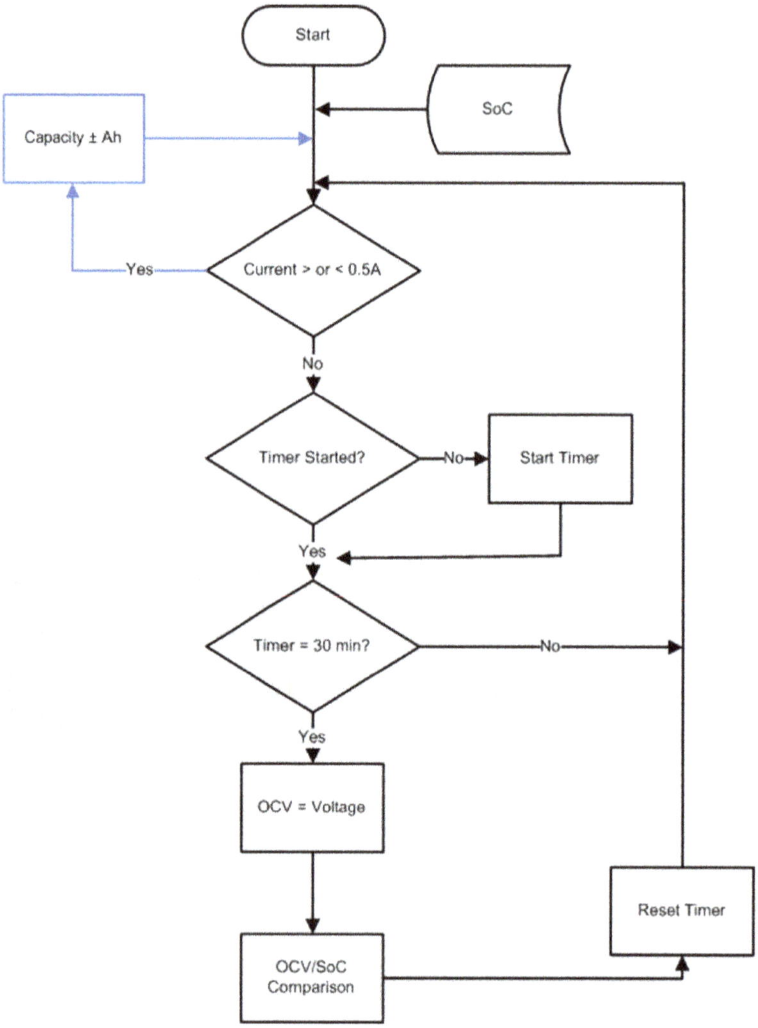

Figure 11. OCV prediction mechanism BMS flow chart.

section of the flow chart that is highlighted blue (dashed) is the coulomb counting section of the code. The SoC block at the top of the flow chart is the SoC value prior to the current iteration of the code.

It should be noted that the BMS is to be placed in a real-world application to monitor performance of a seven-cell lithium module, similar to the module shown in Figure 3. The module can be used as an energy storage system to power a stand-alone system. For example in this case the module is used to power an advertizing sign for a local company, which uses a PV panel as its primary energy source. Figure 12 shows a block diagram of the test rig for this particular case.

As the application for the prediction mechanism and seven-cell module is to be remote, monitoring of the battery condition and performance is difficult. As has been

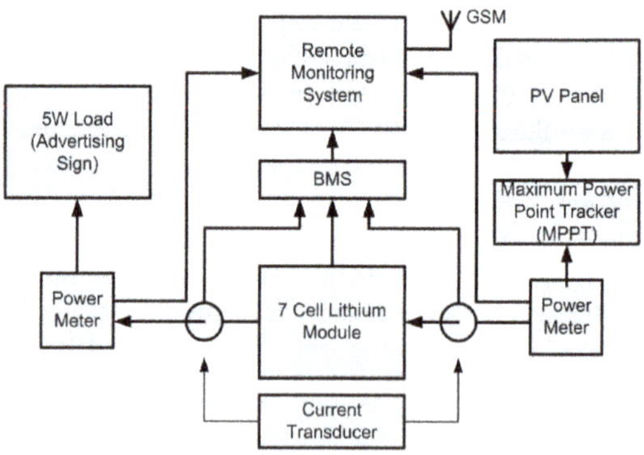

Figure 12. Block diagram of PV powered advertizing sign application for OCV prediction mechanism.

shown in the Introduction, regular monitoring of the battery parameters can lead to improved cycle life and energy storage efficiency. For this purpose, a remote management system is used to provide up-to-date information. The remote management system makes use of a micro-controller as the metre control system, which can read measurements from the BMS using a Modbus connection and sends information to a server using a global system for mobile communications (GSM) connection. The metre is a low cost design and has been implemented at the university for use in this research.

7. Discussion and concluding remarks

The work in this paper shows the advantages of the OCV prediction technique in both a theoretical and practical application. The results achieved from the testing work in Section 3 concur greatly with the tests conducted previously in Stockley et al. (2013, 2014a, 2014b). This is a positive feat as not only were the tests conducted on a seven-cell-connected module, but also on a different cell type to previous cell samples. The cell tested here was a 40-Ah (double the capacity of the previous pouch cell) $LiNiMnCoO_2$ cell. The maximum error from the module OCV prediction tests was well within the acceptable tolerance range of 1%SoC error, by producing a maximum error of just 45, 15 and 5 mV for the mixed, discharge- and charge-based tests as described in Section 3. Added to these low prediction errors is the fact that the OCV has been predicted at a measurement interval of 5 minutes.

While the OCV prediction mechanism was being tested on the module, single-cell measurements were being carried out on the individual cells in the module. This allowed the prediction mechanism to be applied on a module and cell level. Promising results were also seen in this step of the research as a maximum error of 6 mV at a measurement interval of 5 minutes was attained. Excluding the mixed tests would give a further reduced prediction error of 5 mV. The errors seen in this work can be put into perspective by considering the fact that a 1%SoC value has been attributed to a voltage of approximately 6.5 mV. Therefore, the single-cell tests would have equalled a maximum error of approximately 0.8%SoC. A noteworthy point in this work is that the value of the parameter K_V, that is used in Equation (1), has been set to 0.055 for a discharge and 0.035 for a charge relaxation curve. As the voltage for the module is the sum of the cell voltages, the K_V used in the single-cell tests was calculated as 0.0078 and 0.005 for the discharge and charge tests, respectively, which is in fact the module K_V values divided by the number of cells (7).

The prediction mechanism has been modelled in Section 5 which allowed simulations to be run using the results obtained from the full module tests. From Figure 9, it can be seen that the model follows the voltage profile throughout the discharge and until the measurement interval of 5 minutes when the open-circuit period is entered. When the measurement interval is triggered after an open-circuit period of 5 minutes the voltage is predicted. The measured voltage tapers up to the predicted value towards the 3-hour mark, where the cell is judged to be in an equilibrated state. Swapping the module's charge and discharge values of K_V with the cell level K_V allows the single-cell relaxation curve to be modelled, and the OCV to be predicted. Alternatively, using several single-cell models would allow the single cell and module OCV to be predicted, an advantage that is made possible by the simplicity of Equation (2).

The real-world application for the BMS which has the OCV prediction mechanism inbuilt is shown in Figure 10 in Section 6. The system would allow a very simple, low cost BMS to be used to protect the battery system in the stand-alone PV system. An energy management system is included in the system setup which can monitor all battery parameters and then upload the information using a GSM signal. This will allow the company responsible for the battery to be alerted very quickly if the module is being used in an unsafe way or inefficient manor. The flow chart is provided to give a brief understanding of how the software works and how the OCV prediction mechanism and coulomb counting method can be used in conjunction.

Acknowledgements

The authors would also like to thank the staff of CAPSE for their assistance.

Disclosure statement

No potential conflict of interest was reported by the authors.

Funding

The first author would like to acknowledge the financial support from KESS, RUMM Ltd and the University of South Wales.

References

Aylor, J. H., Thieme, A., & Johnso, B. W. (1992). A battery state-of-charge indicator for electric wheelchairs. *IEEE Transactions on Industrial Electronics*, *39*(5), 398–409.

Chiang, Y.-H., Sean, W.-Y., & Ke, J.-C. (2011). Online estimation of internal resistance and open-circuit voltage of lithium-ion batteries in electric vehicles. *Journal of Power Sources*, *196*(8), 3921–3932.

Daowd, M., Omar, N., Bossche, P., & Mierlo, J. (2011). *Passive and active battery balancing comparison based on MATLAB simulation*. The 7th IEEE vehicle power and propulsion conference.

European Photovoltaic Industry Association. Global market outlook for photovoltaics 2014–2018. Retrieved from February 27, 2015. http://files.ctctcdn.com/15d8d5a7001/3f338a6a-eece-4303-b8c4-c007181a59ad.pdf/

European Union. The European Commission. (2007). *The climate and energy package* [Online]. Retrieved from February 27, 2015. http://ec.europa.eu/clima/policies/package/index_en.htm

Great Briton. Department of Energy and Climate Change. (2012). *Smart metering implementation programme* [Online]. Retrieved from February 27, 2015. https://www.gov.uk/government/publications/smart-metering-implementation-programme-first-annual-progress-report-on-the-roll-out-of-smart-meters

Guena, T., & Leblanc, P. (2006). *How depth of discharge affects the cycle life of lithium–metal–polymer batteries*. The 28th annual international telecommunications energy conference 2006 (INTELEC '06) (pp. 1–8).

Hadjipaschalis, I., Poullikkas, A., & Efthimiou, V. (2008). Overview of current and future energy storage technologies for electric power applications. *Renewable and Sustainable Energy Reviews*, *13*, 1513–1522.

Houseman, D. (2005). Smart metering: The holy grail of demand-side energy management? *Refocus Journal*, *6*(5), 50–51.

Hu, X., Li, S., Peng, H., & Sun, F. (2012). Robustness analysis of state-of-charge estimation methods for two types of Li-ion batteries. *Journal of Power Sources*, *217*, 209–219.

Kutluay, K., Cadirci, Y., Ozkazanc, Y. S., & Cadirci, I. (2005). A new online state-of-charge estimation and monitoring system for sealed leadacid batteries in telecommunication power supplies. *IEEE Transactions on Industrial Electronics*, *52*(5), 1315–1327.

Nair Nirmal-Kumar, C., & Garimella, N. (2010). Battery energy storage systems: Assessment for small-scale renewable energy integration. *Energy and Buildings*, *42*(11), 2124–2130.

Ng, K. S., Moo, C.-S., Chen, Y.-P., & Hsieh, Y.-C. (2009). Enhanced coulomb counting method for estimating state-of-charge and state- of-health of lithium-ion batteries. *Applied Energy*, *86*(9), 1506–1511.

Pei, L., Wang, T., Lu, R., & Zhu, C. (2014). Development of a voltage relaxation model for rapid open-circuit voltage prediction in lithium-ion batteries. *Journal of Power Sources*, *253*, 412–418.

Piller, S., Perrin, M., & Jossen, A. (2001). Methods for state-of-charge determination and their applications. *Proceedings of the 22nd International Power Sources Symposium*, *96*(1), 113–120.

Pop, V., Bergveld, H. J., Danilov, D., Regiten, P. P. L., & Notten, P. H. L. (2008). Battery management systems. In V. Pop, H. J. Bergveld, D. Danilov, P. P. L. Regiten, & P. H. L. Notten (Eds.), *Methods for measuring and modelling a battery's electro-motive force* (Vol. *9*, pp. 63–94). Eindhoven: Springer.

Salas, V., Olías, E., Barrado, A., & Lázaro, A. (2006). Review of the maximum power point tracking algorithms for stand-alone photovoltaic systems. *Solar Energy Materials and Solar Cells*, *90*(11), 1555–1578.

Shim, J., Kostecki, R., Richardson, T., Song, X., & Striebel, K. A. (2002). Electrochemical analysis for cucle performance and capacity fading of a lithium-ion battery cycled at elevated temperature. *Journal of Power Sources*, *112*(1), 222–230.

Sima, A. (2006). *Sony exploding batteries – the chronicles*. Softpedia, October 13. Retrieved from February 25, 2015. http://news.softpedia.com/news/Sony-Exploding-Batteries-Chronicles-37848.shtml

Stefanakos, E. K., & Thexton, A. S. (1997). *Remote battery monitoring and management field trial*. The 19th international telecommunications energy conference (INTELEC) (pp. 653–657).

Stockley, T., Thanapalan, K., Bowkett, M., & Williams, J. (2013). *Development of an OCV prediction mechanism for lithium-ion battery system*. The 19th international conference on automation and computing (ICAC) (pp. 48–53).

Stockley, T., Thanapalan, K., Bowkett, M., & Williams, J. (2014a). *Design and implementation of OCV prediction mechanism for PV-lithium ion battery system*. The 20th international conference on automation and computing (ICAC) (pp. 48–54).

Stockley, T., Thanapalan, K., Bowkett, M., & Williams, J. (2014b). Design and implementation of an open circuit voltage prediction mechanism for lithium-ion battery systems. *Systems Science & Control Engineering: An Open Access Journal*, *2*(1), 707–717.

Thanapalan, K., Wang, B., Williams, J., Liu, G., & Rees, D. (2008). *Modelling, parameter estimation and validation of a 300W pem fuel cell system*. Proceedings of the UKACC international conference on control, Manchester (pp. 1–6).

Weng, C., Sun, J., & Peng, H. (2013, October 21–23). *An open-circuit-voltage model of lithium-ion batteries for effective incremental capacity analysis*. Proceedings of the ASME 2013 dynamic systems control conference, CA, USA.

Wolfe, P. (2008). The implications of an increasingly decentralised energy system. *Energy Policy Journal*, *36*(12), 4509–4513.

Yu, F. R., Zhang, P.Xiao, W., & Choudhury, P. (2011). Communication systems for grid integration of renewable energy resources. *IEEE Network*, *25*(5), 22–29.

Zezhong, X., Hongliang, S., & Ting, L. (2010). *Remote monitoring system of lead-acid battery group based on GPRS*. International conference on electrical and control engineering (ICECE) (pp. 4023–4026).

5

Gradient-based step response identification of overdamped processes with time delay

Rui Yan, Tao Liu*, Fengwei Chen and Shijian Dong

Institute of Advanced Control Technology, Dalian University of Technology, Dalian 116024, People's Republic of China

In this paper, a step response identification method is proposed for overdamped industrial processes with time delay from sampled data by developing a gradient searching approach to minimize the output prediction error. Based on establishing a least-squares fitting of the time domain expression of a low-order process model, i.e. first-order-plus-dead-time (FOPDT) or second-order-plus-dead-time (SOPDT), with respect to a step change, the rational model parameters together with the delay parameter can be simultaneously estimated, while the computation effort can be significantly reduced compared to the existing step identification methods based on using the time integral approach for model fitting. Both cases of repetitive poles and distinct poles are considered for the identification of an SOPDT model, along with a guideline for a suitable choice of the model structure for practical applications. The convergence and accuracy of the proposed algorithms are analysed with a strict proof. Four illustrative examples from recent references are used to show the effectiveness of the proposed method.

Keywords: step response identification; overdamped process; time delay; second-order-plus-dead-time model; gradient searching

1. Introduction

For control-oriented model identification in industrial engineering applications, step response tests have been widely used for identifying a transfer function model between the process input and output, owing to its implemental simplicity and economy. In the past decades, the identification of process models with time delay has received increasing attention in view of the fact that time delay is usually involved with industrial processes and system operations (Liu, Wang, & Huang, 2013). Although linear model identification methods have been extensively explored by Rake (1980) and Richard (2003), the extension to identifying a transfer function model with time delay is not straightforward and could become very difficult due to the identification nonlinearity introduced by time delay (Ljung, 1999).

Early references (Gawthrop & Nihtilä, 1985) proposed the approximation of time delay by using Padé approximation and Laguerre expansion, resulting in a higher order rational transfer function model that contains more parameters to be estimated and may cause unacceptable fitting error when the process response has a long time delay. By fitting a few representative points in the transient output response to a step change, Rangaiah and Krishnaswamy (1996) and Huang, Lee, and Chen (2001) reported alternative identification methods for obtaining a first-order-plus-dead-time (FOPDT) or second-order-plus-dead-time (SOPDT) model. Furthermore, based on numerical integral to the time domain expression of step response, Bi et al. (1999) proposed a time integral method for the identification of an FOPDT model, and it has been extended to identify a nth order process model by introducing n-fold multiple integrals in the lecture, e.g. Wang, Guo, and Zhang (2001) and Wang and Zhang (2001). To cope with load disturbance or nonzero initial conditions, modified step response tests were proposed by Wang et al. (2008), Hwang and Lai (2004) and Wang, Liu, and Hang (2005) to develop robust identification algorithms. Liu, Wang, Huang, and Hang (2007) suggested the use of multiple piecewise step tests to establish LS fitting conditions. In contrast, another robust identification algorithm, proposed by Liu and Gao (2008), was developed for obtaining a low-order transfer function model with time delay by using transient system response data from both adding and removing the step change. The idea was further extended to identify the dynamics of inherent-type load disturbance by Liu, Zhou, Yang, and Gao (2010). Note that these multiple-integral-based identification methods can give good accuracy and robustness in comparison with previous methods, but the computation load is relatively high and the use of multiple time integrals for parameter estimation may be sensitive to the test data length. To avoid the use of multiple time integrals for model identification, Ahmed, Huang, and Shah (2006, 2007) developed an iterative identification method based on using a linear filter. By comparison, a frequency domain step identification method was proposed

*Corresponding author. Email: liurouter@ieee.org

by Liu and Gao (2010) to relieve the computation effort by introducing a damping factor to the step response for the computation of the Laplace transform.

For the identification of time delay systems, adaptive identification methods for iteratively estimating the linear model parameters and the delay parameter were proposed in the references, e.g. Orlov, Belkoura, Richard, and Dambrine (2003), Ren, Rad, Chen, and Lo (2005), and Na, Ren, and Xia (2014), obtaining good convergence against measurement noise. A generalized expectation–maximization algorithm was proposed for robust global identification of linear parameter-varying systems with fixed input delay (Yang, Lu, & Yan, 2015). For identifying a canonical state space model with state delay, a recursive least-squares parameter identification algorithm was presented by Gu, Ding, and Li (2014) based on using a state filter. To tackle the identification nonlinearity arising from the time delay, a few numerical optimization methods were developed for estimating all the linear model parameters and the delay parameter, e.g. hierarchical identification strategies (Chen, Garnier, & Gilson, 2015; Previdi & Lovera, 2004; Yang, Iemura, Kanae, & Wada, 2007), Gradient searching algorithm (Ding, Liu, & Chu, 2013), Newton iteration approach (Ji, Xu, Xiong, & Chen, 2015), and the Levenberg–Marquardt optimization method (Sung & Lee, 2001; Baysse, Carrillo, & Habbadi, 2011). Note that these numerical optimization methods were developed based on using persistent excitation tests such as the pseudo-random binary sequence (PRBS), and thus are not suitable for step response tests as preferred in many industrial applications.

In this paper, model identification for overdamped industrial processes with time delay is studied to facilitate model-based control design for such a process in industrial applications. A gradient-based identification method is proposed which can simultaneously identify the linear model parameters together with the delay parameter from the step response data. Based on the time domain expression of the process transfer function model with respect to a step change of the input, a modified Gauss–Newton iteration algorithm for parameter estimation is established with computation efficiency for minimizing the output prediction error. Owing to the use of a step test, it is verified that the cost function of prediction error becomes convex with respect to all the model parameters, such that global convergence can be guaranteed. Moreover, no time integral is required for computation, significantly reducing the computation effort compared with recently developed step response identification methods based on using the time integral approach. For clarity, the paper is organized as follows. Section 2 presents the overdamped process models to be identified. The corresponding identification methods are detailed in Section 3. The convergence of the proposed method is analysed in Section 4, together with some guidelines for the choice of a suitable model structure. Four numerical examples are shown in Section 5 to demonstrate the effectiveness of the proposed identification method. Finally, some conclusions are drawn in Section 6.

2. Overdamped process model

For industrial overdamped processes, e.g. multi-component blending reactors and fermentation tanks, a low-order model of FOPDT or SOPDT is widely used for control system design and tuning (Seborg, Mellichamp, Edgar, & Doyle, 2010). Without loss of generality, two model forms are studied in this paper, one has a single or repetitive poles,

$$G_{m1}(s) = \frac{k_p}{(\tau_p s + 1)^m} e^{-Ls} \tag{1}$$

and the other has two distinct poles,

$$G_{m2}(s) = \frac{k_p}{(\tau_1 s + 1)(\tau_2 s + 1)} e^{-Ls}, \tag{2}$$

where k_p denotes the process static gain, L is the process time delay, τ_p, τ_1 and τ_2 are time constants, and m is the number of repetitive poles or the model order.

It should be noted that higher-order overdamped processes can be effectively described by the above models (Seborg et al., 2010; Liu & Gao, 2012).

Under a step test, the input excitation with a magnitude of h can be described by

$$u(t) = \begin{cases} 0 & \text{if } t < 0 \\ h & \text{if } t \geq 0. \end{cases} \tag{3}$$

Correspondingly, the output response subject to measurement noise may be written as

$$y(t) = y_r(t) + v(t), \tag{4}$$

where $y_r(t)$ denotes the true output, and $v(t)$ is usually assumed to be a white noise with zero mean.

Generally, the process static gain, k_p, can be directly computed from

$$k_p = \frac{\bar{y}(\infty)}{h}, \tag{5}$$

where $\bar{y}(\infty)$ is obtained by averaging 20–30 measured output values after the process response moves into a steady state in a step test.

Define the model prediction error and cost function, respectively, as

$$\delta(t, \hat{\theta}) = y(t) - \hat{y}(t), \tag{6}$$

$$J(\hat{\theta}) = \frac{1}{2}\varepsilon^T \varepsilon, \tag{7}$$

where $\varepsilon = [\delta(t_1, \hat{\theta}), \ldots, \delta(t_N, \hat{\theta})]^T$, $\hat{\theta}$ denotes the estimation of unknown parameters, $\hat{y}(t)$ is the output prediction, N denotes the collected data length, and t_i ($i = 1, 2, \ldots, N$.) are the sampled instant which should be chosen such as $L \leq t_1 < t_2 < \cdots < t_N$. Denote by a column vector θ all the unknown parameters to be identified.

Hence, the identification objective is to estimate the unknown parameters in Equation (1) or Equation (2) from the input–output observations $\{u(t_i), y(t_i)\}$ $(i = 1, 2, \ldots, N)$, satisfying

$$\hat{\theta} = \arg\min_{\theta} J(\hat{\theta}). \tag{8}$$

3. Proposed identification algorithms

Two gradient-based identification algorithms are developed to identify overdamped process models in Equations (1) and (2), as detailed in the following two subsections.

3.1. Time delay model with single or repetitive poles

Letting $m = 1$ in Equation (1) leads to an FOPDT model. Correspondingly, the time domain step response under zero initial conditions can be expressed by

$$y(t) = \begin{cases} v(t) & \text{if} \quad t < L, \\ k_p h(1 - e^{-\alpha(t-L)}) + v(t) & \text{if} \quad t \geq L. \end{cases} \tag{9}$$

where $\alpha = 1/\tau_p$, and the static gain k_p is computed by Equation (5).

Similarly, for the case of $m = 2$ in Equation (1), the time domain step response can be derived as

$$y(t) = \begin{cases} v(t) & \text{if} \quad t < L, \\ k_p h(1 - e^{-\alpha(t-L)} - \alpha(t - L)e^{-\alpha(t-L)}) + v(t) & \text{if} \quad t \geq L. \end{cases} \tag{10}$$

For brevity, the identification procedure is detailed for the case of $m = 1$, which can be simply extended to a case of $m \geq 2$.

To estimate the unknown model parameters, we let

$$\theta = [\alpha, L]^T. \tag{11}$$

The prediction error can be computed from Equation (9) as

$$\delta(t, \theta) = y(t) - \bar{y}(\infty)(1 - e^{-\alpha(t-L)}). \tag{12}$$

To minimize the cost function in Equation (7), we take the first-order derivative for Equation (12) with respect to α and L, respectively, obtaining

$$\frac{\partial \delta(t_i, \theta)}{\partial \alpha} = -\bar{y}(\infty)(t_i - L)e^{-\alpha(t_i-L)}, \tag{13}$$

$$\frac{\partial \delta(t_i, \theta)}{\partial L} = \bar{y}(\infty)\alpha e^{-\alpha(t_i-L)}. \tag{14}$$

The Jacobian matrix of Equation (12) is defined by

$$\Xi(t_i, \theta) = \frac{\partial \delta(t_i, \theta)}{\partial \theta} = \left[\frac{\partial \delta(t_i, \theta)}{\partial \alpha}, \frac{\partial \delta(t_i, \theta)}{\partial L}\right]^T. \tag{15}$$

Denote

$$\Psi(\theta) = [\Xi(t_1, \theta), \Xi(t_2, \theta), \ldots, \Xi(t_N, \theta)]^T. \tag{16}$$

The gradient matrix of $J(\theta)$ with respect to θ is therefore defined by

$$g[J(\theta)] = \Psi^T(\theta)\varepsilon. \tag{17}$$

The corresponding Hessian matrix is formulated by

$$H(\theta) = \Psi^T(\theta)\Psi(\theta). \tag{18}$$

Hence, the parameter vector θ can be estimated using a Gauss–Newton iteration approach, i.e.

$$\hat{\theta}_k = \hat{\theta}_{k-1} - H_k^{-1} g_k, \tag{19}$$

where H_k and g_k can be computed from Equations (17) and (18) in terms of $\hat{\theta}_{k-1}$ estimated from the previous iteration step. The initial estimation of $\hat{\theta}$ may be taken as a vector composed of small positive real numbers while the delay parameter may be taken roughly about the minimal output response delay as can be observed from a step test.

It is well known that the standard Gauss–Newton iteration method may give a very slow convergence rate in the initial stage of the iterative process for parameter estimation, causing a considerable computation effort. To overcome the deficiency, we propose the use of an adaptive step length for iteration based on the Armijo line searching strategy (see, e.g. Wright & Nocedal, 1999), i.e.

$$J(\hat{\theta}_{k-1} + \rho^m d_k) \leq J(\hat{\theta}_{k-1}) + \sigma \rho^m g_k^T d_k, \tag{20}$$

where $d_k = -H_k{}^{-1} g_k$ denotes the searching direction, $\rho \in (0, 1)$ and $\sigma \in (0, 1)$ are generally required for computation, and m is the minimal nonnegative integer satisfying (20).

To accomplish the identification objective in Equation (8) using less computation effort, it is suggested to take the searching step,

$$\rho = 1 - \frac{1}{J_k/J_{k-1} + J_{k-1}/J_k}, \tag{21}$$

$$\rho = \min\{\rho, 0.8\} \tag{22}$$

and $\sigma = \rho$ for the implementation of (20). Note that ρ is step-wise varying with respect to the prediction error. Initially, there is likely to be a larger difference between J_k/J_{k-1} and J_{k-1}/J_k, which may turn out to be a larger value of ρ from Equation (21) that results in a larger step for searching and thus expedites the convergence rate (i.e. reducing the number of external loop iteration relating to the searching step). The lower limit in Equation (22) is set to avoid an over large value of ρ that may cause severe increment of the iteration number of the internal loop iteration for determining m. As the iteration goes on,

J_k/J_{k-1} and J_{k-1}/J_k gradually approach the identity, resulting in $\rho \to 0.5$ that guarantees the searching step not very small to maintain the convergence rate. Owing to the fact that $J_i > 0\,(i = 1,\ 2,\ \ldots,\ N)$, there follows $\rho \in [0.5,\ 0.8]$ which satisfies the general requirement of the Armijo line searching method.

Another issue involved with the Gauss–Newton iteration method is that a suitable initial value is required for iteration, since a local minimum of solution might be turned out for an arbitrary choice of the initial value. To address the issue, numerical simulations are explored to study the relationship between $J(\theta)$ and the model parameters based on using different input excitations for identification. Consider a second-order process with time delay described by $G(s) = \mathrm{e}^{-Ls}/(s^2 + \alpha_1 s + 1)$, where $\alpha_1 = 2$ and $L = 5$. The input excitation is taken as a unity step change or a PRBS sequence with the variance of $\sigma^2 = 1$, respectively, for comparison. The number of measured output data is $N = 3000$ and the sampling period is $T_s = 0.01(\mathrm{s})$. Assuming that the searching ranges of the model parameters are $\alpha_1 \in [0.1, 4]$ and $L \in [0.1, 30]$, the corresponding $J(\theta)$ between the process response and model response are plotted in Figures 1 and 2 to the input excitations of the step change and PRBS sequence, respectively. It is seen from Figure 1 that there is a unique global minimum, owing to the fact that a step change mainly includes low-frequency components that provoke the output response basically in the low-frequency range. This indicates that using a step test for model identification is not sensitive to the initial choice of model parameters for iteration. In contrast, it is seen from Figure 2 that there are multiple local minimums, which implies that different choices of initial model parameters for iteration will turn out different identification results converging to any local minimums of $J(\theta)$.

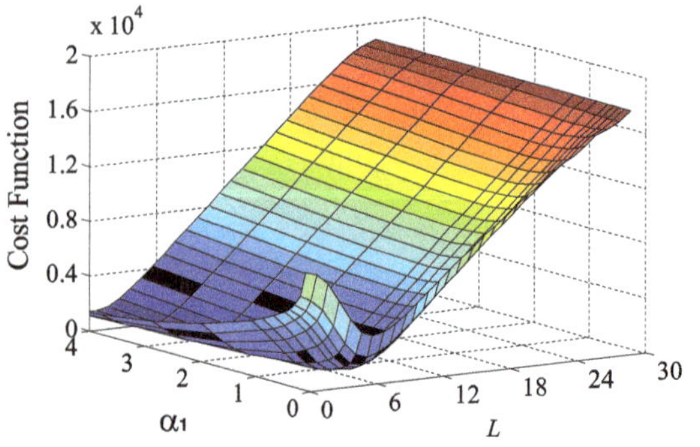

Figure 1. Plot of the identification cost function under a step test.

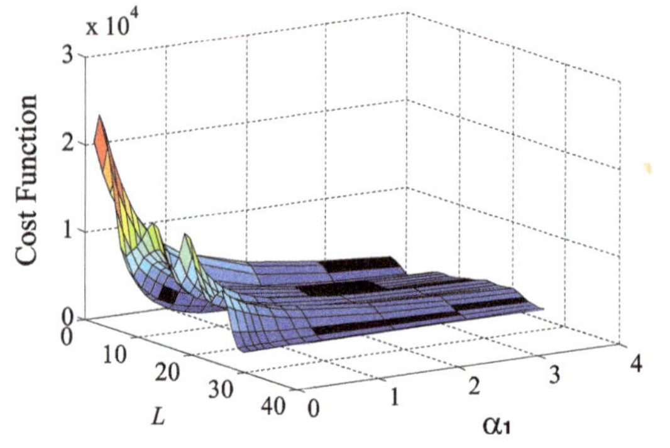

Figure 2. Plot of the identification cost function under a PRBS excitation.

Based on the above analysis, it is suggested to take the initial values of all the model parameters in Equation (1) or Equation (2) as a small positive real number for iteration. To improve the convergence rate, the delay parameter may be taken roughly about the minimum of the possible range as can be observed from a step test.

Hence, the above algorithm named Algorithm I for obtaining a time delay model with a single or repetitive poles can be summarized as

(i) Collect the input–output observations $\{u(t_i), y(t_i)\}$ $(i = 1,\ 2,\ \ldots,\ N)$ from a step test, and compute the process static gain in terms of Equation (5).
(ii) Take the initial estimation of $\hat{\theta}_0$ as a vector composed of small positive real numbers while the delay parameter may be taken roughly about the minimal output response delay as can be observed from a step test.
(iii) Compute the prediction error $\delta(t,\ \hat{\theta}_{k-1})$ and $J(\hat{\theta}_{k-1})$ using Equations (12) and (7).
(iv) Determine the searching direction $d_k = -H_k{}^{-1} g_k$ by computing the gradient matrix g_k in Equation (17) and the Hessian matrix H_k in Equation (18).
(v) Take the searching step in terms of Equation (22).
(vi) Update $\hat{\theta}_k = \hat{\theta}_{k-1} + \mu_k d_k$, where $\mu_k = \rho^{m_k}$, and m is the minimal nonnegative integer that satisfies (20) with the choice of Equation (22).
(vii) End the algorithm if the fitting condition, $1/N \sum_{i=1}^{N} [y(t_i) - \hat{y}(t_i)]^2 < \mathrm{err}$, is satisfied, where err is a user specified fitting threshold, or the maximum iteration number is attained. Otherwise, let $k = k + 1$ and go to step (iii).

3.2. *Second-order time delay model with distinct poles*

The time domain step response of the process model in Equation (2) is in the form of

$$y(t) = k_{\mathrm{p}} h\left[1 + \frac{\alpha_2}{\alpha_1 - \alpha_2}\mathrm{e}^{-\alpha_1(t-L)} + \frac{\alpha_1}{\alpha_2 - \alpha_1}\mathrm{e}^{-\alpha_2(t-L)}\right] + v(t), t > L, \tag{23}$$

where $\alpha_1 = 1/\tau_1$, $\alpha_2 = 1/\tau_2$, and $k_{\mathrm{p}} = \bar{y}(\infty)/h$.

Denote the parameter vector for estimation,

$$\theta = [\alpha_1,\ \alpha_2,\ L]^{\mathrm{T}}. \tag{24}$$

The prediction error can be computed based on the static gain estimated by Equation (5) as

$$\delta(t,\theta) = y(t) - \bar{y}(\infty)\left[1 + \frac{\alpha_2}{\alpha_1 - \alpha_2}\mathrm{e}^{-\alpha_1(t-L)} + \frac{\alpha_1}{\alpha_2 - \alpha_1}\mathrm{e}^{-\alpha_2(t-L)}\right]. \quad (25)$$

To minimize the cost function in Equation (7), we take the first-order derivative for Equation (25) with respect to α_1, α_2 and L, respectively, obtaining

$$\frac{\partial\delta(t_i,\theta)}{\partial\alpha_1} = \bar{y}(\infty)\frac{\alpha_2}{(\alpha_1-\alpha_2)^2}\left[\mathrm{e}^{-\alpha_1(t-L)} + (\alpha_1-\alpha_2)(t-L)\mathrm{e}^{-\alpha_1(t-L)} - \mathrm{e}^{-\alpha_2(t-L)}\right],$$

$$\frac{\partial\delta(t_i,\theta)}{\partial\alpha_2} = \bar{y}(\infty)\frac{\alpha_1}{(\alpha_1-\alpha_2)^2}\left[\mathrm{e}^{-\alpha_2(t-L)} + (\alpha_2-\alpha_1)(t-L)\mathrm{e}^{-\alpha_2(t-L)} - \mathrm{e}^{-\alpha_1(t-L)}\right],$$

$$\frac{\partial\delta(t_i,\theta)}{\partial L} = \bar{y}(\infty)\frac{\alpha_1\alpha_2}{\alpha_1-\alpha_2}\left[\mathrm{e}^{-\alpha_2(t-L)} - \mathrm{e}^{-\alpha_1(t-L)}\right]. \quad (26)$$

Correspondingly, the gradient matrix of $J(\theta)$ and the Hessian matrix can be computed in terms of Equations (17) and (18), respectively.

Hence, an identification algorithm named Algorithm II for obtaining an SOPDT model with distinct poles can be summarized as

(i) Collect the input–output observations $\{u(t_i), y(t_i)\}$ $(i = 1, 2, \ldots, N)$ from a step test, and compute the process static gain in terms of Equation (5).
(ii) Take the initial estimation of $\hat{\theta}_0$ as a vector composed of small positive real numbers while the delay parameter may be taken roughly about the minimal output response delay as can be observed from a step test.
(iii) Compute the prediction error $\delta(t, \hat{\theta}_{k-1})$ and cost function $J(\hat{\theta}_{k-1})$ using Equations (25) and (7).
(iv) Determine the searching direction $d_k = -H_k^{-1}g_k$ by computing the gradient matrix g_k in Equation (17) and the Hessian matrix H_k in Equation (18).
(v) Take the searching step in terms of Equation (22).
(vi) Update $\hat{\theta}_k = \hat{\theta}_{k-1} + \mu_k d_k$, where $\mu_k = \rho^{m_k}$, and m is the minimal nonnegative integer that satisfies (20) with the choice of Equation (22).
(vii) End the algorithm if the fitting condition, $1/N\sum_{i=1}^{N}[y(t_i) - \hat{y}(t_i)]^2 < \text{err}$, is satisfied, where err is a user specified fitting threshold, or the maximum iteration number is attained. Otherwise, let $k = k + 1$ and go to step (iii).

Remark 1 It can be seen from the above algorithms that the identification procedure is relatively independent of the time length of the step response and use no time integral for computation, significantly reducing the computation effort compared with the recently developed step identification methods based on using single or multiple time integrals to the step response data (Ahmed, Huang, & Shah, 2008; Liu et al., 2007, 2010).

4. Analysis on consistency estimation

For practical application in the presence of measurement noise, it should be clarified if consistent estimation could be guaranteed by the proposed method. To address this issue, the following proposition is given accordingly.

PROPOSITION 1 *Assuming that $\delta(t, \theta)$ is differentiable with respect to θ in a neighbourhood of the bounded level set $S = \{\theta | J(\theta) \le J(\theta_0)\}$, and the Jacobian matrix $\Psi(\theta)$ in Equation (16) is full row rank together with the positive-definite Hessian matrix $H(\theta)$ in (18), both Algorithms I and II guarantee uniform convergence satisfying*

$$\lim_{k\to\infty} g_k(\theta) = \lim_{k\to\infty} \Psi^{\mathrm{T}}(\theta_k)\varepsilon(\theta_k) = 0, \quad (27)$$

where $\varepsilon(\theta_k) = [\delta(t_1, \theta_k), \ldots, \delta(t_N, \theta_k)]^{\mathrm{T}}$.

Proof By assuming $\delta(t, \theta)$ is differentiable with respect to θ in the bounded set S, there follows for some positive constants β_1 and β_2 that

$$\|\delta(t,\theta)\| \le \beta_1, \quad \forall\theta \in S \text{ and } \forall t > L, \quad (28)$$

$$\|\Xi(t,\theta)\| \le \beta_2, \quad \forall\theta \in S \text{ and } \forall t > L, \quad (29)$$

where $\|\cdot\|$ denotes the matrix 2-norm.

It can be verified that there exists a constant $q > 0$ such that

$$\|\Psi(\theta)\| = \|\Psi^{\mathrm{T}}(\theta)\| \le q. \quad (30)$$

Owing to that $\Psi(\theta)$ is full column rank, there stands for a sufficient large constant, $p > 0$, such that

$$\|\Psi(\theta)z\| \ge p\|z\|. \quad (31)$$

Denote by ω_k the intersection angle between the negative gradient direction, $-g_k$, and the Gauss–Newton searching direction, d_k^{GN}. It can be derived using Equations (30) and (31) that

$$\cos(\omega_k) = -\frac{g_k^{\mathrm{T}}d_k^{\mathrm{GN}}}{\|g_k\|\|d_k^{\mathrm{GN}}\|} = -\frac{(d_k^{\mathrm{GN}})^{\mathrm{T}}\Psi^{\mathrm{T}}(\theta_k)\Psi(\theta_k)d_k^{\mathrm{GN}}}{\|\Psi^{\mathrm{T}}(\theta_k)\Psi(\theta_k)d_k^{\mathrm{GN}}\|\|d_k^{\mathrm{GN}}\|}$$
$$= \frac{\|\Psi(\theta_k)d_k^{\mathrm{GN}}\|^2}{\|\Psi^{\mathrm{T}}(\theta_k)\Psi(\theta_k)d_k^{\mathrm{GN}}\|\|d_k^{\mathrm{GN}}\|} \ge \frac{p^2\|d_k^{\mathrm{GN}}\|^2}{q^2\|d_k^{\mathrm{GN}}\|^2}$$
$$= \frac{p^2}{q^2} > 0. \quad (32)$$

From the Armijo condition in (20), we obtain

$$\begin{aligned} J(\hat{\theta}_k) &\le J(\hat{\theta}_{k-1}) + \sigma\rho^{m_k} g_k^{\mathrm{T}} d_k \\ &= J(\hat{\theta}_{k-1}) - c_k \|g_k\| \|d_k^{GN}\| \cos(\omega_k), \end{aligned} \tag{33}$$

where $c_k = \sigma\rho^{m_k}$.

Summing (33) over all indices of k yields

$$J(\hat{\theta}_k) \le J(\hat{\theta}_0) - \sum_{j=0}^{k-1} c_j \|g_j\| \|d_j^{\mathrm{GN}}\| \cos(\omega_j). \tag{34}$$

Considering that $J(\theta)$ is a bounded value, we ensure that $J(\hat{\theta}_0) - J(\hat{\theta}_k)$ is finite for all k, and therefore take the limit to both sides of (34), obtaining

$$\sum_{k=0}^{\infty} c_k \|g_k\| \|d_k^{\mathrm{GN}}\| \cos(\omega_k) < \infty, \tag{35}$$

which implies that

$$\lim_{k\to\infty} c_k \|g_k\| \|d_k^{\mathrm{GN}}\| \cos(\omega_k) = 0 \tag{36}$$

With $d_k^{\mathrm{GN}} = -H_k^{-1} g_k$, H_k is positive definite, $c_k = \sigma\rho^{m_k} > 0$ and $\cos(\omega_k) > 0$, it follows that

$$\lim_{k\to\infty} g_k(\theta) = \lim_{k\to\infty} \Psi^T(\theta_k)\varepsilon(\theta_k) = 0.$$

This completes the proof.

Note that the convergence result shown in Equation (27) indicates that the cost function $J(\theta)$ converges to a steady value when $k \to \infty$, that is to say, the estimated model parameters converge to the true values for perfect model match with the plant, leading to $\lim_{k\to\infty} \tilde{\theta}_k = 0$.

According to Proposition 1, the proposed algorithm guarantees convergence in the presence of measurement noise, owing to the fact that

$$\begin{aligned} \lim_{N\to\infty} g(\theta) &= \lim_{N\to\infty} \Psi^{\mathrm{T}}(\theta)\varepsilon \\ &= \lim_{N\to\infty} \sum_{i=1}^{N} \frac{\partial[y(t_i) - \hat{y}(t_i)]}{\partial\theta}[y(t_i) - \hat{y}(t_i)] \\ &= -\lim_{N\to\infty} \sum_{i=1}^{N} \frac{\partial\hat{y}(t_i)}{\partial\theta}[x(t_i) + v(t_i) - \hat{y}(t_i)] \\ &= -\lim_{N\to\infty} \sum_{i=1}^{N} \frac{\partial\hat{y}(t_i)}{\partial\theta}[x(t_i) - \hat{y}(t_i)] \\ &\quad - \lim_{N\to\infty} \sum_{i=1}^{N} v(t_i)\frac{\partial\hat{y}(t_i)}{\partial\theta} \\ &= -\lim_{N\to\infty} \sum_{i=1}^{N} \frac{\partial\hat{y}(t_i)}{\partial\theta}[x(t_i) - \hat{y}(t_i)] \end{aligned} \tag{37}$$

and

$$H(\theta) = \Psi^{\mathrm{T}}(\theta)\Psi(\theta) = \sum_{i=1}^{N} \frac{\partial\hat{y}(t_i)}{\partial\theta}\frac{\partial\hat{y}(t_i)}{\partial\theta^T}. \tag{38}$$

Therefore, the iterative process of the proposed algorithm will not be affected by the measurement noise, guaranteeing good robustness for convergence.

Note that it is often encountered in practical applications that the true model structure of the process to be identified cannot be known exactly. It is desirable to determine an optimal model structure for representing the process dynamic response characteristics from a step test. The following hypothesis testing condition, which was introduced in Liu and Gao (2010), can be taken for the choice of optimal model order,

$$\frac{\sum_{i=1}^{N} [\hat{y}(t_i) - y(t_i)]^2|_{m_1}}{\sum_{i=1}^{N} [\hat{y}(t_i) - y(t_i)]^2|_{m_2}} \le 0.1, \tag{39}$$

where $\hat{y}(t)$ and $y(t)$ denote, respectively, the measured process output and the model output to the step change, m_1 denotes the current model order and m_2 a higher order to be verified.

It can be seen from Equation (39) that a higher model order m_2 should be accepted only if the output prediction error is larger than one-tenth of that of the current model with m_1. To determine if an SOPDT model with repetitive poles or distinct poles should be chosen, we define the curvature (see, e.g. Rynne & Youngson, 2000) of such a model response to a step change by

$$c = \frac{|y''(t)|}{[1 + y'^2(t)]^{3/2}}, \tag{40}$$

where $y'(t)$ and $y''(t)$ denote the first-order and second-order derivatives, respectively. The maximal curvature for the normalized ranges of $\tau_p \in (0, 1]$, $\tau_1 \in (0, 0.5)$ and $\tau_2 \in (0.5, 1]$ in these two models is plotted in Figures 3 and 4, respectively. It is seen from Figure 3 that an evidently larger curvature occurs for an SOPDT model with repetitive poles compared with Figure 4 for an SOPDT model with distinct poles. Therefore, it is suggested to

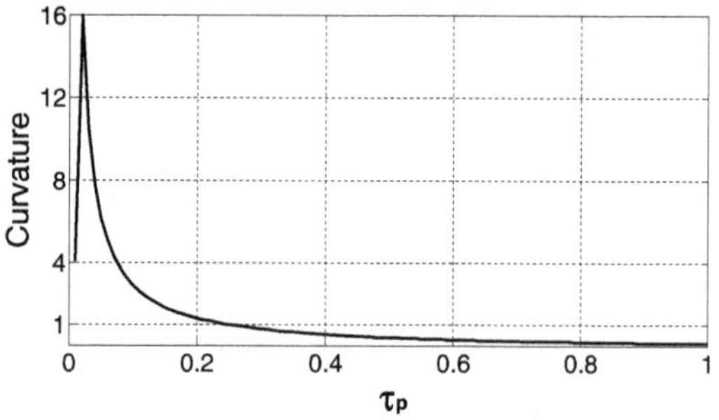

Figure 3. The maximal curvature of an SOPDT model with repetitive poles.

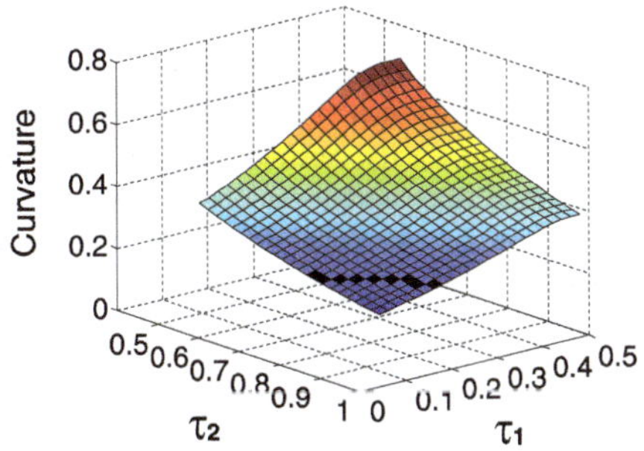

Figure 4. The maximal curvature of an SOPDT model with distinct poles.

take an SOPDT model with repetitive poles if the process response to a step change has an obviously larger curvature.

5. Illustration

Four examples from the recent literatures are studied here to illustrate the performance of the proposed method. Examples 1–3 are given to show good accuracy of the proposed algorithms for identifying low-order processes in terms of the exact model structures, together with measurement noise tests to demonstrate identification robustness. Example 4 is used to show the effectiveness of the proposed algorithm for identification of higher order processes. In the following examples, the measurement noise level is evaluated in terms of the noise-to-signal ratio,

$$\mathrm{NSR} = \frac{\mathrm{mean(abs(noise))}}{\mathrm{mean(abs(signal))}} \times 100\%. \tag{41}$$

The following time domain fitting criterion is used to assess identification accuracy,

$$\mathrm{err} = \frac{1}{N}\sum_{i=1}^{N}[y(t_i) - \hat{y}(t_i)]^2, \tag{42}$$

where $y(t_i)$ is the actual process output and $\hat{y}(t_i)$ is the response of the estimated model.

The parameter estimation error is assessed by the following criterion:

$$\mathrm{ERR} = \sqrt{\frac{\|\hat{\theta}_k - \theta\|_2}{\|\theta\|_2}} \times 100\%, \tag{43}$$

where $\hat{\theta}_k$ is an estimation of the true parameter vector θ.

Example 1 Consider the first-order inertial system with time delay studied by Bi et al. (1999),

$$G(s) = \frac{1}{s+1}\mathrm{e}^{-s}.$$

Based on a unity step response test, Bi et al. (1999) gave a FOPDT model, $G(s) = 1.00\mathrm{e}^{-1.00s}/(0.997s+1)$, by using a time domain integral of the output response for LS fitting. For illustration, the same step test is performed with the sampling interval $T_s = 0.01(\text{ s})$ and the sampling time $T = 50(\text{s})$. According to the proposed guideline for choosing the initial values for iteration, $\theta_0 = [0.01, 0.01]^{\mathrm{T}}$ is taken to apply the proposed Algorithm I, resulting in the exact model parameters based on any data length longer than the transient step response time.

Then assume that the process output measurement is corrupted by a Gaussian white noise with NSR = 10% and NSR = 20%, respectively. The estimation results are listed in Table 1 together with the iteration results. It can be seen from Table 1 that the proposed algorithm converges quickly regardless of poor initial estimation for iteration. Moreover, 100 Monte Carlo tests are conducted under a variety of random measurement noise level (NSR = 10%, 20%) for using the proposed algorithm. Table 2 lists the simulation results where the internal loop indicates the iteration number for determining m in (20), and the external loop indicates the iteration number of searching step for converging to the optimal estimation. It is seen from Table 2 that the proposed algorithm guarantees a fast convergent speed and maintain good robustness against measurement noise. The computation load of the proposed method is listed in Table 3 in comparison with that of the time integral method given by Bi et al. (1999), where N denotes the number of sampled data for identification and K is the number of iterative steps. Owing to the fact that N is much larger than K, the computation load is significantly reduced by the proposed method.

Table 1. Identification results under different noise levels for Example 1.

NSR	Iterative step k	$\hat{\tau}_p$	$\hat{L}$	ERR (%)
10%	1	77.3769	−90.85	8446.80
	5	30.8218	−49.83	4167.30
	10	10.5482	−13.45	1224.50
	20	1.2032	0.91	15.73
	30	0.9816	0.99	1.35
20%	1	77.8236	−91.11	8481.00
	5	32.1390	−52.4817	4376.00
	10	12.1918	−17.06	1502.49
	20	1.5376	0.72	42.97
	30	0.9638	0.99	2.67
	35	0.9633	0.99	2.70
True value		1	1	0

Table 2. Averaged iteration numbers under 100 Monte Carlo tests for Example 1.

NSR	External loop	Internal loop	Success ratio
10%	48	81	100
20%	49	84	99

Table 3. Comparison of the computation load for Example 1.

Methods	Addition	Multiplication
Proposed	$K(11N + 18)$	$K(17N + 19)$
Bi et al. (1999)	$N^2 + 8N + 8$	$N^2 + 13N + 48$

Example 2 Consider a second-order process with repetitive poles studied by Liu et al. (2007),

$$G(s) = \frac{e^{-5s}}{s^2 + 2s + 1}. \quad (44)$$

A square wave with an amplitude 1 and a period $T_p = 15(s)$ was adopted as the input excitation in lecture Liu et al. (2007), obtaining an estimated model,

$$G(s) = \frac{0.9958(\pm 0.083)e^{-5.09(\pm 0.066)s}}{0.9958(\pm 0.083)s^2 + 1.9935(\pm 0.118)s + 1}.$$

For illustration, a unity step response test is used in the proposed method where the sampling period is taken as $T_s = 0.01(s)$ and the overall sampling time is $T = 80$ (s).

To test identification robustness against measurement noise, assume that the noise level is NSR = 10% in the step test. By performing 100 Monte Carlo tests in terms of randomly varying the 'seed' of the noise generator, the proposed algorithm gives

$$G(s) = \frac{1.0000(\pm 0.0016)e^{-5.00(\pm 0.049)s}}{1.0058(\pm 0.068)s^2 + 2.0047(\pm 0.067)s + 1},$$

which indicates improved identification accuracy.

To demonstrate consistent estimation against measurement noise, the identification results for different noise levels (NSR = 5%, 10%, 15%, 20%, 25%, 30%) are shown in Figure 5, where the result for each model parameter to a given noise level is shown as a vertical linear segment along with the sample standard deviation in parentheses for 100 Monte Carlo tests. The square solid point in each linear segment denotes the mean of 100 identification results, and the upper and lower bars correspond to the maximum and the minimum of parameter estimation, respectively. It is seen that good identification accuracy and consistent estimation are obtained against different noise levels. Moreover, it can be easily verified that, given a measurement noise level, the sample standard deviation of the proposed estimation error is much smaller than that of the measurement noise.

Figure 6 shows the estimation results of the proposed method for different lengths (N) of data in a range from 165 to 8000 with NSR = 10%. For data collection with respect to different data lengths, the sampling interval is taken as $T_s = T/N$. It is seen that good identification accuracy can be obtained for a wide range of data lengths and the estimation results become better as the number of data points increases or the sampling interval decreases. Moreover, the computation load is listed in Table 4 in comparison with that of Liu et al (2007) based on using the time integral approach, indicating that the computation effort is significantly reduced by the proposed method.

Example 3 Consider a second-order process with distinct poles studied by Wang, Liu, Hang, and Tang (2006),

$$G(s) = \frac{e^{-4.5s}}{(5s + 1)(s + 1)}. \quad (45)$$

By performing a step test with a sampling period $T_s = 0.01(s)$ and the sampling time $T = 250(s)$, the proposed Algorithm II gives an exact estimation of the process model based on any data length longer than the transient step response time.

Then assume that the process output measurement is corrupted by a white noise causing NSR = 10%. The iteration process of the proposed Algorithm II is listed in Table 5 with an initial choice of $\theta_0 = [0.1,\ 0.3,\ 0.1]^T$, indicating good convergence rate and identification accuracy.

Example 4 Consider the fifth-order system with time delay studied by Wang et al. (2001)

$$G(s) = \frac{1.08}{(s + 1)^2(2s + 1)^3}e^{-10s}.$$

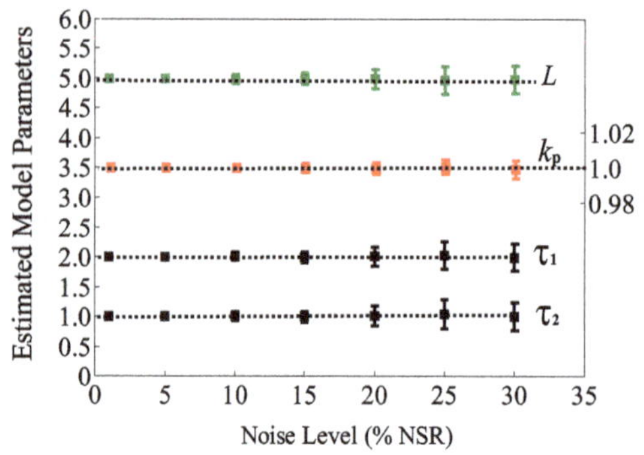

Figure 5. The results of 100 Monte Carlo Tests for Example 2 against measurement noise.

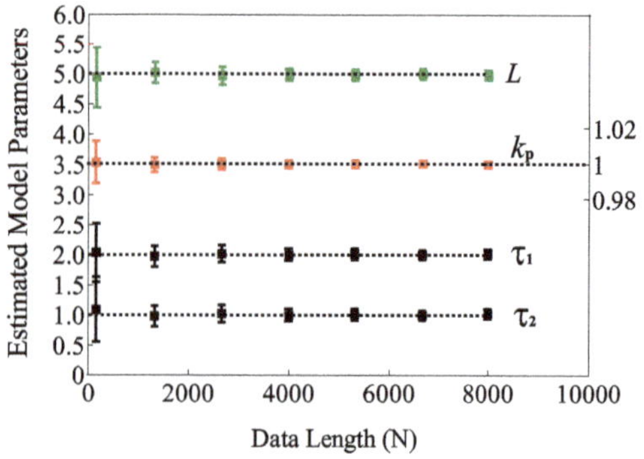

Figure 6. Parameter estimation with respect to data length for Example 2.

Table 4. Comparison of the computation load for Example 2.

Methods	Addition	Multiplication
Proposed	$K(11N+18)$	$K(21N+19)$
Liu et al. (2007)	$\frac{1}{6}N(N+1)(2N+1)+N^2+55N+639$	$\frac{1}{6}N(N+1)(2N+1)+\frac{1}{2}N(N+1)+N^2+66N+1735$

Table 5. Parameter estimation results for Example 3.

Iterative step k	$\hat{\tau}_1$	$\hat{\tau}_2$	$\hat{L}$	ERR (%)
1	10.0000	3.3333	0.10	103.78
10	7.0847	0.6124	10.83	98.21
20	3.6830	0.6836	8.36	60.10
30	5.0930	0.5387	5.68	18.68
40	4.9999	1.0274	4.57	1.20
True value	5	1	4.5	0

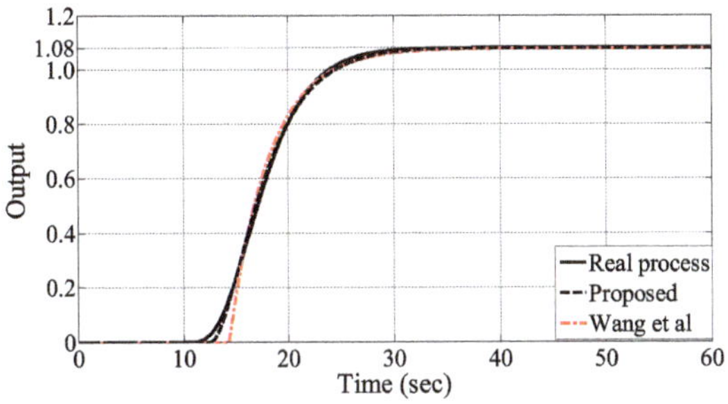

Figure 7. Step response fitting for Example 4.

Based on a step test, Wang et al. (2001) derived an SOPDT model, $G(s) = 1.0800e^{-12.71s}/7.8910s^2 + 5.3231s + 1$, corresponding to the mean-squared output fitting error, err $= 5.71 \times 10^{-4}$. For comparison, the same step test is performed for the sampling time $T = 150$(s) with the sampling interval $T_s = 0.01$(s). The proposed Algorithm II gives an SOPDT model, $G(s) = 1.0800e^{-12.59s}/7.7510s^2 + 5.5679s + 1$, corresponding to err $= 4.88 \times 10^{-4}$.

Then assume that the output measurement noise level is NSR = 10%, the proposed method gives $G(s) = 1.0791e^{-12.49s}/7.8539s^2 + 5.6051s + 1$, corresponding to err $= 3.17 \times 10^{-5}$, compared with that proposed by Wang et al. (2001) $G(s) = 1.0789e^{-14.20s}/0.1111s^2 + 3.8731s + 1$, corresponding to err $= 2.34 \times 10^{-4}$. The step responses of these models are shown in Figure 7, well demonstrating good fitting of the proposed method.

6. Conclusions

A step response identification method has been proposed for overdamped industrial processes with time delay, along with a proof on the convergence and consistency in the presence of measurement noise. By developing a gradient-based searching approach to minimizing the output prediction error, the linear model parameters together with the delay parameter can be simultaneously identified from the time domain expression of a low-order model response to a step change, specifically for the identification of an FOPDT or SOPDT model. A guideline of model structure selection has been given for identifying an SOPDT model with repetitive poles or distinct poles. The computation effort can be significantly reduced compared to recently developed step response identification methods (e.g. Ahmed et al., 2008; Liu et al., 2007, 2010) based on using the time integral to step response data and, moreover, the proposed algorithms are not sensitive to the data length for model identification, therefore facilitating practical applications. Four illustration examples from the literature have well demonstrated that the effectiveness and accuracy of the proposed identification algorithms. It should be noted that the proposed method cannot be applied to an underdamped process with complex poles that hinder the use of a gradient searching strategy for minimizing the output prediction error, which will be studied in our future work.

Funding

This work is supported in part by the National Thousand Talents Program of China, NSF China Grants 61473054, and the Fundamental Research Funds for the Central Universities of China.

References

Ahmed, S., Huang, B., & Shah, S. L. (2006). Parameter and delay estimation of continuous-time models using a linear filter. *Journal of Process Control*, *16*(4), 323–331.

Ahmed, S., Huang, B., & Shah, S. L. (2007). Novel identification method from step response. *Control Engineering Practice*, *15*(5), 545–556.

Ahmed, S., Huang, B., & Shah, S. L. (2008). Identification from step responses with transient initial conditions. *Journal of Process Control*, *18*(2), 121–130.

Baysse, A., Carrillo, F. J., & Habbadi, A. (2011). Time domain identification of continuous-time systems with time delay using output error method from sampled data. The 18th IFAC World Congress, Milano, Italy.

Bi, Q., Cai, W. J., Lee, E. L., Wang, Q. G., Hang, C. C., & Zhang, Y. (1999). Robust identification of first-order plus dead-time model from step response. *Control Engineering Practice*, *7*(1), 71–77.

Chen, F., Garnier, H., & Gilson, M. (2015). Robust identification of continuous-time models with arbitrary time-delay from irregularly sampled data. *Journal of Process Control*, *25*, 19–27.

Ding, F., Liu, X., & Chu, J. (2013). Gradient-based and least-squares-based iterative algorithms for Hammerstein systems using the hierarchical identification principle. *IET Control Theory & Applications*, *7*(2), 176–184.

Gawthrop, P. J., & Nihtilä, M. T. (1985). Identification of time delays using a polynomial identification method. *Systems & Control Letters*, *5*(4), 267–271.

Gu, Y., Ding, F., & Li, J. (2014). State filtering and parameter estimation for linear systems with d-step state-delay. *IET Signal Processing*, *8*(6), 639–646.

Huang, H. P., Lee, M. W., & Chen, C. L. (2001). A system of procedures for identification of simple models using transient step response. *Industrial & Engineering Chemistry Research*, *40*(8), 1903–1915.

Hwang, S. H., & Lai, S. T. (2004). Use of two-stage least-squares algorithms for identification of continuous systems with time delay based on pulse responses. *Automatica*, *40*(9), 1561–1568.

Ji, K., Xu, L., Xiong, W., & Chen, L. (2015). Newton iterative algorithm based modeling and proportional derivative controller design for second-order systems. *Journal of Applied Mathematics and Computing*, *49*(1), 557–572.

Liu, M., Wang, Q. G., Huang, B., & Hang, C. C. (2007). Improved identification of continuous-time delay processes from piecewise step tests. *Journal of Process Control*, *17*(1), 51–57.

Liu, T., & Gao, F. (2008). Robust step-like identification of low-order process model under nonzero initial conditions and disturbance. *IEEE Transactions on Automatic Control*, *53*(11), 2690–2695.

Liu, T., & Gao, F. (2010). A frequency domain step response identification method for continuous-time processes with time delay. *Journal of Process Control*, *20*(7), 800–809.

Liu, T., & Gao, F. (2012). *Industrial process identification and control design: Step-test and Relay-experiment-based Methods*. London: Springer.

Liu, T., Wang, Q. G., & Huang, H. P. (2013). A tutorial review on process identification from step or relay feedback test. *Journal of Process Control*, *23*(10), 1597–1623.

Liu, T., Zhou, F., Yang, Y., & Gao, F. (2010). Step response identification under inherent-type load disturbance with application to injection molding. *Industrial & Engineering Chemistry Research*, *49*(22), 11572–11581.

Ljung, L. (1999). *System identification: Theory for the user*. 2nd ed. Englewood Cliff, NJ: Prentice-Hall.

Na, J., Ren, X. M., & Xia, Y. (2014). Adaptive parameter identification of linear SISO systems with unknown time-delay. *Systems & Control Letters*, *66*, 43–50.

Orlov, Y., Belkoura, L., Richard, J. P., & Dambrine, M. (2003). Adaptive identification of linear time- delay systems. *International Journal of Robust and Nonlinear Control*, *13*(9), 857–872.

Previdi, F., & Lovera, M. (2004). Identification of non-linear parametrically varying models using separable least squares. *International Journal of Control*, *77*(16), 1382–1392.

Rake, H. (1980). Step response and frequency response methods. *Automatica*, *16*, 519–526.

Rangaiah, G. P., & Krishnaswamy, P. R. (1996). Estimating second-order dead time parameters from underdamped process transients. *Chemical Engineering Science*, *51*(7), 1149–1155.

Ren, X. M., Rad, A. B., Chen, P. T., & Lo, W. L. (2005). Online identification of continuous-time systems with unknown time delay. *IEEE Transactions on Automatic Control*, *50*(9), 1418–1422.

Richard, J. P. (2003). Time-delay systems: an overview of some recent advances and open problems. *Automatica*, *39*(10), 1667–1694.

Rynne, B. P., & Youngson, M. A. (2000). *Linear functional analysis*. London: Springer.

Seborg, D. E., Mellichamp, D. A., Edgar, T. F., & Doyle, III F. J. (2010). *Process dynamics and control*. 3rd ed. Hoboken, NJ: John Wiley & Sons.

Sung, S. W., & Lee, I. B. (2001). Prediction error identification method for continuous-time processes with time delay. *Industrial & Engineering Chemistry Research*, *40*(24), 5743–5751.

Wang, Q. G., Guo, X., & Zhang, Y. (2001). Direct identification of continuous time delay systems from step responses. *Journal of Process Control*, *11*(5), 531–542.

Wang, Q. G., Liu, M., & Hang, C. C. (2005). Simplified identification of time-delay systems with nonzero initial conditions from pulse tests. *Industrial & Engineering Chemistry Research*, *44*(19), 7591–7595.

Wang, Q. G., Liu, M., Hang, C. C., & Tang, W. (2006). Robust process identification from relay tests in the presence of nonzero initial conditions and disturbance. *Industrial & Engineering Chemistry Research*, *45*(12), 4063–4070.

Wang, Q. G., Liu, M., Hang, C. C., Zhang, Y., Zhang, Y., & Zheng, W. X. (2008). Integral identification of continuous-time delay systems in the presence of unknown initial conditions and disturbances from step tests. *Industrial & Engineering Chemistry Research*, *47*(14), 4929–4936.

Wang, Q. G., & Zhang, Y. (2001). Robust identification of continuous systems with dead-time from step responses. *Automatica*, *37*(3), 377–390.

Wright, S. J., & Nocedal, J. (1999). *Numerical optimization*. New York, NY: Springer.

Yang, X., Lu, Y., & Yan, Z. (2015). Robust global identification of linear parameter varying systems with generalised expectation–maximisation algorithm. *IET Control Theory & Applications*, *9*(7), 1103–1110.

Yang, Z. J., Iemura, H., Kanae, S., & Wada, K. (2007). Identification of continuous-time systems with multiple unknown time delays by global nonlinear least-squares and instrumental variable methods. *Automatica*, *43*(7), 1257–1264.

6

Dynamical observer-based fault detection and isolation for linear singular systems

G.-L. Osorio-Gordillo[a,b]*, M. Darouach[b], L. Boutat-Baddas[b] and C.-M. Astorga-Zaragoza[a]

[a] *Tecnológico Nacional de México, Centro Nacional de Investigación y Desarrollo Tecnológico, CENIDET, Interior Internado Palmira S/N, Col. Palmira, 62490 Cuernavaca, Mor. Mexico;* [b] *CRAN-CNRS (UMR 7039), Université de Lorraine, IUT de Longwy, 186, Rue de Lorraine, 54400 Cosnes et Romain, France*

This paper is concerned with the fault detection and isolation (FDI) for singular systems by using a dynamical observer having a new structure. The goal set in the task of FDI is to obtain a transfer function from fault to residual of the error dynamics system equal to a diagonal transfer function, to allow multiple faults isolation. Sufficient conditions for the existence and stability of the observer are given. A numerical example is given to illustrate our approach.

Keywords: fault detection and isolation; dynamical observer; singular systems

1. Introduction

This paper concerns the dynamical observer-based fault detection and isolation (FDI) for singular systems. Singular systems also known as descriptor or differential-algebraic systems can be considered as a generalization of dynamical systems. The singular system representation is a powerful modeling tool since they can describe processes governed by both, differential equations (dynamic) and algebraic equations (static). So that represents the physical phenomena that the model by ordinary differential equation cannot describe. These systems were introduced by Luenberger (1977) from a control theory point of view and since, great efforts have been made to investigate singular systems theory and its applications (Araujo, Barros, and Dorea, 2012, Boulkroune, Darouach, Zasadzinski, Gillé, and Fiorelli, 2009, Darouach, 2009, 2012, Liu, Zhang, Yang, and Yang, 2008, Müller, 2005, Müller and Hou, 1993, Zhou and Lu, 2009).

The general goal of fault detection is to determine the fault presence into a system, whereas fault isolation is used to determine the location of the fault, after detection. During the last two decades, FDI has been of considerable interest (Hamdi, Rodriguez, Mechmeche, Theilliol, and Braiek, 2012, Li and Jaimoukha, 2009, Li and Yang, 2011). In Theilliol, Noura, and Ponsart (2002), the fault diagnosis for linear systems is treated. Rodrigues, Theilliol, Adam-Medina, and Sauter (2008) address the FDI for multi-models representation and in Bokor and Szabo (2009), the fault detection for nonlinear systems is presented.

In Liu and Si (1997), the problem of isolating multiples faults in linear systems, is presented by using an approach based on eigenstructure assignment to generate structured residuals, then the observer matrices are determined so that the ith residual represents the ith fault.

In Li and Jaimoukha (2009), an approach which generalizes the results of Liu and Si (1997) is presented. The authors construct an H_∞ FDI observer for linear systems, with the constraint that the transfer function from faults to residual of the error dynamics is equal to a preassigned diagonal transfer matrix. All these results use the proportional observer (PO).

By using a singular system representation, it is shown that the remainder kinds of FDI problems, that is, fault detection, FDI, and disturbance decoupled FDI are generally equivalent to the FDI problem for singular systems (Patton, Frank, and Clark, 2000). To our knowledge, few works have considered the approach of FDI based on an observer for singular systems, due to the structural complexity and strong constraints in designing procedure. Fault detection using unknown input observers were presented in Kim, Yeu, and Kawaji (2001), and in Duan, Howe, and Patton (2002) where the residuals are not affected by the unknown input. In Hamdi et al. (2012), Yeu, Kim, and Kawaji (2005), and Astorga, Theilliol, Ponsart, and Rodrigues (2012), fault diagnosis problem is treated by a PO.

In the estimation by a PO, there always exists a static error estimation. In order to deal with the inconveniences

*Corresponding author. Email: gloriaosorio@cenidet.edu.mx

of PO, proportional-integral observers (PIO) were introduced with an integral gain of the output error in their structure. This change in the structure achieve steady-state accuracy in their estimations. Also a new structure of the observers was developed by Goodwin and Middleton (1989) and Marquez (2003), known as dynamic observers. This structure presents an alternative state estimation which can be considered as more general than PO and PIO. These last can be only considered as particular cases of this structure. The idea of including additional dynamics in the observer was presented by Goodwin and Middleton (1989).

In this paper, we consider FDI problem for singular systems with actuator faults. Residual signals are determined from properly weighted output errors between measurements and estimated outputs.

The main contribution of this paper is that the designed observer is presented in a more general form than the existing dynamical observers: the PO and PIO which are only particular cases of the structure of our observer. This observer is used for actuator FDI in singular systems, which are a generalization of the standard systems. The proposed method is based on the directional residual generation in order to locate simultaneous faults. Finally, the effectiveness of this approach is shown through a numerical example simulation.

2. Preliminaries

In the present paper, the set of real matrices $n \times m$ is denoted by $\mathbb{R}^{n\times m}$. A^{T} denotes the transpose of the matrix A, A^{+} is the generalized inverse of A, i.e. $AA^{+}A = A$. I_n denotes an $n \times n$ identity matrix, I denotes an identity matrix with appropriate dimension, 0 denotes a zero scalar or matrix with appropriate dimension. The notation $A = \mathsf{diag}(a_1, \ldots, a_n)$ denotes that A is a diagonal matrix with elements $(a_1, \ldots, a_n)$ in its diagonal, $\mathsf{ones}_{n,m}$ denotes an $n \times m$ matrix with all elements one.

In Section 4.3, we use the following lemma to solve Linear matrix inequalities (LMIs).

LEMMA 1 (Skelton, Iwasaki, and Grigoriadis, 1998). *Let matrices $\mathcal{B}, \mathcal{C}, \mathcal{D} = \mathcal{D}^{\mathrm{T}}$ be given, then the following statements are equivalent:*

(1) There exists a matrix $\mathcal{X}$ satisfying

$$\mathcal{B}\mathcal{X}\mathcal{C} + (\mathcal{B}\mathcal{X}\mathcal{C})^{\mathrm{T}} + \mathcal{D} < 0.$$

(2) The following two conditions hold:

$$\mathcal{B}^{\perp}\mathcal{D}\mathcal{B}^{\perp\mathrm{T}} < 0 \quad \text{or} \quad \mathcal{B}\mathcal{B}^{\mathrm{T}} > 0,$$
$$\mathcal{C}^{\mathrm{T}\perp}\mathcal{D}\mathcal{C}^{\mathrm{T}\perp\mathrm{T}} < 0 \quad \text{or} \quad \mathcal{C}^{\mathrm{T}}\mathcal{C} > 0.$$

Suppose that the statement 2 holds. Let r_b and r_c be the ranks of $\mathcal{B}$ and $\mathcal{C}$, respectively, and $(\mathcal{B}_l, \mathcal{B}_r)$ and $(\mathcal{C}_l, \mathcal{C}_r)$ be any full rank factors of $\mathcal{B}$ and $\mathcal{C}$, i.e. $\mathcal{B} = \mathcal{B}_l\mathcal{B}_r$, $\mathcal{C} = \mathcal{C}_l\mathcal{C}_r$. Then, the matrix $\mathcal{X}$ in statement 1 is given by

$$\mathcal{X} = \mathcal{B}_r^{+}\mathcal{K}\mathcal{C}_l^{+} + \mathcal{Z} - \mathcal{B}_r^{+}\mathcal{B}_r\mathcal{Z}\mathcal{C}_l\mathcal{C}_l^{+},$$

where $\mathcal{Z}$ is an arbitrary matrix and

$$\mathcal{K} = -\mathcal{R}^{-1}\mathcal{B}_l^{\mathrm{T}}\vartheta\mathcal{C}_r^{\mathrm{T}}(\mathcal{C}_r\vartheta\mathcal{C}_r^{\mathrm{T}})^{-1} + \mathcal{S}^{1/2}\mathcal{L}(\mathcal{C}_r\vartheta\mathcal{C}_r^{\mathrm{T}})^{-1/2},$$
$$\mathcal{S} = \mathcal{R}^{-1} - \mathcal{R}^{-1}\mathcal{B}_l^{\mathrm{T}}[\vartheta - \vartheta\mathcal{C}_r^{\mathrm{T}}(\mathcal{C}_r\vartheta\mathcal{C}_r^{\mathrm{T}})^{-1}\mathcal{C}_r\vartheta]\mathcal{B}_l\mathcal{R}^{-1},$$

where $\mathcal{L}$ is an arbitrary matrix such that $\|\mathcal{L}\| < 1$ and $\mathcal{R}$ is an arbitrary positive-definite matrix such that

$$\vartheta = (\mathcal{B}_r\mathcal{R}^{-1}\mathcal{B}_l^{\mathrm{T}} - \mathcal{D})^{-1} > 0.$$

3. Problem formulation

Consider the following singular system subject to actuator fault:

$$E\dot{x}(t) = Ax(t) + Bu(t) + Gf(t),$$
$$y(t) = Cx(t), \tag{1}$$

where $x(t) \in \mathbb{R}^n$ is the semi state vector, $u(t) \in \mathbb{R}^m$ is the input, $f(t) \in \mathbb{R}^{n_f}$ is the fault vector and $y(t) \in \mathbb{R}^{n_y}$ represents the measured output vector. Matrices $E \in \mathbb{R}^{n\times n}$, $A \in \mathbb{R}^{n\times n}$, $B \in \mathbb{R}^{n\times m}$, $G \in \mathbb{R}^{n\times n_f}$ and $C \in \mathbb{R}^{n_y\times n}$. Let $\mathsf{rank}(E) = \varrho \leq n$ and $E^{\perp} \in \mathbb{R}^{\varrho_1\times n}$ be a full row rank matrix such that $E^{\perp}[E \;\; G] = 0$, in this case $\varrho_1 = n - \varrho$.

In the sequel, we assume that

Assumption 1

$$\mathsf{rank}\begin{bmatrix} E \\ E^{\perp}A \\ C \end{bmatrix} = n$$

Remark 1 Assumption 1 is equivalent to the impulse observability for singular systems. This condition is more general than $\mathsf{rank}\left[\begin{smallmatrix} E \\ C \end{smallmatrix}\right] = n$, generally considered, see, for example, Verhaegen and Dooren (1986), Darouach and Boutayeb (1995), and Hou and Müller (1995).

Now, let us consider the following fault isolation dynamical observer for the system (1):

$$\dot{\zeta}(t) = N\zeta(t) + Hv(t) + F\begin{bmatrix} -E^{\perp}Bu(t) \\ y(t) \end{bmatrix} + Ju(t), \tag{2}$$

$$\dot{v}(t) = S\zeta(t) + Lv(t) + M\begin{bmatrix} -E^{\perp}Bu(t) \\ y(t) \end{bmatrix}, \tag{3}$$

$$\hat{x}(t) = P\zeta(t) + Q\begin{bmatrix} -E^{\perp}Bu(t) \\ y(t) \end{bmatrix}, \tag{4}$$

$$r(t) = W(C\hat{x}(t) - y(t)), \tag{5}$$

where $\zeta \in \mathbb{R}^{q_0}$ represents the state vector of the observer, $v(t) \in \mathbb{R}^{q_1}$ is an auxiliary vector, $\hat{x}(t) \in \mathbb{R}^n$ is the estimate of $x(t)$ and $r(t) \in \mathbb{R}^{n_f}$ is the residual vector.

Now, the following lemma is considered.

LEMMA 2 *There exist a fault isolation observer of the form Equations* (2)–(5) *for the system* (1) *if the following two statements hold.*

I. There exist a matrix T of appropriate dimension such that the following conditions are satisfied:
(a) $NTE + F\begin{bmatrix} E^{\perp}A \\ C \end{bmatrix} - TA = 0$,
(b) $J = TB$,
(c) $STE + M\begin{bmatrix} E^{\perp}A \\ C \end{bmatrix} = 0$, and
(d) $[P \;\; Q]\begin{bmatrix} TE \\ E^{\perp}A \\ C \end{bmatrix} = I_n$.

II. The matrix $\begin{bmatrix} N & H \\ S & L \end{bmatrix}$ is a stability matrix, when $f(t) = 0$.

Proof Let $T \in \mathbb{R}^{q_0 \times n}$ be a parameter matrix and define $\varepsilon(t) = \zeta(t) - TEx(t)$, then its dynamic is given by

$$\dot{\varepsilon}(t) = N\varepsilon(t) + Hv(t) + (J - TB)u(t) + \left(NTE + F\begin{bmatrix} E^{\perp}A \\ C \end{bmatrix} - TA\right)x(t) - TGf(t) \quad (6)$$

by using the definition of $\varepsilon(t)$, Equations (3) and (4), can be written as

$$\dot{v}(t) = S\varepsilon(t) + Lv(t) + \left(STE + M\begin{bmatrix} E^{\perp}A \\ C \end{bmatrix}\right)x(t), \quad (7)$$

$$\hat{x}(t) = P\varepsilon(t) + [P \quad Q]\begin{bmatrix} TE \\ E^{\perp}A \\ C \end{bmatrix}x(t). \quad (8)$$

Now, if the conditions (a)–(d) of Lemma 2 are satisfied, the following observer error dynamics is obtained from Equations (6) and (7)

$$\begin{bmatrix} \dot{\varepsilon}(t) \\ \dot{v}(t) \end{bmatrix} = \begin{bmatrix} N & H \\ S & L \end{bmatrix}\begin{bmatrix} \varepsilon(t) \\ v(t) \end{bmatrix} - \begin{bmatrix} TG \\ 0 \end{bmatrix}f(t) \quad (9)$$

and from Equation (4)

$$\hat{x}(t) - x(t) = P\varepsilon(t), \quad (10)$$

$$e(t) = P\varepsilon(t) \quad (11)$$

so, if $f(t) = 0$ and matrix $\begin{bmatrix} N & H \\ S & L \end{bmatrix}$ is a stability matrix, then $\lim_{t\to\infty} e(t) = 0$.

The residual equation is obtained from Equation (5)

$$r(t) = WCP\varepsilon(t). \quad (12)$$

Remark 2

- The observer (2)–(4) is in general form and generalizes the existing ones. In fact:
  - For $H = 0$, $S = 0$ and $M = 0$ with L a stability matrix, the observer reduces to the PO for singular systems (see, for example, Darouach, 2012 and references therein).
  - For $S = 0$, $H = 0$, $P = I$, $M = 0$ and L a stability matrix, let matrices F and Q be partitioned according to the partition of $\begin{bmatrix} -E^{\perp}Bu(t) \\ y(t) \end{bmatrix}$ as $F = [0 \;\; F_a]$ and $Q = [0 \;\; Q_a]$, respectively, then we obtain the following observer:

$$\dot{\zeta}(t) = N\zeta(t) + F_a y(t) + Ju(t)$$
$$\hat{x}(t) = \zeta(t) + Q_a y(t)$$

    Which is the form used for the unknown input PO for singular systems (Darouach, Zasadzinski, and Hayar, 1996).
  - For $P = I$, $L = 0$ and let $S = -C$ and $M = -CQ + [0 \;\; I]$, then we obtain the following observer:

$$\dot{\zeta}(t) = N\zeta(t) + Hv(t) + F\begin{bmatrix} -E^{\perp}Bu(t) \\ y(t) \end{bmatrix} + Ju(t),$$
$$\dot{v}(t) = y(t) - C\hat{x}(t),$$
$$\hat{x}(t) = \zeta(t) + Q\begin{bmatrix} -E^{\perp}Bu(t) \\ y(t) \end{bmatrix}.$$

  Which is the form used for the unknown input PIO for singular systems.
- The order of the observer is $q_0 \leq n$, when $q_0 = n - p$, we obtain the reduced order observer and for $q_0 = n$, we obtain the full order one.

Now, the problem of the fault isolation observer is reduced to determine the matrices $N, H, F, S, L, M, P, Q, W$, and T such that conditions (a)–(d) from Lemma 2 are satisfied.

4. Observed-based FDI design

4.1. *Parameterization of the observers matrices*

Before giving the solution to the dynamical observer design, the parameterization of the solutions to the algebraic constraints (a)–(d) of Lemma 2 is presented. Let $R \in \mathbb{R}^{q_0 \times n}$ be a full row rank matrix such that the matrix $\Sigma = \begin{bmatrix} R \\ E^{\perp}A \\ C \end{bmatrix}$ is of full column rank, and let $\Omega = \begin{bmatrix} E \\ E^{\perp}A \\ C \end{bmatrix}$, and define the following matrices: $T_1 = R\Omega^{+}\begin{bmatrix} I_n \\ 0 \end{bmatrix}$, $T_2 = (I_{n+Q_1+n_y} - \Omega\Omega^{+})\begin{bmatrix} I_n \\ 0 \end{bmatrix}$, $N_1 = T_1 A\Sigma^{+}\begin{bmatrix} I_{q0} \\ 0 \end{bmatrix}$, $N_2 = T_2 A\Sigma^{+}\begin{bmatrix} I_{q0} \\ 0 \end{bmatrix}$, $N_3 = (I_{q_0+Q_1+n_y} - \Sigma\Sigma^{+})\begin{bmatrix} I_{q0} \\ 0 \end{bmatrix}$, and $P_1 = \Sigma^{+}\begin{bmatrix} I_{q0} \\ 0 \end{bmatrix}$.

The following lemma gives the general form of matrices T, S, M, P, Q, N, and F.

LEMMA 3 *The general form of T, S, M, P, Q, N, and F are*

$$T = T_1 - Z_1 T_2, \tag{13}$$
$$S = -Y_1 N_3, \tag{14}$$
$$M = -Y_1 F_3, \tag{15}$$
$$P = P_1 - Y_2 N_3, \tag{16}$$
$$Q = Q_1 - Y_2 F_3, \tag{17}$$
$$N = N_1 - Z_1 N_2 - Y_3 N_3, \tag{18}$$
$$F = F_1 - Z_1 F_2 - Y_3 F_3, \tag{19}$$

where Z_1, Y_1, Y_2, and Y_3 are arbitrary matrices of appropriate dimensions.

Proof Conditions (c) and (d) of Lemma 2 can be rewritten as

$$\begin{bmatrix} S & M \\ P & Q \end{bmatrix} \begin{bmatrix} TE \\ E^{\perp}A \\ C \end{bmatrix} = \begin{bmatrix} 0 \\ I_n \end{bmatrix}. \tag{20}$$

The necessary and sufficient conditions for (20) to have a solution is

$$\text{rank} \begin{bmatrix} TE \\ E^{\perp}A \\ C \end{bmatrix} = \text{rank} \begin{bmatrix} TE \\ E^{\perp}A \\ C \\ 0 \\ I_n \end{bmatrix} = n. \tag{21}$$

Now, since $\text{rank}\begin{bmatrix} TE \\ E^{\perp}A \\ C \end{bmatrix} = \text{rank}(\Omega) = n$, there exist matrices $T \in \mathbb{R}^{q_0 \times n}$, and $K \in \mathbb{R}^{q_0 \times (\varrho_1 + p)}$ such that

$$TE + K \begin{bmatrix} E^{\perp}A \\ C \end{bmatrix} = R, \tag{22}$$

which can also be written as

$$[T \quad K]\Omega = R \tag{23}$$

and since $\text{rank}\begin{bmatrix} \Omega \\ R \end{bmatrix} = \text{rank}(\Omega)$, or equivalently

$$R\Omega^{+}\Omega = R \tag{24}$$

the general solution to Equation (23) is given by

$$[T \quad K] = R\Omega^{+} - Z_1(I_{n_1+\varrho_1+p} - \Omega\Omega^{+}) \tag{25}$$

or equivalently

$$T = T_1 - Z_1 T_2, \tag{26}$$
$$K = K_1 - Z_1 K_2, \tag{27}$$

where $K_1 = R\Omega^{+}\begin{bmatrix} 0 \\ I_{\varrho_1+n_y} \end{bmatrix}$, $K_2 = (I_{n+\varrho_1+n_y} - \Omega\Omega^{+})\begin{bmatrix} 0 \\ I_{\varrho_1+n_y} \end{bmatrix}$, and Z_1 is an arbitrary matrix of appropriate dimension.

Now, define the following matrices: $F_1 = T_1 A\Sigma^{+}\begin{bmatrix} K \\ I_{\varrho_1+n_y} \end{bmatrix}$, $F_2 = T_2 A\Sigma^{+}\begin{bmatrix} K \\ I_{\varrho_1+n_y} \end{bmatrix}$, $F_3 = (I_{q_0+\varrho_1+n_y} - \Sigma\Sigma^{+})\begin{bmatrix} K \\ I_{\varrho_1+n_y} \end{bmatrix}$, $Q_1 = \Sigma^{+}\begin{bmatrix} K \\ I_{\varrho_1+n_y} \end{bmatrix}$, $\tilde{K}_1 = T_1 A\Sigma^{+}\begin{bmatrix} 0 \\ I_{\varrho_1+n_y} \end{bmatrix}$, $\tilde{K}_2 = T_2 A\Sigma^{+}\begin{bmatrix} 0 \\ I_{\varrho_1+n_y} \end{bmatrix}$, and $\tilde{K}_3 = (I_{q_0+\varrho_1+n_y} - \Sigma\Sigma^{+})\begin{bmatrix} 0 \\ I_{n_y} \end{bmatrix}$.

From Equation (22), we obtain

$$\begin{bmatrix} TE \\ E^{\perp}A \\ C \end{bmatrix} = \begin{bmatrix} I_{q_0} & -K \\ 0 & I_p \end{bmatrix} \Sigma \tag{28}$$

by replacing Equation (28) into Equation (20), it leads to

$$\begin{bmatrix} S & M \\ P & Q \end{bmatrix} \begin{bmatrix} I_{q_0} & -K \\ 0 & I_p \end{bmatrix} \Sigma = \begin{bmatrix} 0 \\ I_n \end{bmatrix}. \tag{29}$$

Since Σ is a full column rank matrix, and $\begin{bmatrix} I_{q_0} & -K \\ 0 & I_p \end{bmatrix}^{-1} = \begin{bmatrix} I_{q_0} & K \\ 0 & I_p \end{bmatrix}$, the general solution of Equation (29) is given by

$$\begin{bmatrix} S & M \\ P & Q \end{bmatrix} = \left(\begin{bmatrix} 0 \\ I_n \end{bmatrix} \Sigma^{+} - \begin{bmatrix} Y_1 \\ Y_2 \end{bmatrix} (I_{q_0+\varrho_1+p} - \Sigma\Sigma^{+}) \right) \times \begin{bmatrix} I_{q_0} & K \\ 0 & I_p \end{bmatrix}, \tag{30}$$

where Y_1 and Y_2 are arbitrary matrices of appropriate dimensions. Then matrices S, M, P, and Q can be determined as

$$S = -Y_1 N_3, \tag{31}$$
$$M = -Y_1 F_3, \tag{32}$$
$$P = P_1 - Y_2 N_3, \tag{33}$$
$$Q = Q_1 - Y_2 F_3. \tag{34}$$

By inserting TE from Equation (22) into condition (a) of Lemma 2 leads to

$$N\left(R - K\begin{bmatrix} E^{\perp}A \\ C \end{bmatrix}\right) + F\begin{bmatrix} E^{\perp}A \\ C \end{bmatrix} = TA,$$
$$NR + \tilde{K}\begin{bmatrix} E^{\perp}A \\ C \end{bmatrix} = TA. \tag{35}$$

where $\tilde{K} = F - NK$. Equation (35) can be written as

$$[N \quad \tilde{K}]\Sigma = TA. \tag{36}$$

The general solution of Equation (36) is given by

$$[N \quad \tilde{K}] = TA\Sigma^{+} - Y_3(I_{q_0+\varrho_1+p} - \Sigma\Sigma^{+}). \tag{37}$$

By replacing T from Equation (26) into Equation (37) it gives

$$N = N_1 - Z_1 N_2 - Y_3 N_3, \tag{38}$$
$$\tilde{K} = \tilde{K}_1 - Z_1 \tilde{K}_2 - Y_3 \tilde{K}_3, \tag{39}$$

where Y_3 is an arbitrary matrix of appropriate dimension.

As N, T, K, and $\tilde{K}$ are known, we can deduce the form of F as follows:

$$F = F_1 - Z_1F_2 - Y_3F_3 \tag{40}$$

Remark 3 From the above results, we can see that the determination of the matrices of the observer (2)–(4) can be done as follows: Matrices N, T, K, and $\tilde{K}$ are known matrices, matrix $J = TB$, and matrices S, M, P, and Q can be deduced from Equations (31)–(34). On the other hand, parameter matrices H, L, Z_1, Y_1, and Y_3 can be obtained from the stability of Equation (9).

Now, let $\bar{T}_2 = T_2G$ and $Z_1 = Z(I_{n+n_y} - \bar{T}_2\bar{T}_2^+)$, where Z is an arbitrary matrix of appropriate dimension. Then, matrices $T, K, N, \tilde{K}$, and F can be expressed as

$$T = T_1 - Z\mathcal{T}_2, \tag{41}$$

$$K = K_1 - Z\mathcal{K}_2, \tag{42}$$

$$N = N_1 - Z\mathcal{N}_2 - Y_3N_3, \tag{43}$$

$$\tilde{K} = \tilde{K}_1 - Z\tilde{\mathcal{K}}_2 - Y_3\tilde{K}_3, \tag{44}$$

$$F = F_1 - Z\mathcal{F}_2 - Y_3F_3, \tag{45}$$

where $\mathcal{T}_2 = (I_{n+n_y} - \bar{T}_2\bar{T}_2^+)T_2$, $\mathcal{K}_2 = (I_{n+n_y} - \bar{T}_2\bar{T}_2^+)K_2$, $\mathcal{N}_2 = (I_{n+n_y} - \bar{T}_2\bar{T}_2^+)N_2$, $\tilde{\mathcal{K}}_2 = (I_{n+n_y} - \bar{T}_2\bar{T}_2^+)\tilde{K}_2$, and $\mathcal{F}_2 = (I_{n+n_y} - \bar{T}_2\bar{T}_2^+)F_2$.

The observer error dynamics (9) can be rewriting as

$$\underbrace{\begin{bmatrix} \dot{\varepsilon}(t) \\ \dot{v}(t) \end{bmatrix}}_{\dot{\varphi}(t)} = \Bigg(\underbrace{\begin{bmatrix} N_1 - Z\mathcal{N}_2 & 0 \\ 0 & 0 \end{bmatrix}}_{\mathbb{A}_1} - \underbrace{\begin{bmatrix} Y_3 & H \\ Y_1 & L \end{bmatrix}}_{\mathbb{Y}} \underbrace{\begin{bmatrix} N_3 & 0 \\ 0 & -I_{q_1} \end{bmatrix}\Bigg)}_{\mathbb{A}_2} \times \begin{bmatrix} \varepsilon(t) \\ v(t) \end{bmatrix} + \underbrace{\begin{bmatrix} -T_1G \\ 0 \end{bmatrix}}_{\mathbb{B}_1} f(t) \tag{46}$$

$$r(t) = W\underbrace{[CP_1 \quad 0]}_{\mathbb{C}_1}\begin{bmatrix} \varepsilon(t) \\ v(t) \end{bmatrix} \tag{47}$$

without lost of generality, $Y_2 = 0$ was taken for simplicity.

4.2. *FDI design*

The objective of the fault detection is to build a residual, which shows the presence of a fault into a system. The mathematical definition of a residual is

$$\lim_{t\to\infty} r(t) = 0 \quad \text{for } f(t) = 0,$$

$$r(t) \neq 0 \quad \text{for } f(t) \neq 0.$$

The fault isolation objective in this paper is to obtain a transfer function from faults to residual equal to a diagonal, in order to deal with faults that may occur simultaneously.

From Equations (46) and (47), the residual dynamics are given by

$$\dot{\varphi}(t) = (\mathbb{A}_1 - \mathbb{Y}\mathbb{A}_2)\varphi(t) + \mathbb{B}_1 f(t), \tag{48}$$

$$r(t) = W\mathbb{C}_1\varphi(t). \tag{49}$$

Let $G_{fr}(s)$ be the transfer function from the fault $f(t)$ to the residual $r(t)$, such that

$$G_{fr}(s) = \left[\begin{array}{c|c} \mathbb{A}_1 - \mathbb{Y}\mathbb{A}_2 & \mathbb{B}_1 \\ \hline W\mathbb{C}_1 & 0 \end{array}\right]. \tag{50}$$

The objective is to render $G_{fr}(s)$ diagonal, i.e.

$$G_{fr}(s) = \mathrm{diag}\{g_{1,1}(s), \ldots, g_{nf,nf}(s)\}, \tag{51}$$

while the stability of the observer is guaranteed. Since $G_{fr}(s)$ has a diagonal structure each residual is affected just by one fault. Considering this, it is possible to isolate simultaneous faults.

PROPOSITION 1 *The transfer function* (50) *can be diagonalized if and only if* $(\mathbb{C}_1\mathbb{B}_1)$ *has full column rank is that* $n_y \geq n_f$.

Proposition 1 is also called output separability condition (White and Speyer, 1987). To isolate n_f faults in Equation (1) the rank of $(\mathbb{C}_1\mathbb{B}_1)$ must be n_f, which in turn requires n_y measured outputs.

The following theorem shows how to design an observer of the form Equations (2)–(4) to perform FDI.

THEOREM 1 *Consider that* $n_y \geq n_f$ *and let*

$$\Lambda = \mathrm{diag}(\lambda_1, \ldots, \lambda_{n_f}) \in \mathbb{R}^{n_f \times n_f}, \lambda_i < 0, \tag{52}$$

$$\Gamma = \mathrm{diag}(\gamma_1, \ldots, \gamma_{n_f}) \in \mathbb{R}^{n_f \times n_f}, |\gamma_i| > 0, \quad \forall i, \{i = 1, \ldots, n_f\} \tag{53}$$

be given. Then, there exist matrices $\mathbb{Y}$ *and* W *such that*

$$(\mathbb{A}_1 - \mathbb{Y}\mathbb{A}_2)\mathbb{B}_1 = \mathbb{B}_1\Lambda, \tag{54}$$

$$W\mathbb{C}_1\mathbb{B}_1 = \Gamma. \tag{55}$$

If $(\mathbb{A}_2\mathbb{B}_1)$ *has full column rank, then matrices* $\mathbb{Y}$ *and* W *are given by*

$$\mathbb{Y} = (\mathbb{A}_1\mathbb{B}_1 - \mathbb{B}_1\Lambda)(\mathbb{A}_2\mathbb{B}_1)^+ - \tilde{Z}(I - (\mathbb{A}_2\mathbb{B}_1)(\mathbb{A}_2\mathbb{B}_1)^+), \tag{56}$$

$$W = \Gamma(\mathbb{C}_1\mathbb{B}_1)^+, \tag{57}$$

where $\tilde{Z}$ *is an arbitrary matrix of appropriate dimension. Finally, if there exist matrices* $\mathbb{Y}$ *and* W *satisfying Equations* (54) *and* (55), *then*

$$G_{fr}(s) = \left[\begin{array}{c|c} \Lambda & I \\ \hline \Gamma & 0 \end{array}\right] = \mathrm{diag}\left(\frac{\gamma_1}{s - \lambda_1}, \ldots, \frac{\gamma_{n_f}}{s - \lambda_{n_f}}\right). \tag{58}$$

Proof Since $\mathbb{B}_1$ has full column rank there exist a matrix completion $\mathbb{B}_1^{\perp} \in \mathbb{R}^{(q_0+q_1)\times(q_0+q_1-n_f)}$ such that $\tilde{\mathcal{B}} = [\mathbb{B}_1 \; \mathbb{B}_1^{\perp}] \in \mathbb{R}^{(q_0+q_1)\times(q_0+q_1)}$ is nonsingular. Let $\tilde{\mathcal{B}}^{-1} = [\tilde{\mathcal{B}}_1 \; \tilde{\mathcal{B}}_2]^{\mathrm{T}}$ with $\tilde{\mathcal{B}}_1 \in \mathbb{R}^{(q_0+q_1)\times n_f}$. Then, we obtain

$$G_{fr}(s) = \left[\begin{array}{c|c} \tilde{\mathcal{B}}^{-1}(\mathbb{A}_1 - \mathbb{Y}\mathbb{A}_2)\tilde{\mathcal{B}} & \tilde{\mathcal{B}}^{-1}\mathbb{B}_1 \\ \hline W\mathbb{C}_1\tilde{\mathcal{B}} & 0 \end{array}\right]$$

$$= \left[\begin{array}{c|c} \begin{bmatrix}\tilde{\mathcal{B}}_1^{\mathrm{T}} \\ \tilde{\mathcal{B}}_2^{\mathrm{T}}\end{bmatrix}(\mathbb{A}_1 - \mathbb{Y}\mathbb{A}_2)\begin{bmatrix}\mathbb{B}_1 & \mathbb{B}_1^{\perp}\end{bmatrix} & \begin{bmatrix}\tilde{\mathcal{B}}_1^{\mathrm{T}} \\ \tilde{\mathcal{B}}_2^{\mathrm{T}}\end{bmatrix}\mathbb{B}_1 \\ \hline W\mathbb{C}_1\begin{bmatrix}\mathbb{B}_1 & \mathbb{B}_1^{\perp}\end{bmatrix} & 0 \end{array}\right]$$

$$= \left[\begin{array}{cc|c} \tilde{\mathcal{B}}_1^{\mathrm{T}}(\mathbb{A}_1 - \mathbb{Y}\mathbb{A}_2)\mathbb{B}_1 & \tilde{\mathcal{B}}_1^{\mathrm{T}}(\mathbb{A}_1 - \mathbb{Y}\mathbb{A}_2)\mathbb{B}_1^{\perp} & I \\ \tilde{\mathcal{B}}_2^{\mathrm{T}}(\mathbb{A}_1 - \mathbb{Y}\mathbb{A}_2)\mathbb{B}_1 & \tilde{\mathcal{B}}_2^{\mathrm{T}}(\mathbb{A}_1 - \mathbb{Y}\mathbb{A}_2)\mathbb{B}_1^{\perp} & 0 \\ \hline W\mathbb{C}_1\mathbb{B}_1 & W\mathbb{C}_1\mathbb{B}_1^{\perp} & 0 \end{array}\right]$$

consider $[\tilde{\mathcal{B}}_1 \; \tilde{\mathcal{B}}_2]^{\mathrm{T}}[\mathbb{B}_1 \; \mathbb{B}_1^{\perp}] = I$, then we have

$$G_{fr}(s) = \left[\begin{array}{cc|c} \tilde{\mathcal{B}}_1^{\mathrm{T}}(\mathbb{A}_1 - \mathbb{Y}\mathbb{A}_2)\mathbb{B}_1 & \tilde{\mathcal{B}}_1^{\mathrm{T}}(\mathbb{A}_1 - \mathbb{Y}\mathbb{A}_2)\mathbb{B}_1^{\perp} & I \\ 0 & \tilde{\mathcal{B}}_2^{\mathrm{T}}(\mathbb{A}_1 - \mathbb{Y}\mathbb{A}_2)\mathbb{B}_1^{\perp} & 0 \\ \hline W\mathbb{C}_1\mathbb{B}_1 & W\mathbb{C}_1\mathbb{B}_1^{\perp} & 0 \end{array}\right]$$

now, removing an uncontrollable subspace, we get

$$G_{fr}(s) = \left[\begin{array}{c|c} \Lambda & I \\ \hline \Gamma & 0 \end{array}\right] = \mathrm{diag}\left(\frac{\gamma_1}{s-\lambda_1}, \ldots, \frac{\gamma_{nf}}{s-\lambda_{nf}}\right). \tag{59}$$

From Equation (59), we found that

$$(\mathbb{A}_1 - \mathbb{Y}\mathbb{A}_2)\mathbb{B}_1 = \mathbb{B}_1\Lambda, \tag{60}$$

$$W\mathbb{C}_1\mathbb{B}_1 = \Gamma \tag{61}$$

the general form of $\mathbb{Y}$ from Equation (60) is given by

$$\mathbb{Y} = (\mathbb{A}_1\mathbb{B}_1 - \mathbb{B}_1\Lambda)(\mathbb{A}_2\mathbb{B}_1)^+ - \tilde{Z}(I - (\mathbb{A}_2\mathbb{B}_1)(\mathbb{A}_2\mathbb{B}_1)^+), \tag{62}$$

where $\tilde{Z}$ is an arbitrary matrix of appropriate dimension.

And the particular form of W in Equation (61) is

$$W = \Gamma(\mathbb{C}_1\mathbb{B}_1)^+, \tag{63}$$

Replacing Equations (62) and (63) in Equations (48) and (49), we obtain

$$\dot{\varphi}(t) = [\underbrace{\mathbb{A}_1 - (\mathbb{A}_1\mathbb{B}_1 - \mathbb{B}_1\Lambda)(\mathbb{A}_2\mathbb{B}_1)^+\mathbb{A}_2}_{\bar{\mathbb{A}}_1} + \tilde{Z}\underbrace{(I - (\mathbb{A}_2\mathbb{B}_1)(\mathbb{A}_2\mathbb{B}_1)^+)\mathbb{A}_2}_{\bar{\mathbb{A}}_2}]\varphi(t) + \mathbb{B}_1 f(t), \tag{64}$$

$$r(t) = \Gamma(\mathbb{C}_1\mathbb{B}_1)^+\mathbb{C}_1\varphi(t). \tag{65}$$

Now, is necessary to study the stability of the observer and determine the remainder of matrices of the observer.

4.3. *Observer design*

The following theorem gives the LMI conditions that allow the determination of the dynamical observer matrices.

THEOREM 2 *There exist matrices $\tilde{Z}$ and Z such that system (64)–(65) is asymptotically stable if and only if there exist a matrix $X = X^{\mathrm{T}} > 0$ such that the following LMIs are satisfied:*

$$X = \begin{bmatrix} X_1 & X_2 \\ X_2^{\mathrm{T}} & X_3 \end{bmatrix} > 0, \tag{66}$$

where $X_1 = X_1^{\mathrm{T}} > 0$, $X_3 = X_3^{\mathrm{T}} > 0$ *and* $X_3 - X_2^{\mathrm{T}}X_1^{-1}X_2 > 0$.

$$\begin{aligned} &(N_3 - N_3T_1G(N_3T_1G)^+N_3)^{\mathrm{T}\perp}(\Pi), \\ &(N_3 - N_3T_1G(N_2T_1G)^+N_3)^{\mathrm{T}\perp\mathrm{T}} < 0, \end{aligned} \tag{67}$$

where

$$\Pi = {\Pi_1}^{\mathrm{T}}X_1 + X_1\Pi_1 - {\Pi_2}^{\mathrm{T}}W_1^{\mathrm{T}} - W_1\Pi_2, \tag{68}$$

$$\Pi_1 = N_1 + (T_1G\Lambda - N_1T_1G)(N_3T_1G)^+N_3, \tag{69}$$

$$\Pi_2 = \mathcal{N}_2 - \mathcal{N}_2T_1G(N_3T_1G)^+N_3. \tag{70}$$

In this case matrix, $Z = X_1^{-1}W_1$ *and the matrix* $\tilde{Z}$ *is parameterized as follows:*

$$\tilde{Z} = X^{-1}(\mathcal{K}\mathcal{C}_l^+ + \mathcal{Z}(I - \mathcal{C}_l\mathcal{C}_l^+)), \tag{71}$$

where

$$\mathcal{K} = -\mathcal{R}^{-1}\vartheta\mathcal{C}_r^{\mathrm{T}}(\mathcal{C}_r\vartheta\mathcal{C}_r^{\mathrm{T}})^{-1} + \mathcal{S}^{-1/2}\mathcal{L}(\mathcal{C}_r\vartheta\mathcal{C}_r^{\mathrm{T}})^{-1/2}, \tag{72}$$

$$\vartheta = (\mathcal{R}^{-1} - \mathcal{D})^{-1} > 0, \tag{73}$$

$$\mathcal{S} = \mathcal{R}^{-1} - \mathcal{R}^{-1}[\vartheta - \vartheta\mathcal{C}_r^{\mathrm{T}}(\mathcal{C}_r\vartheta\mathcal{C}_r^{\mathrm{T}})^{-1}\mathcal{C}_r\vartheta]\mathcal{R}^{-1} \tag{74}$$

with $\mathcal{C} = \begin{bmatrix} N_3 - N_3T_1G(N_3T_1G)^+N_3 & 0 \\ 0 & -I \end{bmatrix}$ *and* $\mathcal{D} = \begin{bmatrix} \Pi & {\Pi_1}^{\mathrm{T}}X_2 - {\Pi_2}^{\mathrm{T}}W_2^{\mathrm{T}} \\ X_2^{\mathrm{T}}\Pi_1 - W_2\Pi_2 & 0 \end{bmatrix}$, *where* Π, Π_1 *and* Π_2 *are defined in Equations (68)–(70), respectively, and* $W_2 = X_2^{\mathrm{T}}Z$.

Matrices $\mathcal{R}, \mathcal{L}$, and $\mathcal{Z}$ are arbitrary matrices of appropriate dimensions satisfying $\mathcal{R} > 0$ and $\|\mathcal{L}\| < 1$. Matrices $\mathcal{C}_l$ and $\mathcal{C}_r$ are full rank matrices such that $\mathcal{C} = \mathcal{C}_l\mathcal{C}_r$.

Proof Consider a matrix $X = X^{\mathrm{T}} > 0$ such that

$$(\bar{\mathbb{A}}_1 + \tilde{Z}\bar{\mathbb{A}}_2)^{\mathrm{T}}X + X(\bar{\mathbb{A}}_1 + \tilde{Z}\bar{\mathbb{A}}_2) < 0. \tag{75}$$

This last inequality can be rewritten as

$$\mathcal{B}\mathcal{X}\mathcal{C} + (\mathcal{B}\mathcal{X}\mathcal{C})^{\mathrm{T}} + \mathcal{D} < 0, \tag{76}$$

where $\mathcal{X} = X\tilde{Z}$, $\mathcal{D} = \bar{\mathbb{A}}_1^{\mathrm{T}}X + X\bar{\mathbb{A}}_1$ its equivalence is defined in Theorem 2. Also matrix $\mathcal{B}$ is taken as $\mathcal{B} = I$, then $\mathcal{B}_l = I$, $\mathcal{B}_r = I$ and $\mathcal{B}^{\perp} = 0$.

The solvability conditions of Lemma 1 applied to Equation (76) are reduced to

$$\mathcal{C}^{\mathrm{T}\perp}\mathcal{D}\mathcal{C}^{\mathrm{T}\perp\mathrm{T}} < 0 \quad (77)$$

with $\mathcal{C}^{\mathrm{T}\perp} = [(N_3 - N_3T_1G(N_3T_1G)^{+}N_3)^{\mathrm{T}\perp}\ \ 0]$. By using the definition of $\mathcal{D}$ and W_1 we obtain Equation (67).

From Theorem 2 if condition (67) is satisfied, the matrix $\tilde{Z}$ is obtained as in Equations (71)–(74).

Remark 4 The dynamical observer application to standard systems can be obtained directly form our results by setting $E = I$, then we have $E^{\perp} = 0$, $\Sigma = \left[\begin{smallmatrix} R \\ C \end{smallmatrix}\right]$, and $\Omega = \left[\begin{smallmatrix} E \\ C \end{smallmatrix}\right]$.

The following algorithm summarize the procedure to compute all the observer matrices.

ALGORITHM 1

Step 1. Select the observer order q_0 and a matrix $R \in \mathbb{R}^{q_0\times n}$ such that $\mathrm{rank}(\Sigma) = n$.

Step 2. Compute the matrices $T_1, \mathcal{T}_2, K_1, \mathcal{K}_2, N1, \mathcal{N}_2, N_3, \tilde{K}_1, \tilde{\mathcal{K}}_2, \tilde{K}_3$, and P_1 defined in Section 4.

Step 3. Select the matrices Λ and Γ as were defined in Theorem 1.

Step 4. Compute matrix W as in Equation (63).

Step 5. Find $\mathcal{R} > 0$ such that Equation (73) be positive definite.

Step 6. Find the matrices $\mathcal{L}$ and $\mathcal{Z}$ such that $\|\mathcal{L}\| < 1$ to solve Equations (66) and (67), then obtain the matrix $\tilde{Z}$ as in (71).

Step 7. Compute the matrices of the dynamical observer (2)–(3): N, H, F, J, S, L, M, P and Q, by using (43) to compute N, (56) to compute H and L, (31)–(34) to compute S, M, P and Q, F is defined in (45) and J is defined by Lemma 2.

5. Illustrative example

In order to illustrate our results, let us consider the following singular system:

$$\begin{bmatrix} 1 & 0 & 0 \\ 0 & 1 & 0 \\ 0 & 0 & 0 \end{bmatrix}\dot{x}(t) = \begin{bmatrix} -2.7 & 0 & 0.3 \\ -0.2 & -3 & 0 \\ -0.11 & 1.74 & -1 \end{bmatrix}x(t) + \begin{bmatrix} 1 \\ 0.5 \\ 1 \end{bmatrix}u(t) + \begin{bmatrix} 1 & 1 \\ 0 & 1 \\ 0 & 0 \end{bmatrix}f(t),$$

$$y(t) = \begin{bmatrix} 1 & 0 & 1 \\ 0 & 0 & 1 \end{bmatrix}x(t).$$

By following Algorithm 1, an observer with order $q_0 = 3$ was selected and $R = I_3$, such that $\mathrm{rank}(\Sigma) = 3$. Matrices $\mathcal{R}$, $\mathcal{L}$ and $\mathcal{Z}$ were selected as $\mathcal{R} = I_6$, $\mathcal{L} = \mathrm{ones}_{6,4} \times 0.1$, and

$$\mathcal{Z} = \begin{bmatrix} 9 & 3 & 2 & 1 & 8 & 9 & 9 & 0 & 3 \\ 9 & 4 & 1 & 3 & 8 & 7 & 8 & 2 & 8 \\ 9 & 2 & 8 & 4 & 1 & 7 & 7 & 4 & 2 \\ 9 & 4 & 4 & 5 & 7 & 8 & 3 & 8 & 3 \\ 9 & 1 & 8 & 4 & 7 & 2 & 8 & 4 & 1 \\ 9 & 4 & 8 & 2 & 8 & 4 & 9 & 3 & 8 \end{bmatrix}.$$

By using the LMI toolbox of MATLAB, we solved the inequalities (66) and (67). The observer gains are constructed using Theorem 2.

$$\begin{aligned}\dot{\zeta}(t) &= \begin{bmatrix} -3.11 & -2.89 & -3.67 \\ -1.10 & -2.04 & -2.08 \\ 0.21 & 0.36 & -0.66 \end{bmatrix}\zeta(t) + \begin{bmatrix} 0.87 & 0.87 & 0.87 \\ 0.36 & 0.36 & 0.36 \\ 0.20 & 0.20 & 0.20 \end{bmatrix} \times 0.01v(t) \\ &+ \begin{bmatrix} 14.56 & -0.97 & 2.23 \\ 6.13 & 0.10 & -0.10 \\ 2.55 & 0.42 & -0.88 \end{bmatrix}\begin{bmatrix} -E^{\perp}Bu(t) \\ y(t) \end{bmatrix} + \begin{bmatrix} 14.09 \\ 6.17 \\ 2.75 \end{bmatrix}u(t),\end{aligned}$$

$$\begin{aligned}\dot{v}(t) &= \begin{bmatrix} 0.44 & -0.69 & 1.18 \\ 0.44 & -0.69 & 1.18 \\ 0.44 & -0.69 & 1.18 \end{bmatrix} \times 0.01\zeta(t), + \begin{bmatrix} -2.03 & 0.22 & 0.22 \\ 0.22 & -2.03 & 0.22 \\ 0.22 & 0.22 & -2.03 \end{bmatrix} \times 0.01v(t) \\ &+ \begin{bmatrix} 0 & 0 & 0 \\ 0 & 0 & 0 \\ 0 & 0 & 0 \end{bmatrix}\begin{bmatrix} -E^{\perp}Bu(t) \\ y(t) \end{bmatrix}\end{aligned}$$

$$\hat{x}(t) = \begin{bmatrix} 0.59 & -0.05 & -0.19 \\ -0.05 & 0.31 & 0.15 \\ -0.19 & 0.15 & 0.37 \end{bmatrix}\zeta(t) + \begin{bmatrix} 0.07 & 0.56 & -0.49 \\ 0.49 & 0.16 & 0.33 \\ -0.09 & 0.24 & 0.67 \end{bmatrix}\begin{bmatrix} -E^{\perp}Bu(t) \\ y(t) \end{bmatrix},$$

$$r(t) = \begin{bmatrix} -2.20 & -5.85 \\ 6.05 & 4.97 \end{bmatrix}(C\hat{x}(t) - y(t)),$$

In order to evaluate the observer performance a measurement noise $n(t)$ was considered in the measured output, then the noise-corrupted outputs become $y_1(t) = x_1(t) + x_3(t) + n(t)$, and $y_2(t) = x_3(t) + n(t)$.

The results are depicted in Figure 1–5, which show two cases of simulation. The first case shows step faults with different time of apparition, and the second case shows simultaneous faults, and one of these is time variant.

Figure 1 shows the measurement noise, the system input $u(t)$ was considered as a constant $u(t) = 2$.

Case 1 Step actuator faults: In this case, the actuators faults were considered as a step, each one applied at different time, see Figure 2.

Figure 3 gives the residual where each fault can be readily distinguished from the other, which illustrates that the proposed observer satisfies the requirement of FDI.

Once the residuals were generated, the next step is the evaluation of the residuals by assigning a symptom.

symptom	
1	if residue > threshold,
0	if residue < threshold.

With these symptoms we can generate the following signature table.

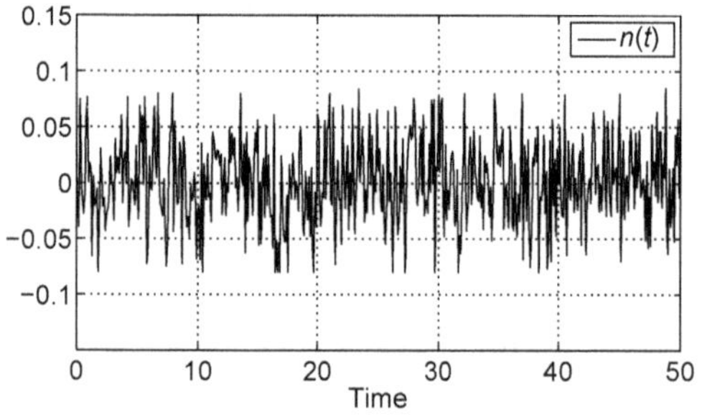

Figure 1. The measurement noise $n(t)$.

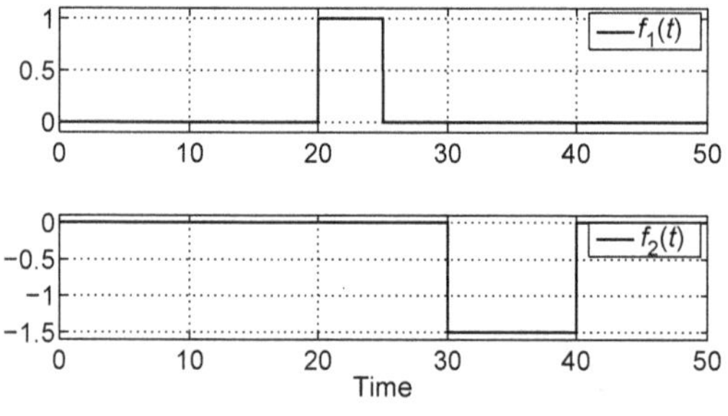

Figure 2. Case 1: step actuator faults.

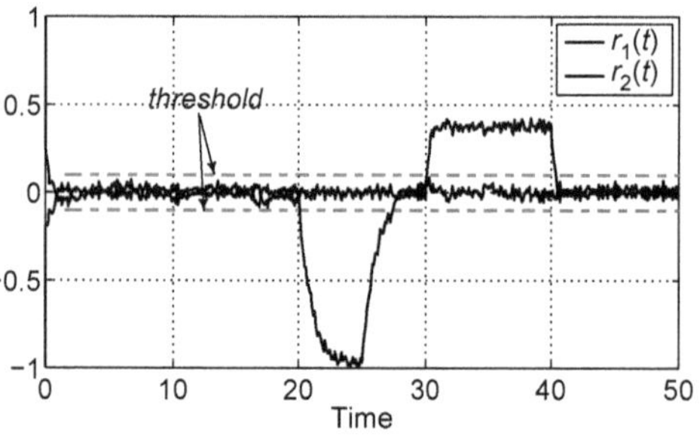

Figure 3. Case 1: residuals fault detection.

Table 1. Case 1: residual evaluation.

	Time		
Residue	20.12–27.37	30.09–40.38	Other time
$r_1(t)$	1	0	0
$r_2(t)$	0	1	0

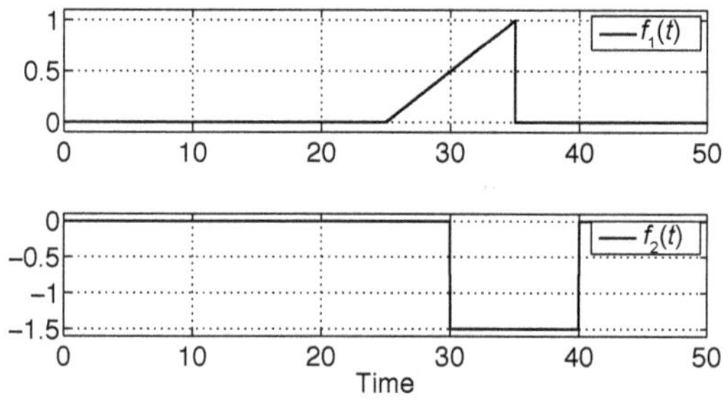

Figure 4. Case 2: actuator faults($f_1(t)$ time variant and $f_2(t)$ step).

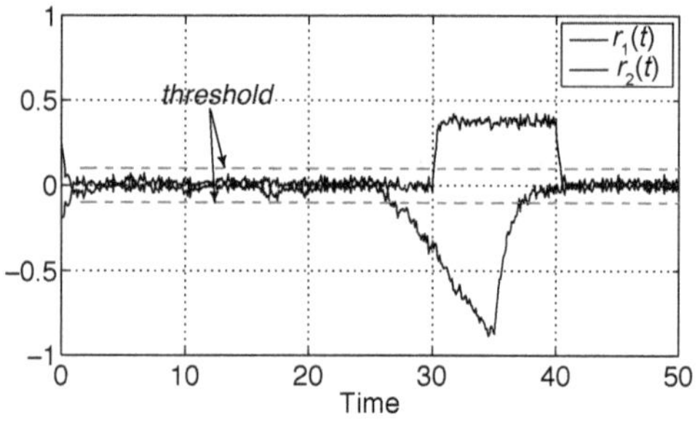

Figure 5. Case 2: residuals simultaneous faults.

Table 2. Case 2: residual evaluation.

	Time			
Residue	26.57–30.09	30.1–37.78	37.79–40.38	Other time
$r_1(t)$	1	1	0	0
$r_2(t)$	0	1	1	0

From Table 1, we observe that the signature to represent the presence of the fault $f_1(t)$ is different from the signature representing the fault $f_2(t)$, so that we can isolate each fault.

Case 2 Simultaneous faults: Figure 4 shows the faults, where $f_1(t)$ is time variant with a ramp behavior. Between 30 and 35 both faults are present into the system.

From Table 2, we observer that the signature in the case of simultaneous faults is different from the other cases, so that even simultaneous faults can be isolated.

The example makes clear that the actual residual response reflects the fault presence in the system. With a good choice of threshold, the observer designed satisfies the performance requirements of FDI.

6. Conclusion

In this paper, a dynamical observer-based FDI for singular systems has been presented. The conditions for the existence of this observer were given in terms of a set of LMIs. The obtained observer satisfies the constraint to obtain a diagonal transfer function between the fault $f(t)$ and the residual $r(t)$ to isolate faults. The approach presented here permits to parameterize the others dynamical observers for fault detection (Remark 2). The case of standard systems can be directly obtained from our results (Remark 4).

Disclosure statement

No potential conflict of interest was reported by the author(s).

References

Araujo J. M., Barros P. R., & Dorea C. E. T. (2012). Design of observers with error limitation in discrete-time descriptor systems: A case study of a hydraulic tank system. *IEEE Transactions on Control Systems Technology*, *20*, 1041–1047.

Astorga-Zaragoza C. M., Theilliol D., Ponsart J. C., & Rodrigues M. (2012). Fault diagnosis for a class of descriptor linear parameter-varying systems. *International Journal of Adaptive Control and Signal Processing*, *26*, 208–223.

Bokor J., & Szabó Z. (2009). Fault detection and isolation in nonlinear systems. *Annual Reviews in Control*, *33*(2), 113–123.

Boulkroune B., Darouach M., Zasadzinski M., Gillé S., & Fiorelli D. (2009). A nonlinear observer design for an activated sludge wastewater treatment process. *Journal of Process control*, *19*, 1558–1565.

Darouach M. (2009). H_∞ unbiased filtering for linear descriptor systems via LMI. *IEEE Transactions on Automatic Control*, *54*, 1966–1972.

Darouach M. (2012). On the functional observers for linear descriptor systems. *Systems & Control Letters*, *61*, 427–434.

Darouach M., & Boutayeb M. (1995). Design of observers for descriptor systems. *IEEE Transactions on Automatic Control*, *40*, 1323–1327.

Darouach M., Zasadzinski M., & Hayar M. (1996). Reduced-order observer design for descriptor systems with unknown inputs. *IEEE Transaction on Automatic Control*, (41), 1068–1072.

Duan G. R., Howe D., & Patton R. J. (2002). Robust fault detection in descriptor linear systems via generalized unknown input observers. *International Journal of Systems Science*, *33*, 369–377.

Goodwin G. C., & Middleton R. H. (1989). The class of all stable unbiased state estimators. *Systems & Control Letters*, *13*, 161–163.

Hamdi H., Rodriguez M., Mechmeche C., Theilliol D., & Benhadj Braiek N. (2012). Fault detection and isolation in linear parameter-varying descriptor systems via proportional integral observer. *International Journal of Adaptive Control and Signal Processing*, *26*, 208–223.

Hou M., & Müller P. C. (1995). Design of a class of Luenberger observers for descriptor systems. *IEEE Transactions on Automatic Control*, *40*, 133–136.

Kim H. S., Yeu T. K., & Kawaji S. (2001). Fault detection in linear descriptor systems via unknown input PI observer. *Transactions on Control, Automation and Systems Engineering*, *3*, 77–82.

Li Z., & Jaimoukha I. M. (2009). Observer-based fault detection and isolation filter design for linear time-invariant systems. *International Journal of Control*, *82*, 171–182.

Li X. J., & Yang G. H. (2011). Dynamic observer-based robust control and fault detection for linear systems. *IET Control Theory and Applications*, *6*, 2657–2666.

Liu B., & Si J. (1997). Fault isolation filter design for linear time-invariant systems. *IEEE Transaction on Automatic Control*, *42*, 704–707.

Liu P., Zhang Q., Yang X., & Yang L. (2008). Passivity and optimal control of descriptor biological complex systems. *IEEE Transactions on Automatic Control*, *53*, 122–125.

Luenberger D. G. (1977). Dynamic equations in descriptor form. *IEEE Transactions on Automatic Control*, *AC-22*, 312–321.

Marquez H. J. (2003). A frequency domain approach to state estimation. *Journal of the Franklin Institute*, *340*, 147–157.

Müller P. C. (2005). Modelling and control of mechatronic systems by the descriptor approach. *Journal of Theoretical and Applied Mechanics*, *43*, 593–607.

Müller P. C., & Hou M. (1993). On the observer design for descriptor systems. *IEEE Transactions on Automatic Control*, *38*, 1666–1671.

Patton R. J., Frank P. M., & Clark R. N. (2000). *Issues of fault detection for dynamic systems*. London: Springer.

Rodrigues M., Theilliol D., Adam-Medina M., & Sauter D. (2008). A fault detection and isolation scheme for industrial systems based on multiple operating models. *Control Engineering Practice*, (16), 225–239.

Skelton R. E., Iwasaki T., & Grigoriadis K. (1998). *A unified algebraic approach to linear control design*. London: Taylor & Francis.

Theilliol D., Noura H., & Ponsart J. C. (2002). Fault diagnosis and accommodation of three-tank system based on analytical redundancy. *ISA Transactions*, (41), 365–382.

Verhaegen M. H., & Dooren P. V. (1986). A reduced order observer for descriptor systems. *Systems & Control Letters*, *8*, 29–37.

White J. E., & Speyer J. L. (1987). Detection filter design: Spectral theory and algorithms. *IEEE Transactions on Automatic Control*, *AC-32*, 593–603.

Yeu T. K., Kim H. S., & Kawaji S. (2005). Fault detection, isolation and reconstruction for descriptor systems. *Asian Journal of Control*, *7*, 356–367.

Zhou L., & Lu G. (2009). Detection and stabilization for discrete-time descriptor systems via a limited capacity communication channel. *Automatica*, *45*, 2272–2277.

7

Switch actuators in process control: constraint problems and corrections

Ricardo J. Mantz[a,b*]

[a]*Laboratorio de Electrónica Industrial, Control e Instrumentación (LEICI), FI. Universidad Nacional de La Plata UNLP., CC 91., La Plata 1900, Argentina;* [b]*Comisión de Investigaciones Científicas de la Provincia de Buenos Aires (CIC), Argentina*

The paper deals with switch actuators in process control. The inclusion of a signal modulator required for commanding the actuator in proportional-integral-derivative (PID) controller structures is discussed. In particular, it is analysed how the restrictions of this modulator may worsen the well-known problems of process-input saturation, such as windup or loss of control directionality among others. From this analysis a simple correction methodology is proposed for simultaneously addressing constraints in both modulator and actuator. This methodology is easily implemented from feeding back a modulator signal and takes advantage of the wide knowledge existing in anti-windup techniques. Two examples (SISO and MIMO) are presented to assess the effectiveness of the proposed correction.

Keywords: PID controller; switch actuator; actuator constraint; sliding mode; reach mode

1. Introduction

Proportional-integral-derivative (PID) controllers are widely used in industrial applications, being commonly preferred over controllers obtained from more sophisticated techniques. Their popularity has also encouraged the formulation of a large number of methods for tuning the controller gains (O'Dwyer, 2006). Like other industrial controllers, PIDs must address corrective actions to overcome the constraints of the process and power actuators which cause, among other problems, controller windup, plant windup, loss of control directionality in MIMO systems, loss of decoupling properties, etc. (Åström & Hagglund, 2006; Hippe, 2006).

Depending on the characteristics of the process and the type of control action, actuators can be continuous or switched (relays, electronic devices operating as power switches, etc.). For efficiency reasons, switching actuators are commonly used in processes where the control action is an electrical variable. In these cases, the control action switches between two values at high frequency with respect to the process dynamics in such a way that produces, on average, the same effects that the continuous control action (Garelli, Mantz, & De Battista, 2011). To command this type of actuators is also necessary to modulate the controller output in two levels (Sira-Ramírez & Villeda, 2004). Although the signal modulator can be connected to the output of the continuous controller, it is common to incorporate it into the controller structure which simplifies the implementation (Al-Hosani, Utkin, & Malinin, 2011). In greater or lesser degree, the insertion of the modulator tends to reduce the closed loop performance with respect to the case of continuous actuators. Processes with fast dynamics are more sensitive to this loss of performance (e.g. electrolysis processes for clean hydrogen production, controls of fuel cell converters, energy conversion systems for process excitation, etc.). Although the effects of the restrictions of both actuator and modulator are dependent on each other, they have been treated independently. This is probably due to the fact that in the nonlinearity used in the modulator (relay type) it is not possible to define the saturation error used in most of the methods that address the constrained input problems. From a practical point of view, the problem has been usually addressed with a more conservative controller tuning.

The paper is organized as follows. Section 2 presents how the modulator can be incorporated in the structure of PID controllers and suggests a structure 2DOF/PID to command switching actuators. In Section 3, it is shown how the restrictions of the modulator can degrade the closed loop response with respect to the case of continuous actuator. The following section proposes to use a common framework to solve the problems inherent to the constraints of both the modulator and the actuator, that is, reach mode (RM) and saturation problems, respectively. In this context, well-proven anti-reset-windup (ARW) algorithms can be adapted to improve the RM of the modulator at the same time that the closed loop performance in presence of constrains is increased. In order to clarify the advantages of using a common framework, a simple case of RM correction, based on ARW observer ideas, is considered.

*Email: mantz@ing.unlp.edu.ar

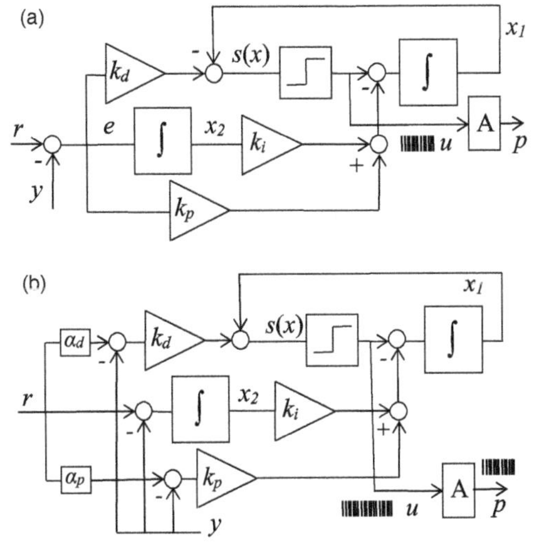

Figure 1. (a) Classical and (b) 2DOF-PID controllers with output to command switch actuators.

The main features of the proposal are validated through two examples. Finally, conclusions are summarized.

2. PID controller with output for switching actuators

Although the problems addressed in this paper can appear with different kinds of controllers and techniques of modulation, for reasons of clarity, the analysis is restricted to PID controllers (for being the most commonly accepted in industrial applications) and to modulation by a sliding mode (SM) regimens (because SM theory allows a more precise and easier analysis than other techniques). Figure 1(a) shows a PID controller structure, implemented via a sliding regime that is suitable to command switching actuators. Indeed, while the values of the output $u(t)$ enable or not the power flow $p(t)$ into the plant, its mean value preserves the PID controller actions. An attractive property of this structure is that it avoids the implementation of a derivator and therefore the problems associated with the measurement noises. The circuit can be modelled by

$$\begin{bmatrix} \dot{x}_1 \\ \dot{x}_2 \end{bmatrix} = f(x) + g(x)u + d = \begin{bmatrix} -k_i x_2 \\ 0 \end{bmatrix} + \begin{bmatrix} 1 \\ 0 \end{bmatrix} u + \begin{bmatrix} -k_p e \\ e \end{bmatrix}. \quad (1)$$

In a normal operation, the controller states $x^T = [x_1, x_2]$ evolve inside an invariant control surface defined by:

$$s(x) = k_d e - x_1 = 0, \quad (2)$$

that verifies two necessary conditions for achieving the SM regime (its relative degree with respect to the discontinuous signal $u(t)$ is 1, and the Lie derivative $L_g s(x) < 0$), being the sliding range associated to the limit values u^+ and u^- (Sira-Ramírez, 1988). Once the circuit operates in SM, the invariance conditions are satisfied

$$s(x) = k_d e - x_1 = 0$$
$$\dot{s}(x) = k_d \dot{e} - (u_{eq} - k_i x_2 - k_p e) = 0 \quad (3)$$

being u_{eq} the equivalent control, that is, a fictitious continuous signal that produces the same effect on the process than the actual discontinuous signal $u(t)$. From Equation (3), it can be verified that contains the PID action

$$u_{eq} = k_p e + k_i \int e \, dt + k_d \dot{e}. \quad (4)$$

Here a small modification is suggested to the PID structure of Figure 1(a) which, without modifying the SM surface, allows us to obtain the control action of a 2DOF/PID

$$u_{eq} = k_p(\alpha_p r - y) + k_i \int e \, dt + k_d \frac{d(\alpha_d r - y)}{dt}. \quad (5)$$

The additional parameters (α_p, α_d) permit to meet simultaneous specifications (basically, set-point tracking and perturbation rejecting) and, even, contribute to reduce windup risks (Åström & Hagglund, 2006; Bianchi, Mantz, & Christiansen, 2008; Puleston & Mantz, 1995).

3. Effects of the modulator constraints

As previously mentioned, the discontinuous signal $u(t)$ can be used to command the switching actuator in such a way that the mean value of the process input preserve the PID actions. However, it is important to keep in mind that the equivalent controls (4) and (5) are only defined on the surface $s(x) = 0$, that is, during the SM. Throughout the time that precedes the SM, that is, during the RM, the modulator output only takes one of the limit values u^+ or u^-. From a practical standpoint, the control loop is opened as the process-input loses the information of the PID action. In greater or lesser degree, this deteriorates the behaviour of the system with respect to the case in which the actuator is continuous. Moreover, this fact can induce integrator windup in the PID, or vice versa, the actuator saturation can produce, besides of integrator windup in the PID, an inadequate charging of the integrator of the modulator and as a consequence the extending of the RM. In extreme cases, the interaction of these effects (RM and windup) could cause system instability.

3.1. *Example*

Consider the system

$$P(s) = \frac{1}{(1 + sT_E)sT_I}, \quad (6)$$

being $T_E = 0.01$ y $T_I = 1$, controlled with a 2DOF/PI where $k_p = 50$, $k_i = 25k_p$, $\alpha_p = 0$ and a continuous actuator (Garelli et al., 2011; Hippe, 2006). Assuming that there

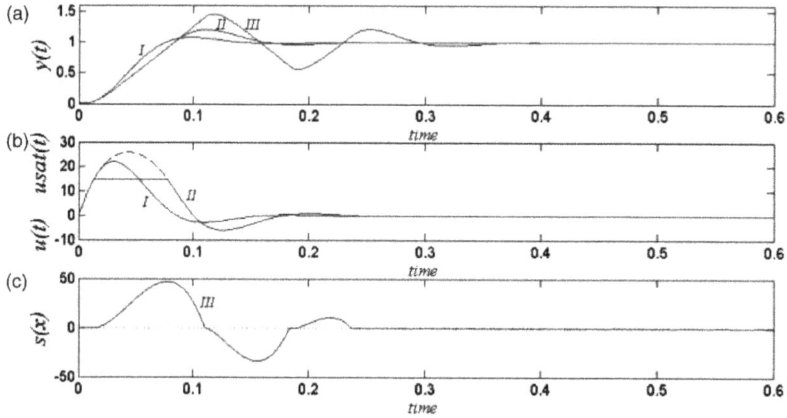

Figure 2. (a) Set-point tracking responses: (I) without constraints, (II) continuous actuator with constrains, (III) switch case with considers both modulator and actuator constraints; (b) control actions; (c) $s(x)$.

are no restrictions, this linear control allows obtaining suitable set-point tracking and disturbance rejection. However, the set-point tracking is affected when the actuator saturates. This fact can be corroborated in Figure 2(a) where the curve I corresponds to the case without restrictions and the curve II to the case where the actuator saturates at ± 15. Figure 2(b) displays the corresponding controller outputs and process inputs.

Consider now that, due to requirements of energy efficiency, the system is controlled using a switch actuator and a 2DOF/PI controller with structure shown in Figure 1(b) and with the same gain values than the continuous 2DOF/PI. The system response is shown in curve III of Figure 2(a), being observed a severe deterioration compared with the two previous cases of continuous actuators. As was mentioned above, this worsening can be attributed to the mutual interaction between two problems: (1) the PI controller windup and (2) the RM preceding the sliding regime during which the PI action is not present in the mean value of $u(t)$.

As can be seen in Figure 2(c), the SM is just reached at $t = 0.237$ (when $s(x) = 0$) which is substantially greater than the time during which the actuator remains saturated in the continuous case (see Figure 2(b)). It is important to note that while the windup and RM are different problems, they manifest in the same way by sending the limit signal at process input, which does not correspond to the ideal PI action.

4. Proposal of a common framework for RM and windup

Conventional methods to reduce the RM are based on the increase of extreme values u^+ and u^-. However, these values are constrained by the DC electrical sources used for implementing the PID circuit. To overcome this shortcoming, and even though that the RM and the actuator saturation could be addressed independently; it is proposed to use a unique theoretical framework for both problems.

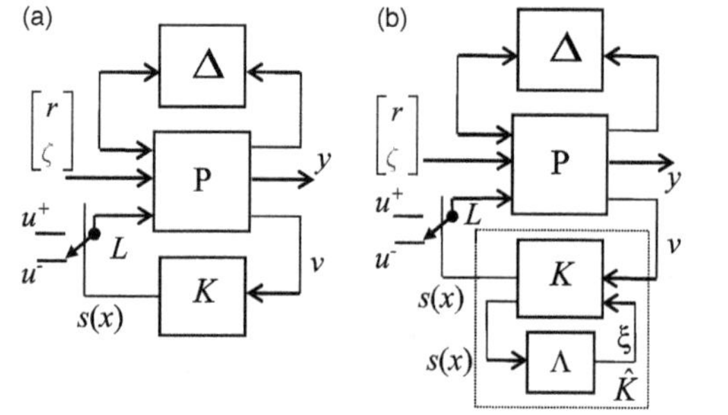

Figure 3. (a) Schematic diagram of generic VSS; (b) RM correction as a problem of windup in a wide sense.

4.1. *Similarities between the RM in variable structure systems and the windup problems. Correction*

The purpose of this section is to show that the effects of RM in variable structure systems (VSS) can be analysed as problems of windup in a wide sense. Figure 3(a) shows a diagram of a generic VSS, where P is the system to be controlled and L represents a switching device which is governed by the output of the block K. Controller K is often conformed by static gains, but in same cases it also has its own dynamics (e.g. it is common practice to expand the states to ensure disturbance rejection at steady state). Then, considering that K is linear time invariant (LTI) system, it can be expressed as $K(s) = C(sI - A)^{-1}B + D$ or schematically:

$$K = \left[\begin{array}{c|c} A & B \\ \hline C & D \end{array}\right] = \left[\begin{array}{c|c|c|c} A & B_r & B_y & B_{x_s} \\ \hline C & D & D_y & D_{x_s} \end{array}\right]$$

being its input $v = [r, y, x_s]^{\mathrm{T}}$. Obviously, in the static case, only D must be contemplated. The goal of an SM control is to force the states to evolve on the surface $s(x) = 0$ where the system has the required dynamic properties and the necessary robustness to parameter uncertainties and state disturbances $\zeta(t)$. When this target is met, namely:

$$s(x) = 0 \quad \dot{s}(x) = 0 \tag{7}$$

it is possible to define the equivalent continuous control action $u_{\mathrm{eq}}(x_s, x_k)$ that produces the same effect on the states than the discontinuous signal $u(t)$ which switches ideally with infinite frequency.

It is necessary and sufficient condition for the existence of SM that:

$$u^- \leq u_{\mathrm{eq}} \leq u^+ \tag{8}$$

being u^- and u^+ the extreme values of the available control action (Sira-Ramírez, 1988).

Since all the benefits of robustness of the VSS reside in the SM, it is important that this operation mode predominates over the RM. To this end, the states should reach the surface $s(x) = 0$ in minimum time and, once it is achieved, verify the condition (8). In this paper, the delay in reaching the surface and in establishing of the SM is attributed to lack of correspondence between the states of K and the P input, which is due to the opening of the loop (if K is static, we can speak of no correspondence between the $s(x) = 0$ and the input of P).

In order to recover the lost correspondence, we propose adding a correction Λ on K for conditioning its states and/or its input (Figure 3(b)). For the new controller to perform as an LTI system is established that:

$$\Lambda : v \rightarrow \xi = \begin{bmatrix} \xi_1 & \xi_2 \end{bmatrix}^{\mathrm{T}}, \tag{9}$$

being ξ_1 and ξ_2 the correction terms of the states and input of K, respectively; it must be casual, lineal and time invariant. Furthermore, assuming that the correction only has to act during the RM,

$$\text{if} \quad s(x) = 0 \Rightarrow \xi(t) = 0. \tag{10}$$

Note: the previous reasoning has led us to a statement of the RM problem that keeps extreme similarity with the theoretical framework proposed by Kothare, Campo, Morari, and Nett (1994) to address the problem of control with input constraints and on which significant contributions have been made in recent years (Garelli et al., 2011). To overcome the main difference, referred to the discontinuous action of sliding regimes, the switching device can assimilated to an actuator with saturation and gain tending to infinity in the linear region. In this context, the error of saturation commonly used in anti-windup algorithms can be modified to:

$$e = \lim_{k \to \infty} \left(s(x) - \frac{u}{k} \right) = s(x) \tag{11}$$

and applied for conditioning K according to

$$\xi = \begin{bmatrix} \Lambda_1 \\ \Lambda_2 \end{bmatrix} \lim_{k \to \infty} \left(s(x) - \frac{u}{k} \right) = \begin{bmatrix} \Lambda_1 \\ \Lambda_2 \end{bmatrix} s(x), \tag{12}$$

where, to verify Equation (10), the elements Λ_1 and Λ_2 of Λ must be static gains.

Note that in order to select Λ for correcting the RM, we can take advantage of the discussed similarity. Then, proposing small modifications to the well-proven ARW methods it is possible to introduce different algorithms to RM improvement. Next, we evaluate a RM correction based on one among many possible ARW algorithms.

4.2. A RM correction based on the conventional ARW observer method

From the previous analysis and accepting the interpretation that the RM is consequence of the mismatch between K controller output and the plant input, in this section is introduced a RM correction deduced from the observer-based ARW method. For this purpose it is assumed that during the open-loop operation, $\widehat{K}$ acts as a state observer of a fictitious controller that provides the actual input to the plant

$$\dot{\widehat{x}}_K = A\widehat{x}_K + B_r r + B_y y + B_{x_s} x_s + L\left(\lim_{k \to \infty} \frac{u}{k} - s(x) \right), \tag{13}$$

being the controller output

$$s(x) = C\widehat{x}_k + Dv = C\widehat{x}_k + D_r r + D_y y + D_{x_s} x_s. \tag{14}$$

Then, the state equations of $\widehat{K}$, that is, of the controller with RM correction from the conditioning of the state vector x_k, results:

$$\begin{aligned} \dot{\widehat{x}}_K &= (A - LC)\widehat{x}_K + (B_r - LD_r)r + (B_y - LD_y)y \\ &\quad + (B_{x_s} - LD_{x_s})x_s, \\ s(x) &= C\widehat{x}_k + Dv = C\widehat{x}_k + D_r r + D_y y + D_{x_s} x_s. \end{aligned} \tag{15}$$

Note that in the context of the previous analysis, the correction of RM by the method of the observer corresponds to:

$$\xi = \begin{bmatrix} L \\ 0 \end{bmatrix} s(x). \tag{16}$$

For the purposes of assessing the effects of the proposed correction, we can compare the rates of change of $s(x)$ in both the conventional RM and the RM with correction from the observer ideas, Equations (17) and (18) respectively:

$$\dot{s}(x) = C(A\widehat{x}_k + Bv) + D\dot{v}, \tag{17}$$

$$\dot{s}(x) = -CLs(x) + C(A\widehat{x}_k + Bv) + D\dot{v}, \tag{18}$$

where it can be seen that during the RM the proposed correction has introduced a stabilization term $(-CLs(x))$ in the differential equation $s(x)$.

Observation. It is important to note that $s(x)$, beside of containing information about the operation mode of the

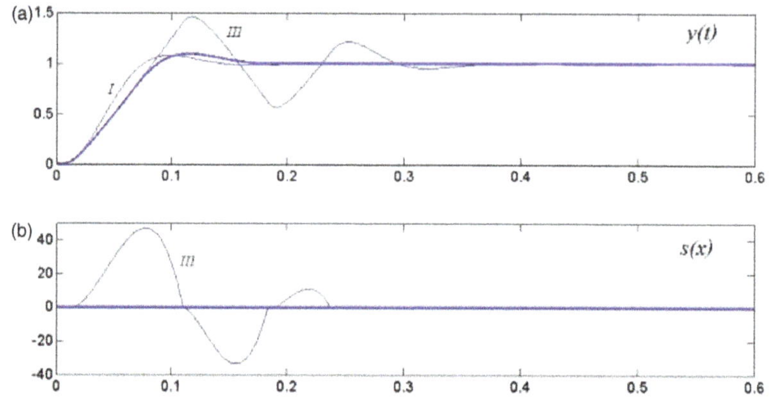

Figure 4. (a) Set-point tracking responses: (I) ideal case, (II) with constraints, and (thick line) with constraints and the propose correction; (b) $s(x)$ without and with correction.

modulator, also has information about the process-input saturation, which happens when the equivalent control exceeds the limit values. Then the signal $s(x)$ can also be employed to correct the state of the controller in order to overcome constraint problems such as windup or loss of control directionality. This is an interesting advantage of addressing in a common way both RM and input saturation. Effectively, the use of $s(x)$ avoids filtering $u(t)$ for calculating the saturation error (most of the ARW algorithms evade dynamics in the correction loop because this fact degrades their performance).

Consider again the example in Section 3.1, where now the correction proportional to $s(x)$ based on concepts of ARW observer method is applied. The simulation has been obtained from adding correction terms $-s(x)$ in the weighted errors of the 2DOF/PI that affects both modulator and PI states with $L = [L_1\ L_2]^T = [k_p\ 1]^T$. The effects of the correction are presented in the thick trace curves of Figure 4 and compared with the curves I and III of the Figure 2(a) that correspond to the ideal case (without considering restrictions) and the case that considers both modulator and actuator constraints. The response and the settling time of the controlled variable greatly improve with respect to the non-compensated case. The effectiveness of the proposed correction is also revealed in Figure 4(b), which shows a drastic reduction of the RM time (thick trace). From a practical point of view the RM has been eliminated.

4.3. *Example*

Consider the benchmark MIMO problem (Campo & Morari, 1990; Mantz & De Battista, 2002) with strong coupling between variables

$$\begin{bmatrix} y_1 \\ y_2 \end{bmatrix} = \frac{1}{10s+1}\begin{bmatrix} 4 & -5 \\ -3 & 4 \end{bmatrix}\begin{bmatrix} u_1 \\ u_2 \end{bmatrix}, \tag{19}$$

and a PI-MIMO centralized controller that give nominal decoupling under close loop operation

$$\mathrm{PI}(s) = \frac{10s+1}{s}\begin{bmatrix} 4 & 5 \\ 3 & 4 \end{bmatrix},$$

$$\begin{bmatrix} u_1 \\ u_2 \end{bmatrix} = \frac{10s+1}{s}\begin{bmatrix} 4 & 5 \\ 3 & 4 \end{bmatrix}\begin{bmatrix} e_1 \\ e_2 \end{bmatrix}. \tag{20}$$

Figures 5–8 show the closed loop MIMO tracking responses when set points $r_1 = 0.6$ and $r_2 = 0.8$ are applied, considering different cases. Simulations in Figure 5 consider continuous and ideal actuators. Curves in Figure 6 correspond to actuators with saturation at $u_i^+ = +40$ and $u_i^- = -40$. As discussed by Campo and Morari (1990) the deterioration in the system response is not due to a problem of windup but to the loss of control directionality when actuators saturate. Figure 7 shows the response of the same system but considering a PI controller whose output is modulated to command a switching actuator. A deterioration in the responses (which is attributable to the RM of the modulator and to its interaction with the pre-existing problem of control directionality) is clearly observed. Part b of the figure displays the signals $s_1(x)$ and $s_2(x)$ which contain information on the status of the actuator and the operating mode of the modulator (i.e. if the modulator has reached the sliding regime where it can be ensured that the control contains, in average, the actions PI/MIMO (20)). In the box at the top right of this figure it can be seen that the period during which the system inputs are saturated is greater than the period corresponding to the case with continuous controller and actuator.

According to the previous discussion, it is now added a correction $-s_1(x) - s_2(x)$ in each error in order to correct simultaneously the constraint problems of both modulator and actuator. In terms of the observer method this correction is equivalent to $L = [L_{1M}\ L_{1PI}\ L_{2M}\ L_{2PI}]^T = [k_{p1}\ 1\ k_{p2}\ 1]^T$ being k_{pi} the PI proportional gains and the subscripts $_M$ and $_{PI}$ corresponding to the state of the modulator

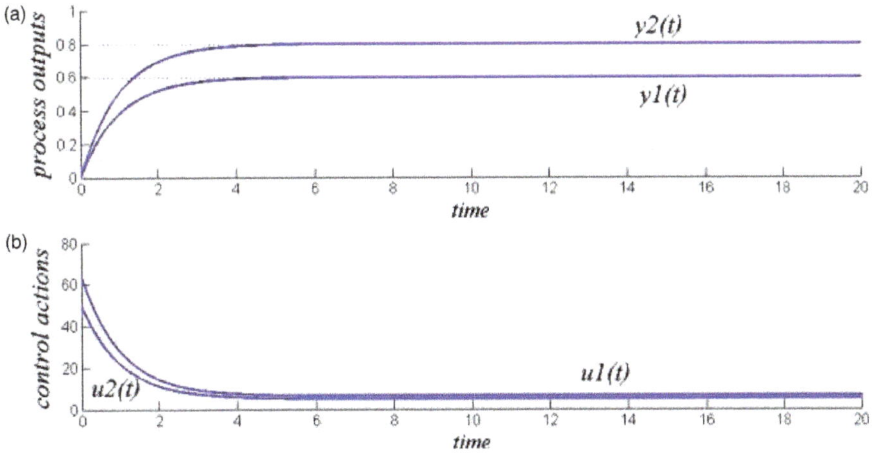

Figure 5. Ideal case without constrains. (a) Set-points tracking responses; (b) unconstrained control actions.

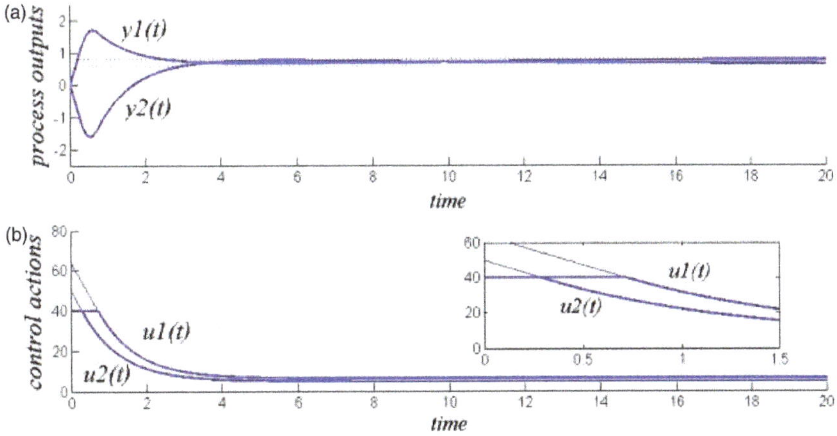

Figure 6. Loss of control directionality due to actuator saturation. (a) Set-points tracking responses; (b) constrained control actions. Box: detail of the actuator saturation.

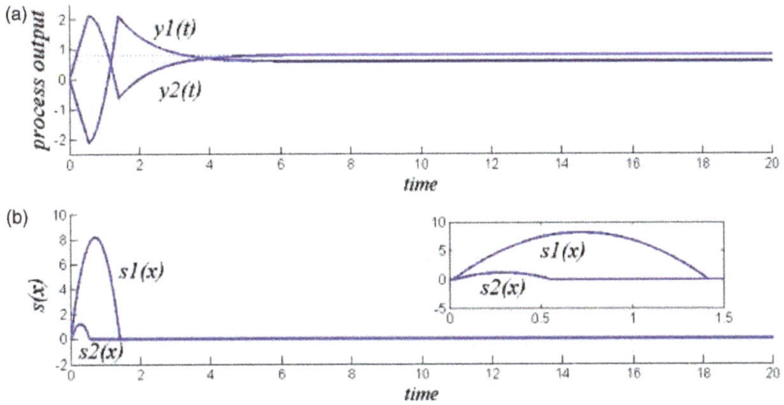

Figure 7. Loss of control directionality with modulator constraints. (a) Set-points tracking responses; (b) $s(x) = [s_1(x)\ s_1(x)]^T$. Box: detail of the period in which $s(x) \neq 0$.

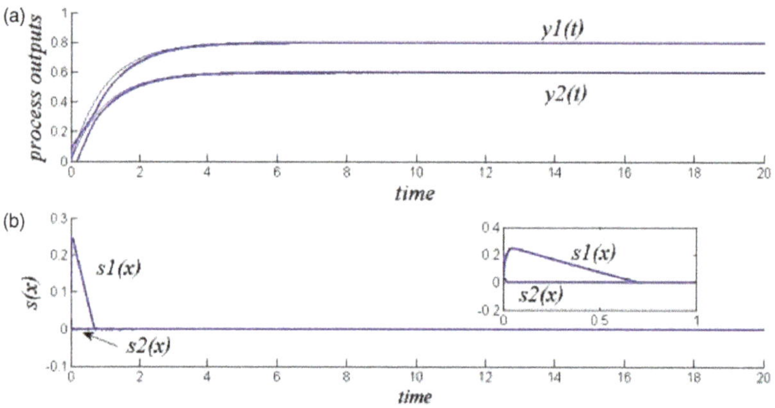

Figure 8. (a) Set-points tracking responses. Thin line: ideal case without constraint. Thick trace: case considering constraints and with the proposed correction. (b) $s(x) = [s_1(x)\ s_1(x)]^T$.

and controller, respectively. The effectiveness of the proposed correction is exhibited in Figure 8. From a practical standpoint the MIMO system outputs remain dynamically decoupled and with a performance that is close to the performance of the ideal unconstrained case. Figure 8(b) shows the drastic reduction in the RM time.

5. Conclusions

The switch actuators, frequently used in industrial applications, require the modulation of the output of the controller. The paper analysed how the insertion of the modulator can reduce the closed loop performance with respect to the case of continuous actuators. Modulator restrictions are manifested through its operation mode known as RM, during which the modulator output does not contain information from the controller output. This can emphasize problems such as windup and loss of control directionality and decoupling among others. Conventional correction methods for constrained systems cannot be directly applied to the modulator because of its type of nonlinearity (relay type) where it is not possible to define a saturation error. Additionally, the measurement of the saturation error in the switching actuators requires to filter the switched control action, fact that introduces dynamics in the correction loop, which loses its effectiveness as a consequence. In this paper these shortcomings are avoided by feeding back a continuous signal of the modulator into the states and/or input of the controller. The paper addresses the constraints of both modulator and switching actuator in a common framework that allows to adapt conventional well-proven ARW techniques to the analysed problem. The effectiveness of the proposal is shown through two examples.

Disclosure statement

No potential conflict of interest was reported by the author.

Funding

This work was supported by ANPCyT, CICpBA, and UNLP.

References

Al-Hosani, K., Utkin, V., & Malinin, A. (2011). Chapter 3: Sliding mode control for industrial controllers. In A. Bartoszewicz (Ed.), *Sliding mode control* (pp. 45–76). Rijeka: InTech.

Åström, K., & Hagglund, T. (2006). *Advanced PID control.* Research Triangle Park: ISA.

Bianchi, F., Mantz, R., & Christiansen, C. (2008). Multivariable PID control with set-point weighting via BMI optimisation. *Automatica, 44*, 472–478.

Campo, P., & Morari, M. (1990). Robust control of processes subject to saturation nonlinearities. *Computers & Chemical Engineering, 14*, 343–358.

Garelli, F., Mantz, R. J., & De Battista, H. (2011). *Advanced control for constrained processes and systems. A unified and practical approach* (Chapter 3). London: IET, Control Engineering Series.

Hippe, P. (2006). *Windup in control. Its effects and their prevention* (Chapter 3). Erlangen: Springer.

Kothare, M., Campo, P., Morari, M., & Nett, C. N. (1994). A unified framework for the study of anti-windup designs. *Automatica, 30*, 1869–1883.

Mantz, R. J., & De Battista, H. (2002). Sliding mode compensation for windup and direction of control problems in two-input-two-output PI controllers. *Industrial and Engineering Chemistry Research, 41*, 3179–3185.

O'Dwyer, A. (2006). *Handbook of PI and PID controller tuning rules.* London: Imperial College Press.

Puleston, P., & Mantz, R. J. (1995). An anti-windup proportional integral structure for controlling time-delayed multiinput-multioutput processes. *Industrial Engineering Chemistry Research, 34*, 2993–3000.

Sira-Ramírez, H. (1988). Differential geometric methods in variable structure control. *International Journal of Control, 48*, 1359–1390.

Sira-Ramírez, H., & Lugo Villeda, L. I. (2004). Sliding modes, delta-modulation and output feedback control of dynamic systems. In A. Sabanovic, L. Fridman, & S. Spurgeon (Eds.), *Variable structure systems from principles to implementation* (Chapter 7, pp. 157–175). London: IET.

8

Global stability in Lagrange sense for BAM-type Cohen–Grossberg neural networks with time-varying delays

Jigui Jian* and Zhihua Zhao

College of Science, China Three Gorges University, Yichang, Hubei 443002, People's Republic of China

In this paper, we investigate the positive invariant sets and global exponential attractive sets for a class of bidirectional associative memory (BAM)-type Cohen–Grossberg neural networks with multiple time-varying delays. By applying inequality techniques, some easily verifiable delay-independent criteria for the ultimate boundedness and global exponential attractive sets of BAM-type Cohen–Grossberg neural networks are obtained by constructing appropriate Lyapunov functions. Finally, one example with numerical simulations is given to illustrate the results obtained in this paper.

Keywords: BAM-type Cohen–Grossberg neural networks; Lagrange stability; globally attractive set; inequality

1. Introduction

Recently, the stability of different types of Cohen–Grossberg neural networks and the bidirectional associative memory (BAM) neural networks has been widely studied by many researchers and various interesting results have been reported (Jiang & Cao, 2008; Li, 2009; Li, Fei, Tan, & Zhang, 2009; Li, Zhang, Zhang, & Li, 2010; Liu & Zong, 2009; Wang, Jian, & Guo, 2008; Zhang, Liu, & Zhou, 2012). In many applications, since BAM-type Cohen–Grossberg neural networks consider the interaction between two neural networks, the studies of the stability behavior of BAM-type Cohen–Grossberg neural networks are of greater interest than the studies of the stability of Cohen–Grossberg neural networks and BAM neural networks.

For BAM-type Cohen–Grossberg neural networks, the existence of periodic solution and an equilibrium point and their stability have been investigated in Jiang and Cao (2008), Li (2009), Li et al. (2010) and Zhang et al. (2012). But in many actual applications, these conclusions are no longer appropriate in the multistable dynamics which have multiple equilibrium and so many of them are unstable (Lu, Wang, & Chen, 2011; Wang & Chen, 2012). Such as the Cohen–Grossberg neural network, when applications are taken into account in biology, it is necessary and important to deal with multistable properties. In this context, it is worth mentioning that the Lagrange stability refers to the stability of the total system which does not require the information of equilibrium points, because the Lagrange stability is considered on the basis of the boundedness of solutions and the existence of global attractive sets (Liao, Luo, & Zeng, 2008). Just as verified in (Liao et al., 2008), outside the globally attracting set, there is no equilibrium point, chaos attractor, periodic state or almost periodic state in the neural networks. Therefore, the research on positive invariant sets and globally attractive sets of the neural networks have been done by many scholars (Liao et al., 2008; Luo, Zeng, & Liao, 2011; Song & Zhao, 2005; Tu, Jian, & Wang, 2011; Tu, Wang, Zha, & Jian, 2013; Wang, Jian, & Jiang, 2010). Song and Zhao (2005) investigated the dissipativity of neural networks with both variable and unbounded delays by constructing proper Lyapunov functions and using some analytic techniques. And the global stability in the Lagrange sense for a class of Cohen–Grossberg neural networks with time-varying delays and finite distributed delays was studied in Tu et al. (2011) and Wang et al. (2010). In Tu et al. (2013), the authors study the global dissipativity of a class of BAM neural networks with both time-varying and unbound delays. To our best knowledge, few authors have discussed the global attractive sets for BAM-type Cohen–Grossberg neural networks.

Motived by the above analysis, the aim of this paper is to study Lagrange stability and global exponential attractive sets for BAM-type Cohen–Grossberg neural networks with time-varying delays and some delay-independent criteria for the ultimate boundedness and global exponential attractive sets of BAM-type Cohen–Grossberg neural networks are obtained. And some results here obtained in this paper are more general than that of the existing reference on the globally exponentially attractive (GEA) set as special cases. The remaining paper is organized as follows: Section 2 describes some preliminaries including some necessary notations, definitions, assumptions and some lemmas. The main results are stated in Section 3. Section 4

*Corresponding author. Email: jiguijian@ctgu.du.cn

gives a numerical example to testify the theoretical analysis. Finally, conclusions are drawn in Section 5.

2. Problem statement

Consider the following BAM-type Cohen–Grossberg neural network model

$$\begin{cases} \dot{x}_i(t) = a_i(x_i(t)) \begin{bmatrix} -c_i(x_i(t)) + \sum_{j=1}^{m} a_{ij} f_j(y_j(t)) \\ + \sum_{j=1}^{m} b_{ij} f_j(y_j(t-\tau_j(t))) + I_i \end{bmatrix}, \\ \quad i = 1,2,\ldots,n, \\ \dot{y}_j(t) = b_j(y_j(t)) \begin{bmatrix} -d_j(y_j(t)) + \sum_{i=1}^{n} m_{ji} g_i(x_i(t)) \\ + \sum_{i=1}^{n} n_{ji} g_i(x_i(t-\sigma_i(t))) + J_j \end{bmatrix}, \\ \quad j = 1,2,\ldots,m, \end{cases} \tag{1}$$

where $x(t) = (x_1(t),\ldots,x_n(t))^{\mathrm{T}}, y(t) = (y_1(t),\ldots,y_m(t))^{\mathrm{T}}$ are the neuron state vectors of the neural network (1), $a_i(x_i(t)) > 0$ for $i \in \Lambda = \{1,2,\ldots,n\}$ and $b_j(y_j(t)) > 0$ for $j \in \Gamma = \{1,2,\ldots,m\}$ represent amplification functions of the ith neurons from the neural field F_x and the jth neurons from the neural field F_y, respectively; $c_i(x_i)$, $d_j(y_j)$ are appropriately behaved functions of the ith neurons from the neural field F_x and the jth neurons from the neural field F_y, respectively; f_j and g_i are the activation functions, I_i and J_j are the exogenous inputs. $A = (a_{ij})_{n\times m}$, $M = (m_{ji})_{m\times n}$; $B = (b_{ij})_{n\times m}$, $N = (n_{ji})_{m\times n}$ are the connection weight matrices and the delayed weight matrices, respectively, $I = (I_1, I_2,\ldots,I_n)^{\mathrm{T}}, J = (J_1, J_2,\ldots,J_m)^{\mathrm{T}}$ are external input vectors. The time-varying delays $\tau_j(t)$, $\sigma_i(t)$ are non-negative and bounded, i.e. $0 \le \tau_j(t) \le \tau_j$, $0 \le \sigma_i(t) \le \sigma_i$. We define the vector functions f, g by

$$g(x(\cdot)) = (g_1(x_1(\cdot)), g_2(x_2(\cdot)),\ldots,g_n(x_n(\cdot)))^{\mathrm{T}} \in R^n,$$

$$f(y(\cdot)) = (f_1(y_1(\cdot), f_2(y_2(\cdot),\ldots,f_m(y_m(\cdot)))^{\mathrm{T}} \in R^m.$$

The initial conditions associated with Equation (1) are given by

$$\begin{cases} x_i(\theta) = \phi_i(\theta), & \theta \in [-\sigma, 0], i \in \Lambda, \\ y_j(\theta) = \psi_j(\theta), & \theta \in [-\tau, 0], j \in \Gamma, \end{cases} \tag{2}$$

where $\phi_i(\theta)$ and $\psi_j(\theta)$ are continuous real valued functions defined on their respective domains, $\sigma = \max_{1\le i\le n}\{\sigma_i\}$, $\tau = \max_{1\le j\le m}\{\tau_j\}$.

Remark 1 It is obvious that system (1) includes neural systems considered in Tu et al. (2013) as its special case. For example, $a_i(x_i(t)) = 1$, $b_j(y_j(t)) = 1$ for $i \in \Lambda$ and $j \in \Gamma$, $c_i(x_i(t)) = \tilde{a}_i x_i(t)$ and $d_j(y_j(t)) = \tilde{c}_j y_j(t)$ with the constants $\tilde{a}_i > 0$ and $\tilde{c}_j > 0$, system (1) reduces to the BAM neural network in Tu et al. (2013).

In order to establish the conditions of main results for the neural networks (1), we have the following assumptions:

(H1) $a_i(u), b_j(u) \in C(R, R^+)$. Furthermore, there exist positive constants $\underline{a}_i$, $\bar{a}_i$, $\underline{b}_j$ and $\bar{b}_j (i \in \Lambda, j \in \Gamma)$ such that $0 < \underline{a}_i \le a_i(u) \le \bar{a}_i$, $0 < \underline{b}_j \le b_j(u) \le \bar{b}_j$, $u \in R$.

(H2) $c_i(u), d_j(u) \in C(R, R^+)$. Moreover, there exist positive constants $\underline{c}_i$, $\bar{c}_i$, $\underline{d}_j$ and $\bar{d}_j (i \in \Lambda, j \in \Gamma)$ such that $\underline{c}_i u^2 \le u c_i(u) \le \bar{c}_i u^2$, $\underline{d}_j u^2 \le u d_j(u) \le \bar{d}_j u^2$, $u \in R$.

The set of bounded activation functions is defined as

$$B = \{p(x) | p_i(x_i) \in C(R,R), \exists k_i > 0, |p_i(x_i)| \le k_i, \forall x_i \in R\}$$

The sigmoid function is defined as

$$S = \left\{ p(x) \,\middle|\, \begin{array}{l} p_i(0) = 0, p_i(x_i) \in C(R,R), \\ D^+ p_i(x_i) \ge 0, |p_i(x_i)| \le k_i, \forall x_i \in R \end{array} \right\}.$$

Remark 2 In this paper, $f(\cdot), g(\cdot) \in B$ represent $|g_i(\cdot)| \le s_i$, $|f_j(\cdot)| \le r_j$ for $i \in \Lambda$ and $j \in \Gamma$, respectively, where s_i, r_j are all positive constants.

Let

$$\tilde{M} = \sum_{i=1}^{n} \bar{a}_i M_i s_i,$$

$$M_i = \bar{a}_i \left(\sum_{j=1}^{m} (|a_{ij}| + |b_{ij}|) r_j + |I_i| \right), \quad i \in \Lambda;$$

$$\tilde{N} = \sum_{j=1}^{m} \bar{b}_j N_j r_j,$$

$$N_j = \bar{b}_j \left(\sum_{i=1}^{n} (|m_{ji}| + |n_{ji}|) s_i + |J_j| \right), \quad j \in \Gamma.$$

Let $\Omega \subset R^{n+m}$ be a compact set in R^{n+m}. Denote the complement of Ω by $R^{n+m} \backslash \Omega$. For any

$$\begin{pmatrix} x(t) \\ y(t) \end{pmatrix} \in R^{n+m}, \ \rho\left(\begin{pmatrix} x \\ y \end{pmatrix}, \Omega \right) = \inf_{(x_1^T y_1^T) \in \Omega} \left\| \begin{pmatrix} x \\ y \end{pmatrix} - \begin{pmatrix} x_1 \\ y_1 \end{pmatrix} \right\|$$

is the distance between $\binom{x}{y}$ and Ω. We call a compact set Ω as a global attractive set of networks (1), if for every solution

$$\begin{pmatrix} x(t) \\ y(t) \end{pmatrix} \in R^{n+m} \backslash \Omega$$

with initial condition (2), we have

$$\lim_{t\to+\infty}\rho\left(\begin{pmatrix}x(t)\\y(t)\end{pmatrix},\Omega\right)=0.$$

Obviously, if the network (1) has global attractive sets, then the solutions are ultimately bounded.

Definition 1 *A compact set $\Omega\in R^{n+m}$ is said to be a global exponential stable (GES) set of system* (1)*, if there exists a constant α and a non-negative bounded continuous functional K such that for every solution $\begin{pmatrix}x(t)\\y(t)\end{pmatrix}\in R^{n+m}\backslash\Omega$ with an initial condition* (2)*, we have*

$$\rho\left(\begin{pmatrix}x(t)\\y(t)\end{pmatrix},\Omega\right)\le K(\phi,\psi)\,e^{-\alpha(t-t_0)}.$$

Definition 2 *If there exists a radially unbounded and positive definite Lyapunov function $V(t)=V(x(t),y(t))$ and positive constants l and α such that for any solution $\begin{pmatrix}x(t)\\y(t)\end{pmatrix}\in R^{n+m}\backslash\Omega$ of* (1)*, $V(t)>l$ for $t\ge t_0$ implies $V(t)-l\le(V(t_0)-l)\,e^{-\alpha(t-t_0)}$, system* (1) *is said to be GEA. The compact set*

$$\Omega=\left\{\begin{pmatrix}x\\y\end{pmatrix}\in R^{n+m}|V(t)\le l.\right\}$$

is called a GEA set of Equation (1).

Definition 3 *Network* (1) *is called GES in the Lagrange sense, if it is both uniformly bounded and GEA.*

Lemma 1 *For $\forall x,y\in R$, $a>0$, the inequality $-ax^2+xy\le-\frac{1}{2}ax^2+y^2/2a$ holds.*

Lemma 2 (Wang et al., 2010) *Let $a\ge0$, $b\ge0$, $p>1$, $q>1$ with $1/p+1/q=1$. Then the inequality $ab\le(1/p)a^p+(1/q)b^q$ holds, and the equality holds if and only if $a^p=b^q$.*

Lemma 3 (Luo et al., 2011) *Let $V(t)\in C[R^n,R^+]$ be a positive definite and radially unbounded function, and suppose there exist two constants $\alpha>0$, $\beta>0$ such that $D^+V(t)\le-\alpha V(t)+\beta$ for $t\ge t_0$, then $V(t)\ge\beta/\alpha$ implies*

$$V(t)-\frac{\beta}{\alpha}\le\left(V(t_0)-\frac{\beta}{\alpha}\right)e^{-\alpha(t-t_0)}.$$

3. Main results

Theorem 1 *If the activation functions $f(\cdot),g(\cdot)\in B$ and (H1), (H2) are also satisfied, then system* (1) *is globally exponentially stable in Lagrange sense and Ω_i for $i=1,2,3,4,5$ are all GES set and the set $\Omega=\bigcap_{i=1}^{5}\Omega_i$ is a better GES set of Equation* (1)*, where*

$$\Omega_1=\left\{\begin{pmatrix}x\\y\end{pmatrix}\in R^{n+m}\;\middle|\;\sum_{i=1}^{n}x_i^2(t)+\sum_{j=1}^{m}y_j^2(t)\le\frac{\sum_{i=1}^{n}M_i^2/\underline{a}_i\underline{c}_i+\sum_{j=1}^{m}N_j^2/\underline{b}_j\underline{d}_j}{\min_{\substack{1\le i\le n\\1\le j\le m}}\{\underline{a}_i\underline{c}_i,\underline{b}_j\underline{d}_j\}}\right\}.$$

$$\Omega_2=\left\{\begin{pmatrix}x\\y\end{pmatrix}\in R^{n+m}\;\middle|\;\sum_{i=1}^{n}|x_i(t)|+\sum_{j=1}^{m}|y_j(t)|\le\frac{\sum_{i=1}^{n}M_i+\sum_{j=1}^{m}N_j}{\min_{\substack{1\le i\le n\\1\le j\le m}}\left\{\underline{a}_i\underline{c}_i,\underline{b}_j\underline{d}_j\right\}}\right\}.$$

$$\Omega_3=\left\{\begin{pmatrix}x\\y\end{pmatrix}\in R^{n+m}\;\middle|\;\begin{array}{l}|x_i(t)|\le\dfrac{M_i}{\underline{a}_i\underline{c}_i},i\in\Lambda;\\[2ex]|y_j(t)|\le\dfrac{N_j}{\underline{b}_j\underline{d}_j},j\in\Gamma\end{array}\right\}.$$

$$\Omega_4=\left\{\begin{pmatrix}x\\y\end{pmatrix}\in R^{n+m}\;\middle|\;\begin{array}{l}\displaystyle\sum_{i=1}^{n}x_i^2(t)\le\frac{\sum_{i=1}^{n}M_i^2/\underline{a}_i\underline{c}_i}{\min_{1\le i\le n}\{\underline{a}_i\underline{c}_i\}},\\ \displaystyle\sum_{j=1}^{m}y_j^2(t)\le\frac{\sum_{j=1}^{m}N_j^2/\underline{b}_j\underline{d}_j}{\min_{1\le i\le n}\{\underline{b}_j\underline{d}_j\}}\end{array}\right\}.$$

$$\Omega_5=\left\{\begin{pmatrix}x\\y\end{pmatrix}\in R^{n+m}\;\middle|\;\begin{array}{l}\displaystyle\sum_{i=1}^{n}|x_i(t)|\le\frac{\sum_{i=1}^{n}M_i}{\min_{1\le i\le n}\{\underline{a}_i\underline{c}_i\}},\\ \displaystyle\sum_{j=1}^{m}|y_j(t)|\le\frac{\sum_{j=1}^{m}N_j}{\min_{1\le j\le m}\{\underline{b}_j\underline{d}_j\}}\end{array}\right\}.$$

Proof (1) Employ a radially unbounded and positive definite Lyapunov function as

$$V(t)=\frac{1}{2}\sum_{i=1}^{n}x_i^2(t)+\frac{1}{2}\sum_{j=1}^{m}y_j^2(t).$$

Calculating the Dini derivative of $V(t)$ along the positive semi-trajectory of Equation (1), and by virtue of Lemma 1, we obtain

$$\left.\frac{\mathrm{d}V(t)}{\mathrm{d}t}\right|_{(1)}$$

$$\le-\sum_{i=1}^{n}\underline{a}_i\underline{c}_ix_i^2(t)$$

$$+\sum_{i=1}^{n}\bar{a}_i\left(\sum_{j=1}^{m}(|a_{ij}|+|b_{ij}|)r_j+|I_i|\right)|x_i(t)|$$

$$-\sum_{j=1}^{m}\underline{b}_j\underline{d}_jy_j^2(t)$$

$$+\sum_{j=1}^{m}\bar{b}_j\left(\sum_{i=1}^{n}(|m_{ji}|+|n_{ji}|)s_i+|J_j|\right)|y_j(t)|$$
$$=-\sum_{i=1}^{n}\underline{a}_i\underline{c}_i x_i^2(t)+\sum_{i=1}^{n}M_i|x_i(t)|-\sum_{j=1}^{m}\underline{b}_j\underline{d}_j y_j^2(t)$$
$$+\sum_{j=1}^{m}N_j|y_j(t)|$$
$$\leq -\frac{1}{2}\sum_{i=1}^{n}\underline{a}_i\underline{c}_i x_i^2(t)+\frac{1}{2}\sum_{i=1}^{n}\frac{M_i^2}{\underline{a}_i\underline{c}_i}$$
$$-\frac{1}{2}\sum_{j=1}^{m}\underline{b}_j\underline{d}_j y_j^2(t)+\frac{1}{2}\sum_{j=1}^{m}\frac{N_j^2}{\underline{b}_j\underline{d}_j}$$
$$\leq -\min_{\substack{1\leq i\leq n\\1\leq j\leq m}}\{\underline{a}_i\underline{c}_i,\underline{b}_j\underline{d}_j\}V(t)+\frac{1}{2}\sum_{i=1}^{n}\frac{M_i^2}{\underline{a}_i\underline{c}_i}+\frac{1}{2}\sum_{j=1}^{m}\frac{N_j^2}{\underline{b}_j\underline{d}_j}$$
$$=-\alpha V(t)+\beta,$$

where

$$\alpha=\min_{\substack{1\leq i\leq n\\1\leq j\leq m}}\left\{\underline{a}_i\underline{c}_i,\underline{b}_j\underline{d}_j\right\},$$
$$\beta=\frac{1}{2}\sum_{i=1}^{n}\frac{M_i^2}{\underline{a}_i\underline{c}_i}+\frac{1}{2}\sum_{j=1}^{m}\frac{N_j^2}{\underline{b}_j\underline{d}_j}.$$

According to Lemma 3, for $V(t)\geq\beta/\alpha$, $V(t_0)\geq\beta/\alpha$, we have

$$V(t)-\frac{\beta}{\alpha}\leq(V(t_0)-\frac{\beta}{\alpha})\mathrm{e}^{-\alpha(t-t_0)}.$$

Hence Ω_1 is a GES set of Equation (1).

(2) Construct another positive definite and radially unbounded Lyapunov function as

$$V(t)=\sum_{i=1}^{n}|x_i(t)|+\sum_{j=1}^{m}|y_j(t)|.$$

So we can get

$$D^+V(t)|_{(1)}$$
$$\leq-\sum_{i=1}^{n}\underline{a}_i\underline{c}_i|x_i(t)|$$
$$+\sum_{i=1}^{n}\left(\bar{a}_i\sum_{j=1}^{m}(|a_{ij}t|+|b_{ij}|)r_j+|I_i|\right)$$
$$-\sum_{j=1}^{m}\underline{b}_j\underline{d}_j|y_j(t)|$$
$$+\sum_{j=1}^{m}\bar{b}_j\left(\sum_{i=1}^{n}(|m_{ji}|+|n_{ji}|)s_i+|J_j|\right)$$
$$\leq-\alpha V(t)+\beta,$$

where

$$\alpha=\min_{\substack{1\leq i\leq n\\1\leq j\leq m}}\{\underline{a}_i\underline{c}_i,\underline{b}_j\underline{d}_j\},\quad \beta=\sum_{i=1}^{n}M_i+\sum_{j=1}^{m}N_j.$$

So we get

$$V(t)-\frac{\beta}{\alpha}\leq\left(V(t_0)-\frac{\beta}{\alpha}\right)\mathrm{e}^{-\alpha(t-t_0)}.$$

And the set Ω_2 is a GES set of Equation (1).

(3) Choose another two positive definite and radially unbounded Lyapunov functions as

$$V_i(t)=|x_i(t)|,\ i\in\Lambda;\quad V_j(t)=|y_j(t)|,\ j\in\Gamma.$$

And we have

$$D^+V_i(t)|_{(1)}\leq-\underline{a}_i\underline{c}_iV_i(t)+M_i,\quad i\in\Lambda,$$
$$D^+V_j(t)|_{(1)}\leq-\underline{b}_j\underline{d}_jV_j(t)+N_j,\quad j\in\Gamma.$$

So we have

$$V_i(t)-\frac{M_i}{\underline{a}_i\underline{c}_i}\leq\left(V_i(t_0)-\frac{M_i}{\underline{a}_i\underline{c}_i}\right)\mathrm{e}^{-\underline{a}_i\underline{c}_i(t-t_0)},\ i\in\Lambda,$$
$$V_j(t)-\frac{N_j}{\underline{b}_j\underline{d}_j}\leq\left(V_j(t_0)-\frac{N_j}{\underline{b}_j\underline{d}_j}\right)\mathrm{e}^{-\underline{b}_j\underline{d}_j(t-t_0)},\ j\in\Gamma.$$

So the Ω_3 is a GES set of Equation (1).

(4) Employ only the following two radially unbounded and positive definite Lyapunov functions as

$$V_x(t)=\frac{1}{2}\sum_{i=1}^{n}x_i^2(t),\quad V_y(t)=\frac{1}{2}\sum_{j=1}^{m}y_j^2(t).$$

The remaining proof is similar to the proof in the previous part (1). Meanwhile, consider only the following other two Lyapunov functions

$$V_x(t)=\sum_{i=1}^{n}|x_i(t)|,\quad V_y(t)=\sum_{j=1}^{m}|y_j(t)|.$$

The remaining proof is similar to that in the previous part (2). So the sets Ω_4 and Ω_5 are also GES sets of (1). According to the definition of intersection set, we know that the set $\Omega=\bigcap_{i=1}^{5}\Omega_i$ is a better GES set of NN (1). The proof of Theorem1 is completed.

Remark 3 When $a_i(x_i(t))=1, b_j(y_j(t))=1$ for $i\in\Lambda$ and $j\in\Gamma$, $c_i(x_i(t))=\tilde{a}_ix_i(t)$ and $d_j(y_j(t))=\tilde{c}_jy_j(t)$ with the constants $\tilde{a}_i > 0$ and $\tilde{c}_j > 0$, the set Ω_5 in Theorem 1 here is just the main result (I) of Theorem 3.2 in Tu et al. (2013).

THEOREM 2 *Let $p > 1$, $q > 1$ and $1/p + 1/q = 1$. Choose $\varepsilon_i > 0$, $\bar{\varepsilon}_j > 0 (i \in \Lambda, j \in \Gamma)$ such that $\mu_i = p\underline{a}_i\underline{c}_i - (p-1)\varepsilon_i > 0$, $\eta_j = q\underline{b}_j\underline{d}_j - (q-1)\bar{\varepsilon}_j > 0$. If the activation functions $f(\cdot), g(\cdot) \in B$ and (H1), (H2) are also satisfied, then NN (1) is globally exponentially stable in Lagrange sense and Ω_6 is a GES set, where*

$$\Omega_6 = \left\{ \begin{pmatrix} x \\ y \end{pmatrix} \in R^{n+m} \;\middle|\; \begin{array}{l} \frac{1}{p}\sum_{i=1}^{n} |x_i(t)|^p + \frac{1}{q}\sum_{j=1}^{m} |y_j(t)|^q \\ \leq \dfrac{\sum_{i=1}^n M_i^p / p\varepsilon_i^{p-1} + \sum_{j=1}^m N_j^q / q\bar{\varepsilon}_j^{q-1}}{\min_{1\leq i\leq n, 1\leq j\leq m} \{\mu_i, \eta_j\}} \end{array} \right\}.$$

Proof We introduce the following Lyapunov function

$$V(t) = \frac{1}{p}\sum_{i=1}^{n} |x_i(t)|^p + \frac{1}{q}\sum_{j=1}^{m} |y_j(t)|^q.$$

Calculating the Dini derivative of $V(t)$ along (1), and by virtue of Lemma 2, we can obtain

$$\begin{aligned}
&D^+V(t)|_{(1)} \\
&\leq -\sum_{i=1}^{n} \underline{a}_i\underline{c}_i |x_i(t)|^p + \sum_{i=1}^{n} M_i |x_i(t)|^{p-1} \\
&\quad - \sum_{j=1}^{m} \underline{b}_j\underline{d}_j |y_j(t)|^q + \sum_{j=1}^{m} N_j |y_j(t)|^{q-1} \\
&\leq -\sum_{i=1}^{n} \underline{a}_i\underline{c}_i |x_i(t)|^p + \sum_{i=1}^{n} \left(\frac{p-1}{p}\varepsilon_i |x_i(t)|^p + \frac{1}{p\varepsilon_i^{p-1}} M_i^p \right) \\
&\quad - \sum_{j=1}^{m} \underline{b}_j\underline{d}_j |y_j(t)|^q + \sum_{j=1}^{m} \left(\frac{q-1}{q}\bar{\varepsilon}_j |y_j(t)|^q + \frac{1}{q\bar{\varepsilon}_j^{q-1}} N_j^q \right) \\
&\leq -\sum_{i=1}^{n} \left(\underline{a}_i\underline{c}_i - \frac{p-1}{p}\varepsilon_i \right) |x_i(t)|^p + \sum_{i=1}^{n} \frac{1}{p\varepsilon_i^{p-1}} M_i^p \\
&\quad - \sum_{j=1}^{m} \left(\underline{b}_j\underline{d}_j - \frac{q-1}{q}\bar{\varepsilon}_j \right) |y_j(t)|^q + \sum_{j=1}^{m} \frac{1}{q\bar{\varepsilon}_j^{q-1}} N_j^q \\
&\leq -\alpha V(t) + \beta,
\end{aligned}$$

where

$$\alpha = \min_{\substack{1\leq i\leq n \\ 1\leq j\leq m}} \{\mu_i, \eta_j\}, \quad \beta = \sum_{i=1}^{n} \frac{M_i^p}{p\varepsilon_i^{p-1}} + \sum_{j=1}^{m} \frac{N_j^q}{q\bar{\varepsilon}_j^{q-1}}.$$

And by Lemma 3, we get

$$V(t) - \frac{\beta}{\alpha} \leq \left(V(t_0) - \frac{\beta}{\alpha} \right) e^{-\alpha(t-t_0)}.$$

So the set Ω_6 is a GES set of Equation (1).

Remark 4 When $a_i(x_i(t)) = 1, b_j(y_j(t)) = 1$ for $i \in \Lambda$ and $j \in \Gamma$, $c_i(x_i(t)) = \tilde{a}_i x_i(t)$ and $d_j(y_j(t)) = \tilde{c}_j y_j(t)$ with the constants $\tilde{a}_i > 0$ and $\tilde{c}_j > 0$ the sets Ω_6 in Theorem 2 here are just the main result of Theorem 3.1 in Tu et al. (2013).

THEOREM 3 *If the activation functions $f(\cdot), g(\cdot) \in S$ and (H1), (H2) are also satisfied, then NN (1) has positive invariant and globally exponential attractive sets*

$$\Omega_7 = \left\{ \begin{pmatrix} x \\ y \end{pmatrix} \in R^{n+m} \;\middle|\; \begin{array}{l} \sum_{i=1}^{n} \int_0^{x_i(t)} g_i(s)\,\mathrm{d}s \\ + \sum_{j=1}^{m} \int_0^{y_j(t)} f_j(\eta)\,\mathrm{d}\eta \\ \leq \dfrac{\tilde{M} + \tilde{N}}{\min_{1\leq i\leq n, 1\leq j\leq m} \{\underline{a}_i\underline{c}_i, \underline{b}_j\underline{d}_j\}} \end{array} \right\},$$

$$\Omega_8 = \left\{ \begin{pmatrix} x \\ y \end{pmatrix} \in R^{n+m} \;\middle|\; \begin{array}{l} \sum_{i=1}^{n} \int_0^{x_i(t)} g_i(s)\,\mathrm{d}s \leq \dfrac{\tilde{M}}{\min_{1\leq i\leq n}\{\underline{a}_i\underline{c}_i\}}, \\ \sum_{j=1}^{m} \int_0^{y_j(t)} f_j(\eta)\,\mathrm{d}\eta \leq \dfrac{\tilde{N}}{\min_{1\leq j\leq m}\{\underline{b}_j\underline{d}_j\}} \end{array} \right\}.$$

And $\Omega = \Omega_7 \cap \Omega_8$ is a better GES set of Equation (1).

Proof Firstly, employ the following Lyapunov function

$$V(t) = \sum_{i=1}^{n} \int_0^{x_i(t)} g_i(s)\,\mathrm{d}s + \sum_{j=1}^{m} \int_0^{y_j(t)} f_j(\eta)\,\mathrm{d}\eta.$$

Calculating the derivative of $V(t)$, we have

$$\begin{aligned}
\left.\frac{\mathrm{d}V(t)}{\mathrm{d}t}\right|_{(1)}
&= \sum_{i=1}^{n} g_i(x_i(t))\dot{x}_i(t) + \sum_{j=1}^{m} f_j(y_j(t))\dot{y}_j(t) \\
&\leq -\sum_{i=1}^{n} \underline{a}_i\underline{c}_i x_i(t) g_i(x_i(t)) + \tilde{M} \\
&\quad - \sum_{j=1}^{m} \underline{b}_j\underline{d}_j y_j(t) f_j(y_j(t)) + \tilde{N} \\
&\leq -\min_{\substack{1\leq i\leq n \\ 1\leq j\leq m}} \{\underline{a}_i\underline{c}_i, \underline{b}_j\underline{d}_j\} V(t) + \tilde{M} + \tilde{N} = -\alpha V(t) + \beta,
\end{aligned}$$

where $\alpha = \min_{\substack{1\le i\le n \\ 1\le j\le m}} \{\underline{a}_i\underline{c}_i, \underline{b}_j\underline{d}_j\}$, $\beta = \tilde{M} + \tilde{N}$. In the light of Lemma 3, we get

$$V(t) - \frac{\beta}{\alpha} \le \left(V(t_0) - \frac{\beta}{\alpha}\right) e^{-\alpha(t-t_0)}.$$

So the set Ω_7 is a GES set of Equation (1).

Secondly, consider the following Lyapunov functions

$$V_1(t) = \sum_{i=1}^{n} \int_0^{x_i(t)} g_i(s)\,ds, \quad V_2(t) = \sum_{j=1}^{m} \int_0^{y_j(t)} f_j(\eta)\,d\eta.$$

Similar to the proof in the previous part, we can obtain that the set Ω_8 is a GES set of Equation (1). Hence, $\Omega = \Omega_7 \cap \Omega_8$ is a better GES set of neural network (1).

4. Illustrative examples

In this section, we will give an example to verify our theoretical results.

Example 4.1 Consider the following example:

$$\begin{cases} \dot{x}_i(t) &= a_i(x_i(t)) \begin{bmatrix} -c_i(x_i(t)) + \sum_{j=1}^{2} a_{ij} f_j(y_j(t)) \\ + \sum_{j=1}^{2} b_{ij} f_j(y_j(t-\tau_j(t))) + I_i \end{bmatrix}, \\ & i = 1,2, \\ \dot{y}_j(t) &= b_j(y_j(t)) \begin{bmatrix} -d_j(y_j(t)) + \sum_{i=1}^{2} m_{ji} g_i(x_i(t)) \\ + \sum_{i=1}^{2} n_{ji} g_i(x_i(t-\sigma_i(t))) + J_j \end{bmatrix}, \\ & j = 1,2, \end{cases} \tag{3}$$

where $a_i(x_i(t)) = 2 + \cos x_i(t)$, $c_i(x_i(t)) = 3x_i(t)$, $g_i(x_i) = 2x_i/(1+x_i^2)$; $b_j(y_j(t)) = 2 + \sin y_j(t)$, $d_j(y_j(t)) = 3y_j(t)$, $f(y_j) = \frac{1}{2}(|y_j+1| - |y_j-1|)$. Let $A = \begin{pmatrix} 1 & 2 \\ 2 & 1 \end{pmatrix}$, $B = \begin{pmatrix} 1 & 0 \\ 0 & 1 \end{pmatrix}$, $M = \begin{pmatrix} 2 & 1 \\ 1 & 0 \end{pmatrix}$, $N = \begin{pmatrix} 2 & 0 \\ 0 & 1 \end{pmatrix}$, $I = (1\ 2)^T$, $J = (2\ 1)^T$. So $\underline{a}_i = \underline{b}_j = 1$, $\bar{a}_i = \bar{b}_j = 3$, $\underline{c}_i = \bar{c}_i = \underline{d}_j = \bar{d}_j = 3$, $s_i = r_j = 1$, $M_1 = 15, M_2 = 18, N_1 = 21, N_2 = 9$. Since $f(\cdot), g(\cdot) \in B$, according to Theorem 1, the neural network model (3) has positive invariant and globally exponential attractive sets as follows:

$$\Omega_1 = \left\{ \begin{pmatrix} x \\ y \end{pmatrix} \in R^4 \,\middle|\, x_1^2(t) + x_2^2(t) + y_1^2(t) + y_2^2(t) \le 119 \right\},$$

$$\Omega_2 = \left\{ \begin{pmatrix} x \\ y \end{pmatrix} \in R^4 \,\middle|\, |x_1(t)| + |x_2(t)| + |y_1(t)| + |y_2(t)| \le 21 \right\}.$$

Meanwhile, let the initial conditions $x_1(t) = 0.7 + y_2(t), x_2(t) = 1 + y_2(t), y_1(t) = 1.2 + y_2(t), y_2(t) = 0.9 + 0.5\sin 2t$, and the delays $\tau_1 = \tau_2 = 100 - \sin t$,

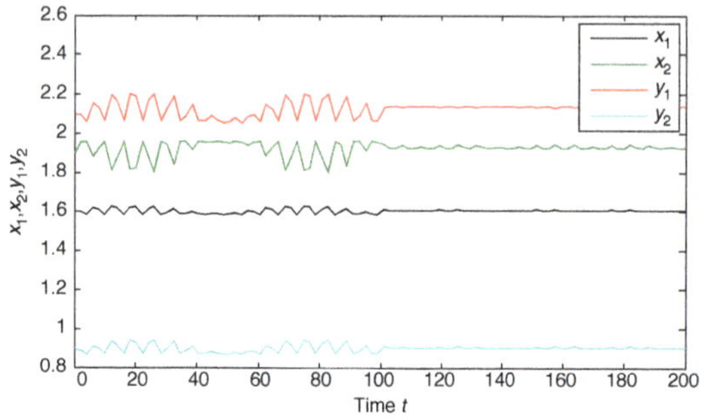

Figure 1. Time response of states $x_1(t)$, $x_2(t)$, $y_1(t)$ and $y_2(t)$ of Equation (3).

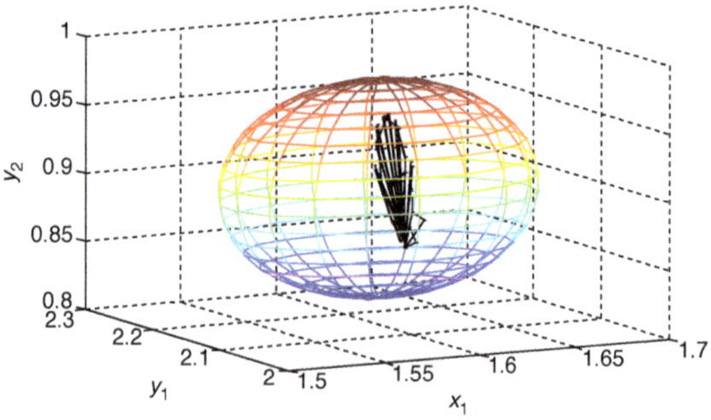

Figure 2. The ultimate bound of Equation (3) in coordinate system (x_1, y_1, y_2).

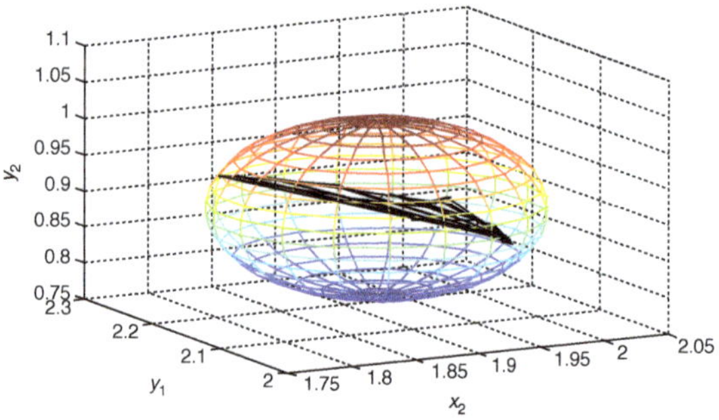

Figure 3. The ultimate bound of Equation (3) in coordinate system (x_2, y_1, y_2).

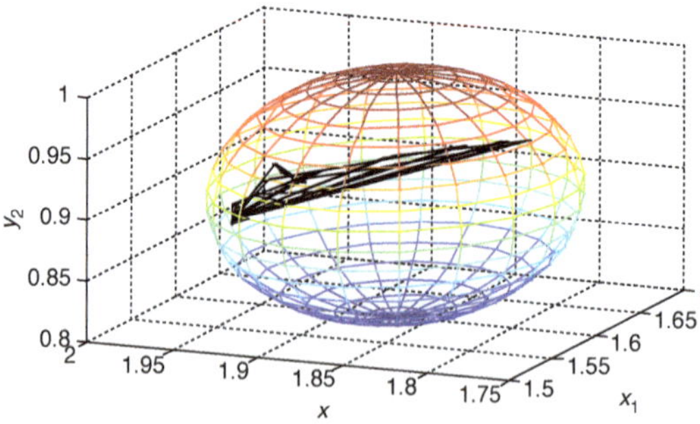

Figure 4. The ultimate bound of Equation (3) in coordinate system (x_1, x_2, y_2).

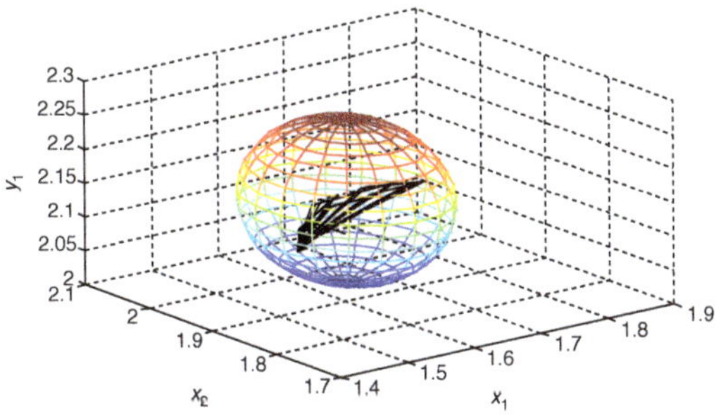

Figure 5. The ultimate bound of Equation (3) in coordinate system (x_1, x_2, y_1).

$\sigma_1 = \sigma_2 = 100 - \sin t$. Figure 1 shows time response of states $x_1(t)$, $x_2(t)$, $y_1(t)$ and $y_2(t)$. Figures 2–5 show the estimations of the ultimate bound of system (3) in the three-dimensional phase space, respectively.

5. Conclusions

Based on the Lyapunov stability theory and some inequalities, this paper has derived some sufficient delay-independent conditions of positive invariant set and globally exponential attractive set for the BAM-type Cohen–Grossberg neural networks with time-varying delays. According to the parameters, the detailed estimations for the positive invariant and globally attractive set of the BAM-type Cohen–Grossberg neural networks have been established without any hypothesis on the existence. Meanwhile, the results obtained in this paper are more general than that of the existing references (Tu et al., 2013) on the GEA set as special cases. Moreover, the proposed methods here can be also applied to nonlinear discrete-time systems with time-varying delays such as that in Dong, Wang, and Gao (2013) and Hu, Wang, Niu, and Stergioulas (2012). Finally, an illustrative example is shown to verify our results.

Acknowledgements

The authors are grateful for the support of the National Natural Science Foundation of China (61174216) and (61273183), the Scientific Innovation Team Project of Hubei Provincial Department of Education (T200809).

Disclosure statement

No potential conflict of interest was reported by the author(s).

References

Dong, H. L., Wang, Z. D., & Gao, H. J. (2013, October). Distributed H_∞ filtering for a class of Markovian jump nonlinear time-delay systems over lossy sensor networks. *IEEE Transactions on Industrial Electronics*, *60*, 4665–4672.

Hu, J., Wang, Z. D., Niu, Y. G., & Stergioulas, L. K. (2012, November). H_∞ sliding mode observer design for a class of nonlinear discrete time-delay systems: A delay-fractioning approach. *International Journal of Robust and Nonliner Control*, *22*, 1806–1826.

Jiang, H. J., & Cao, J. D. (2008, January). BAM-type Cohen–Grossberg neural networks with time delays. *Mathematical and Computer Modelling*, *47*, 92–103.

Li, T., Fei, S. M., Tan, M. C., & Zhang, Y. N. (2009, August). New sufficient conditions for global asymptotic stability of Cohen–Grossberg neural networks with time-varying delays. *Nonlinear Analysis: Real World Applications*, *10*, 2139–2145.

Li, K. L., Zhang, L. P., Zhang, X. H., & Li, Z. A. (2010, February). Stability in impulsive Cohen–Grossberg-type BAM neural networks with distributed delays. *Applied Mathematics and Computation*, *215*, 3970–3984.

Li, X. D. (2009, September). Existence and global exponential stability of periodic solution for impulsive Cohen–Grossberg-type BAM neural networks with continuously distributed delays. *Applied Mathematics and Computation*, *215*, 292–307.

Liao, X. X., Luo, Q., & Zeng, Z. G. (2008, January). Positive invariant and global exponential attractive sets of neural-networks with time-varying delays. *Neurocomputing*, *71*, 513–518.

Liu, J., & Zong, G. D. (2009, June). New delay-dependent asymptotic stability conditions concerning BAM neural networks of neutral type. *Neurocomputing*, *72*, 2549–2555.

Lu, W. L., Wang, L. L., & Chen, T. P. (2011, March). On attracting basins of multiple equilibria of a class of cellular neural networks. *IEEE Transactions on Neural Networks*, *22*, 381–394.

Luo, Q., Zeng, Z. G., & Liao, X. X. (2011, January). Global exponential stability in Lagrange sense for neutral type recurrent neural networks. *Neurocomputing*, *74*, 638–645.

Song, Q. K., & Zhao, Z. J. (2005, July). Global dissipativity of neural networks with both variable and unbounded delays. *Chaos, Solitons and Fractals*, *25*, 393–401.

Tu, Z. W., Jian, J. G., & Wang, W. W. (2011, March). *Global dissipativity for Cohen–Grossberg neural networks with both time-varying delays and infinite distributed delays.* International Conference on Information Science and Technology, Nanjing, 982–985.

Tu, Z. W., Wang, L. W., Zha, Z. W., & Jian, J. G. (2013, September). Global dissipativity of a class of BAM neural networks with time-varying and unbound delays. *Communications in Nonlinear Science and Numerical Simulation*, *18*, 2562–2570.

Wang, B. X., Jian, J. G., & Guo, C. D. (2008, January). Global exponential stability of a class of BAM networks with time-varying delays and continuously distributed delays. *Neurocomputing*, *71*, 495–501.

Wang, B. X., Jian, J. G., & Jiang, M. H. (2010, February). Stability in Lagrange sense for Cohen–Grossberg neural networks with time-varying delays and finite distributed delays. *Nonlinear Analysis: Hybrid Systems*, *4*, 65–78.

Wang, L. L., & Chen, T. P. (2012, December). Complete stability of cellular neural networks with unbounded time-varying delays. *Neural Networks*, *36*, 11–17.

Zhang, Z. Q., Liu, W. B., & Zhou, D. M. (2012, January). T Global asymptotic stability to a generalized Cohen–Grossberg BAM neuralnetworks of neutral type delays. *Journal of Neural Networks*, *25*, 94–105.

9

The paradigm of complex probability and the Brownian motion

Abdo Abou Jaoude*

Department of Mathematics and Statistics, Faculty of Natural and Applied Sciences, Notre Dame University – Louaize, Zouk Mosbeh, Lebanon

Andrey N. Kolmogorov's system of axioms can be extended to encompass the imaginary set of numbers and this by adding to his original five axioms an additional three axioms. Hence, any experiment can thus be executed in what is now the complex probability set $\mathscr{C}$ which is the sum of the real set $\mathscr{R}$ with its corresponding real probability, and the imaginary set $\mathscr{M}$ with its corresponding imaginary probability. The objective here is to evaluate the complex probabilities by considering supplementary new imaginary dimensions to the event occurring in the 'real' laboratory. Whatever the probability distribution of the input random variable in $\mathscr{R}$ is, the corresponding probability in the whole set $\mathscr{C}$ is always one, so the outcome of the random experiment in $\mathscr{C}$ can be predicted totally. The result indicates that chance and luck in $\mathscr{R}$ is replaced now by total determinism in $\mathscr{C}$. This is the consequence of the fact that the probability in $\mathscr{C}$ is got by subtracting the chaotic factor from the degree of our knowledge of the stochastic system. This novel complex probability paradigm will be applied to the classical theory of Brownian motion and to prove as well the law of large numbers in an original way.

Keywords: extended Kolmogorov's axioms; complex set; probability norm; degree of our knowledge; chaotic factor; Gauss–Laplace distribution; diffusion; entropy; resultant complex random vector

Nomenclature

$\mathscr{R}$	real set of events
$\mathscr{M}$	imaginary set of events
$\mathscr{C}$	complex set of events
i	the imaginary number, where $i^2 = -1$
EKA	extended Kolmogorov's axioms
P_{rob}	probability of any event
P_r	probability in the real set $\mathscr{R}$
P_m	probability in the imaginary set $\mathscr{M}$ corresponding to the real probability in $\mathscr{R}$
Pc	probability of an event in $\mathscr{R}$ with its associated event in $\mathscr{M}$ probability in the complex set $\mathscr{C}$
z	complex number = sum of P_r and P_m = complex random vector
DOK	$\lvert z\rvert^2$ = the degree of our knowledge of the random experiment it is the square of the norm of z.
Chf	the chaotic factor of z
MChf	magnitude of the chaotic factor of z
N	number of random variables
Z	the resultant complex random vector
$\mathrm{DOK}_Z = \frac{\lvert Z\rvert^2}{N^2}$	the degree of our knowledge of Z
$\mathrm{Chf}_Z = \frac{\mathrm{Chf}}{N^2}$	the chaotic factor of Z
MChf_Z	magnitude of the chaotic factor of Z
S_R	entropy in $\mathscr{R}$
$\bar{S}_R$	entropy of the complementary real probability set to $\mathscr{R}$
S_M	entropy in $\mathscr{M}$
S_C	entropy in $\mathscr{C}$

1. Introduction

The terms 'pedesis' (πήδησις = pĕ : dε : sis = 'leaping') in Greek or 'Brownian motion' in English, both refer to the random motion of particles suspended in a liquid or a gas. This motion is the result of the particles collision with the rapid atoms or molecules in the fluid. Moreover, the term Brownian motion can also refer to the mathematical model used to describe such random movements. This model is usually called a 'particle theory' (Kuhn, 1970).

This phenomenon, also called a 'transport phenomenon', is named after the botanist Robert Brown. In the nineteenth century and precisely in1827, while Brown was looking through a microscope at particles found in pollen grains in water, he noticed that the particles moved through the water but was not able to determine the mechanisms that caused this motion. It had long been theorized that atoms and molecules are the constituents of matter, and many decades later (precisely in 1905), Albert Einstein published a paper that explained in accurate detail the

*Email: abdoaj@idm.net.lb

Brownian motion. In fact the motion that Brown had perceived was the consequence of the pollen being moved by individual water molecules. Consequently, this explanation of Brownian motion yielded a decisive confirmation of the actual existence of atoms and molecules. Furthermore, in 1908, Jean Perrin experimentally verified the theory. As a result, in 1926, the Nobel Prize in Physics was attributed to Perrin 'for his work on the discontinuous structure of matter', whereas five years earlier, Einstein had received the prize 'for his services to theoretical physics' with specific citation of different research. In fact, the seemingly random nature of the Brownian motion was due to the constantly changing direction of the force of atomic bombardment and therefore the particle is hit more on one side than another at different times (Poincaré, 1968; Stewart, 1996).

Additionally, there are various real-world applications to the Brownian motion mathematical model. As an example, we frequently cite the stock market fluctuations, even though Benoit Mandelbrot rejected to apply it to the movements of stock price in part due to the discontinuous nature of these movements (Stewart, 2012).

We consider Brownian motion among the simplest of the continuous-time stochastic processes. In fact, it is a limit of both simpler and more complicated stochastic processes (random walk and Donsker's theorem). In both cases, Brownian motion is usually more convenient rather than more accurate of the models. Consequently, this motivates the use of the Brownian probabilistic model (Barrow, 1992; Warusfel & Ducrocq, 2004).

Around 60 BC, in his scientific poem 'On the Nature of Things', the Roman Lucretius described remarkably in the verses 113–140 from Book II, the Brownian motion of dust particles. In fact, Lucretius considered this motion as an inevitable proof of the existence of atoms. Even though air currents largely cause the interspersing motion of dust particles, the tumbling motion of small dust particles is certainly and primarily due to the true Brownian dynamics (Greene, 2003; Hawking, 2005).

Moreover, in the eighteenth century, precisely in 1785, Jan Ingenhousz had described the irregular motion of coal dust particles on the surface of alcohol. But the discovery is usually credited to the botanist Robert Brown in 1827. In fact, while studying pollen grains of the plant 'Clarkia pulchella' suspended in water under a microscope, Brown noticed tiny particles which were ejected by the pollen grains and performing a jittery motion. Although the origin of the motion was yet to be explained, Brown repeated the experiment with particles of inorganic matter and concluded that the motion was not life-related (Bogdanov & Bogdanov, 2013; Penrose, 1999).

In 1880, in a paper on the method of least squares, Thorvald N. Thiele was the first person to describe mathematically the Brownian motion. In 1900, Louis Bachelier, and independently, followed Thiele and explained the motion in his Ph.D. thesis entitled 'The theory of speculation' and hence presented a stochastic analysis of the stock market and option markets. In 1905, Albert Einstein's miraculous year, and Marian Smoluchowski in 1906, using Brownian motion, they confirmed indirectly the existence of atoms and molecules. Both scientists' equations that described Brownian motion were verified experimentally in 1908 by Jean Perrin (Bogdanov & Bogdanov, 2012).

Moreover, let us consider the Brownian motion of pollen grain travelling randomly in water and emitting erratic particles. It is well known that a water molecule has the size of about 0.1×0.2 nm; however, Robert Brown observed particles of the order of a few micrometers in size. Notice that they are not to be confused with the actual pollen particle which has a size of about 100 micrometers. The instantaneous imbalance in the combined forces exerted by the collisions of the particle with the much smaller liquid molecules which are in random thermal motion yields the Brownian motion of the surrounded particle in a liquid (Bogdanov & Bogdanov, 2010).

The time evolution of the probability density function (PDF) associated with the position of the particle in Brownian motion is approximated by the diffusion equation, knowing that this approximation is valid on short timescales. In addition, Langevin equation which contains a random force field representing the effect of the thermal fluctuations of the solvent on the particle is the best description of the time evolution of the position of the Brownian particle itself (Bogdanov & Bogdanov, 2009).

Furthermore, by solving the diffusion equation under appropriate boundary conditions and hence by finding its solution, we acquire the displacement of a particle in Brownian motion. Consequently, the solution indicates that the displacement varies as the square root of the time and not linearly. This leads to the explanation why the results concerning the Brownian particles gave nonsensical results. In fact, the assumption of a linear time dependence was incorrect (Davies, 1993).

Additionally, the particle motion is governed by its inertia and its displacement will be linearly dependent on time, and this at very short-time scales. Hence, we obtained $\Delta x = v\Delta t$, and the instantaneous velocity of the Brownian motion is expressed by $v = \Delta x/\Delta t$, when $\Delta t \ll \tau$, where τ is the momentum relaxation time. For a glass microsphere trapped in air with an optical tweezer, we succeeded in 2010 to measure the instantaneous velocity of a Brownian particle. This important experiment, verified the distribution of Maxwell–Boltzmann, and the Brownian particle equipartition theorem (Wikipedia, *Brownian Motion*; Wikipedia, *Entropy*).

Also, a random walk can model the Brownian motion. We must mention that random walks in porous media or fractals are anomalous. In the general case, Brownian motion is considered as a non-Markov random process

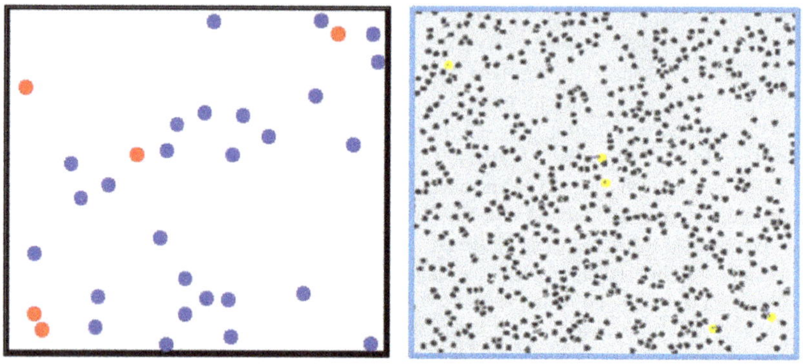

Figure 1. (a) and (b) This is an example of the Brownian motion of five particles (red at left and yellow at right) that collide with a large set of 800 particles (blue at left and black at right) (Wikipedia, *Brownian Motion*).

and can be expressed by stochastic integral equations (Figure 1(a) and 1(b)) (Wikipedia, *Mass Diffusivity*).

2. Albert Einstein's contribution

Einstein's theory is divided into two parts (Aczel, 2000; Balibar, 2002; Hawking, 2002, 2011; Hoffmann, 1975; Pickover, 2008; Reeves, 1988; Ronan, 1988; Wikipedia, *Brownian Motion*; Wikipedia, *Entropy*; Wikipedia, *Mass Diffusivity*):

(1) The first part consists in the formulation of the Brownian particles diffusion equation. The diffusion coefficient is linked to the mean-squared displacement of a Brownian particle.

(2) The second part consists in connecting the diffusion coefficient to measurable physical quantities.

Consequently, Einstein succeeded in determining the size of atoms, the number of atoms in a mole, or the weight of molecules in grams of a gas. As fixed by Avogadro's law, this volume is the same for all ideal gases, and which is 22.414 L at standard temperature and pressure. It is well known that the number of atoms contained in this volume is denoted by Avogadro's number. This leads to determining the mass of an atom which can be computed by dividing the mass of a mole of the gas by Avogadro's number.

In the first part of Einstein's theory, we are able to determine the distance that a Brownian particle travels in a given interval of time. Knowing that due to the enormous number of collisions that a Brownian particle undergoes (approximately of the order of 10^{21} collisions per second), classical mechanics is incapable to determine this distance. Consequently, Einstein considered the Brownian particles collective motion.

Moreover, the type of dynamical equilibrium suggested by Einstein was not new. In fact, in his series of lectures at Yale University in May 1903, J. J. Thomson proposed that

> the dynamic equilibrium between the velocity generated by a concentration gradient given by Fick's law and the velocity due to the variation of the partial pressure due to the ions set in motion gives us a method of determining Avogadro's Constant which is independent of any hypothesis as to the shape or size of molecules, or of the way in which they act upon each other. (Wikipedia, *Brownian Motion*)

Furthermore, in 1888, Walter Nernst found a similar expression to Einstein's formula for the diffusion coefficient. In fact, he formulated that the diffusion coefficient is the ratio of the osmotic pressure to the ratio of the frictional force and the velocity to which it gives rise. The former was equated to the law of Van't Hoff, while the latter was given by Stokes's law. Nernst wrote

$$k' = \frac{p_0}{k},$$

where k' is the diffusion coefficient, p_0 is the osmotic pressure, and k is the ratio of the frictional force to the molecular viscosity which he assumes is given by Stokes's formula for the viscosity.

> By introducing the ideal gas law per unit volume for the osmotic pressure, Nernst formula is consequently identical to Einstein's formula. The use of Stokes's law in Nernst's case, as well as in Einstein and Smoluchowski, is not strictly applicable since it does not apply to the case where the radius of the sphere is small in comparison with the mean free path. (Wikipedia, *Brownian Motion*)

The series of experiments conducted in 1906 and 1907 by Svedberg, refuted seemingly at first Einstein's formula predictions. These experiments gave the displacements value as 4–6 times Einstein's value. In 1908, Henri's experiments yielded displacements three times greater than those predicted by Einstein's formula. Finally, in 1908 and in 1909, Chaudesaigues and Perrin, respectively, confirmed in a series of experiments Einstein's predictions.

> The confirmation of Einstein's theory constituted empirical progress for the kinetic theory of heat. In essence, Einstein showed that the motion can be predicted directly from the kinetic model of thermal equilibrium. The importance of the theory lays in the fact that it confirmed the kinetic

theory's account of the second law of thermodynamics as being an essentially statistical law. (Wikipedia, *Brownian Motion*)

3. The purpose and the advantages of the present work

All our work in classical probability theory is to compute probabilities. The original idea in this paper is to add new dimensions to our random experiment and this will make the work deterministic. In fact, the probability theory is a nondeterministic theory by nature, meaning that the outcome of the events is due to chance and luck. By adding new dimensions to the event in $\mathcal{R}$, we make the work deterministic and hence a random experiment will have a certain outcome in the complex set of probabilities $\mathcal{C}$. It is of great importance that the stochastic system becomes totally predictable since we will be totally knowledgeable to foretell the outcome of chaotic and random events that occur in nature like, for example, in statistical mechanics or in all stochastic processes. Therefore, the work that should be done is to add to the real set of probabilities $\mathcal{R}$, the contributions of $\mathcal{M}$ which is the imaginary set of probabilities, that makes the event in $\mathcal{C} = \mathcal{R} + \mathcal{M}$ deterministic. If this is found to be fruitful, then a new theory in statistical sciences and prognostic is elaborated and this to understand deterministically those phenomena that used to be random phenomena in $\mathcal{R}$. This is what I called 'the complex probability paradigm' that was initiated and elaborated in my five previous papers (Abou Jaoude, 2013a, 2013b, 2014, 2015; Abou Jaoude, El-Tawil, & Kadry, 2010).

Consequently, the purpose and the advantages of the present work are to

(1) Extend classical probability theory to the set of complex numbers, hence to relate probability theory to the field of complex analysis. This task was initiated and elaborated in my five previous papers.
(2) Apply the new probability axioms and paradigm to the Brownian motion.
(3) Prove that all random phenomena can be expressed deterministically in the complex set $\mathcal{C}$.
(4) Quantify both the *d*egree of *o*ur *k*nowledge (DOK for short) and the chaos of the random walk of particles.
(5) Draw and represent graphically the functions and parameters of the novel paradigm associated with the Brownian motion.
(6) Show that the classical concept of entropy is always equal to zero in the complex set; hence, no chaos and disorder exist in $\mathcal{C}$ (complex set) $= \mathcal{R}$ (real set) $+ \mathcal{M}$ (imaginary set).
(7) Prove the very well-known law of large numbers using the newly defined axioms.
(8) Pave the way to apply the original paradigm to other topics in statistical mechanics, in stochastic processes, and to the field of prognostics in engineering. These will be the subjects of my subsequent research papers.

3.1. *The original Andrey N. Kolmogorov set of axioms*

The simplicity of Kolmogorov's system of axioms may be surprising. Let E be a collection of elements $\{E_1, E_2, \ldots\}$ called elementary events and let F be a set of subsets of E called random events. The five axioms for a finite set E are (Benton, 1966a, 1966b; Feller, 1968; Montgomery & Runger, 2003; Walpole, Myers, Myers, & Ye, 2002) as follows:

Axiom 1: F is a field of sets.
Axiom 2: F contains the set E.
Axiom 3: A non-negative real number $P_{\text{rob}}(A)$, called the probability of A, is assigned to each set A in F. We have always $0 \leq P_{\text{rob}}(A) \leq 1$.
Axiom 4: $P_{\text{rob}}(E)$ equals 1.
Axiom 5: If A and B have no elements in common, the number assigned to their union is as follows:

$$P_{\text{rob}}(A \cup B) = P_{\text{rob}}(A) + P_{\text{rob}}(B).$$

Hence, we say that A and B are disjoint; otherwise, we have:

$$P_{\text{rob}}(A \cup B) = P_{\text{rob}}(A) + P_{\text{rob}}(B) - P_{\text{rob}}(A \cap B).$$

And we say also that $P_{\text{rob}}(A \cap B) = P_{\text{rob}}(A) \times P_{\text{rob}}(B|A) = P_{\text{rob}}(B) \times P_{\text{rob}}(A|B)$ which is the conditional probability. If both A and B are independent, then $P_{\text{rob}}(A \cap B) = P_{\text{rob}}(A) \times P_{\text{rob}}(B)$.

In addition, we can generalize and say that, for N disjoint events $A_1, A_2, \ldots, A_j, \ldots, A_N$ (for $1 \leq j \leq N$), we have

$$P_{\text{rob}}\left(\bigcup_{j=1}^{N} A_j\right) = \sum_{j=1}^{N} P_{\text{rob}}(A_j).$$

3.2. *Adding the imaginary part $\mathcal{M}$*

Now, we can add to this system of axioms an imaginary part such that

Axiom 6: Let $P_m = \mathrm{i}(1 - P_r)$ be the probability of an associated event in $\mathcal{M}$ (the imaginary part) to the event A in $\mathcal{R}$ (the real part). It follows that $P_r + P_m/\mathrm{i} = 1$, where i is the imaginary number with $\mathrm{i}^2 = -1$.
Axiom 7: We construct the complex number or vector $Z = P_r + P_m = P_r + \mathrm{i}(1 - P_r)$ having a norm $|Z|$ such that $|Z|^2 = P_r^2 + (P_m/\mathrm{i})^2$.
Axiom 8: Let Pc denote the probability of an event in the universe $\mathcal{C}$ where $\mathcal{C} = \mathcal{R} + \mathcal{M}$.

We say that Pc is the probability of an event A in $\mathscr{R}$ with its associated event in $\mathscr{M}$ such that

$$Pc^2 = \left(P_r + \frac{P_m}{\mathrm{i}}\right)^2 = |Z|^2 - 2\mathrm{i}P_rP_m$$

and is always equal to 1.

We can see that the system of axioms defined by Kolmogorov could be hence expanded to take into consideration the set of imaginary probabilities by adding three new axioms (Abou Jaoude, 2013a, 2013b, 2014, 2015; Abou Jaoude et al., 2010).

3.3. *The purpose of extending the axioms*

It is apparent from the set of axioms that the addition of an imaginary part to the real event makes the probability of the event in $\mathscr{C}$ always equal to 1. In fact, if we begin to see the set of probabilities as divided into two parts, one real and the other imaginary, understanding will directly follow. The random event that occurs in $\mathscr{R}$ (like tossing a coin and getting a head) has a corresponding probability P_r. Now, let $\mathscr{M}$ be the set of imaginary probabilities and let $|Z|^2$ be the DOK of this phenomenon. P_r is always, and according to Kolmogorov's axioms, the probability of an event.

A total ignorance of the set $\mathscr{M}$ makes

$P_r = 0.5$ and $|Z|^2$ in this case is equal to $1 - 2P_r(1 - P_r) = 1 - (2 \times 0.5) \times (1 - 0.5) = 0.5$.

Conversely, a total knowledge of the set in $\mathscr{R}$ makes

$P_{\text{rob}}(\text{event}) = P_r = 1$ and $P_m = P_{\text{rob}}(\text{imaginary part}) = 0$. Here we have $|Z|^2 = 1 - (2 \times 1) \times (1 - 1) = 1$ because the phenomenon is totally known, that is, its laws and variables are completely determined, hence; our DOK of the system is 1 or 100%.

Now, if we can tell for sure that an event will never occur, that is, like 'getting nothing' (the empty set), P_r is accordingly $= 0$, that is the event will never occur in $\mathscr{R}$. P_m will be equal to

$\mathrm{i}(1 - P_r) = \mathrm{i}(1 - 0) = \mathrm{i}$, and $|Z|^2 = 1 - (2 \times 0) \times (1 - 0) = 1$, because we can tell that the event of getting nothing surely will never occur; thus, the DOK of the system is 1 or 100% (Abou Jaoude et al., 2010).

We can infer that we have always

$$0.5 \le |Z|^2 \le 1 \quad \forall P_r\colon \quad 0 \le P_r \le 1$$

and

$$|Z|^2 = \text{DOK} = P_r^2 + \left(\frac{P_m}{\mathrm{i}}\right)^2, \tag{1}$$

where $0 \le P_r, P_m/i \le 1$.

And what is important is that in all cases we have

$$Pc^2 = \left(P_r + \frac{P_m}{\mathrm{i}}\right)^2 = |Z|^2 - 2\mathrm{i}P_rP_m = [P_r + (1 - P_r)]^2 = 1^2 = 1. \tag{2}$$

In fact, according to an experimenter in $\mathscr{R}$, the game is a game of luck: the experimenter does not know the output of the event. He will assign to each outcome a probability P_r and he will say that the output is nondeterministic. But in the universe $\mathscr{C} = \mathscr{R} + \mathscr{M}$, an observer will be able to predict the outcome of the game of chance since he takes into consideration the contribution of $\mathscr{M}$, so we write

$$Pc^2 = \left(P_r + \frac{P_m}{\mathrm{i}}\right)^2.$$

Hence Pc is always equal to 1. In fact, the addition of the imaginary set to our random experiment resulted to the abolition of ignorance and indeterminism. Consequently, the study of this class of phenomena in $\mathscr{C}$ is of great usefulness since we will be able to predict with certainty the outcome of experiments conducted. In fact, the study in $\mathscr{R}$ leads to unpredictability and uncertainty. So instead of placing ourselves in $\mathscr{R}$, we place ourselves in $\mathscr{C}$ then study the phenomena, because in $\mathscr{C}$ the contributions of $\mathscr{M}$ are taken into consideration and therefore a deterministic study of the phenomena becomes possible. Conversely, by taking into consideration the contribution of the set $\mathscr{M}$, we place ourselves in $\mathscr{C}$ and by ignoring $\mathscr{M}$, we restrict our study to nondeterministic phenomena in $\mathscr{R}$ (Bell, 1992; Boursin, 1986; Dacunha-Castelle, 1996; Dalmedico-Dahan, Chabert, & Chemla, 1992; Srinivasan & Mehata, 1988; Stewart, 2002; Van Kampen, 2006).

Moreover, it follows from the above definitions and axioms that (Abou Jaoude et al., 2010):

$$\begin{aligned} 2\mathrm{i}P_rP_m &= 2\mathrm{i} \times P_r \times \mathrm{i} \times (1 - P_r) \\ &= 2\mathrm{i}^2 \times P_r \times (1 - P_r) = -2P_r(1 - P_r) \\ &\Rightarrow 2\mathrm{i}P_rP_m = \text{Chf}. \end{aligned} \tag{3}$$

$2\mathrm{i}P_rP_m$ will be called the *Chaotic factor* in our experiment and will be denoted accordingly by 'Chf'. We will see why we have called this term the chaotic factor; in fact:

In case $P_r = 1$, that is the case of a certain event, then the chaotic factor of the event is equal to 0.

In case $P_r = 0$, that is the case of an impossible event, then $\text{Chf} = 0$. Hence, in both two last cases, there is no chaos since the outcome is certain and is known in advance.

In case $P_r = 0.5$, $\text{Chf} = -0.5$ (Figure 2).

We notice that $-0.5 \le \text{Chf} \le 0 \quad \forall P_r\colon\ 0 \le P_r \le 1$.

What is interesting here is thus we have quantified both the DOK and the chaotic factor of any random event and hence we write now

$$Pc^2 = |Z|^2 - 2\mathrm{i}P_rP_m = \text{DOK} - \text{Chf}. \tag{4}$$

Then, we can conclude that

$Pc^2 = \text{DOK of the system} - \text{chaotic factor} = 1$, therefore $Pc = 1$.

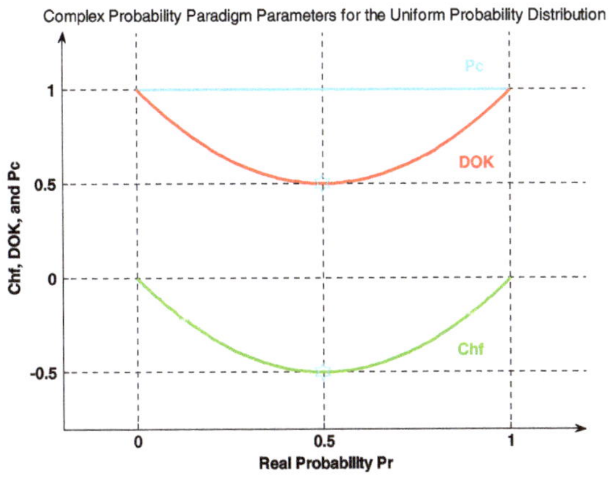

Figure 2. Chf, DOK, and Pc for a uniform probability distribution.

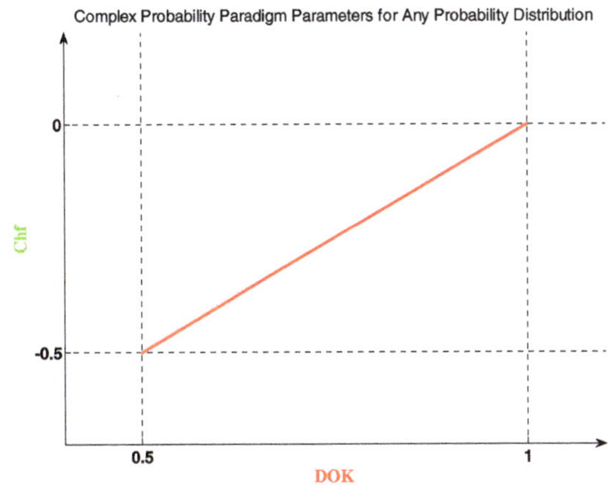

Figure 3. Graph of DOK − Chf = 1.

This directly means that if we succeed to subtract and eliminate the chaotic factor in any random experiment, then the output will always be with the probability equal to 1 (Dalmedico-Dahan & Peiffer, 1986; Ekeland, 1991; Gleick, 1997; Gullberg, 1997; Science et vie, 1999).

The graph in Figure 3 shows the linear relation between both DOK and Chf.

Furthermore, we need in our current study the absolute value of the chaotic factor that will give us the magnitude of the chaotic and random effects on the studied system materialized by the PDF, and which leads to an increasing system chaos in $\mathcal{R}$. This new term will be denoted accordingly MChf or *M*agnitude of the *Ch*aotic *f*actor (Figures 4–6). Hence, we can deduce the following:

$$\begin{aligned} \text{MChf} &= |\text{Chf}| = |2iP_rP_m| = -2iP_rP_m \\ &= 2P_r(1-P_r) \geq 0 \quad \forall P_r:\ 0 \leq P_r \leq 1, \end{aligned} \tag{5}$$

and

$$\begin{aligned} Pc^2 &= \text{DOK} - \text{Chf} \\ &= \text{DOK} + |\text{Chf}|, \text{since} -0.5 \leq \text{Chf} \leq 0 \end{aligned}$$

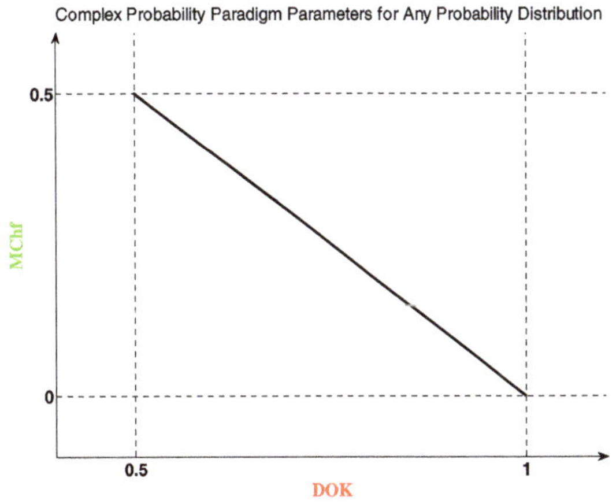

Figure 4. Graph of DOK + MChf = 1.

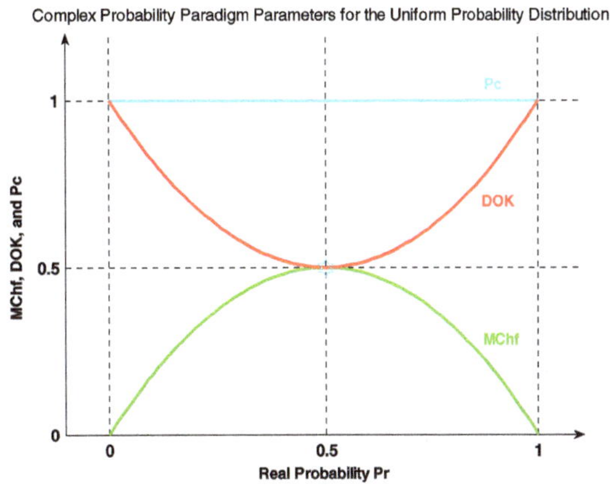

Figure 5. MChf, DOK, and Pc for a uniform probability distribution.

$$\begin{aligned} &= \text{DOK} + \text{MChf} = 1, \\ &\Leftrightarrow 0 \leq \text{MChf} \leq 0.5, \end{aligned}$$

where $0.5 \leq \text{DOK} \leq 1$

The graph in Figure 4 shows the linear relation between both DOK and MChf. Moreover, the graphs in Figures 5 and 6 show the Chf, MChf, DOK, and Pc as functions of the real probability P_r for a uniform probability distribution.

To summarize and to conclude, as the *D*egree of *O*ur certain *K*nowledge or DOK in the real universe $\mathcal{R}$ is unfortunately incomplete, the extension to the complex set $\mathcal{C}$ includes the contributions of both the real set of probabilities $\mathcal{R}$ and the imaginary set of probabilities $\mathcal{M}$. Consequently, this will result in a complete and perfect DOK in $\mathcal{C} = \mathcal{R} + \mathcal{M}$ ($Pc = 1$). In fact, in order to have a certain prediction of any random event, it is necessary to work in the complex set $\mathcal{C}$ in which the chaotic factor is quantified and subtracted from the computed DOK to lead to a probability in $\mathcal{C}$

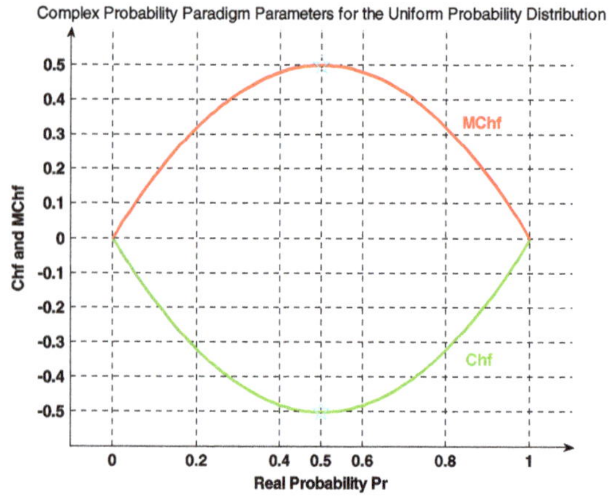

Figure 6. Chf and MChf for a uniform probability distribution.

equal to one ($Pc^2 = \text{DOK} - \text{Chf} = \text{DOK} + \text{MChf} = 1$). This hypothesis is verified in my five previous research papers by the mean of many examples encompassing both discrete and continuous distributions (Abou Jaoude et al., 2010, 2013a, 2013b, 2014, 2015). The *e*xtended *K*olmogorov *a*xioms (EKA for short) or the complex probability paradigm can be illustrated in Figure 7:

4. The new paradigm and the diffusion equation

The continuous law that we will illustrate here is the normal diffusion distribution of one particle starting from the origin at the initial time $t = 0$ (Wikipedia, *Brownian Motion*):

$$\mathrm{d}F = \rho(x,t)\mathrm{d}x = \frac{1}{\sqrt{4\pi D t}} \exp\left(\frac{-x^2}{4Dt}\right)\mathrm{d}x \quad \text{for} \quad -\infty < x < +\infty, \tag{6}$$

where $\rho(x,t)$ = the PDF of diffusion, $F(x,t)$ = the probability cumulative distribution function (CDF) of diffusion, D = the diffusion factor, x = the displacement of the particle, t = the time of displacement, $\bar{x} = 0$ = the mean value of x, $\sigma = \sqrt{2Dt}$ = the standard deviation of x (Abou Jaoude, 2004, 2005, 2007, 2013a, 2013b, 2014, 2015; Abou Jaoude et al., 2010; Bidabad, 1992; Chan Man Fong, De Kee, & Kaloni, 1997; Cox, 1955; Fagin, Halpern, & Megiddo, 1990; Ognjanović, Marković, Rašković, Doder, & Perović, 2012; Stepić, & Ognjanović, 2014; Weingarten, 2002; Wikipedia, *Brownian Motion*; Wikipedia, *Entropy*; Wikipedia, *Mass Diffusivity*; Youssef, 1994).

This bell-shaped density function has been taken from thermodynamics and statistical mechanics and it is the normal law of *Karl Friedrich Gauss* (the Prince of Mathematicians) and the *Marquis Pierre-Simon de Laplace*, or for short, the Gauss–Laplace distribution.

To illustrate the novel probability paradigm, I will consider in the simulations of the Brownian motion throughout the whole paper the diffusion of oxygen gas in air gas; hence, $D = 0.176\ \text{cm}^2/\text{s}$ at a temperature $T = 25\ ^\circ\text{C}$ (Wikipedia, *Mass Diffusivity*).

Figure 8 shows the characteristic bell-shaped curves of the diffusion of Brownian particles. The distribution begins as a Dirac delta function, indicating that all the particles are located at the origin at time $t = 0$ s, and for increasing

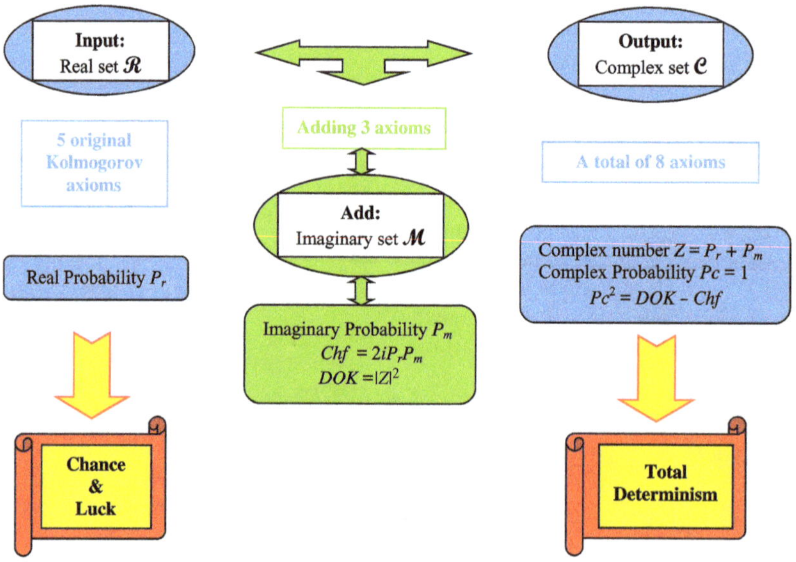

Figure 7. The EKA or the complex probability paradigm diagram.

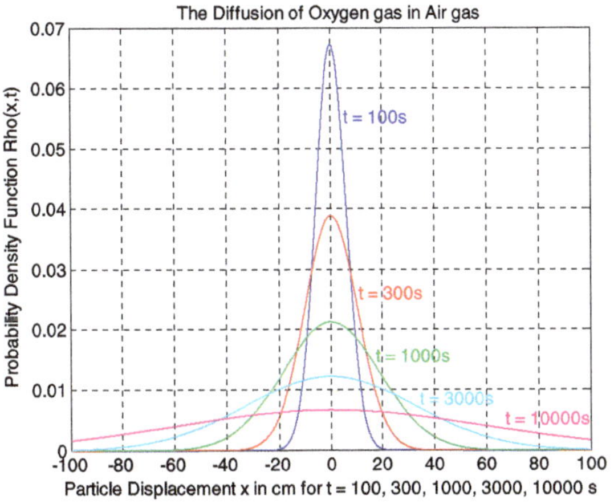

Figure 8. The PDF $\rho(x, t)$ of the diffusion of oxygen gas in air gas for different times t.

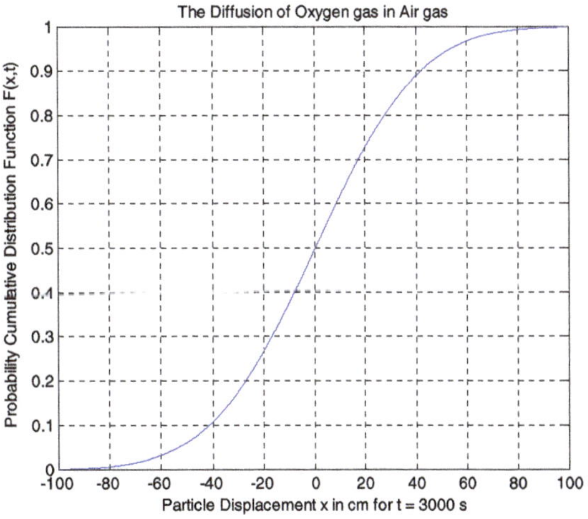

Figure 9. The CDF $F(x,t)$ of the diffusion of oxygen gas in air gas for $t = 3000$ s.

times they become flatter and flatter until the distribution becomes uniform in the asymptotic time limit.

To find the probability of an event following the normal distribution, we use the density function. We should have in the discrete case $\sum_j p_j = 1$, or in the continuous case we should have the integration over the whole domain equals to 1 like in this way:

$$\int_{-\infty}^{+\infty} \mathrm{d}F = \int_{-\infty}^{+\infty} \rho(x,t)\mathrm{d}x = \int_{-\infty}^{+\infty} \frac{1}{\sqrt{4\pi Dt}} \exp\left(\frac{-x^2}{4Dt}\right)\mathrm{d}x$$
$$= \frac{1}{\sqrt{4\pi Dt}} \int_{-\infty}^{+\infty} \exp\left(\frac{-x^2}{4Dt}\right)\mathrm{d}x = \frac{1}{\sqrt{4\pi Dt}} \times \sqrt{4\pi Dt} = 1.$$

Now, the real probability P_{rL} in $\mathcal{R}$ of finding the particle in the interval between $-\infty$ and L is as follows:

$$\int_{-\infty}^{L} \mathrm{d}F = \int_{-\infty}^{L} \rho(x,t)\mathrm{d}x = \int_{-\infty}^{L} \frac{1}{\sqrt{4\pi Dt}} \exp\left(\frac{-x^2}{4Dt}\right)\mathrm{d}x$$
$$= P_{\mathrm{rob}}[x \le L] = F(x = L) = P_{rL} = p_L. \quad (7)$$

We note that P_{rL} is a non-decreasing function (Figure 9).

The associated imaginary probability P_{mL} in $\mathcal{M}$ is as follows:

$$P_{mL} = \mathrm{i} \times (1 - P_{rL}) = \mathrm{i} \times P_{\mathrm{rob}}[x > L] = \mathrm{i}\int_{L}^{+\infty} \mathrm{d}F$$
$$= \mathrm{i}\int_{L}^{+\infty} \rho(x,t)\,\mathrm{d}x = \mathrm{i}\int_{L}^{+\infty} \frac{1}{\sqrt{4\pi Dt}} \exp\left(\frac{-x^2}{4Dt}\right)\mathrm{d}x$$
$$= \mathrm{i} \times [1 - F(x = L)]$$
$$\Rightarrow P_{mL} = \mathrm{i} \times P_{\mathrm{rob}}[\text{to find the particle in the interval between } L \text{ and } +\infty]. \quad (8)$$

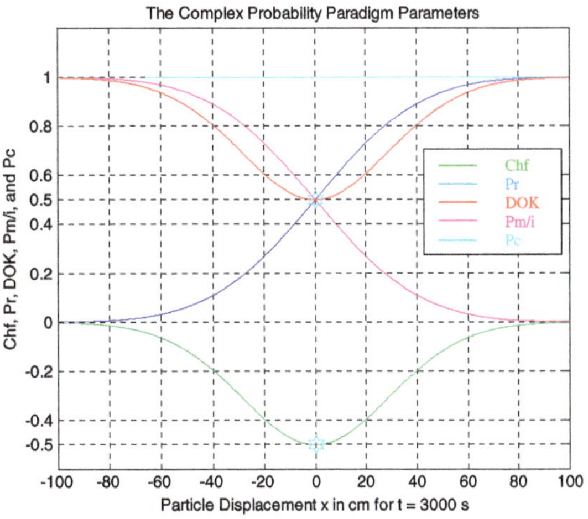

Figure 10. The complex probability parameters of the diffusion of oxygen gas in air gas for $t = 3000$ s.

The complementary probability P_{mL}/i is as follows:

$$\frac{P_{mL}}{\mathrm{i}} = 1 - P_{rL} = 1 - F(x = L)$$
$$= \int_{L}^{+\infty} \frac{1}{\sqrt{4\pi Dt}} \exp\left(\frac{-x^2}{4Dt}\right)\mathrm{d}x. \quad (9)$$

We note that P_{mL}/i is a non-increasing function (Figure 10).

Now, if we compute the norm of the complex number $Z_L = P_{rL} + P_{mL}$, we obtain

$$|Z_L|^2 = P_{rL}^2 + \left(\frac{P_{mL}}{\mathrm{i}}\right)^2 = p_L^2 + (1 - p_L)^2$$
$$= 1 + 2p_L(p_L - 1) = 1 - 2p_L(1 - p_L). \quad (10)$$

This implies that

$$1 = |Z_L|^2 + 2p_L(1-p_L) = |Z_L|^2 - 2\mathrm{i}^2 p_L(1-p_L) = |Z_L|^2 - 2\mathrm{i}p_L\mathrm{i}(1-p_L) = |Z_L|^2 - 2\mathrm{i}P_{rL}P_{mL} = P_{rL}^2 + \left(\frac{P_{mL}}{\mathrm{i}}\right)^2 - 2\mathrm{i}P_{rL}P_{mL} = P_{rL}^2 + \left(\frac{P_{mL}}{\mathrm{i}}\right)^2 + 2P_{rL}\frac{P_{mL}}{\mathrm{i}} = \left(P_{rL} + \frac{P_{mL}}{\mathrm{i}}\right)^2 = Pc_L^2 \Rightarrow Pc_L = 1, \quad (11)$$

where $Z_L = P_{rL} + P_{mL} = \int_{-\infty}^{L} \rho(x,t)\mathrm{d}x + \mathrm{i}\int_{L}^{+\infty} \rho(x,t)\mathrm{d}x$, written for short $Z_L = \int_{-\infty}^{L} + \mathrm{i}\int_{L}^{+\infty}$; and with the imaginary number i having $\mathrm{i}^2 = -1 \Rightarrow (1/i) = -\mathrm{i}$.

We can deduce from the above that

$$Pc_L^2 = \left(P_{rL} + \frac{P_{mL}}{\mathrm{i}}\right)^2 = \left(\int_{-\infty}^{L} + \frac{\mathrm{i}\int_{L}^{+\infty}}{\mathrm{i}}\right)^2 = \left(\int_{-\infty}^{L} + \int_{L}^{+\infty}\right)^2 = \left(\int_{-\infty}^{+\infty}\right)^2 = 1^2 = 1. \quad (12)$$

This is also coherent with the three added axioms previously defined.

The DOK is as follows:

$$\mathrm{DOK}_L = |Z_L|^2 = 1 - 2p_L(1-p_L) = 1 - 2 \times \int_{-\infty}^{L} \times \left(1 - \int_{-\infty}^{L}\right); \quad (13)$$

which is a curve concave upward having a minimum at $p_L = 0.5 \Leftrightarrow \mathrm{At}(L = \bar{x} = 0, 0.5)$ since the diffusion equation considered is a normal distribution symmetric about the mean which is $\bar{x} = 0$ cm (Figure 10).The chaotic factor is as follows:

$$\mathrm{Chf}_L = 2\mathrm{i}P_{rL}P_{mL} = 2\mathrm{i} \times \int_{-\infty}^{L} \times \mathrm{i} \times \int_{L}^{+\infty} = -2 \times \int_{-\infty}^{L} \times \int_{L}^{+\infty} = -2 \times \int_{-\infty}^{L} \times \left(1 - \int_{-\infty}^{L}\right), \quad (14)$$

which is a curve concave upward having a minimum at $p_L = 0.5 \Leftrightarrow \mathrm{At}(L = \bar{x} = 0, -0.5)$ since the diffusion equation considered is a normal distribution symmetric about the mean which is $\bar{x} = 0$ cm (Figure 10).

Moreover, the MChf is as follows:

$$\mathrm{MChf}_L = -2\mathrm{i}P_{rL}P_{mL} = -2\mathrm{i} \times \int_{-\infty}^{L} \times \mathrm{i} \times \int_{L}^{+\infty} = 2 \times \int_{-\infty}^{L} \times \int_{L}^{+\infty} = 2 \times \int_{-\infty}^{L} \times \left(1 - \int_{-\infty}^{L}\right); \quad (15)$$

which is a curve concave downward having a maximum at

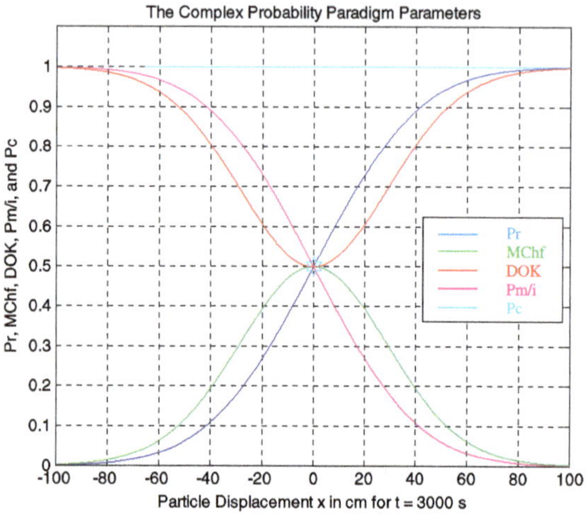

Figure 11. The complex probability parameters of the diffusion of oxygen gas in air gas with MChf for $t = 3000$ s.

$p_L = 0.5 \Leftrightarrow \mathrm{At}(L = \bar{x} = 0, 0.5)$ since the diffusion equation considered is a normal distribution symmetric about the mean which is $\bar{x} = 0$cm (Figure 11).

One can directly see that $\mathrm{DOK}_L = 1$ and $\mathrm{Chf}_L = \mathrm{MChf}_L = 0$ if $L \to -\infty$ or $L \to +\infty$, that means that we will not find the particle anywhere ($P_{rL} = 0$; impossible event) or we will find always the particle somewhere ($P_{rL} = 1$; certain event), respectively (Figures 10 and 11).

Furthermore, the intersection point of the complex probability model functions can be computed as follows:

$$P_{rL} = \frac{P_{mL}}{\mathrm{i}} \Leftrightarrow \int_{-\infty}^{L} \frac{1}{\sqrt{4\pi Dt}} \exp\left(\frac{-x^2}{4Dt}\right) \mathrm{d}x = 1 - \int_{-\infty}^{L} \frac{1}{\sqrt{4\pi Dt}} \exp\left(\frac{-x^2}{4Dt}\right) \mathrm{d}x \Leftrightarrow 2\int_{-\infty}^{L} \frac{1}{\sqrt{4\pi Dt}} \exp\left(\frac{-x^2}{4Dt}\right) \mathrm{d}x = 1 \Leftrightarrow \int_{-\infty}^{L} \frac{1}{\sqrt{4\pi Dt}} \exp\left(\frac{-x^2}{4Dt}\right) \mathrm{d}x = \frac{1}{2} = 0.5 \Leftrightarrow L = \bar{x} = 0 \text{ cm}.$$

Notice that $P_{rL}(L = \bar{x} = 0) = 0.5$ and $P_{mL}(L = \bar{x} = 0)/\mathrm{i} = 1 - 0.5 = 0.5$, so P_{rL} and P_{mL}/i intersect at (0, 0.5).

Moreover, we can deduce mathematically that the minimum of DOK and the maximum of MChf occur at (0, 0.5).

So we conclude that P_{rL}, $P_{\mathrm{mL}}/\mathrm{i}$, DOK, and MChf all intersect at (0, 0.5) (Figure 11).

Finally, we state that

$$Pc_L^2 = |Z_L|^2 - 2\mathrm{i}P_{rL}P_{mL} = \text{degree of our knowledge} - \text{chaotic factor}$$

$$= \text{degree of our knowledge}$$
$$+ \text{magnitude of the chaotic factor}$$
$$\Rightarrow Pc_L^2 = 1 \quad \text{for} \quad -\infty < L < +\infty, \qquad (16)$$

where $|Z_L|$ is the norm of the complex number Z_L and it combines here both the contributions of $\mathcal{R}$ and $\mathcal{M}$. Note that, it is the chaotic factor that makes the study of an event in $\mathcal{R}$ a random process. Hence, any event in $\mathcal{C}$ is deterministic since $Pc_L = 1$. Consequently, we deduce that in the set $\mathcal{C}$, we have a complete knowledge of the random variable and therefore this is the advantage of working in $\mathcal{C} = \mathcal{R} + \mathcal{M}$.

5. The evolution of DOK, Chf, and MChf

Since $Pc^2(x,t) = \text{DOK}(x,t) - \text{Chf}(x,t) = 1$ then $(\partial\text{DOK}/\partial x) - (\partial\text{Chf}/\partial x) = (\partial(1)/\partial x) = 0$

$$\Rightarrow \frac{\partial\text{DOK}}{\partial x} = \frac{\partial\text{Chf}}{\partial x}. \qquad (17)$$

Similarly,

$$\frac{\partial\text{DOK}}{\partial t} = \frac{\partial\text{Chf}}{\partial t}. \qquad (18)$$

That means if DOK increases (or decreases) with displacement and time evolution, then Chf increases (or decreases).

Moreover, since $Pc^2(x,t) = \text{DOK}(x,t) + \text{MChf}(x,t) = 1$ then $(\partial\text{DOK}/\partial x) + (\partial\text{MChf}/\partial x) = (\partial(1)/\partial x) = 0$

$$\Rightarrow \frac{\partial\text{DOK}}{\partial x} = -\frac{\partial\text{MChf}}{\partial x}. \qquad (19)$$

Similarly,

$$\frac{\partial\text{DOK}}{\partial t} = -\frac{\partial\text{MChf}}{\partial t}. \qquad (20)$$

That means if DOK increases (or decreases) with displacement and time evolution, then MChf decreases (or increases).

In addition, since $\text{Chf}(x,t) + \text{MChf}(x,t) = 0$, therefore:

$$\frac{\partial\text{Chf}}{\partial x} = -\frac{\partial\text{MChf}}{\partial x} \qquad (21)$$

and

$$\frac{\partial\text{Chf}}{\partial t} = -\frac{\partial\text{MChf}}{\partial t}. \qquad (22)$$

That means if Chf increases (or decreases) with displacement and time evolution, then MChf decreases (or increases).

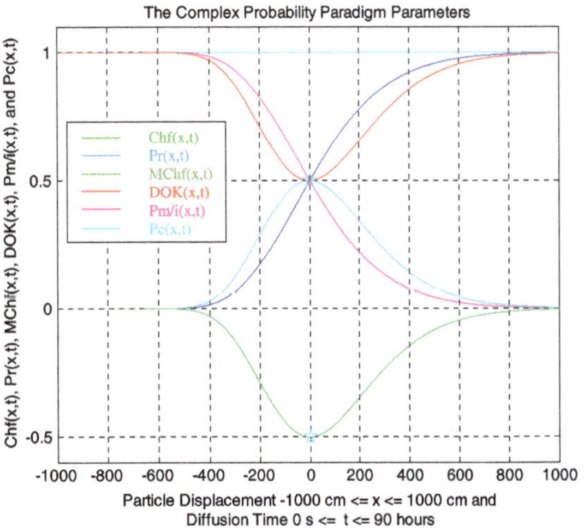

Figure 12. The complex probability parameters of the diffusion of oxygen gas in air gas whenever both x and t vary simultaneously.

Moreover, the mixed partial derivatives are the following:

$$\frac{\partial^2\text{DOK}}{\partial x \partial t} = \frac{\partial^2\text{Chf}}{\partial x \partial t}$$
$$\Rightarrow \text{both DOK and Chf are concave upward,} \qquad (23)$$

$$\frac{\partial^2\text{DOK}}{\partial x \partial t} = -\frac{\partial^2\text{MChf}}{\partial x \partial t}$$
$$\Rightarrow \text{DOK and MChf are of opposite concavity,} \qquad (24)$$

$$\frac{\partial^2\text{Chf}}{\partial x \partial t} = -\frac{\partial^2\text{MChf}}{\partial x \partial t} \Rightarrow$$
$$\text{Chf and MChf are of opposite concavity.} \qquad (25)$$

Figure 12 shows that all the complex probability paradigm functions conserve their characteristics and behaviour whenever both x and t vary simultaneously:

$$-1000\text{ cm} \leq x \leq 1000\text{ cm} \quad \text{and} \quad 0\,\text{s} \leq t \leq 90\,\text{h}.$$

But as we can notice, all the graphs are skewed to the right. Furthermore, we can write

$$\int_{t=0\,\text{s}}^{t=90\,\text{h}} \int_{x=-1000\,\text{cm}}^{x=g(t)} f(x,t)\,\mathrm{d}x\,\mathrm{d}t$$
$$= F(x = 1000\text{ cm}, t = 90\,\text{h}) = 1,$$

where

$$x = g(t) = t/162 - 1000$$

$$\Rightarrow \begin{cases} x = g(t = 0\,\text{s}) = 0/162 - 1000 = -1000\,\text{cm}, \\ x = g(t = 90\,\text{h} = 324,000\,\text{s}) = 324,000/ \\ \quad 162 - 1000 = 1000\,\text{cm}. \end{cases}$$

6. Numerical example

Consider for this numerical example always the diffusion of oxygen gas in air gas (Wikipedia, *Mass Diffusivity*), hence: $\bar{x}$ = mean value of $x = 0$ cm.

The diffusion factor $D = 0.176\,\text{cm}^2/\text{s}$ at a temperature $T = 25\,°\text{C}$.

The diffusion time: $t = 3000$ s.

This implies that the standard deviation of the particles Brownian motion is as follows:

$$\sigma = \sqrt{2Dt} = \sqrt{2 \times 0.176 \times 3000} = 32.4962\ \text{cm}.$$

We can compute from the CDF the following:

$P_{\text{rob}}[7\,\text{cm} \le x \le 21\,\text{cm}] = 0.1557$, $P_{\text{rob}}[-\infty < x \le 21\,\text{cm}] = 0.7409$, and $P_{\text{rob}}[21\,\text{cm} \le x < +\infty] = 0.2591$.

We can see that $P_{\text{rob}}[L_a \le x \le L_b] = F(u_b) - F(u_a)$, where $u_b = \frac{L_b - \bar{x}}{\sigma}$ and $u_a = \frac{L_a - \bar{x}}{\sigma}$.

If $L_a = 7$ cm and $L_b = 21$ cm, then the probability $= F\left(\frac{21-0}{32.4962}\right) - F\left(\frac{7-0}{32.4962}\right) = 0.1557$.

Note that

$$F(u_L) = \int_{-\infty}^{u_L} \frac{1}{\sqrt{2\pi}} \exp\left(\frac{-u^2}{2}\right) du = P_{\text{rob}}[u \le u_L],$$

where $u = \frac{x - \bar{x}}{\sigma}$.

In the real domain $\mathscr{R}$, we have $dF = f(u)\,du = (1/\sqrt{2\pi}) \exp(-u^2/2)\,du$, and which is the standard normal distribution of diffusion having $\bar{u} = 0$ and $\sigma_u = 1$.

And we know that

$$\int_{-\infty}^{+\infty} dF = \int_{-\infty}^{+\infty} \frac{1}{\sqrt{2\pi}} \exp\left(\frac{-u^2}{2}\right) du$$
$$= \int_{-\infty}^{+\infty} \frac{1}{\sqrt{2\pi}\sigma} \exp\left[-\frac{1}{2}\left(\frac{x - \bar{x}}{\sigma}\right)^2\right] dx = 1.$$

Now, as a numerical example, take $L = 21$ cm, then in the real domain $\mathscr{R}$ we have

$$P_{\text{rob}}[-\infty < x \le 21] = P_{rL}$$
$$= \int_{-\infty}^{L=21} \frac{1}{\sqrt{2\pi} \times 32.4962} \times \exp\left[-\frac{1}{2}\left(\frac{x - 0}{32.4962}\right)^2\right] dx$$
$$= 0.7409.$$

The correspondent probability in the imaginary domain $\mathscr{M}$ is as follows:

$$P_{mL} = \mathrm{i}(1 - P_{rL}) = \mathrm{i}P_{\text{rob}}[x > 21]$$
$$= \mathrm{i}\int_{21}^{+\infty} \frac{1}{\sqrt{2\pi} \times 32.4962} \exp\left[-\frac{1}{2}\left(\frac{x - 0}{32.4962}\right)^2\right] dx$$
$$= \mathrm{i} \times 0.2591.$$

And the complementary probability is $P_{mL}/\mathrm{i} = 0.2591$. We note that

$$Z_L = P_{rL} + P_{mL} = \int_{-\infty}^{L=21} f(u)\,du + \mathrm{i}\int_{L=21}^{+\infty} f(u)\,du$$
$$= 0.7409 + \mathrm{i} \times 0.2591.$$

We also have

$$Pc_L^2 = \left(P_{rL} + \frac{P_{mL}}{i}\right)^2 = \left(\int_{-\infty}^{L=21} + \int_{L=21}^{+\infty}\right)^2$$
$$= \left(\int_{-\infty}^{+\infty}\right)^2 = 1^2 = 1.$$

And the DOK is as follows:

$$\text{DOK}_L = |Z_L|^2 = 1 - 2P_{rL}(1 - P_{rL}) = 1 - 2 \times \int_{-\infty}^{L=21}$$
$$\times \left(1 - \int_{-\infty}^{L=21}\right),$$

where

$$\text{DOK}_L = 1 \quad \text{if} \begin{cases} L \to -\infty & \text{hence } P_{rL} = 0, \\ L \to +\infty & \text{hence } P_{rL} = 1. \end{cases}$$

Furthermore, the chaotic factor is as follows:

$$\text{Chf}_L = 2\mathrm{i} \times P_{rL} \times P_{mL} = 2\mathrm{i} \times \int_{-\infty}^{L=21} \times \mathrm{i} \times \int_{L=21}^{+\infty}$$
$$= -2 \times \int_{-\infty}^{21} \times \left(1 - \int_{-\infty}^{21}\right),$$

where

$$\text{Chf}_L = 0 \text{ if} \begin{cases} L \to -\infty & \text{hence } P_{rL} = 0, \\ L \to +\infty & \text{hence } P_{rL} = 1. \end{cases}$$

Moreover, the MChf is as follows:

$$\text{MChf}_L = -2\mathrm{i} \times P_{rL} \times P_{rL} = -2\mathrm{i} \times \int_{-\infty}^{L=21} \times \mathrm{i} \times \int_{L=21}^{+\infty}$$
$$= 2 \times \int_{-\infty}^{21} \times \left(1 - \int_{-\infty}^{21}\right),$$

where

$$\mathrm{MChf}_L = 0 \text{ if } \begin{cases} L \to -\infty & \text{hence } P_{rL} = 0, \\ L \to +\infty & \text{hence } P_{rL} = 1. \end{cases}$$

Numerically, we write

$$\mathrm{DOK}_L = |Z_L|^2 = (0.7409)^2 + (0.2591)^2 = 0.5489 + 0.0671 = 0.6160 \Rightarrow |Z_L| = 0.7849$$
$$\Rightarrow \mathrm{Chf}_L \neq 0 \quad \text{notice that } \frac{1}{2} \le \mathrm{DOK}_L \le 1.$$

Hence,

$$\mathrm{Chf}_L = -2 \times 0.7409 \times 0.2591 = -0.3840 \quad \text{notice that } -\frac{1}{2} \le \mathrm{Chf}_L \le 0.$$

And

$$\mathrm{MChf}_L = 2 \times 0.7409 \times 0.2591 = 0.3840 \quad \text{notice that } 0 \le \mathrm{MChf}_L \le \frac{1}{2}.$$

What is interesting here and throughout the whole original paradigm simulation, is that we have quantified both the DOK and the chaotic factor of the random event in $\mathscr{R}$ which is the stochastic Brownian motion of the oxygen gas particles in air gas.

Moreover, notice that
the DOK − the chaotic factor = 0.6160 − (− 0.3840)

$$= 0.6160 + 0.3840 = 1 = Pc_L.$$

And
the DOK + the MChf = 0.6160 + 0.3840

$$= 1 = Pc_L.$$

Therefore, we state that we have always

$$Pc_L^2 = |Z_L|^2 - 2iP_{rL}P_{mL}$$
$$= \text{degree of our knowledge} - \text{chaotic factor}$$
$$= \text{degree of our knowledge} + \text{magnitude of the chaotic factor}$$
$$= 1.$$

And if $\mathrm{Chf}_L = \mathrm{MChf}_L = 0 \Rightarrow |Z_L|^2 = 1$, in other words, if the chaotic factor or the MChf are zero, then the DOK in $\mathscr{R}$ is 1 or 100%.

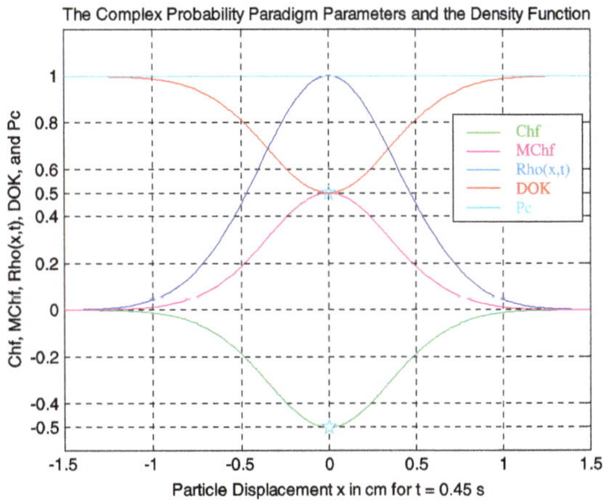

Figure 13. The complex probability parameters of the diffusion of oxygen gas in air gas with $\rho(x, t)$ at $t = 0.45$ s.

Conversely, if we assume that

$$\mathrm{Chf}_L = 0 \Rightarrow |Z_L|^2 = 1 \Rightarrow P_{rL}^2 + \left(\frac{P_{mL}}{i}\right)^2 = 1$$
$$\Rightarrow 2P_{rL}(1 - P_{rL}) = 0 \Rightarrow \begin{cases} P_{rL} = 0 \\ \text{or} \\ P_{rL} = 1 \end{cases}$$
$$\Rightarrow \begin{cases} L \to -\infty \\ \text{or} \\ L \to +\infty \end{cases}$$

And if

$$\mathrm{Chf}_L = -\frac{1}{2} \quad \text{or} \quad \mathrm{MChf}_L = \frac{1}{2}$$
$$\Rightarrow L = \bar{x} = 0 \quad \text{and} \quad \mathrm{DOK}_L = \frac{1}{2}.$$

Now, if L increases from 21 cm to become equal to 50 cm for example, then: DOK_L and Chf_L both increase, and MChf_L decreases.

Therefore, we can deduce that

$$\lim_{L\to\pm\infty} \mathrm{Chf}_L = \lim_{L\to\pm\infty} \mathrm{MChf}_L = 0 \text{ and } \lim_{L\to\pm\infty} \mathrm{DOK}_L = 1,$$

where

$$Pc_L^2 = \mathrm{DOK}_L - \mathrm{Chf}_L = \mathrm{DOK}_L + \mathrm{MChf}_L = 1$$

for every value of L in the real set $\mathscr{R}$ and for any diffusion time $0 \le t < +\infty$, specifically for $t = 0.45$ s as in Figure 13, and as we will see in Section 8 for other values of t.

7. Flowchart of the complex probability paradigm

The following flowchart summarizes all the procedures of the proposed complex probability prognostic model:

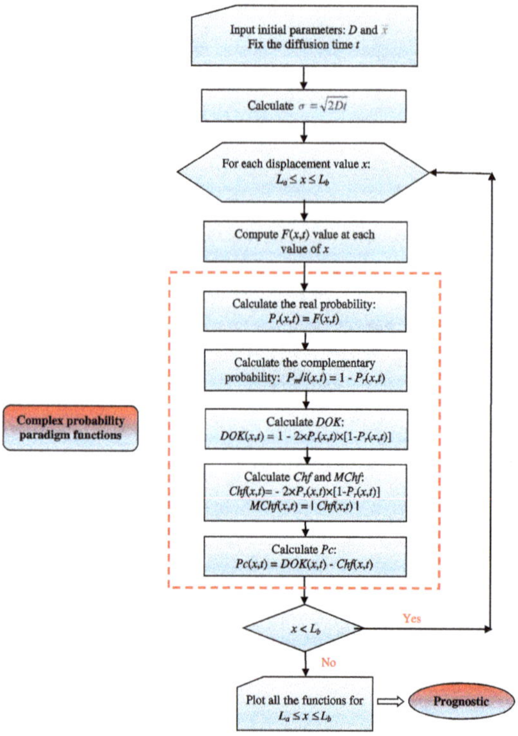

8. Simulation of the new paradigm

Note that all the numerical values found in the paradigm functions analysis for $t = 3000$ s or $t = 1000$ s or $t = 100$ s were computed using the MATLAB version 2015 software.

8.1. *The paradigm functions analysis for t = 3000 s*

We notice from Figures 14–16 that the DOK is maximum (DOK = 1) when MChf is minimum (MChf = 0) (points J & H) and that means when the MChf decreases our certain knowledge in $\mathcal{R}$ increases.

At the beginning $P_r(x < -100\,\text{cm}) \approx 0$, the system is nearly intact and has nearly zero chaotic factor (Chf = MChf ≈ 0) before any diffusion, hence at this instant DOK ≈ 1. Here $P_m/\text{i} \approx 1$ with $Pc = P_r + P_m/\text{i} \approx 0 + 1 = 1$ and $Pc = \text{DOK} + \text{MChf} \approx 1 + 0 = 1$.

Note that if $x \to -\infty$ cm, then DOK → 1, Chf → 0, MChf → 0, $P_r \to 0$, $P_m/\text{i} \to 1$, with $Pc = P_r + P_m/\text{i} \to 0 + 1 = 1$ and $Pc = \text{DOK} + \text{MChf} \to 1 + 0 = 1$.

Afterward, with diffusion, x starts to increase with $-\infty < x < +\infty$, $P_r(L) = \int_{-\infty}^{L} \rho(x,t)\,dx \neq 0$ and we have always $Pc(L) = P_r(L) + P_m/\text{i}(L) = \text{DOK}(L) + \text{MChf}(L)$

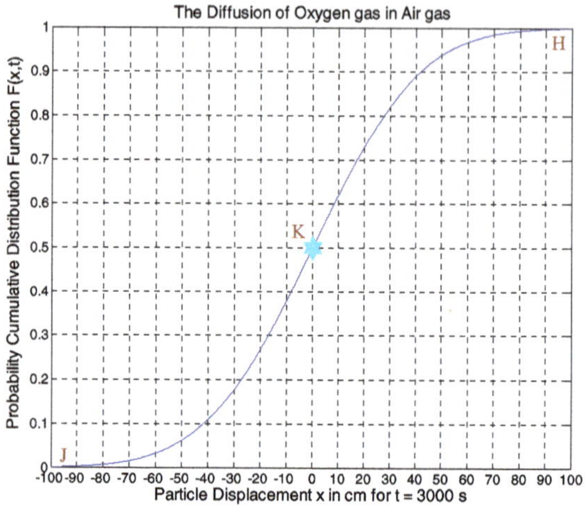

Figure 14. The CDF of the diffusion of oxygen gas in air gas for $t = 3000$ s.

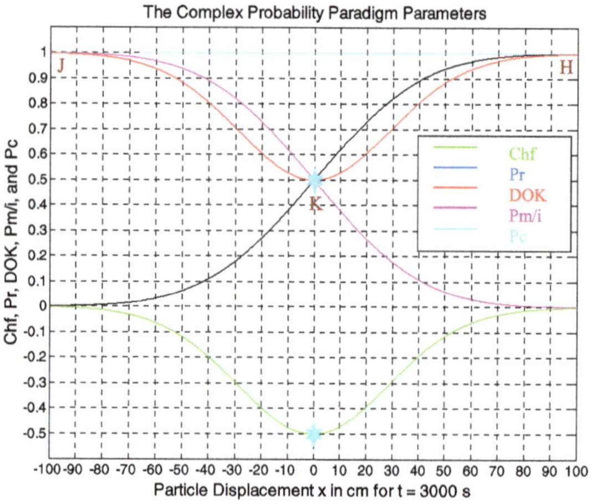

Figure 15. The complex probability parameters for $t = 3000$ s.

= 1, thus MChf starts to increase also during the diffusion due to the environment and intrinsic conditions thus leading to a decrease in DOK.

If $x = -100$ cm (point J), then DOK = 0.997913, Chf = −0.002087, MChf = 0.002087, $P_r = 0.001044$, $P_m/\text{i} = 0.998956$, with $Pc = P_r + P_m/\text{i} = 0.001044 + 0.998956 = 1$ and $Pc = \text{DOK} + \text{MChf} = 0.997913 + 0.002087 = 1$.

If $x = -60$ cm, then DOK = 0.93726, Chf = −0.06274, MChf = 0.06274, $P_r = 0.03242$, $P_m/\text{i} = 0.96758$, with $Pc = P_r + P_m/\text{i} = 0.03242 + 0.96758 = 1$ and $Pc = \text{DOK} + \text{MChf} = 0.93726 + 0.06274 = 1$. We can see that with the increase of x, the real probability P_r also increases.

If $x = -20$ cm, then DOK = 0.6066, Chf = −0.3934, MChf = 0.3934, $P_r = 0.2691$, $P_m/\text{i} = 0.7309$, with $Pc = P_r + P_m/\text{i} = 0.2691 + 0.7309 = 1$ and $Pc = \text{DOK} + \text{MChf} = 0.6066 + 0.3934 = 1$. We can see here

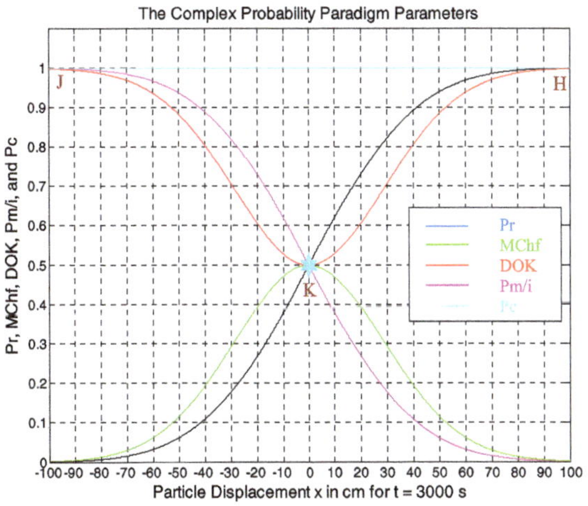

Figure 16. The complex probability parameters with MChf for $t = 3000\,\text{s}$.

that with the increase of x, the MChf also increases, whereas DOK decreases.

If $x = 0\,\text{cm}$ (point K), both DOK (minimum) and MChf (maximum) reach 0.5, $P_r = 0.5$, $P_m/\text{i} = 0.5$, and $\text{Chf} = -0.5$ with $Pc = P_r + P_m/\text{i} = 0.5 + 0.5 = 1$ and $Pc = \text{DOK} + \text{MChf} = 0.5 + 0.5 = 1$, as always. Hence, all the EKA parameters will intersect at the point K. We have here maximum chaos and the minimum of the system knowledge in $\mathscr{R}$; therefore, the real probability is $P_r = 1/2 =$ probability half way to full diffusion. Furthermore, notice in Figure 16 the complete symmetry at the vertical axis $x = 0\,\text{cm} =$ half way to complete dispersion.

If $x = 30\,\text{cm}$, then $\text{DOK} = 0.7074$, $\text{Chf} = -0.2926$, $\text{MChf} = 0.2926$, $P_r = 0.822$, $P_m/\text{i} = 0.178$, with $Pc = P_r + P_m/\text{i} = 0.822 + 0.178 = 1$ and $Pc = \text{DOK} + \text{MChf} = 0.7074 + 0.2926 = 1$. We can see that with the increase of x and consequently the decrease of MChf, the real probability P_r also increases.

If $x = 70\,\text{cm}$, then $\text{DOK} = 0.96926$, $\text{Chf} = -0.03074$, $\text{MChf} = 0.03074$, $P_r = 0.98438$, $P_m/\text{i} = 0.01562$, with $Pc = P_r + P_m/\text{i} = 0.98438 + 0.01562 = 1$ and $Pc = \text{DOK} + \text{MChf} = 0.96926 + 0.03074 = 1$.

If $x = 100\,\text{cm}$ (point H), then $\text{DOK} = 0.997913$, $\text{Chf} = -0.002087$, $\text{MChf} = 0.002087$, $P_r = 0.998956$, $P_m/\text{i} = 0.001044$, with $Pc = P_r + P_m/\text{i} = 0.998956 + 0.001044 = 1$ and $Pc = \text{DOK} + \text{MChf} = 0.997913 + 0.002087 = 1$. Here, since $x = 100\,\text{cm}$ which is relatively very big in this case, then the real probability P_r is very near to 1 since we will reach total and full dispersion very soon.

With the increase of displacement beyond 100 cm, MChf and Chf are nearly zero, DOK returns to 1; hence we have the total dispersion of the system of the oxygen gas particles in air gas. At this last point, $P_r \approx 1$, $P_m/\text{i} \approx 0$, with $Pc = P_r + P_m/\text{i} \approx 1 + 0 = 1$ and $Pc = \text{DOK} + \text{MChf} \approx 1 + 0 = 1$; thus, the logical explanation of the value of $\text{DOK} \approx 1$ follows.

Note that if $x \to +\infty\,\text{cm}$, then $\text{DOK} \to 1$, $\text{Chf} \to 0$, $\text{MChf} \to 0$, $P_r \to 1$, $P_m/\text{i} \to 0$, with $Pc = P_r + P_m/\text{i} \to 1 + 0 = 1$ and $Pc = \text{DOK} + \text{MChf} \to 1 + 0 = 1$.

Moreover, at each value of x, the probability to find the particle $Pc(x,t)$ is certainly predicted in the complex set $\mathscr{C}$ with Pc maintained as equal to one through a continuous compensation between DOK and Chf. This compensation is from $x = -\infty$, where $P_r = 0$ until $x = +\infty$ where $P_r = 1$, keeping always $Pc = \text{DOK} - \text{Chf} = \text{DOK} + \text{MChf} = 1$. Furthermore, what is truly interesting for all the particle displacement simulations is that we have quantified and visualized both the DOK and the chaotic factor of the Brownian motion.

It becomes clear from the simulations that DOK is the measure of our certain knowledge in $\mathscr{R}$ (100% probability) about the expected event and it does not include any uncertain knowledge (with a probability less than 100%).

We note that the same logic and analysis concerning the diffusion as well as all the EKA parameters apply for all the three instants of Brownian motion.

8.1.1. The complex probability cubes

In Figure 17, we can see the simulation of DOK and Chf as functions of x and of each other for $t = 3000\,\text{s}$. The line in cyan is $Pc^2(x,t) = \text{DOK}(x,t) - \text{Chf}(x,t) = 1 = Pc(x,t)$. This line, projected on the $x = -100\,\text{cm}$ plane, starts at the point ($\text{DOK} = 1$, $\text{Chf} = 0$) when $x = -100\,\text{cm}$ (point J in Figures 14–16), reaches the point ($\text{DOK} = 0.5$, $\text{Chf} = -0.5$) when $x = 0\,\text{cm}$ (point K in Figures 14–16), and returns at the end to the point ($\text{DOK} = 1$, $\text{Chf} = 0$) when $x = 100\,\text{cm}$ (point H in Figures 14–16). The other curves are the graphs of $\text{DOK}(x,t)$ and $\text{Chf}(x,t)$ in different planes. Notice that they all have a minimum at $x = 0\,\text{cm}$ (point K in Figures 14–16), as explained previously.

In Figure 18, we can notice the simulation of the real probability $P_r(x,t)$ and its complementary probability $P_m/\text{i}(x,t)$ as functions of x for $t = 3000\,\text{s}$. The line in cyan is $Pc^2(x,t) = P_r(x,t) + P_m/\text{i}(x,t) = 1 = Pc(x,t)$. This line, projected on the $x = -100$ plane, starts at the point ($P_r = 0$, $P_m/\text{i} = 1$) (point J in Figures 14–16) and ends at the point ($P_r = 1$, $P_m/\text{i} = 0$) (point H in Figures 14–16). The blue curve represents $P_r(x,t)$ in the plane $P_r = P_m/\text{i}$ and the red curve represents $P_m/\text{i}(x,t)$ in the plane $P_r + P_m/\text{i} = 1$. Notice the importance of the point ($P_r = 0.5$, $P_m/\text{i} = 0.5$) corresponding to $x = 0\,\text{cm}$ (point K in Figures 14–16).

Note that similar cubes can be drawn for the instants $t = 1000\,\text{s}$ and for $t = 100\,\text{s}$ with their corresponding points J, K, and H.

8.2. The paradigm functions analysis for t = 1000 s

We notice from Figures 19–21 that the DOK is maximum ($\text{DOK} = 1$) when MChf is minimum ($\text{MChf} = 0$) (points

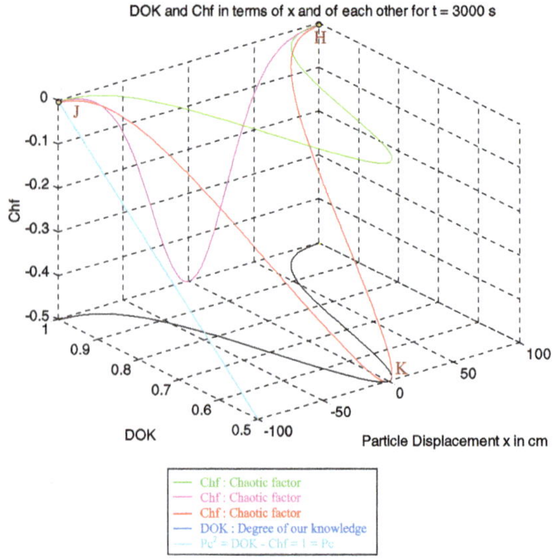

Figure 17. DOK Chf and Pc, in terms of x and of each other for the diffusion time $t = 3000$ s.

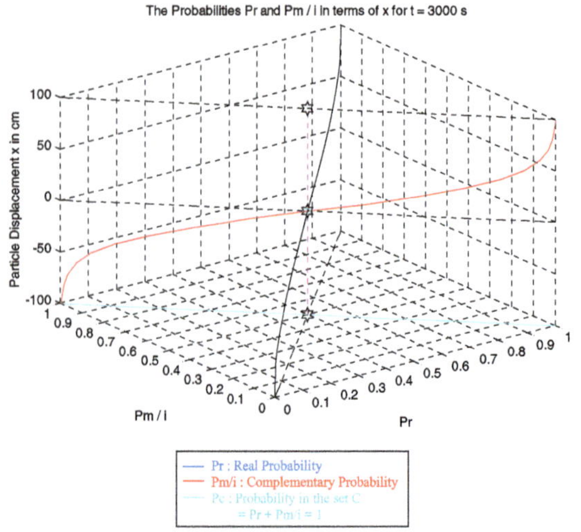

Figure 18. P_r, P_m/i, and Pc, in terms of x for the diffusion time $t = 3000$ s.

J & H) and that means when the MChf decreases our certain knowledge in $\mathcal{R}$ increases.

At the beginning $P_r(x < -60\,\text{cm}) \approx 0$, the system is nearly intact and has nearly zero chaotic factor (Chf = MChf ≈ 0) before any diffusion, hence at this instant DOK ≈ 1. Here $P_m/\text{i} \approx 1$ with $Pc = P_r + P_m/\text{i} \approx 0 + 1 = 1$ and $Pc = \text{DOK} + \text{MChf} \approx 1 + 0 = 1$.

Note that if $x \to -\infty$ cm, then DOK $\to 1$, Chf $\to 0$, MChf $\to 0$, $P_r \to 0$, $P_m/\text{i} \to 1$, with $Pc = P_r + P_m/\text{i} \to 0 + 1 = 1$ and $Pc = \text{DOK} + \text{MChf} \to 1 + 0 = 1$.

Afterward, with diffusion, x starts to increase with $-\infty < x < +\infty$, $P_r(L) = \int_{-\infty}^{L} \rho(x,t)\,\text{d}x \neq 0$ and we have always $Pc(L) = P_r(L) + P_m/\text{i}(L) = \text{DOK}(L) + \text{MChf}(L)$

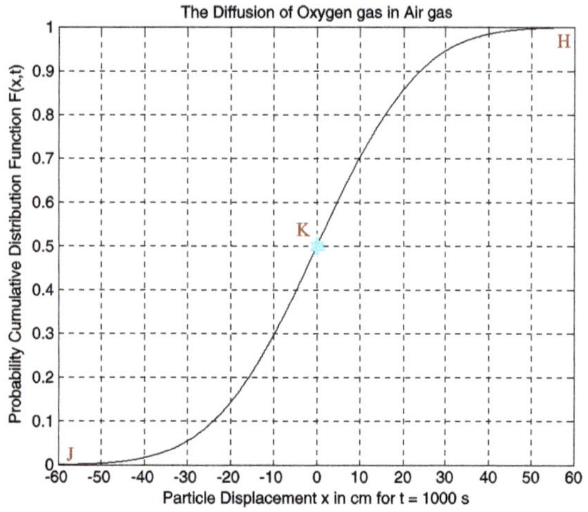

Figure 19. The CDF of the diffusion of oxygen gas in air gas for $t = 1000$ s.

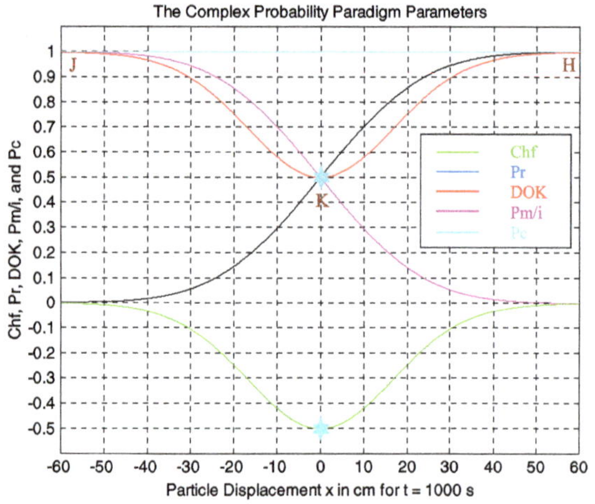

Figure 20. The complex probability parameters for $t = 1000$ s.

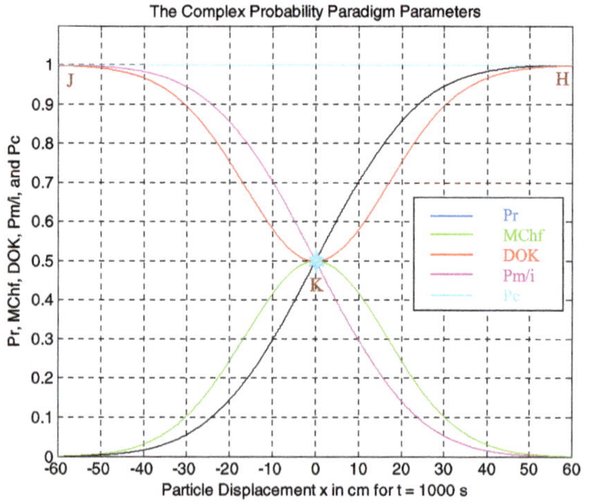

Figure 21. The complex probability parameters with MChf for $t = 1000$ s.

$= 1$, thus MChf starts to increase also during the diffusion due to the environment and intrinsic conditions thus leading to a decrease in DOK.

If $x = -60\,\text{cm}$ (point J), then DOK = 0.998617, Chf = −0.001383, MChf = 0.001383, $P_r = 0.0006919$, $P_m/\text{i} = 0.9993081$, with $Pc = P_r + P_m/\text{i} = 0.0006919 + 0.9993081 = 1$ and $Pc = \text{DOK} + \text{MChf} = 0.998617 + 0.001383 = 1$.

If $x = -40\,\text{cm}$, then DOK = 0.96754, Chf = −0.03246, MChf = 0.03246, $P_r = 0.0165$, $P_m/\text{i} = 0.9835$, with $Pc = P_r + P_m/\text{i} = 0.0165 + 0.9835 = 1$ and $Pc = \text{DOK} + \text{MChf} = 0.96754 + 0.03246 = 1$. We can see that with the increase of x, the real probability P_r also increases.

If $x = -20\,\text{cm}$, then DOK = 0.7546, Chf = −0.2454, MChf = 0.2454, $P_r = 0.1432$, $P_m/\text{i} = 0.8568$, with $Pc = P_r + P_m/\text{i} = 0.1432 + 0.8568 = 1$ and $Pc = \text{DOK} + \text{MChf} = 0.7546 + 0.2454 = 1$. We can see here that with the increase of x, the MChf also increases, whereas DOK decreases.

If $x = 0\,\text{cm}$ (point K), both DOK (minimum) and MChf (maximum) reach 0.5, $P_r = 0.5$, $P_m/\text{i} = 0.5$, and Chf = −0.5 with $Pc = P_r + P_m/\text{i} = 0.5 + 0.5 = 1$ and $Pc = \text{DOK} + \text{MChf} = 0.5 + 0.5 = 1$, as always. Hence, all the EKA parameters will intersect at the point K. We have here maximum chaos and the minimum of the system knowledge in $\mathcal{R}$; therefore, the real probability is $P_r = 1/2$ = probability half way to full diffusion. Furthermore, notice in the last figure (Figure 21) the complete symmetry at the vertical axis $x = 0\,\text{cm}$ = half way to complete dispersion.

If $x = 10\,\text{cm}$, then DOK = 0.5824, Chf = −0.4176, MChf = 0.4176, $P_r = 0.703$, $P_m/\text{i} = 0.297$, with $Pc = P_r + P_m/\text{i} = 0.703 + 0.297 = 1$ and $Pc = \text{DOK} + \text{MChf} = 0.5824 + 0.4176 = 1$. We can see here that with the increase of x and consequently the decrease of MChf, the real probability P_r also increases.

If $x = 30\,\text{cm}$, then DOK = 0.8962, Chf = −0.1038, MChf = 0.1038, $P_r = 0.94509$, $P_m/\text{i} = 0.05491$, with $Pc = P_r + P_m/\text{i} = 0.94509 + 0.05491 = 1$ and $Pc = \text{DOK} + \text{MChf} = 0.8962 + 0.1038 = 1$.

If $x = 60\,\text{cm}$ (point H), then DOK = 0.998617, Chf = −0.001383, MChf = 0.001383, $P_r = 0.9993081$, $P_m/\text{i} = 0.0006919$, with $Pc = P_r + P_m/\text{i} = 0.9993081 + 0.0006919 = 1$ and $Pc = \text{DOK} + \text{MChf} = 0.998617 + 0.001383 = 1$. Here, since $x = 60\,\text{cm}$ which is relatively very big in this case, then the real probability P_r is very near to 1 since we will reach total and full dispersion very soon.

With the increase of displacement beyond 60 cm, MChf and Chf are nearly zero, DOK returns to 1, hence we have the total dispersion of the system of the oxygen gas particles in air gas. At this last point, $P_r \approx 1$, $P_m/\text{i} \approx 0$, with $Pc = P_r + P_m/\text{i} \approx 1 + 0 = 1$ and $Pc = \text{DOK} + \text{MChf} \approx 1 + 0 = 1$; thus, the logical explanation of the value of DOK ≈ 1 follows.

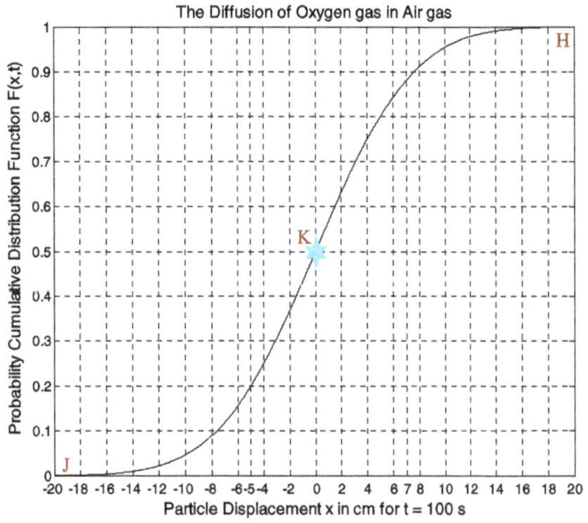

Figure 22. The CDF of the diffusion of oxygen gas in air gas for $t = 100\,\text{s}$.

Note that, if $x \to +\infty\,\text{cm}$, then DOK → 1, Chf → 0, MChf → 0, $P_r \to 1$, $P_m/\text{i} \to 0$, with $Pc = P_r + P_m/\text{i} \to 1 + 0 = 1$ and $Pc = \text{DOK} + \text{MChf} \to 1 + 0 = 1$.

Moreover, at each value of x, the probability to find the particle $Pc(x,t)$ is certainly predicted in the complex set $\mathcal{C}$ with Pc maintained as equal to one through a continuous compensation between DOK and Chf. This compensation is from $x = -\infty$ where $P_r = 0$ until $x = +\infty$ where $P_r = 1$, keeping always $Pc = \text{DOK} - \text{Chf} = \text{DOK} + \text{MChf} = 1$. We can understand now that DOK is the measure of our certain knowledge in $\mathcal{R}$ (100% probability) about the expected event, it does not include any uncertain knowledge (with a probability less than 100%). Furthermore, what is truly interesting in the particle displacement simulations is that we have quantified and visualized both the DOK and the chaotic factor of the Brownian motion.

We note that the same logic and analysis for the first instant of Brownian motion were applied to the second instant concerning the diffusion as well as all the EKA parameters.

8.3. The paradigm functions analysis for t = 100 s

We notice from Figures 22–24 that the DOK is maximum (DOK = 1) when MChf is minimum (MChf = 0) (points J & H) and that means when the MChf decreases our certain knowledge in $\mathcal{R}$ increases.

At the beginning $P_r(x < -20\,\text{cm}) \approx 0$, the system is nearly intact and has nearly zero chaotic factor (Chf = MChf ≈ 0) before any diffusion, hence at this instant DOK ≈ 1. Here $P_m/\text{i} \approx 1$ with $Pc = P_r + P_m/\text{i} \approx 0 + 1 = 1$ and $Pc = \text{DOK} + \text{MChf} \approx 1 + 0 = 1$.

Note that, if $x \to -\infty\,\text{cm}$, then DOK → 1, Chf → 0, MChf → 0, $P_r \to 0$, $P_m/\text{i} \to 1$, with $Pc = P_r + P_m/\text{i} \to 0 + 1 = 1$ and $Pc = \text{DOK} + \text{MChf} \to 1 + 0 = 1$.

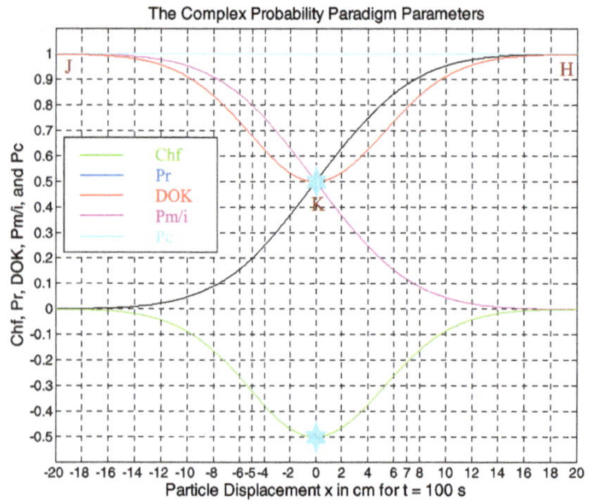

Figure 23. The complex probability parameters for $t = 100$ s.

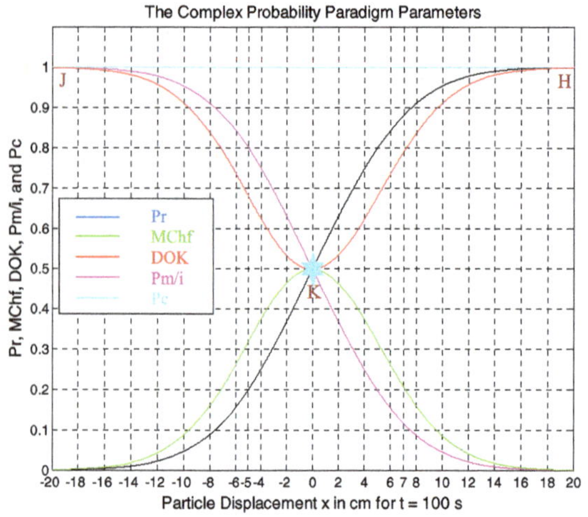

Figure 24. The complex probability parameters with MChf for $t = 100$ s.

Afterward, with diffusion, x starts to increase with $-\infty < x < +\infty, P_r(L) = \int_{-\infty}^{L} \rho(x,t)\, dx \neq 0$ and we have always $Pc(L) = P_r(L) + P_m/i(L) = \text{DOK}(L) + \text{MChf}(L) = 1$, thus MChf also starts to increase during the diffusion due to the environment and intrinsic conditions thus leading to a decrease in DOK.

If $x = -20$ cm (point J), then DOK = 0.9992513, Chf = − 0.0007487, MChf = 0.0007487, $P_r = 0.0003745$, $P_m/i = 0.9996255$, with $Pc = P_r + P_m/i = 0.0003745 + 0.9996255 = 1$ and $Pc = \text{DOK} + \text{MChf} = 0.9992513 + 0.0007487 = 1$.

If $x = -10$ cm, then DOK = 0.91233, Chf = − 0.08767, MChf = 0.08767, $P_r = 0.04595$, $P_m/i = 0.95405$, with $Pc = P_r + P_m/i = 0.04595 + 0.95405 = 1$ and $Pc = \text{DOK} + \text{MChf} = 0.91233 + 0.08767 = 1$. We can see that with the increase of x, the real probability P_r also increases.

If $x = -5$ cm, then DOK = 0.6804, Chf = − 0.3196, MChf = 0.3196, $P_r = 0.1997$, $P_m/i = 0.8003$, with $Pc = P_r + P_m/i = 0.1997 + 0.8003 = 1$ and $Pc = \text{DOK} + \text{MChf} = 0.6804 + 0.3196 = 1$. We can see here that with the increase of x, the MChf also increases, whereas DOK decreases.

If $x = 0$ cm (point K), both DOK (minimum) and MChf (maximum) reach 0.5, $P_r = 0.5$, $P_m/i = 0.5$, and Chf = − 0.5 with $Pc = P_r + P_m/i = 0.5 + 0.5 = 1$ and $Pc = \text{DOK} + \text{MChf} = 0.5 + 0.5 = 1$, as always. Hence, all the EKA parameters will intersect at the point K. We have here maximum chaos and the minimum of the system knowledge in $\mathcal{R}$; therefore, the real probability is $P_r = 1/2$ = probability half way to full diffusion. Furthermore, notice in Figure 24 the complete symmetry at the vertical axis $x = 0$ cm = half way to complete dispersion.

If $x = 7$ cm, then DOK = 0.7903, Chf = − 0.2097, MChf = 0.2097, $P_r = 0.881$, $P_m/i = 0.119$, with $Pc = P_r + P_m/i = 0.881 + 0.119 = 1$ and $Pc = \text{DOK} + \text{MChf} = 0.7903 + 0.2097 = 1$. We can see here that with the increase of x and consequently the decrease of MChf, the real probability P_r also increases.

If $x = 12$ cm, then DOK = 0.95781, Chf = − 0.04219, MChf = 0.04219, $P_\text{r} = 0.97844$, $P_\text{m}/i = 0.02156$, with $Pc = P_\text{r} + P_\text{m}/i = 0.97844 + 0.02156 = 1$ and $Pc = \text{DOK} + \text{MChf} = 0.95781 + 0.04219 = 1$.

If $x = 20$ cm (point H), then DOK = 0.9992513, Chf = − 0.0007487, MChf = 0.0007487, $P_r = 0.9996255$, $P_m/i = 0.0003745$, with $Pc = P_r + P_m/i = 0.9996255 + 0.0003745 = 1$ and $Pc = \text{DOK} + \text{MChf} = 0.9992513 + 0.0007487 = 1$. Here, since $x = 20$ cm which is relatively very big in this case, then the real probability P_r is very near to 1 since we will reach total and full dispersion very soon.

With the increase of displacement beyond 20 cm, MChf and Chf are nearly zero, DOK returns to 1; hence, we have the total dispersion of the system of the oxygen gas particles in air gas. At this last point, $P_r \approx 1$, $P_m/i \approx 0$, with $Pc = P_r + P_m/i \approx 1 + 0 = 1$ and $Pc = \text{DOK} + \text{MChf} \approx 1 + 0 = 1$; thus, the logical explanation of the value of DOK ≈ 1 follows.

Note that if $x \to +\infty$ cm, then DOK → 1, Chf → 0, MChf → 0, $P_r \to 1$, $P_m/i \to 0$, with $Pc = P_r + P_m/i \to 1 + 0 = 1$ and $Pc = \text{DOK} + \text{MChf} \to 1 + 0 = 1$.

Moreover, at each value of x, the probability to find the particle $Pc(x,t)$ is certainly predicted in the complex set $\mathcal{C}$ with Pc maintained as equal to one through a continuous compensation between DOK and Chf. This compensation is from $x = -\infty$, where $P_r = 0$ until $x = +\infty$, where $P_r = 1$, keeping always throughout the whole process $Pc = \text{DOK} - \text{Chf} = \text{DOK} + \text{MChf} = 1$.

It is clear from the particle displacements simulations that DOK is the measure of our certain knowledge in $\mathcal{R}$ (100% probability) about the expected event and it does not include any uncertain knowledge (with a probability less than 100%).

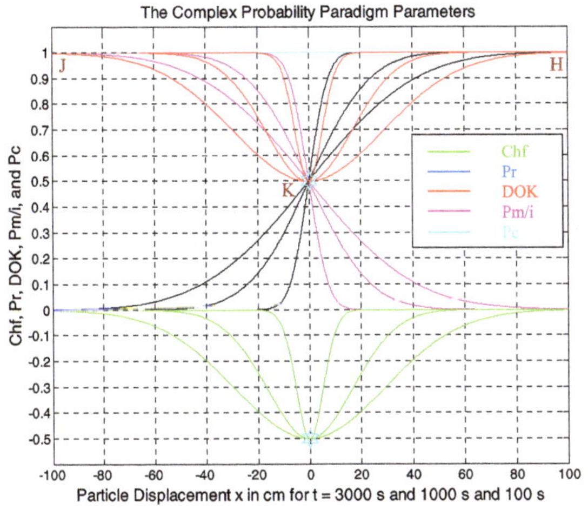

Figure 25. The complex probability parameters for $t = 3000$ s, $t = 1000$ s, and $t = 100$ s.

Furthermore, what is truly interesting in three instants simulations is that we have quantified and visualized both the DOK and the chaotic factor of the Brownian motion.

We note that the same methodology and analysis for the first and second instants were applied to the third instant concerning the diffusion, as well as all the EKA parameters. Thus, we can consequently conclude that whatever the instant is, both the logic and the method implemented are similar. This proves the validity of the new axioms developed and of the novel prognostic model adopted.

Figure 25 summarizes what has been previously explained.

9. The new paradigm and entropy

In the nineteenth century, specifically in the 1870s, Ludwig Boltzmann developed the statistical definition of entropy as a consequence of analysing the statistical behaviour of the microscopic components of a system (Boltzmann, 1995; Cercignani, 2010; Planck, 1969; Wikipedia, *Brownian Motion*; Wikipedia, *Entropy*; Wikipedia, *Mass Diffusivity*). The definition of entropy proposed by Boltzmann proved that entropy was equivalent to the thermodynamic entropy to within a constant number which is designated by Boltzmann's constant. This result was ascertained by Boltzmann himself. To summarize, the experimental definition of entropy was provided by the definition of the thermodynamic definition, whereas the definition of the statistical entropy extends the concept and yields a deeper explanation and a profound understanding of its nature.

Moreover, in statistical mechanics, the interpretation of entropy is the measure of uncertainty. Gibbs used the phrase *mixedupness* to designate entropy. When we determine the system set of macroscopic variables, the entropy measures the degree to which the probability of the system is spread out over different microstates. A macrostate characterizes simply observable average quantities, whereas a microstate expresses all the details of molecules in a system and this includes the velocity and the position of every molecule, knowing that the greater the available system states probability, the bigger the system entropy. In addition, we state that in statistical mechanics, entropy is a measure of the number of ways in which a system may be arranged, usually it is considered as a measure of 'disorder', that means that the higher the entropy, the greater the disorder. This definition describes the entropy as being proportional to the natural logarithm of the number of possible microscopic configurations of the individual atoms and molecules of the system (microstates) which could give rise to the observed macroscopic state (macro-state) of the system. The constant of proportionality is the Boltzmann constant.

More specifically, entropy is a logarithmic measure of the number of states with significant probability of being occupied; so mathematically we write

$$S = -k_B \sum_j p_j \operatorname{Ln}(p_j), \tag{26}$$

where k_B is the Boltzmann constant, equal to 1.38065×10^{-23} J/K or 8.6173324×10^{-5} eV/K. The summation is over all the possible microstates of the system, and p_j is the probability that the system is in the jth microstate. This definition assumes that the basis set of states has been picked so that there is no information on their relative phases.

In a different basis set, the more general expression is as follows:

$$S = -k_B \operatorname{Tr}(\hat{\rho} \operatorname{Ln}(\hat{\rho})), \tag{27}$$

where $\hat{\rho}$ is the density matrix and Ln is the matrix logarithm.

> This density matrix formulation is not needed in cases of thermal equilibrium so long as the basis states are chosen to be energy eigen-states. For most practical purposes, this can be taken as the fundamental definition of entropy since all other formulas for S can be mathematically derived from it, but not vice versa. (Wikipedia, *Entropy*)

In addition, the *fundamental assumption of statistical thermodynamics* or *the fundamental postulate in statistical mechanics* states that the occupation of any microstate is assumed to be equally probable, that is: $p_j = 1/\Omega$, where Ω is the number of microstates. It is important to mention that this assumption is usually justified for an isolated system in equilibrium. Then, Equation (26) becomes equal to

$$S = k_B \operatorname{Ln} \Omega. \tag{28}$$

Furthermore,

> the most general interpretation of entropy is as a measure of our uncertainty about a system. The equilibrium state

of a system maximizes the entropy because we have lost all information about the initial conditions except for the conserved variables; maximizing the entropy maximizes our ignorance about the details of the system. This uncertainty is not of the everyday subjective kind, but rather the uncertainty inherent to the experimental method and interpretative model. In addition, entropy can be defined for any Markov processes with reversible dynamics and the detailed balance property. (Wikipedia, *Entropy*)

In 1896, in his book entitled 'Lectures on Gas Theory', Ludwig Boltzmann showed that this expression yields a measure of the entropy for systems of atoms and molecules in the gas phase, and consequently it provided a measure for the classical thermodynamics entropy.

Also, the very well-known second law of thermodynamics states that in general any system total entropy will not decrease other than by increasing the entropy of some other system. Therefore, in a system isolated from its environment, the entropy of that system will tend not to decrease. So mathematically we write

$$dS \geq 0. \tag{29}$$

It follows that

heat will not flow from a colder body to a hotter body without the application of work (the imposition of order) to the colder body. Secondly, it is impossible for any device operating on a cycle to produce network from a single temperature reservoir; the production of network requires flow of heat from a hotter reservoir to a colder reservoir, or a single expanding reservoir undergoing adiabatic cooling, which performs adiabatic work. As a result, there is no possibility of a perpetual motion system. It follows that a reduction in the increase of entropy in a specified process, such as a chemical reaction, means that it is energetically more efficient. (Wikipedia, *Entropy*)

Additionally, statistical mechanics shows that entropy is governed by probability. In fact, it allows a disorder decrease even in an isolated system. Although this is mathematically possible, such an event has a small probability of occurring, making it unlikely.

Hence, in the novel complex probability paradigm, we can deduce the following consequences.

In the set $\mathscr{R}$, we denote the corresponding real entropy by S_R and $\Omega =$ the number of microstates. We have for an isolated system in equilibrium the real probability equals to

$$p_j = P_r = \frac{1}{\Omega} \Rightarrow S_R = -k_B \sum_{j=1}^{\Omega} \frac{1}{\Omega} \mathrm{Ln}\left(\frac{1}{\Omega}\right)$$
$$= -k_B \left(\Omega \times \frac{1}{\Omega}\right)(-\mathrm{Ln}\,\Omega)$$
$$\Rightarrow S_R = k_B \mathrm{Ln}\,\Omega \tag{30}$$

and is a divergent non-decreasing series

$$\Rightarrow dS_R \geq 0 \quad \text{and} \quad \lim_{\Omega \to +\infty} dS_R = +\infty \tag{31}$$

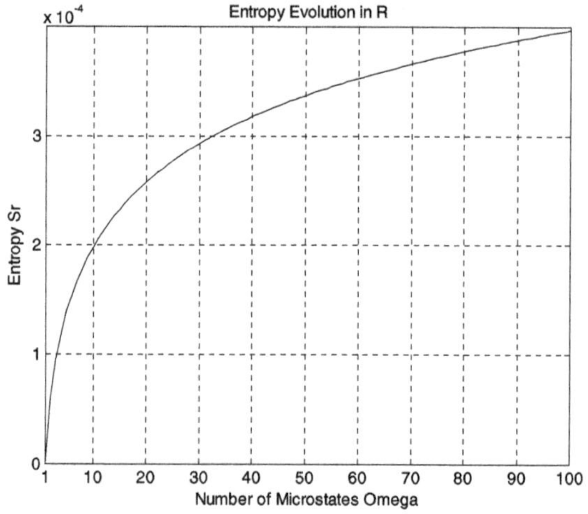

Figure 26. The real entropy S_R in $\mathscr{R}$ as function of the number of microstates Ω.

that means in $\mathscr{R}$, chaos and disorder are increasing with time (Figure 26).

In the set $\mathscr{M}$, we denote the corresponding imaginary entropy by S_M. We have for an isolated system in equilibrium the imaginary probability equals to

$$p_j = P_m = \mathrm{i}(1 - P_r) = \mathrm{i}\left(1 - \frac{1}{\Omega}\right)$$
$$\Rightarrow S_M = -k_B \sum_{j=1}^{\Omega} \mathrm{i}\left(1 - \frac{1}{\Omega}\right) \mathrm{Ln}\left[\mathrm{i}\left(1 - \frac{1}{\Omega}\right)\right]. \tag{32}$$

In the complementary real probability set to $\mathscr{R}$, we denote the corresponding real entropy by $\bar{S}_R$.

The meaning of $\bar{S}_R$ is the following: it is the real entropy in the real set $\mathscr{R}$ and which is related to the complementary real probability $p_j = P_m/\mathrm{i} = 1 - P_r$. We have for an isolated system in equilibrium the complementary probability equals to

$$p_j = \frac{P_m}{\mathrm{i}} = 1 - P_r = 1 - \frac{1}{\Omega} \Rightarrow$$
$$\bar{S}_R = -k_B \sum_{j=1}^{\Omega} \left(1 - \frac{1}{\Omega}\right) \mathrm{Ln}\left(1 - \frac{1}{\Omega}\right) = -k_B \times \Omega$$
$$\times \left(1 - \frac{1}{\Omega}\right) \times \left[\mathrm{Ln}\left(\frac{\Omega - 1}{\Omega}\right)\right]$$
$$= -k_B \times (\Omega - 1) \times \left[\mathrm{Ln}\left(\frac{\Omega - 1}{\Omega}\right)\right]$$
$$\Rightarrow \bar{S}_R = k_B \times (\Omega - 1) \times \left[\mathrm{Ln}\left(\frac{\Omega}{\Omega - 1}\right)\right] \tag{33}$$

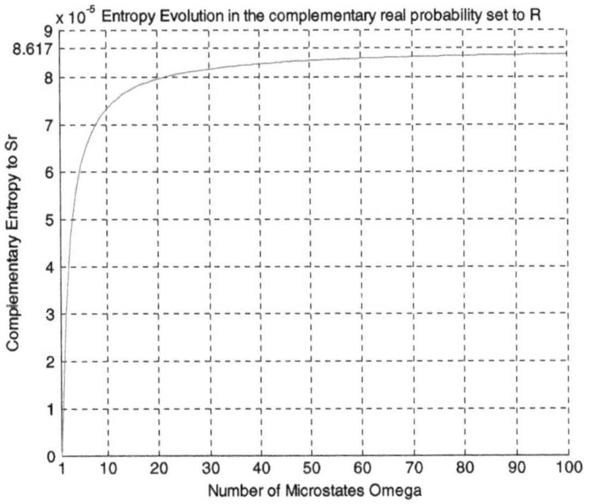

Figure 27. The complementary entropy $\bar{S}_R$ to S_R in $\mathscr{R}$ as function of the number of microstates Ω.

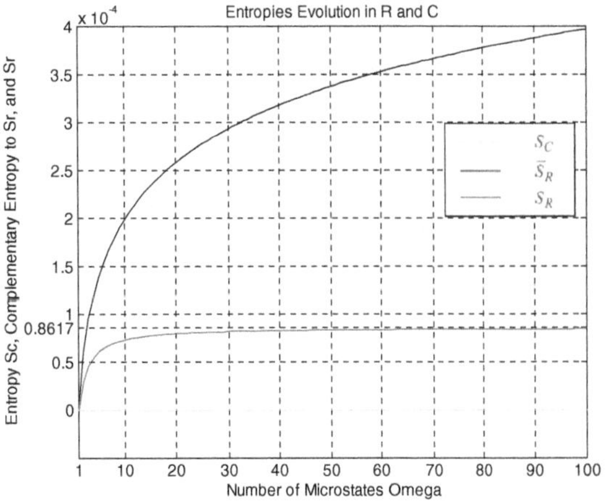

Figure 28. The entropies S_C, $\bar{S}_R$, and S_R as functions of the number of microstates Ω.

and is a convergent non-decreasing series. In fact,

$$\lim_{\Omega\to+\infty}\sum_{j=1}^{\Omega}\left(1-\frac{1}{\Omega}\right)\mathrm{Ln}\left(1-\frac{1}{\Omega}\right)=-1\Rightarrow\bar{S}_R$$
$$\to[-k_B\times(-1)]=k_B$$
$$\Rightarrow d\bar{S}_R\geq 0\quad\text{and}\quad\lim_{\Omega\to+\infty}d\bar{S}_R=0,\tag{34}$$

that means in the complementary real probability set to $\mathscr{R}$, chaos is increasing with time, and its corresponding entropy is converging to the Boltzmann constant $k_B = 8.6173324\times 10^{-5}$ eV/K (Figure 27).

In the set $\mathscr{C}$, we denote the corresponding real entropy by S_C. We have for an isolated system in equilibrium the real probability equals to

$$p_j = Pc = 1\Rightarrow S_C = -k_B\sum_j p_j\mathrm{Ln}(p_j)$$
$$= -k_B\sum_j 1\times\mathrm{Ln}1\Rightarrow S_C = 0\tag{35}$$

$$\text{and is a constant series}\Rightarrow dS_C = 0\quad\forall\Omega\in[1,+\infty)\tag{36}$$

that means, in $\mathscr{C}=\mathscr{R}+\mathscr{M}$, we have complete order, no chaos, and no ignorance since all measurements are completely deterministic (Figure 28).

10. The resultant complex random vector *Z*

I will describe in this section a powerful tool, developed in a personal previous paper, based on the concept of a complex random vector which is a vector representing the real and the imaginary probabilities of an outcome, defined in the added axioms by the term $z = P_r + P_m$ (Abou Jaoude, 2013a, 2013b, 2014, 2015; Abou Jaoude et al., 2010). Then express the resultant complex random vector Z as the vector which is the sum of all the complex random vectors in the complex probability space $\mathscr{C}$. I will illustrate this methodology by considering a general Bernoulli distribution first, then a discrete distribution with N random variables as a general case. Afterward, I will prove the very well-known law of large numbers using this new powerful concept. In fact, if z represents one particle in Brownian motion, then Z represents the whole system of particles in a gas or liquid that means the whole random distribution in the complex probability space $\mathscr{C}$. So it follows directly that a Bernoulli distribution can be understood as a simplified system with two random particles (paragraph 10.1), whereas the general case is a random system with N particles (paragraph 10.2).

10.1. The resultant complex random vector Z of a general Bernoulli distribution

First, let us define the complex random vectors and their resultant by considering the following general Bernoulli distribution (Table 1):

We have

$$\sum_{j=1}^{2}P_{rj} = P_{r1}+P_{r2} = p+q = 1$$

and

$$\sum_{j=1}^{2}P_{mj} = P_{m1}+P_{m2} = iq+ip = i(1-p)+ip$$
$$= i-ip+ip = i = i(2-1) = i(N-1),$$

where N is the number of random variables which is equal to 2 for a Bernoulli distribution.

Table 1. A general Bernoulli distribution.

Outcome	x_j	x_1	x_2
In $\mathcal{R}$	P_{rj}	$P_{r1} = p$	$P_{r2} = q$
In $\mathcal{M}$	P_{mj}	$P_{m1} = \mathrm{i}(1-p) = \mathrm{i}q$	$P_{m2} = \mathrm{i}(1-q) = \mathrm{i}p$
In $\mathcal{C} = \mathcal{R} + \mathcal{M}$	z_j	$z_1 = P_{r1} + P_{m1}$	$z_2 = P_{r2} + P_{m2}$

Notes: x_1 and x_2 are the outcomes of the first and second random variables, respectively. P_{r1} and P_{r2} are the real probabilities of x_1 and x_2, respectively. P_{m1} and P_{m2} are the imaginary probabilities of x_1 and x_2, respectively.

The complex random vector corresponding to the random outcome x_1 is as follows:

$$z_1 = P_{r1} + P_{m1} = p + \mathrm{i}(1-p) = p + \mathrm{i}q.$$

The complex random vector corresponding to the random outcome x_2 is as follows:

$$z_2 = P_{r2} + P_{m2} = q + \mathrm{i}(1-q) = q + \mathrm{i}p.$$

The resultant complex random vector is defined as follows:

$$\begin{aligned} Z = z_1 + z_2 &= \sum_{j=1}^{2} P_{rj} + \sum_{j=1}^{2} P_{mj} \\ &= (p + \mathrm{i}q) + (q + \mathrm{i}p) = (p+q) + \mathrm{i}(p+q) \\ &= 1 + \mathrm{i} = 1 + \mathrm{i}(2-1) \\ &\Rightarrow Z = 1 + \mathrm{i}(N-1). \end{aligned} \tag{37}$$

The probability Pc_1 in the complex space $\mathcal{C} = \mathcal{R} + \mathcal{M}$ which corresponds to the complex random vector z_1 is computed as follows:

$$|z_1|^2 = P_{r1}^2 + \left(\frac{P_{m1}}{\mathrm{i}}\right)^2 = p^2 + q^2,$$

$$\begin{aligned} \mathrm{Chf}_1 &= -2P_{r1}\frac{P_{m1}}{\mathrm{i}} = -2pq, \Rightarrow Pc_1^2 = |z_1|^2 - \mathrm{Chf}_1 \\ &= p^2 + q^2 + 2pq = (p+q)^2 = 1^2 = 1 \Rightarrow Pc_1 = 1. \end{aligned}$$

This is coherent with the three new complementary axioms defined for the extended Kolmogorov's system.

Similarly, Pc_2 corresponding to z_2 is as follows:

$$|z_2|^2 = P_{r2}^2 + \left(\frac{P_{m2}}{\mathrm{i}}\right)^2 = q^2 + p^2,$$

$$\begin{aligned} \mathrm{Chf}_2 &= -2P_{r2}\frac{P_{m2}}{\mathrm{i}} = -2qp \Rightarrow Pc_2^2 = |z_2|^2 - \mathrm{Chf}_2 \\ &= q^2 + p^2 + 2qp = (q+p)^2 = 1^2 = 1 \Rightarrow Pc_2 = 1. \end{aligned}$$

The probability Pc in the complex space $\mathcal{C}$ which corresponds to the resultant complex random vector $Z = 1 + \mathrm{i}$ is computed as follows:

$$|Z|^2 = \left(\sum_{j=1}^{2} P_{rj}\right)^2 + \left(\sum_{j=1}^{2} \frac{P_{mj}}{\mathrm{i}}\right)^2 = 1^2 + 1^2 = 2,$$

$$\mathrm{Chf} = -2\sum_{j=1}^{2} P_{rj} \sum_{j=1}^{2} \frac{P_{mj}}{\mathrm{i}} = -2(1)(1) = -2.$$

$$\begin{aligned} \text{Let } s^2 &= |Z|^2 - \mathrm{Chf} = 2 + 2 = 4 \Rightarrow s = 2 \Rightarrow Pc^2 = \frac{s^2}{N} \\ &= \frac{|Z|^2 - \mathrm{Chf}}{N^2} = \frac{|Z|^2}{N^2} - \frac{\mathrm{Chf}}{N^2} \Rightarrow Pc = \frac{s}{N} = \frac{2}{2} = 1, \end{aligned}$$

where s is an intermediary quantity used in our computation of Pc.

Pc is the probability corresponding to the resultant complex random vector Z in the universe $\mathcal{C} = \mathcal{R} + \mathcal{M}$ and is also equal to 1. In fact, Z represents both z_1 and z_2 that means the whole distribution of random variables in the complex space $\mathcal{C}$ and its probability Pc is computed in the same way as Pc_1 and Pc_2.

By analogy, for the case of one random variable z_j we have

$$Pc_j^2 = |z_j|^2 - \mathrm{Chf}_j \quad \text{with } (N = 1).$$

In general, for the vector Z we have

$$Pc^2 = \frac{|Z|^2}{N^2} - \frac{\mathrm{Chf}}{N^2}; \quad (N \geq 1), \tag{38}$$

where the DOK of the whole distribution is equal to $(|Z|^2/N^2)$ and its relative chaotic factor is (Chf/N^2).

Notice, if $N = 1$ in Equation (38), then:

$$\begin{aligned} Pc^2 &= \frac{|Z|^2}{N^2} - \frac{\mathrm{Chf}}{N^2} = \frac{|Z|^2}{1^2} - \frac{\mathrm{Chf}}{1^2} = |Z|^2 - \mathrm{Chf} = |z_j|^2 \\ &- \mathrm{Chf}_j = Pc_j^2, \end{aligned}$$

which is coherent with the calculations already done.

To illustrate the concept of the resultant complex random vector Z, I will use Figure 29.

10.2. *The general case: a discrete distribution with N random variables*

As a general case, let us consider then this discrete probability distribution with N equiprobable random variables (Table 2).

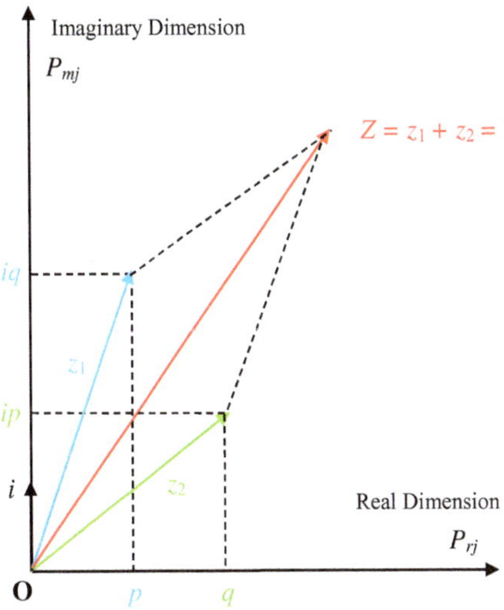

Figure 29. The resultant complex random vector $Z = z_1 + z_2$ for a general Bernoulli distribution in the complex probability space $\mathscr{C}$.

We have here in $\mathscr{C} = \mathscr{R} + \mathscr{M}$:

$$z_j = P_{rj} + P_{mj} \quad \forall j: \quad 1 \le j \le N$$

and

$$z_1 = z_2 = \cdots = z_N = \frac{1}{N} + \frac{\mathrm{i}(N-1)}{N} \Rightarrow Z = z_1 + z_2 + \cdots + z_N = Nz_j = N\left(\frac{1}{N} + \frac{\mathrm{i}(N-1)}{N}\right) = 1 + \mathrm{i}(N-1).$$

Moreover, we can notice that $|z_1| = |z_2| = \cdots = |z_N|$, hence,

$$|Z| = |z_1 + z_2 + \cdots + z_N| = N|z_1| = N|z_2| = \cdots = N|z_N| \Rightarrow |Z|^2 = N^2|z_j|^2 = N^2\left(\frac{1}{N^2} + \frac{(N-1)^2}{N^2}\right) = 1 + (N-1)^2 \quad \text{where } 1 \le j \le N$$

and

$$Chf = N^2 \times Chf_j = -2 \times P_{r_j} \times \frac{P_{m_j}}{i} \times N^2 = -2N^2 \times \left(\frac{1}{N}\right)\left(\frac{N-1}{N}\right) = -2(1)(N-1) = -2(N-1)$$

$$= -2(N-1) \Rightarrow s^2 = |Z|^2 - \mathrm{Chf} = 1 + (N-1)^2 + 2(N-1) = N^2 \Rightarrow Pc^2 = \frac{s^2}{N^2} = \frac{N^2}{N^2} = 1 \Rightarrow Pc = 1,$$

where s is an intermediary quantity used in our computation of Pc.

Therefore, the DOK corresponding to the resultant complex vector Z representing the whole distribution is as follows:

$$\mathrm{DOK}_Z = \frac{|Z|^2}{N^2} = \frac{1 + (N-1)^2}{N^2} \tag{39}$$

and its relative chaotic factor is as follows:

$$\mathrm{Chf}_Z = \frac{\mathrm{Chf}}{N^2} = -\frac{2(N-1)}{N^2}. \tag{40}$$

Similarly, its relative MChf is as follows:

$$\mathrm{MChf}_Z = |\mathrm{Chf}_Z| = \left|\frac{\mathrm{Chf}}{N^2}\right| = \left|-\frac{2(N-1)}{N^2}\right| = \frac{2(N-1)}{N^2}. \tag{41}$$

Thus, we can verify that we always have

$$Pc^2 = \frac{|Z|^2}{N^2} - \frac{\mathrm{Chf}}{N^2} = \mathrm{DOK}_Z - \mathrm{Chf}_Z = \mathrm{DOK}_Z + \mathrm{MChf}_Z = 1. \tag{42}$$

What is important here is that we can notice the following: Take for example:

$$N = 2 \Rightarrow \frac{|Z|^2}{N^2} = \frac{1 + (2-1)^2}{2^2} = 0.5 \quad \text{and} \quad \frac{\mathrm{Chf}}{N^2} = \frac{-2(2-1)}{2^2} = -0.5,$$

$$N = 4 \Rightarrow \frac{|Z|^2}{N^2} = \frac{1 + (4-1)^2}{4^2} = 0.625 \ge 0.5 \quad \text{and} \quad \frac{\mathrm{Chf}}{N^2} = \frac{-2(4-1)}{4^2} = -0.375 \ge -0.5,$$

$$N = 5 \Rightarrow \frac{|Z|^2}{N^2} = \frac{1 + (5-1)^2}{5^2} = 0.68 \ge 0.625 \quad \text{and} \quad \frac{\mathrm{Chf}}{N^2} = \frac{-2(5-1)}{5^2} = -0.32 \ge -0.375,$$

Table 2. A discrete distribution with N equiprobable random variables.

Outcome	x_j	x_1	x_2	$\dots$	x_N
In $\mathscr{R}$	P_{rj}	$P_{r1} = \frac{1}{N}$	$P_{r2} = \frac{1}{N}$	$\dots$	$P_{rN} = \frac{1}{N}$
In $\mathscr{M}$	P_{mj}	$P_{m1} = \mathrm{i}\left(1 - \frac{1}{N}\right)$	$P_{m2} = \mathrm{i}\left(1 - \frac{1}{N}\right)$	$\dots$	$P_{mN} = \mathrm{i}\left(1 - \frac{1}{N}\right)$

$$N = 10 \Rightarrow \frac{|Z|^2}{N^2} = \frac{1+(10-1)^2}{10^2} = 0.82 \geq 0.68 \quad \text{and}$$

$$\frac{\text{Chf}}{N^2} = \frac{-2(10-1)}{10^2} = -0.18 \geq -0.32,$$

$$N = 100 \Rightarrow \frac{|Z|^2}{N^2} = \frac{1+(100-1)^2}{100^2}$$

$$= 0.9802 \geq 0.82 \quad \text{and} \quad \frac{\text{Chf}}{N^2} = \frac{-2(100-1)}{100^2}$$

$$= -0.0198 \geq -0.18,$$

$$N = 1000 \Rightarrow \frac{|Z|^2}{N^2} = \frac{1+(1000-1)^2}{1000^2}$$

$$= 0.998002 \geq 0.9802, \quad \text{and}$$

$$\frac{\text{Chf}}{N^2} = \frac{-2(1000-1)}{1000^2} = -0.001998 \geq -0.0198.$$

We can deduce mathematically that

$$\lim_{N\to+\infty} \frac{|Z|^2}{N^2} = \lim_{N\to+\infty} \frac{1+(N-1)^2}{N^2} = 1 \qquad (43)$$

and

$$\lim_{N\to+\infty} \frac{\text{Chf}}{N^2} = \lim_{N\to+\infty} -\frac{2(N-1)}{N^2} = 0. \qquad (44)$$

From the above, we can also deduce this conclusion.

As much as N increases, as much as the DOK in $\mathcal{R}$ corresponding to the resultant complex vector is perfect, that is, it is equal to 1, and as much as the chaotic factor that forbids us from predicting exactly the result of the random experiment in $\mathcal{R}$ approaches 0. Mathematically, we say: if N tends to infinity then the DOK in $\mathcal{R}$ tends to 1 and the chaotic factor tends to 0.

Moreover,

$$\text{For } N = 1 \Rightarrow \frac{|Z|^2}{N^2} = \frac{1+(1-1)^2}{1^2} = 1 \quad \text{and}$$

$$\frac{\text{Chf}}{N^2} = \frac{-2(1-1)}{1^2} = 0.$$

This means that we have a random experiment with only one outcome, hence, either $P_r = 1$ or $P_r = 0$, that means we have, respectively, either a sure event or an impossible event in $\mathcal{R}$. For this, we have surely the DOK is 1 and the chaotic factor is 0 since the experiment is either certain or impossible, which is absolutely logical.

10.3. The resultant complex random vector Z and the law of large numbers

The law of large numbers states that:

'As N increases, then the probability that the value of sample mean to be close to population mean approaches 1.'

We can deduce now the following conclusion related to the law of large numbers.

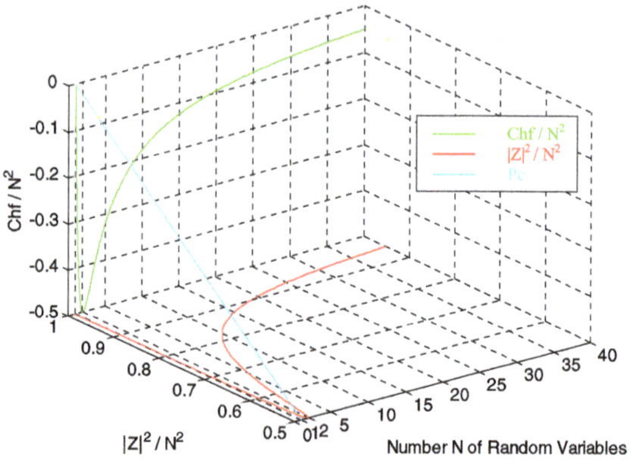

Figure 30. $\text{Chf}_Z = (\text{Chf}/N^2)$, $\text{DOK}_Z = (|Z|^2/N^2)$, and Pc, as functions of N in 3D.

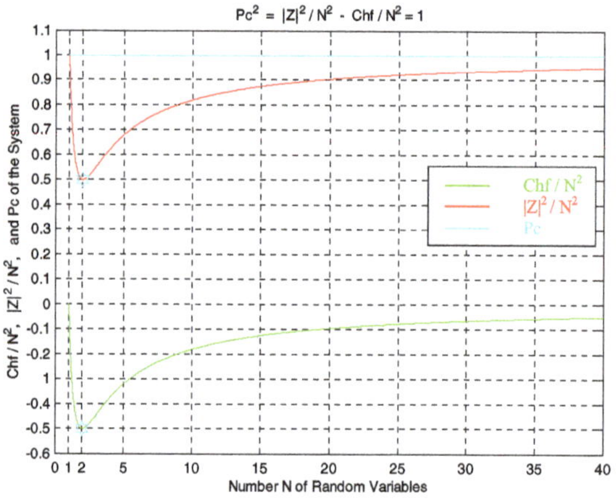

Figure 31. $\text{Chf}_Z = (\text{Chf}/N^2)$, $\text{DOK}_Z = (|Z|^2/N^2)$, and Pc, as functions of N in 2D.

We can see, as we have proved, that as much as N increases, as much as the DOK of the resultant complex vector $\text{DOK}_Z = (|Z|^2/N^2)$ tends to 1 and its relative chaotic factor $\text{Chf}_Z = (\text{Chf}/N^2)$ tends to 0. Assume now that the random variables x_j's correspond to the particles or molecules moving randomly in a gas or a liquid. So if we study a gas or a liquid with billions of such particles, N is big enough (e.g. Avogadro number $\approx 6.02214 \times 10^{23}$/mole in the International System of Units) to allow that its corresponding temperature, pressure, energy, etc. tend to the mean of these quantities corresponding to the whole system. This is because the chaotic factor of the whole gas or liquid, that is, of the resultant complex random vector Z representing all the random particles or vectors, tends to 0; thus, the behaviour and characteristics of the whole system in $\mathcal{R}$ is predictable with great precision since the DOK of the whole gas or liquid tends to 1. Figures 30 and 31 illustrate this result.

Hence, we have joined here two different key concepts which are as follows: the law of large numbers and the

Table 3. Computation of Pc for different of values of z_1 and z_2 which are the complex random vectors of a Bernoulli distribution and which are chosen at random.

The complex random vectors and their probabilities	z_1	Pc_1	z_2	Pc_2	Z	Pc
Simulation #1	0.8106 + i(0.1894)	1	0.1894 + i(0.8106)	1	1 + i	1
Simulation #2	0.0084 + i(0.9916)	1	0.9916 + i(0.0084)	1	1 + i	1
Simulation #3	0.4558 + i(0.5442)	1	0.5442 + i(0.4558)	1	1 + i	1
Simulation #4	0.5225 + i(0.4775)	1	0.4775 + i(0.5225)	1	1 + i	1
Simulation #5	0.3723 + i(0.6277)	1	0.6277 + i(0.3723)	1	1 + i	1
Simulation #6	0.1908 + i(0.8092)	1	0.8092 + i(0.1908)	1	1 + i	1
Simulation #7	0.208 + i(0.792)	1	0.792 + i(0.208)	1	1 + i	1

Notes: In this case, the resultant complex random vector is $Z = z_1 + z_2$ and is always equal to $1 + \mathrm{i}$. The corresponding probability of Z in $\mathscr{C}$ is always 1, just as postulated and expected.

Table 4. Computation of Pc for different of values of z_1, z_2, z_3 which are the complex random vectors of the distribution and which are chosen at random.

The complex random vectors and their probabilities	z_1	Pc_1	z_2	Pc_2	z_3	Pc_3	Z	Pc
Simulation #1	0.636 + i(0.364)	1	0.136 + i(0.864)	1	0.228 + i(0.772)	1	1 + i(2)	1
Simulation #2	0.8393 + i(0.1607)	1	0.0402 + i(0.9598)	1	0.1205 + i(0.8795)	1	1 + i(2)	1
Simulation #3	0.7802 + i(0.2198)	1	0.0220 + i(0.978)	1	0.1978 + i(0.8022)	1	1 + i(2)	1
Simulation #4	0.3619 + i(0.6381)	1	0.1381 + i(0.8619)	1	0.5 + i(0.5)	1	1 + i(2)	1
Simulation #5	0.9909 + i(0.0091)	1	0.0015 + i(0.9985)	1	0.0076 + i(0.9924)	1	1 + i(2)	1
Simulation #6	0.5205 + i(0.4795)	1	0.0533 + i(0.9467)	1	0.4262 + i(0.5738)	1	1 + i(2)	1
Simulation #7	0.0651 + i(0.9349)	1	0.2349 + i(0.7651)	1	0.7 + i(0.3)	1	1 + i(2)	1

Notes: In this case, the resultant complex random vector is $Z = z_1 + z_2 + z_3$ and is always equal to $1 + 2\mathrm{i}$. The corresponding probability of Z in $\mathscr{C}$ is always 1, just as postulated and expected.

Table 5. The resultant complex random vector $Z = z_1 + z_2 + \cdots + z_j + \cdots + z_N = Nz_j = 1 + i(N-1)$ with $1 \leq j \leq N$ and the verification of the law of large numbers.

Complex random vectors and their characteristics	N	z_j	Z	$\mathrm{DOK}_Z = \frac{\lvert Z\rvert^2}{N^2}$	$\mathrm{Chf}_Z = \frac{\mathrm{Chf}}{N^2}$	Pc
Simulation #1	1	1 + i(0)	1 + i(0)	1	0	1
Simulation #2	2	0.5 + i(0.5)	1 + i(1)	0.5	– 0.5	1
Simulation #3	3	0.3333 + i(0.6667)	1 + i(2)	0.5556	– 0.4444	1
Simulation #4	5	0.2 + i(0.8)	1 + i(4)	0.68	– 0.32	1
Simulation #5	10	0.1 + i(0.9)	1 + i(9)	0.82	– 0.18	1
Simulation #6	100	0.01 + i(0.99)	1 + i(99)	0.9802	– 0.0198	1
Simulation #7	1000	0.001 + i(0.999)	1 + i(999)	0.998002	– 0.001998	1
Simulation #8	10000	0.0001 + i(0.9999)	1 + i(9999)	0.99980002	– 0.00019998	1
Simulation #9	100000	1e – 005 + i(0.99999)	1 + i(99999)	0.9999800002	– 1.99998e – 005	1
Simulation #10	1000000	1e – 006 + i(0.999999)	1 + i(999999)	0.999998	– 1.999998e – 006	1
Simulation #11	1000000000	1e – 009 + i(0.999999999)	1 + i(999999999)	0.999999998	– 1.999999998e – 009	1
Simulation #12	10^{12} = 1e + 12	≈ 1e – 012 + i(1)	≈ 1 + i(1e + 012)	≈ 1	≈ – 2e – 012 ≈ 0	1

resultant complex random vector. The first concept comes from ordinary statistics, classical probability theory, and statistical mechanics, whereas the second concept comes from the new theory of complex probability and statistics. This looks very interesting and fruitful and shows the validity and the benefits of extending Kolmogorov's axioms to the complex set $\mathscr{C}$.

10.4. Numerical simulations

Numerical simulations verify what has been found earlier (Gentle, 2003). We will use Monte Carlo simulation method with the help of the programming language C++ with its predefined pseudorandom function rand() that generates random numbers with a uniform distribution. Table 3 is a simulation of a Bernoulli distribution where the complex random vectors are chosen randomly by C++. Table 4 is a simulation of a uniform distribution with three random variables having their complex random vectors also chosen randomly by C++. Table 5 is a simulation that confirms the direct relation between the resultant complex vector Z and the law of large numbers.

11. Conclusion and perspectives

In the current paper, I applied the theory of EKA to the classical theory of Brownian motion. Hence, I established a tight link between the new paradigm and diffusion. Thus, I developed the theory of 'Complex Probability' beyond the scope of my previous five papers on this topic.

As it was proved and illustrated, before any diffusion and after the full particles dispersion, the DOK is one and the chaotic factor (Chf and MChf) is 0 since the state of the system is totally known. During the process of gas diffusion ($-\infty < x < +\infty, 0 < t < +\infty$,) we have $0.5 \leq \text{DOK} < 1$, $-0.5 \leq \text{Chf} < 0$, and $0 < \text{MChf} \leq 0.5$. Notice that during the whole process of the Brownian motion, we have $Pc = \text{DOK} - \text{Chf} = \text{DOK} + \text{MChf} = 1$, that means that the phenomenon which seems to be random and stochastic in $\mathscr{R}$ is now deterministic and certain in $\mathscr{C} = \mathscr{R} + \mathscr{M}$, and this after adding to $\mathscr{R}$ the contributions of $\mathscr{M}$ and hence after subtracting the chaotic factor from the DOK. Moreover, for each value of the diffusion instant t, I have determined its corresponding probability of finding the gas particle in the interval $(-\infty, \text{L}]$. Therefore, at each instant, this probability is certainly predicted in the complex set $\mathscr{C}$ with Pc maintained as equal to one through a continuous compensation between DOK and Chf. This compensation is from the instant $t = 0$ (before any diffusion) until the instant of full gas dispersion when $t \to +\infty$. Furthermore, what is important, is that using all these graphs and simulations illustrated throughout the whole paper, we can visualize and quantify both the system chaos (Chf and MChf) and the system certain knowledge (DOK and Pc). I applied this novel methodology in the current research paper to the diffusion of oxygen gas in air gas, knowing that this novel paradigm can be applied to any diffusion phenomenon not discussed here.

Additionally, I have used in this paper a new powerful tool, already defined in a personal previous paper, which is the concept of the complex random vector that is a vector representing the real and the imaginary probabilities of an outcome, identified in the added axioms as being the term $z = P_r + P_m$. Then I have defined and expressed the resultant complex random vector as the vector which is the sum of all the complex random vectors and representing the whole distribution and system in the complex space $\mathscr{C}$. I have illustrated this methodology by considering a general Bernoulli distribution, then a discrete distribution with N random variables as a general case. Afterward, I have proven that there is a direct correlation between the concept of the resultant complex random vector and the very well-known law of large numbers. In fact, using this original concept and tool, I have succeeded to demonstrate the law of large numbers in a new way.

All the results found are certainly very interesting and fruitful and show once again the benefits of extending Kolmogorov's axioms and thus the originality and usefulness of this new field in applied mathematics and prognostic that can be called verily: 'The Complex Probability Paradigm.'

Additional development of this new paradigm will be done in subsequent works. And as prospective and future works, it is planned to more elaborate the novel proposed prognostic methodology and to apply it to a wide set of random dynamic systems and stochastic processes.

Disclosure statement

No potential conflict of interest was reported by the author.

References

Abou Jaoude, A. (2004). *Numerical methods and algorithms for applied mathematicians* (Ph.D. thesis in applied mathematics). Bircham International University. Retrieved August 1, 2004, from http://www.bircham.edu

Abou Jaoude, A. (2005). *Computer simulation of Montle Carlo methods and random phenomena* (Ph.D. thesis in computer science). Bircham International University. Retrieved October 14, 2005, from http://www.bircham.edu

Abou Jaoude, A. (2007). *Analysis and algorithms for the statistical and stochastic paradigm* (Ph.D. thesis in applied statistics and probability). Bircham International University. Retrieved April 27, 2007, from http://www.bircham.edu

Abou Jaoude, A. (2013a). The complex statistics paradigm and the law of large numbers. *Journal of Mathematics and Statistics*, *9*(4), 289–304.

Abou Jaoude, A. (2013b). The theory of complex probability and the first order reliability method. *Journal of Mathematics and Statistics*, *9*(4), 310–324.

Abou Jaoude, A. (2014). Complex probability theory and prognostic. *Journal of Mathematics and Statistics*, *10*(1), 1–24.

Abou Jaoude, A. (2015). The complex probability paradigm and analytic linear prognostic for vehicle suspension systems. *American Journal of Engineering and Applied Sciences*, *8*(1), 147–175.

Abou Jaoude, A., El-Tawil, K., & Kadry, S. (2010). Prediction in complex dimension using Kolmogorov's set of axioms. *Journal of Mathematics and Statistics*, *6*(2), 116–124.

Aczel, A. (2000). *God's equation*. New York: Dell.

Balibar, F. (2002). *Albert Einstein: Physique, Philosophie, Politique* (1st ed.). Paris: Le Seuil.

Barrow, J. (1992). *Pi in the sky*. London: Oxford University Press.

Bell, E. T. (1992). *The development of mathematics*. New York: Dover.

Benton, W. (1966a). *Probability, Encyclopedia Britannica* (Vol. 18, pp. 570–574). Chicago, IL: Encyclopedia Britannica.

Benton, W. (1966b). *Mathematical probability, Encyclopedia Britannica* (Vol. 18, pp. 574–579). Chicago, IL: Encyclopedia Britannica.

Bidabad, B. (1992). *Complex probability and Markov stochastic processes*. Proceedings first Iranian statistics conference, Isfahan University of Technology, Tehran.

Bogdanov, I., & Bogdanov, G. (2009). *Au Commencement du Temps*. Paris: Flammarion.
Bogdanov, I., & Bogdanov, G. (2010). *Le Visage de Dieu*. Paris: Editions Grasset et Fasquelle.
Bogdanov, I., & Bogdanov, G. (2012). *La Pensée de Dieu*. Paris: Editions Grasset et Fasquelle.
Bogdanov, I., & Bogdanov, G. (2013). *La Fin du Hasard*. Paris: Editions Grasset et Fasquelle.
Boltzmann, L. (1995). *Lectures on gas theory*. New York, NY: Dover.
Boursin, J.-L. (1986). *Les structures du Hasard*. Paris: Editions du Seuil.
Cercignani, C. (2010). *Ludwig Boltzmann, the man who trusted atoms*. Oxford: Oxford University Press.
Chan Man Fong, C. F., De Kee, D., & Kaloni, P. N. (1997). *Advanced mathematics for applied and pure sciences*. Amsterdam: Gordon and Breach Sciences.
Cox, D. R. (1955). A use of complex probabilities in the theory of stochastic processes. *Mathematical Proceedings of the Cambridge Philosophical Society*, *51*, 313–319.
Dacunha-Castelle, D. (1996). *Chemins de l'Aléatoire*. Paris: Flammarion.
Dalmedico-Dahan, A., Chabert, J-L., & Chemla, K. (1992). *Chaos Et Déterminisme*. Paris: Edition du Seuil.
Dalmedico-Dahan, A., & Peiffer, J. (1986). *Une Histoire des Mathématiques*. Paris: Edition du Seuil.
Davies, P. (1993). *The mind of god*. London: Penguin Books.
Ekeland, I. (1991). *Au Hasard. La Chance, la Science et le Monde*. Paris: Editions du Seuil.
Fagin, R., Halpern, J., & Megiddo, N. (1990). A logic for reasoning about probabilities. *Information and Computation*, *87*, 78–128.
Feller, W. (1968). *An introduction to probability theory and its applications* (3rd ed.). New York, NY: Wiley.
Gentle, J. (2003). *Random number generation and Monte Carlo methods* (2nd ed.). Sydney: Springer.
Gleick, J. (1997). *Chaos, making a new science*. New York: Penguin Books.
Greene, B. (2003). *The elegant universe*. New York: Vintage.
Gullberg, J. (1997). *Mathematics from the birth of numbers*. New York: W.W. Norton.
Hawking, S. (2002). *On the shoulders of giants*. London: Running Press.
Hawking, S. (2005). *God created the integers*. London: Penguin Books.
Hawking, S. (2011). *The dreams that stuff is made of*. London: Running Press.
Hoffmann, B. (1975). *In collaboration with Helen Dukas, Albert Einstein, Créateur et Rebelle* (1st ed.). Paris: Editions du Seuil.
Kuhn, T. (1970). *The structure of scientific revolutions* (2nd ed.). Chicago: Chicago Press.
Montgomery, D. C., & Runger, G. C. (2003). *Applied statistics and probability for engineers* (3rd ed.). New York: John Wiley & Sons.
Ognjanović, Z., Marković, Z., Rašković, M., Doder, D., & Perović, A. (2012). A connection between the Cantor–Bendixson derivative and the well-founded semantics of finite logic programs. *Annals of Mathematics and Artificial Intelligence*, *65*, 1–24.
Penrose, R. (1999). *traduction Française: Les Deux Infinis et L'Esprit Humain*. Paris: Roland Omnès, Flammarion.
Pickover, C. (2008). *Archimedes to Hawking*. Oxford: Oxford University Press.
Planck, M. (1969). *Treatise on thermodynamics*. New York, NY: Dover.
Poincaré, H. (1968). *La Science et l'Hypothèse* (1st ed.). Paris: Flammarion.
Reeves, H. (1988). *Patience dans L'Azur, L'Evolution Cosmique*. Paris: Le Seuil.
Ronan, C. (1988). *traduction Française: Histoire Mondiale des Sciences*. Paris: Claude Bonnafont, Le Seuil.
Science et vie. (1999). *Le Mystère des Mathématiques. Numéro*, *984*, 1–1.
Srinivasan, S. K., & Mehata, K. M. (1988). *Stochastic processes* (2nd ed.). New Delhi: McGraw-Hill.
Stepić, A. I., & Ognjanović, Z. (2014). Complex valued probability logics. *Publications De L'institut Mathématique, Nouvelle Série, tome*, *95*(109), 73–86. doi:10.2298/PIM1409073I
Stewart, I. (1996). *From here to infinity* (2nd ed.). Oxford: Oxford University Press.
Stewart, I. (2002). *Does god play dice?* (2nd ed.). Oxford: Blackwell.
Stewart, I. (2012). *In pursuit of the unknown*. Oxford: Basic Books.
Van Kampen, N. G. (2006). *Stochastic processes in physics and chemistry* (Revised and Enlarged ed.). Sydney: Elsevier.
Walpole, R., Myers, R., Myers, S., & Ye, K. (2002). *Probability and statistics for engineers and scientists* (7th ed.). Upper Saddle River, NJ: Prentice Hall.
Warusfel, A., & Ducrocq, A. (2004). *Les Mathématiques, Plaisir et Nécessité* (1st ed.). Paris: Edition Vuibert.
Weingarten, D. (2002). Complex probabilities on RN as real probabilities on CN and an application to path integrals. *Physical Review Letters*, *89*, 1–1.
Wikipedia, the free encyclopedia, *Brownian Motion*. Retrieved from https://en.wikipedia.org/
Wikipedia, the free encyclopedia, *Entropy*. Retrieved from https://en.wikipedia.org/
Wikipedia, the free encyclopedia, *Mass Diffusivity*. Retrieved from https://en.wikipedia.org/
Youssef, S. (1994). Quantum mechanics as complex probability theory. *Modern Physics Letters A*, *9*, 2571–2586.

10

Robust H_∞ control of time delayed power systems

Mohammed Jamal Alden and Xin Wang*

Department of Electrical and Computer Engineering, Southern Illinois University Edwardsville, Edwardsville, IL 62026, USA

Power system is the backbone of our society. The purpose of this work is to design a stable, robust and efficient controller for a power-generation system with time delays, model uncertainties and disturbances. Based on the practical dynamics of generator, prime mover, exciter and automatic voltage regulator, a mathematical power-generation system model is developed with state space dynamical equations involving time delays in the feedback. A novel robust H_∞ control framework based on linear matrix inequalities is proposed in the paper, which controls the energy system effectively. Computer simulations are used to show the efficacy of the proposed control algorithm.

Keywords: time delay; linear matrix inequalities; power system; H_∞

Nomenclature

δ	rotor angle
ω_s	synchronous speed
ν	normalized frequency $\nu = \omega/\omega_s$
T'_{do}	equivalent transient rotor time constant
P	total number of poles
S_B	rated three phase voltage ampere
ω_B	rated speed in electrical radians per second, $\omega_B = \omega_s$
H	the shaft inertia constant is scaled by defining $H = \frac{1}{2}J(\omega_B(2/P))^2/S_B$
K_d	the damping factor
I_d	direct axis current
I_q	quadrature axis current
X_d	direct axis reactance
X_q	quadrature axis reactance
X'_d	direct axis transient reactance
E_q	quadrature axis voltage
E'_q	quadrature axis transient voltage
E_{fd}	excitation voltage
E'_{fd}	transient excitation voltage
T_{mech}	mechanical torque
R_s	stator resistance
V_T	terminal bus voltage
V_{ref}	reference voltage
R_e	transmission line resistance
X_e	transmission linear reactance
z_1, z_2, z_3	state variables of power system stabilizer
U_{pss}	power system stabilizer control signal
$\star$	transposed value of the corresponding element

1. Introduction

Power systems are the basic infrastructure of modern civilization in our society. Stability analysis and control system development of the smart power grid are becoming more and more important, due to the rapid deployment of the distributed energy resources. In practical engineering applications, time delay plays a significant role in performance and stability of the overall power systems. Severe delays can even lead to catastrophic breakdown of the entire energy system due to instability (Alrifai, Zribi, Rayan, & Mahmoud, 2013; Bayrak & Tatlicioglu, 2012; Mahmoud, 2000; Okuno & Nakabayashi, 2006; Scorletti & Fromion, 1998; Wu, Ni, & Heydt, 2002; Zribi, Mahmoud, Karkoub, & Lie, 2000). For this reason, extensive research of transient and steady-state stability analysis and controller design have been conducted during the past decade (Jiang, 2007). System design engineers should consider time delays in designing and implementing practical power systems due to their significance (Alrifai et al., 2013; Jiang, Cai, Dorsey, & Qu, 1997; Mahmoud & Zribi, 1999; Wu et al., 2002; Yu, Jia, & Zhao, 2008).

Many different control approaches have been studied in the literature for effectively controlling the time delayed power systems Zhang et al. (2012). Based on an optimal control approach, the effect of time delays on the region of stability for small signals variation is studied in Jia, Yu, Yu, and Wang (2008). Zhang, Jiang, Wu, and Wu (2013) present a robust control method to design a PID type load frequency control of power systems considering time delays. Snyder, Ivanescu, Hadjsaid, Georges,

*Corresponding author. Email: xwang@siue.edu

and Margotin (2000) introduce a robust controller for a wide-area power system involving input time delays. The controller is developed based on the model reduction and linear matrix inequalities (LMIs). An adaptive wide-area damping controller based on generalized predictive control and model identification for time delayed power system is proposed in Yao, Jiang, Wen, Cheng, and Wu (2009). Yu, Zhang, Xie, and Wang (2007) propose a nonlinear robust control algorithm for power system considering signal delays and measurement incompleteness. Yu et al. (2008) discuss the maximal allowable time delay margin for a stable power systems based on the Lyapunov method involving three generators and nine buses. Chowdhury, Kulhare, and Raina (2011) present the nonlinear limit cycle effect of time delays on local stability of the single machine infinite bus system. In Liu, Zhu, and Jiang (2008), the cluster treatment of eigenvalues is introduced to analyze the stability of a power system with time delays in the feedback loop.

This paper presents a general robust control framework based on linear matrix inequality for time delayed power systems. The mathematical dynamics is modeled as a seventh-order nonlinear system with bounded model uncertainties and external disturbances. By formulating the nonlinear control as a convex optimization problem, the linear matrix inequality can provide the optimal and robust solution satisfying the Lyapunov stability and the robust H_∞ performance objective.

This paper is organized as follows: Section 2 discusses the mathematical modeling of the power-generation system. Section 3 presents the novel H_∞ controller design with linear matrix inequality. Computer simulations conducted with MATLAB is given in Section 4. In Section 5, the conclusion is reached and future work is discussed.

2. Mathematical model of power-generation system

In this section, the mathematical model of an infinite bus power system involving a synchronous generator is developed.

2.1. Synchronous generator model

A model for the synchronous generator is given as follows (Wang & Gu, 2014):

$$\dot{E}'_{\mathrm{q}} = -\frac{1}{T'_{\mathrm{do}}}(E'_{\mathrm{q}} - (X_{\mathrm{d}} - X'_{\mathrm{d}})I_{\mathrm{d}} - E_{\mathrm{fd}}), \tag{1}$$

$$\dot{\delta} = \omega - \omega_{\mathrm{s}}, \tag{2}$$

$$\dot{\omega} = \frac{\omega_{\mathrm{s}}}{2H}[T_{\mathrm{mech}} - (E'_{\mathrm{q}}I_{\mathrm{q}} + (X_{\mathrm{q}} - X'_{\mathrm{d}})I_{\mathrm{d}}I_{\mathrm{q}} + K_{\mathrm{d}}(\omega - \omega_{\mathrm{s}}))]. \tag{3}$$

The stator algebraic equations are

$$V_{\mathrm{T}}\sin(\delta - \theta) + R_{\mathrm{s}}I_{\mathrm{d}} - X_{\mathrm{q}}I_{\mathrm{q}} = 0, \tag{4}$$

$$E'_{\mathrm{q}} - V_{\mathrm{T}}\cos(\delta - \theta) - R_{\mathrm{s}}I_{\mathrm{q}} - X'_{\mathrm{d}}I_{\mathrm{d}} = 0. \tag{5}$$

Neglecting stator resistance by assuming $R_{\mathrm{s}} = 0$, we can write the stator dynamical equations as

$$V_{\mathrm{T}}\sin(\delta - \theta) - X_{\mathrm{q}}I_{\mathrm{q}} = 0, \tag{6}$$

$$E'_{\mathrm{q}} - V_{\mathrm{T}}\cos(\delta - \theta) - X'_{\mathrm{d}}I_{\mathrm{d}} = 0. \tag{7}$$

Since we have

$$(V_{\mathrm{d}} + jV_{\mathrm{q}})\,e^{j(\delta - \pi/2)} = V_{\mathrm{T}}\,e^{j\theta}, \tag{8}$$

therefore,

$$V_{\mathrm{d}} = V_{\mathrm{T}}\sin(\delta - \theta), \tag{9}$$

$$V_{\mathrm{q}} = V_{\mathrm{T}}\cos(\delta - \theta). \tag{10}$$

Substituting in Equations (6) and (7), we get

$$V_{\mathrm{d}} - X_{\mathrm{q}}I_{\mathrm{q}} = 0, \tag{11}$$

$$E'_{\mathrm{q}} - V_{\mathrm{q}} - X'_{\mathrm{d}}I_{\mathrm{d}} = 0. \tag{12}$$

Assuming zero degree phase angle for the infinite bus voltage, we have

$$(I_{\mathrm{d}} + \mathrm{j}I_{\mathrm{q}})\,\mathrm{e}^{\mathrm{j}(\delta - \pi/2)} = \frac{(V_{\mathrm{d}} + \mathrm{j}V_{\mathrm{q}})\,\mathrm{e}^{\mathrm{j}(\delta - \pi/2)} - V_\infty\angle 0^\circ}{R_e + \mathrm{j}X_e}, \tag{13}$$

By separating the imaginary and real parts of Equation (13), we get

$$\begin{aligned} R_{\mathrm{d}}I_{\mathrm{d}} - X_eI_{\mathrm{q}} &= V_{\mathrm{d}} - V_\infty\sin(\delta), \\ X_eI_{\mathrm{d}} - R_eI_{\mathrm{q}} &= V_{\mathrm{q}} - V_\infty\cos(\delta). \end{aligned} \tag{14}$$

By linearizing Equations (11) and (12), we get

$$\begin{bmatrix} \Delta V_{\mathrm{d}} \\ \Delta V_{\mathrm{q}} \end{bmatrix} = \begin{bmatrix} 0 & X_{\mathrm{q}} \\ -X'_{\mathrm{d}} & 0 \end{bmatrix}\begin{bmatrix} \Delta I_{\mathrm{q}} \\ \Delta I_{\mathrm{q}} \end{bmatrix} + \begin{bmatrix} 0 \\ \Delta E'_{\mathrm{q}} \end{bmatrix}. \tag{15}$$

And by linearizing Equation (14), we get

$$\begin{bmatrix} \Delta V_{\mathrm{d}} \\ \Delta V_{\mathrm{q}} \end{bmatrix} = \begin{bmatrix} R_e & -X_e \\ X_e & R_e \end{bmatrix}\begin{bmatrix} \Delta I_{\mathrm{q}} \\ \Delta I_{\mathrm{q}} \end{bmatrix} + \begin{bmatrix} V_\infty\sin(\delta) \\ -V_\infty\cos(\delta) \end{bmatrix}\Delta\delta. \tag{16}$$

By equating the right-hand sides of Equations (15) and (16), we have

$$\begin{bmatrix} R_e & -(X_e + X_{\mathrm{q}}) \\ (X_e + X'_{\mathrm{d}}) & R_e \end{bmatrix}\begin{bmatrix} \Delta I_{\mathrm{d}} \\ \Delta I_{\mathrm{q}} \end{bmatrix} = \begin{bmatrix} 0 \\ \Delta E'_{\mathrm{q}} \end{bmatrix} + \begin{bmatrix} -V_\infty\cos\delta \\ V_\infty\sin\delta \end{bmatrix}\Delta\delta, \tag{17}$$

$\Delta I_\mathrm{d}, \Delta I_\mathrm{q}$ can be obtained from Equations (15) and (16) as

$$\begin{bmatrix}\Delta I_\mathrm{d}\\ \Delta I_\mathrm{q}\end{bmatrix} = \frac{1}{\Delta}\begin{bmatrix}(X_e+X_\mathrm{q}) & -R_eV_\infty\cos\delta + V_\infty\sin\delta(X_\mathrm{q}+X_e)\\ R_e & R_eV_\infty\sin\delta + V_\infty\cos\delta(X'_\mathrm{d}+X_e)\end{bmatrix} \times \begin{bmatrix}\Delta E'_\mathrm{q}\\ \Delta\delta\end{bmatrix}, \tag{18}$$

where

$$\Delta = R_e^2 + (X_e+X_\mathrm{q})(X_e+X'_\mathrm{d}). \tag{19}$$

Denote the normalized frequency $\nu = \omega/\omega_\mathrm{s}$. The linearized synchronous generator model of Equations (1)–(3) is given as follows:

$$\begin{bmatrix}\Delta\dot{E}'_\mathrm{q}\\ \Delta\dot{\delta}\\ \Delta\dot{\nu}\end{bmatrix} = \begin{bmatrix}-\frac{1}{T'_\mathrm{do}} & 0 & 0\\ 0 & 0 & \omega_\mathrm{s}\\ -\frac{I^o_\mathrm{q}}{2H} & 0 & -\frac{K_\mathrm{d}\omega_\mathrm{s}}{2H}\end{bmatrix}\begin{bmatrix}\Delta E'_\mathrm{q}\\ \Delta\delta\\ \Delta\nu\end{bmatrix} + \begin{bmatrix}-\frac{1}{T'_\mathrm{d}}(X_\mathrm{d}-X'_\mathrm{d}) & 0\\ 0 & 0\\ \frac{1}{2H}(X'_\mathrm{d}-X_\mathrm{q})I^o_\mathrm{q} & \frac{1}{2H}(X'_\mathrm{d}-X_\mathrm{q})I^o_\mathrm{d} - \frac{1}{2H}E'^o_\mathrm{q}\end{bmatrix} \times \begin{bmatrix}\Delta I_\mathrm{d}\\ \Delta I_\mathrm{q}\end{bmatrix} + \begin{bmatrix}\frac{1}{T'_\mathrm{do}} & 0\\ 0 & 0\\ 0 & \frac{1}{2H}\end{bmatrix}\begin{bmatrix}\Delta E_\mathrm{fd}\\ \Delta T_\mathrm{mech}\end{bmatrix}. \tag{20}$$

Substitute for $\Delta I_\mathrm{d}, \Delta I_\mathrm{q}$, we have

$$\Delta\dot{E}'_\mathrm{q} = -\frac{1}{K_3T'_\mathrm{do}}\Delta E'_\mathrm{q} - \frac{K_4}{T'_\mathrm{do}}\Delta\delta + \frac{1}{T'_\mathrm{do}}\Delta E_\mathrm{fd}, \tag{21}$$

$$\Delta\dot{\delta} = \omega_\mathrm{s}\Delta\nu, \tag{22}$$

$$\Delta\dot{\nu} = -\frac{K_2}{2H}\Delta E'_\mathrm{q} - \frac{K_1}{2H}\Delta\delta - \frac{K_\mathrm{d}\omega_\mathrm{s}}{2H}\Delta\nu + \frac{1}{2H}\Delta T_\mathrm{mech}. \tag{23}$$

where

$$\frac{1}{K_3} = 1 + \frac{(X_\mathrm{d}-X'_\mathrm{d})(X_\mathrm{q}+X_e)}{\Delta}, \tag{24}$$

$$K_4 = \frac{V_\infty(X_\mathrm{d}-X'_\mathrm{d})}{\Delta}[(X_\mathrm{q}+X_e)\sin\delta - R_e\cos\delta], \tag{25}$$

$$K_2 = \frac{1}{\Delta}[I^o_\mathrm{q}\Delta - I^o_\mathrm{q}(X'_\mathrm{d}-X_\mathrm{q})(X_\mathrm{q}+X_e) - R_e(X'_\mathrm{d}-X_\mathrm{q})I^o_\mathrm{d} + R_eE'^o_\mathrm{q}], \tag{26}$$

$$K_1 = -\frac{1}{\Delta}[I^o_\mathrm{q}V_\infty(X'_\mathrm{d}-X_\mathrm{q})\{(X_\mathrm{q}+X_e)\sin\delta - R_e\cos\delta\} + V_\infty\{(X'_\mathrm{d}-X_\mathrm{q})I^o_\mathrm{d} - E'^o_\mathrm{q}\}\{(X'_\mathrm{d}+X_e)\cos\delta + R_e\sin\delta\}]. \tag{27}$$

Since

$$V_\mathrm{T}^2 = V_\mathrm{d}^2 + V_\mathrm{q}^2,$$

the differential terms is given as follows:

$$\Delta V_\mathrm{T} = \frac{V^o_\mathrm{d}}{V_\mathrm{T}}\Delta V_\mathrm{d} + \frac{V^o_\mathrm{q}}{V_\mathrm{T}}\Delta V_\mathrm{q}, \tag{28}$$

Substituting Equation (18) into Equation (15), we obtain

$$\begin{pmatrix}\Delta V_\mathrm{d}\\ \Delta V_\mathrm{q}\end{pmatrix} = \frac{1}{\Delta}\begin{pmatrix}X_\mathrm{q}R_e & X_\mathrm{q}(R_eV_\infty\sin\delta + V_\infty\cos\delta(X'_\mathrm{d}+X_e))\\ -X'_\mathrm{d}(X_\mathrm{q}+X_e) & -X'_\mathrm{d}(-R_eV_\infty\cos\delta + V_\infty(X_\mathrm{q}+X_e)\sin\delta)\end{pmatrix}\begin{pmatrix}\Delta E'_\mathrm{q}\\ \Delta\delta\end{pmatrix} + \begin{pmatrix}0\\ \Delta E'_\mathrm{q}\end{pmatrix}, \tag{29}$$

Based on Equations (28) and (29), we get

$$\Delta V_\mathrm{T} = K_5\Delta\delta + K_6\Delta E'_\mathrm{q}, \tag{30}$$

where

$$K_5 = \frac{1}{\Delta}\left\{\frac{V^o_\mathrm{d}}{V_\mathrm{T}}X_\mathrm{q}[R_eV_\infty\sin\delta + V_\infty\cos\delta(X'_\mathrm{d}+X_e)] + \frac{V^o_\mathrm{q}}{V_\mathrm{T}}[X'_\mathrm{d}(R_eV_\infty\cos\delta) - V_\infty(X_\mathrm{q}+X_e)\sin\delta]\right\}, \tag{31}$$

$$K_6 = \frac{1}{\Delta}\left\{\frac{V^o_\mathrm{d}}{V_\mathrm{T}}X_\mathrm{q}R_e - \frac{V^o_\mathrm{q}}{V_\mathrm{T}}X'_\mathrm{d}(X_\mathrm{q}+X_e)\right\} + \frac{V^o_\mathrm{q}}{V_\mathrm{T}}. \tag{32}$$

2.2. *Automatic voltage regulator (AVR) and exciter circuit dynamics*

The following dynamical equations for AVR and excitation control system are adopted:

$$\dot{E}_\mathrm{fd} = \frac{K_A}{T_A}\left(V_\mathrm{ref} - V_\mathrm{T} + U_\mathrm{pss} - \frac{E_\mathrm{fd}}{T_A}\right). \tag{33}$$

By linearizing Equation (33), we get

$$\Delta\dot{E}_\mathrm{fd} = \frac{K_A}{T_A}(\Delta V_\mathrm{ref} - \Delta V_\mathrm{T} + \Delta U_\mathrm{pss}) - \frac{\Delta E_\mathrm{fd}}{T_A}. \tag{34}$$

Based on (30), (34) can be rewritten as

$$\Delta\dot{E}_\mathrm{fd} = -\frac{\Delta E_\mathrm{fd}}{T_A} - \frac{K_AK_5}{T_A}\Delta\delta - \frac{K_AK_6}{T_A}\Delta E'_\mathrm{q} + \frac{K_A}{T_A}\Delta U_\mathrm{pss} + \frac{K_A}{T_A}\Delta V_\mathrm{ref}. \tag{35}$$

A typical power system stabilizer (PSS) control scheme include a washout filter and two lead-lag blocks. The retarded measure of v propagates in the PSS equations. The linearized form of PSS can be modeled as

$$\Delta\dot{z}_1 = \frac{-(K_w\Delta v(t-\tau)+\Delta z_1)}{T_w},$$

$$\Delta\dot{z}_2 = \frac{[(1-T_1/T_2)(K_w\Delta v(t-\tau)+\Delta z_1)-\Delta z_2]}{T_2},$$

$$\Delta\dot{z}_3 = \frac{\{(1-T_3/T_4)[\Delta z_2+(T_1/T_2)(K_w\Delta v(t-\tau)+\Delta z_1)]-\Delta z_3\}}{T_4},$$

$$\Delta U_{pss} = \Delta z_3 + \frac{T_3}{T_4}\left[\Delta z_2 + \frac{T_1}{T_2}(K_w\Delta v(t-\tau)+\Delta z_1)\right]. \quad (36)$$

Hence, the overall linearized model of the power-generation system is given as follows:

$$\Delta\dot{\delta} = \omega_s\Delta v, \quad (37)$$

$$\Delta\dot{v} = -\frac{K_2}{2H}\Delta E'_q - \frac{K_1}{2H}\Delta\delta - \frac{K_d\omega_s}{2H}\Delta v + \frac{1}{2H}\Delta T_{mech}, \quad (38)$$

$$\Delta\dot{E}'_q = -\frac{1}{K_3T'_{do}}\Delta E'_q - \frac{K_4}{T'_{do}}\Delta\delta + \frac{1}{T'_{do}}\Delta E_{fd}, \quad (39)$$

$$\Delta\dot{E}_{fd} = -\frac{\Delta E_{fd}}{T_A} - \frac{K_AK_5}{T_A}\Delta\delta - \frac{K_AK_6}{T_A}\Delta E'_q + \frac{K_A}{T_A}\Delta V_{ref} + \frac{K_A}{T_A}\left\{\Delta z_3 + \frac{T_3}{T_4}\left[\Delta z_2 + \frac{T_1}{T_2}(K_w\Delta v(t-\tau)+\Delta z_1)\right]\right\}, \quad (40)$$

$$\Delta\dot{z}_1 = \frac{-(K_w\Delta v(t-\tau)+\Delta z_1)}{T_w}, \quad (41)$$

$$\Delta\dot{z}_2 = \frac{[(1-T_1/T_2)(K_w\Delta v(t-\tau)+\Delta z_1)-\Delta z_2]}{T_2}, \quad (42)$$

$$\Delta\dot{z}_3 = \frac{\{(1-T_3/T_4)[\Delta z_2+((T_1/T_2)(K_w\Delta v(t-\tau)+\Delta z_1)]-\Delta z_3\}}{T_4}. \quad (43)$$

Denote $x = [\Delta\delta, \Delta v, \Delta E'_q, \Delta E_{fd}, \Delta z_1, \Delta z_2, \Delta z_3]^t$ and $u = [\Delta T_{mech}, \Delta V_{ref}]^t$, the linearized model becomes

$$\dot{x} = Ax(t) + A_dx(t-\tau) + Bu(t), \quad (44)$$

where

$$A = \begin{pmatrix} 0 & \omega_s & 0 & 0 & 0 & 0 & 0 \\ -\frac{K_1}{2H} & -\frac{K_d\omega_s}{2H} & -\frac{K_2}{2H} & 0 & 0 & 0 & 0 \\ -\frac{K_4}{T'_{do}} & 0 & -\frac{1}{K_3T'_{do}} & -\frac{1}{T'_{do}} & 0 & 0 & 0 \\ -\frac{K_AK_5}{T_A} & 0 & -\frac{K_AK_6}{T_A} & -\frac{1}{T_A} & \frac{K_AT_3T_1}{T_AT_4T_2} & \frac{K_AT_3}{T_AT_4} & \frac{K_A}{T_A} \\ 0 & 0 & 0 & 0 & -\frac{1}{T_w} & 0 & 0 \\ 0 & 0 & 0 & 0 & \left(1-\frac{T_1}{T_2}\right)\frac{1}{T_2} & -\frac{1}{T_2} & 0 \\ 0 & 0 & 0 & 0 & \left(1-\frac{T_3}{T_4}\right)\frac{T_1}{T_2}\frac{1}{T_4} & \left(1-\frac{T_3}{T_4}\right)\frac{1}{T_4} & -\frac{1}{T_4} \end{pmatrix}, \quad (45)$$

$$A_d = \begin{pmatrix} 0 & 0 & 0 & 0 & 0 & 0 & 0 \\ 0 & 0 & 0 & 0 & 0 & 0 & 0 \\ 0 & 0 & 0 & 0 & 0 & 0 & 0 \\ 0 & \frac{T_3T_1K_w}{T_4T_2} & 0 & 0 & 0 & 0 & 0 \\ 0 & -\frac{K_w}{T_w} & 0 & 0 & 0 & 0 & 0 \\ 0 & \left(1-\frac{T_1}{T_2}\right)K_w\frac{1}{T_2} & 0 & 0 & 0 & 0 & 0 \\ 0 & \left(1-\frac{T_3}{T_4}\right)\frac{T_1}{T_2}K_w\frac{1}{T_4} & 0 & 0 & 0 & 0 & 0 \end{pmatrix}, \quad (46)$$

$$B = \begin{pmatrix} 0 & 0 \\ \frac{1}{2H} & 0 \\ 0 & 0 \\ 0 & \frac{K_A}{T_A} \\ 0 & 0 \\ 0 & 0 \\ 0 & 0 \end{pmatrix}. \quad (47)$$

3. Robust H_∞ controller design

In this section, we propose the novel design of the robust controller satisfying H_∞ performance objective. The system with state and input delays, uncertainties and disturbances is considered. The system is of the form:

$$\dot{x}(t) = (A + \delta A)x(t) + (A_d + \delta A_d)x(t - \tau_s) + (B + \delta B)u(t) + (B_d + \delta B_d)u(t - \tau_i) + Dw(t), \quad (48)$$

the performance output is chosen as

$$z(t) = Ex(t) \quad (49)$$

and

$$x(t) = \phi(t) \quad \text{for } t \in [-\tau_s, 0]. \quad (50)$$

Assume that the state variables are available for feedback. Otherwise, estimators can be developed for state estimation purposes. Then, we have state feedback control input as

$$u(t) = Kx(t).$$

Therefore, the closed-loop system becomes:

$$\dot{x}(t) = (A + \delta A + BK + \delta B)x(t) + (A_d + \delta A_d)x(t - \tau_s) + (B_d + \delta B_d)Kx(t - \tau_i) + Dw(t). \quad (51)$$

Rearranging Equation (51), we get

$$\dot{x}(t) = A_c x(t) + \delta A_c x(t) + (A_d + \delta A_d)x(t - \tau_s) + (B_d + \delta B_d)Kx(t - \tau_i) + Dw(t), \quad (52)$$

where

$$A_c = A + BK,$$
$$\delta A_c = \delta A + \delta BK.$$

Before proceeding to the theorem derivation, Assumption 1 and Lemma 1 are introduced (Wang, Yaz, & Yaz, 2010).

ASSUMPTION 1 *The general form of unstructured L_2 bounded uncertainties is used in this work:*

$$\delta A \delta A^t \le \gamma_A I,$$
$$\delta A_d \delta A_d^t \le \gamma_{A_d} I,$$
$$\delta B \delta B^t \le \gamma_B I,$$
$$\delta B_d \delta B_d^t \le \gamma_{B_d} I.$$

LEMMA 1

$$AB^t + BA^t \le \alpha AA^t + \alpha^{-1} BB^t.$$

To prove this inequality, we can consider the following equivalent inequality which always holds, given arbitrary $\alpha > 0$:

$$(\alpha^{1/2} A - \alpha^{-1/2} B)(\alpha^{1/2} A - \alpha^{-1/2} B)^t \ge 0.$$

Furthermore, if A and B are chosen to be $\begin{bmatrix} a^t \\ 0 \end{bmatrix}$ and $\begin{bmatrix} 0 \\ b^t \end{bmatrix}$, respectively, we get

$$\begin{bmatrix} 0 & a^t b \\ b^t a & 0 \end{bmatrix} \le \begin{bmatrix} \zeta a^t a & 0 \\ 0 & \zeta^{-1} b^t b \end{bmatrix}.$$

Based on Assumption 1 and Lemma 1, the main theorem of the paper is summarized as follows:

THEOREM 1 *Under the feedback control law $u(t) = Kx(t)$, the system of Equation (52) is asymptotically stable for all delays satisfying τ_s, $\tau_i \ge 0$. And the H_∞ performance objective $\|T_{zw}\|_\infty \le \gamma, \gamma > 0$ can be satisfied. If there exist matrices $Y, Q_t^t = Q_t > 0$ and $Q_s^t = Q_s > 0$ satisfying the following LMI:*

$$\begin{bmatrix} m_1 & A_d X & B_d Y & D & X & Y^t \\ XA_d^t & m_2 & 0 & 0 & 0 & 0 \\ Y^t B_d^t & 0 & m_3 & 0 & 0 & 0 \\ D^t & 0 & 0 & m_4 & 0 & 0 \\ X & 0 & 0 & 0 & m_5 & 0 \\ Y & 0 & 0 & 0 & 0 & m_6 \end{bmatrix} < 0, \quad (53)$$

where

$$\begin{aligned} m_1 &= AX + BY + XA^t + Y^t B + Q_t + Q_s + \alpha_1(\gamma_A + \gamma_B)I, \\ m_2 &= \alpha_2^{-1} I - Q_t, \\ m_3 &= \alpha_3^{-1} I - Q_s, \\ m_4 &= -\gamma^2 I, \\ m_5 &= -[\alpha_1^{-1} I + \alpha_2 \gamma_{A_d} I + E^t E]^{-1}, \\ m_6 &= -[\alpha_1^{-1} I + \alpha_3 \gamma_{B_d} I]^{-1}. \end{aligned} \quad (54)$$

Proof A Lyapunov–Krasovskii function is chosen as follows:

$$V(x,t) = x^t(t)Px(t) + \int_{t-\tau_s}^{t} x^t(v)Q_1 x(v)\,dv + \int_{t-\tau_i}^{t} x^t(v)Q_2 x(v)\,dv \quad (55)$$

where $V(x,t)$ is a positive semi-definite functional and the matrices P, Q_1, Q_2 are all positive definite.

By taking derivative, we have

$$\dot{V}(x,t) = \dot{x}^t(t)Px(t) + x^t(t)P\dot{x}(t) + x^t(t)Q_1 x(t) - x^t(t - \tau_s)Q_1 x(t - \tau_s) + x^t(t)Q_2 x(t) - x^t(t - \tau_i)Q_2 x(t - \tau_i). \quad (56)$$

Based on LaSalle's theorem, in order to achieve the asymptotic stability, the conditions $V > 0$ and $\dot{V} < 0$ need to be satisfied.

In order to satisfy H_∞ performance objective, the following H_∞ performance inequality needs be employed.

$$J = \int_0^\infty (z^t z - \gamma^2 \omega^t \omega)\,dt < 0. \quad (57)$$

The sufficient condition to achieve both the asymptotic stability and H_∞ performance objective is

$$J = \int_0^\infty (z^t z - \gamma^2 \omega^t \omega + \dot{V})\,dt < 0. \quad (58)$$

Condition (58) implies

$$z^t z - \gamma^2 \omega^t \omega + \dot{V} < 0. \tag{59}$$

By substituting Equations (52) into (56), we get

$$\begin{aligned}\dot{V}(x,t) = {}& [A_c x(t) + \delta A_c x(t) + (A_d + \delta A_d)x(t-d) \\ & + (B_d + \delta B_d)Kx(t-\tau_i) + Dw(t)]^t Px \\ & + x^t P[A_c x(t) + \delta A_c x(t) + (A_d + \delta A_d)x(t-d) \\ & + (B_d + \delta B_d)Kx(t-\tau_i) + Dw(t)] + x^t(t)Q_1 x(t) \\ & - x^t(t-\tau_s)Q_1 x(t-\tau_s) + x^t(t)Q_2 x(t) \\ & - x^t(t-\tau_i)Q_2 x(t-\tau_i) < 0.\end{aligned} \tag{60}$$

Based on condition (58), we have

$$\begin{aligned}& [A_c x(t) + \delta A_c x(t) + (A_d + \delta A_d)x(t-d) \\ & + (B_d + \delta B_d)Kx(t-\tau_i) + Dw(t)]^t Px \\ & + x^t P[A_c x(t) + \delta A_c x(t) + (A_d + \delta A_d)x(t-d) \\ & + (B_d + \delta B_d)Kx(t-\tau_i) + Dw(t)] + x^t(t)Q_1 x(t) \\ & - x^t(t-\tau_s)Q_1 x(t-\tau_s) + x^t(t)Q_2 x(t) \\ & - x^t(t-\tau_i)Q_2 x(t-\tau_i) + z^t z - \gamma^2 I < 0.\end{aligned} \tag{61}$$

Denote $\zeta(t) = [x^t(t) \quad x^t(t-\tau_d) \quad x^t(t-\tau_i) \quad \omega^t(t)]^t$, then Equation (61) can be written as

$$\dot{V}(x,t) = \zeta^t(t) W_o \zeta(t),$$

where

$$W_o = \begin{bmatrix} P(A_c + \delta A_c) + (A_c + \delta A_c)^t P + Q_1 + Q_2 + E^t E & P(A_d + \delta A_d) & P(B_d + \delta B_d)K & PD \\ (A_d + \delta A_d)^t P & -Q_1 & 0 & 0 \\ K^t(B_d + \delta B_d)^t P & 0 & -Q_2 & 0 \\ D^t P & 0 & 0 & -\gamma^2 I \end{bmatrix} < 0. \tag{62}$$

By pre- and post-multiplying Equation (62) with the diagonal matrix diag(X, I, I, I) and denote

$$X = P^{-1}, \quad Y = KX, \quad Q_t = XQ_1X, \quad Q_s = XQ_2X,$$

we get

$$\begin{bmatrix} \phi_{11} & \phi_{12} & \phi_{13} & D \\ \star & \phi_{22} & 0 & 0 \\ \star & 0 & \phi_{33} & 0 \\ \star & 0 & 0 & \phi_{44} \end{bmatrix} < 0, \tag{63}$$

where

$$\begin{aligned}\phi_{11} &= (A + BK + \delta A + \delta BK)X + X(A + BK + \delta A + \\ &\quad \delta BK)^t + Q_s + Q_t + XE^t EX, \\ \phi_{12} &= (A_d + \delta A_d)X, \\ \phi_{13} &= (B_d + \delta B_d)Y, \\ \phi_{22} &= -Q_t, \\ \phi_{33} &= -Q_s, \\ \phi_{44} &= -\gamma^2 I.\end{aligned}$$

Applying Lemma 1, we get

$$\begin{aligned}& (\delta A + \delta BK)X + X(\delta A + \delta BK)^t \\ &= X[I \quad K^t]^t \begin{bmatrix} \delta A^t \\ \delta B^t \end{bmatrix} + [\delta A \quad \delta B] \begin{bmatrix} I \\ K \end{bmatrix} X \\ &\leq \alpha_1 [\delta A \quad \delta B] \begin{bmatrix} \delta A^t \\ \delta B^t \end{bmatrix} + \alpha_1^{-1} X[I \quad K^t] \begin{bmatrix} I \\ K \end{bmatrix} X.\end{aligned}$$

By applying Assumption 1, we have

$$\begin{aligned}(\delta A + \delta BK)X + X(\delta A + \delta BK)^t &\leq \alpha_1(\gamma_A I + \gamma_B I) \\ &\quad + \alpha_1^{-1} X[I \quad K^t] \begin{bmatrix} I \\ K \end{bmatrix} X.\end{aligned} \tag{64}$$

Based on Lemma 1 and Assumption 1, the following matrix inequality is reached:

$$\begin{bmatrix} 0 & \delta A_d X & \delta B_d Y & 0 \\ X^t \delta A_d^t & 0 & 0 & 0 \\ Y^t \delta B_d^t & 0 & 0 & 0 \\ 0 & 0 & 0 & 0 \end{bmatrix} \leq \begin{bmatrix} \alpha_2 \gamma_{A_d} X^t X + \alpha_3 \gamma_{B_d} Y^t Y & 0 & 0 & 0 \\ 0 & \alpha_2^{-1} I & 0 & 0 \\ 0 & 0 & \alpha_3^{-1} I & 0 \\ 0 & 0 & 0 & 0 \end{bmatrix}. \tag{65}$$

Now, by substituting Equations (64) and (65) into Equation (63) and applying Schur complement, we obtain the following LMI result:

$$\begin{bmatrix} \zeta_{11} & \zeta_{12} & \zeta_{13} & D & X & Y^t \\ \star & \zeta_{22} & 0 & 0 & 0 & 0 \\ \star & 0 & \zeta_{33} & 0 & 0 & 0 \\ \star & 0 & 0 & \zeta_{44} & 0 & 0 \\ \star & 0 & 0 & 0 & \zeta_{55} & 0 \\ \star & 0 & 0 & 0 & 0 & \zeta_{66} \end{bmatrix} < 0, \tag{66}$$

where

$\zeta_{11} = AX + BY + XA^t + Y^tB + Q_t + Q_s + \alpha_1(\gamma_A + \gamma_B)I,$

$\zeta_{12} = A_dX,$

$\zeta_{13} = B_dY,$

$\zeta_{22} = \alpha_2^{-1}I - Q_t,$

$\zeta_{33} = \alpha_3^{-1}I - Q_s,$

$\zeta_{44} = -\gamma^2 I,$

$\zeta_{55} = -[\alpha_1^{-1}I + \alpha_2\gamma_{A_d}I + E^tE]^{-1},$

$\zeta_{66} = -[\alpha_1^{-1}I + \alpha_3\gamma_{B_d}I]^{-1}.$

4. Simulation and results

The following parameters are used for simulations. Assuming that $R_e = 0, X_e = 0.5pu, V_T\angle\theta = 1\angle 15^\circ pu$, and $V_\infty\angle 0^\circ = 1.05\angle 0^\circ pu$.

The generator, automatic voltage regulator and exciter parameters are $H = 3.2$ s, $T'_{do} = 9.6$ s, $K_A = 400$, $T_A = 0.2$ s, $R_s = 0$ pu, $X_q = 2.1$ pu, $X_d = 2.5$ pu, $X'_d = 0.39$ pu, $K_d = 0$, and $\omega_s = 377$.

The PSS parameters are $K_w = 0.5, T_1 = 0.5, T_2 = 0.01, T_3 = 1, T_4 = 0.1, T_w = 10$.

Based on Equations (24)–(27), (31) and (32), we can calculate the values: $K_1 = 0.9224$, $K_2 = 1.0739$, $K_3 = 0.296667$, $K_4 = 2.26555$, $K_5 = 0.005$, $K_6 = 0.3572$.

The L_2 of disturbance is chosen as $w(t) = 5 \times 0.9^t$, notice that the disturbance energy is finite.

MATLAB robust control toolbox provide the capability to design the optimal control feedback with LMI. Computer simulation shows that our proposed controller effectively stabilizes the time response of rotor angle in Figure 1, normalized frequency in Figure 2, quadrature axis transient voltage in Figure 3 and excitation voltage in Figure 4. The control input is shown in Figure 5. Simulation results have demonstrated the effectiveness and robustness of our proposed approach (Figure 5).

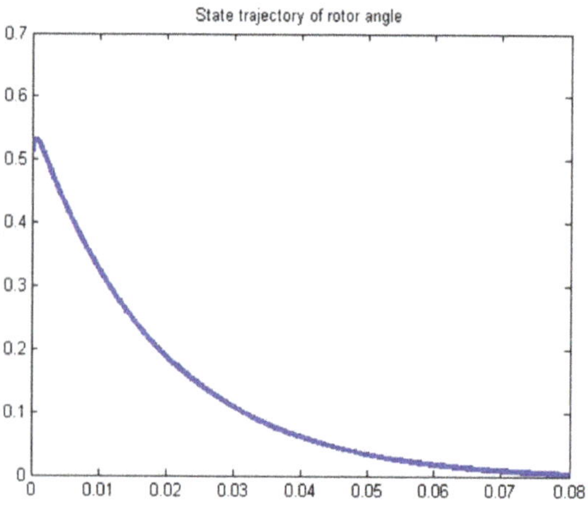

Figure 1. Time response of rotor angle.

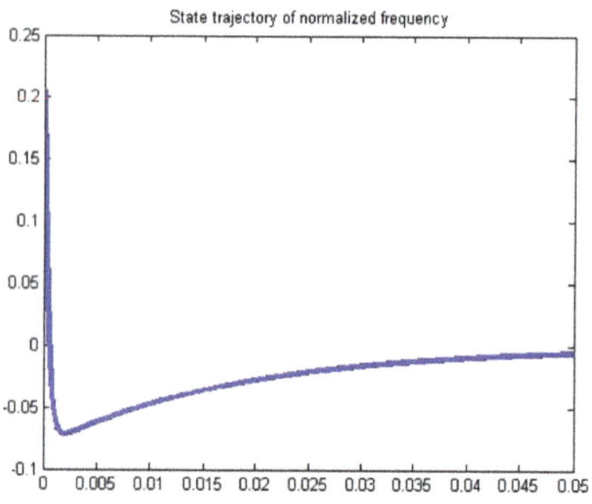

Figure 2. Time response of normalized frequency.

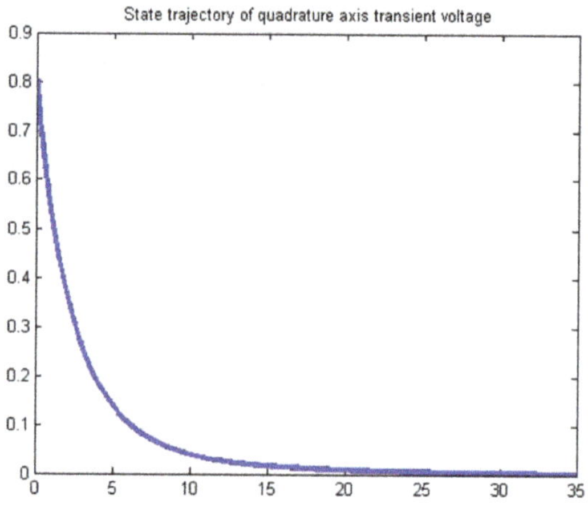

Figure 3. Time response of quadrature axis transient voltage.

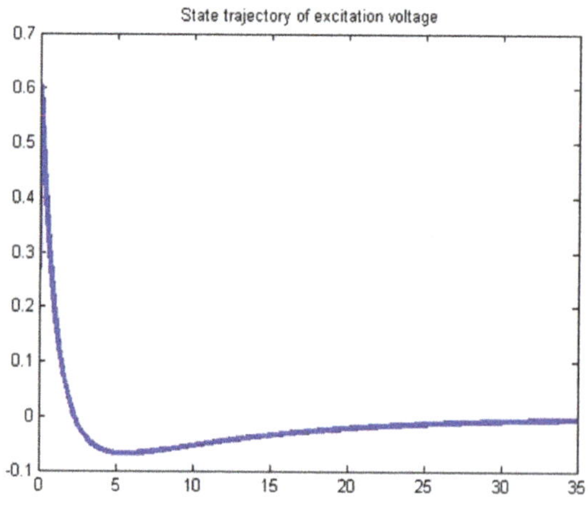

Figure 4. Time response of excitation voltage.

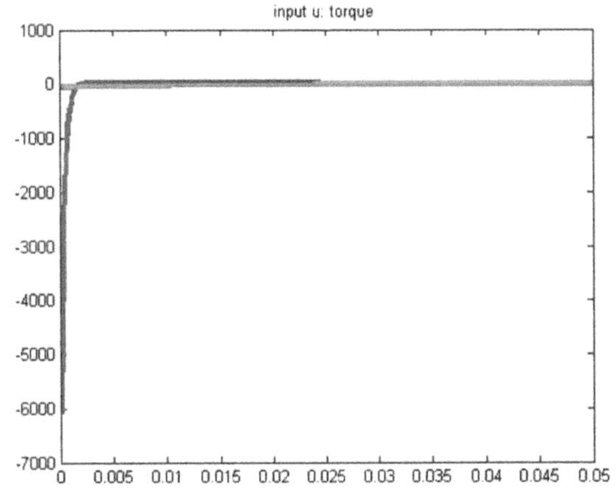

Figure 5. Control input.

5. Conclusion

A general robust H_∞ control approach is proposed in this paper for a power-generation system with state and input delays, disturbances and model uncertainties. The uncertainties are assumed to be bounded and unstructured. A novel seventh-order state space model for the power-generation system is developed. Computer simulation studies conducted through the use of MATLAB show the robustness and effectiveness of the novel approach. Notice that our LMI control solution only applies for delay independent power system control applications. In the future development, we would investigate the delay-dependent cases based on the Discretized Lyapunov–Krasovskii functional method.

Disclosure statement

No potential conflict of interest was reported by the authors.

References

Alrifai M., Zribi M., Rayan M., & Mahmoud M. (2013). On the control of time delay power systems. *International Journal of Innovative Computing, Information and Control*, *9*(2), 762–792.

Bayrak A., & Tatlicioglu E. (2012). *Online time delay identification and control for general classes of nonlinear systems*. Proceedings of the IEEE 51st annual conference on decision and control, Maui, HI, USA (pp. 1591–1596).

Chowdhury B., Kulhare A., & Raina G. (2011). *A study of the SMIB power system model with delayed feedback*. Proc. of international conference on power and energy systems, Chennai, India (pp. 1–6).

Jia H., Yu X., Yu Y., & Wang C. (2008). Power system small signal stability region with time delay. *International Journal of Electrical Power and Energy Systems*, *30*(1), 16–22.

Jiang Z. (2007). *Design of power system stabilizers using synergetic control theory*. Proceedings of IEEE power engineering society general meeting, Tampa, FL, USA (pp. 1–8).

Jiang H., Cai H., Dorsey J. F., & Qu Z. (1997). Toward a globally robust decentralized control for large-scale power systems. *IEEE Transactions on Control Systems Technology*, *5*(3), 309–319.

Liu Z., Zhu C., & Jiang Q. (2008). *Stability analysis of time delayed power system based on cluster treatment of characteristic roots method*. Proceeding of IEEE power and energy society general meeting, Pittsburgh, PA, USA (pp. 1–6).

Mahmoud M. (2000). *Robust control and filtering for time-delay systems*. New York, NY: Marcel Dekker Control Engineering.

Mahmoud M. S., & Zribi M. (1999). H_∞-controllers for time-delay systems using linear matrix inequalities. *Journal of Optimization Theory and Applications*, *100*(1), 89–122.

Okuno H., & Nakabayashi T. (2006). *Basin of attraction and controlling chaos of five-synchronous-generator infinite-bus system*. Proc. of the IEEE international symposium on industrial electronics, Montreal, Quebec, Canada (pp. 1818–1823).

Saad M. S., Hassouneh M. A., Abed E. H., & Edris A. (2005). *Delaying instability and voltage collapse in power systems using SVCs with washout filter-aided feedback*. Proceedings of the 2005 American control conference, Portland, OR, USA (pp. 4357–4362).

Scorletti G., & Fromion V. (1998). *A unified approach to time-delay system control: Robust and gain-scheduled*. Proceedings of the American control conference, Philadelphia, PA (pp. 2391–2395).

Snyder A. F., Ivanescu D., Hadjsaid N., Georges D., & Margotin T. (2000). *Delayed-input wide-area stability control with synchronized phasor measurements and linear matrix inequalities*. Proc. of IEEE power engineering society summer meeting, Seattle, WA (pp. 1009–1014).

Wang X., & Gu K. (2014). *Time-delay power systems control and stability with discretized Lyapunov functional method*. Proceeding of the 2014 Chinese control conference. Invited Session Paper, Nanjing, Jiangsu, China.

Wang X., Yaz E. E., & Yaz Y. I. (2010). *Robust and resilient state dependent control of continuous-time nonlinear systems with general performance criteria*. Proceedings of the 49th IEEE conference on decision and control (CDC) (pp. 603–608).

Wu H., Ni H., & Heydt G. T. (2002). *The impact of time delay on robust control design in power systems*. Proc. of power engineering society winter meeting, New York, NY, USA (pp. 1511–1516).

Yao W., Jiang L., Wen J. Y., Cheng S. J., & Wu Q. H. (2009). *An adaptive wide-area damping controller based on generalized predictive control and model identification*. Proc. of IEEE power and energy society general meeting, Calgary, AB, Canada (pp. 1–7).

Yu X., Jia H., & Zhao J. (2008). *A LMI based approach to power system stability analysis with time delay*. Proceedings of IEEE region 10 conference (TENCON), Hyderabad, India (pp. 1–6).

Yu G. L., Zhang B. H., Xie H., & Wang C. G. (2007). *Wide-area measurement-based nonlinear robust control of power system considering signals delay and incompleteness*. Proceedings of IEEE power engineering society general meeting, Tampa, FL, USA (pp. 1–8) .

Zhang C., Jiang L., Wu Q. H., & Wu M. (2013). Delay-dependent robust load frequency control for time delay power systems. *IEEE Transactions on Power Systems*, *28*(3), 2192–2201.

Zhang W., Xu F., Hu W., Li M., Ge W., & Wang Z. (2012). *Research of coordination control system between nonlinear robust excitation control and governor power system stabilizer in multi-machine power system*. Proceedings of the IEEE international conference on power system technology, Auckland, New Zealand (pp. 1–5).

Zribi M., Mahmoud M. S., Karkoub M., & Lie T. T. (2000). H_∞-controllers for linearised time-delay power systems. *IEEE Transaction on Generation, Transmission and Distribution*, *147*(6), 401–408.

11

Adaptive internal model-based suppression of torque ripple in brushless DC motor drives

Yoni Mandel[a]* and George Weiss[b]

[a]*Core Photonics, Tel Aviv 6971035, Israel;* [b]*School of EE, Tel Aviv University, Tel Aviv 6997801, Israel*

Permanent-magnet synchronous motors (PMSMs) are widely used as high-performance variable-speed drives. Ripple in the electric torque of such motors is often a source of vibration and tracking errors, especially at low speeds. We study the torque characteristics of PMSMs and propose a method to minimize the torque ripple. First, we establish a detailed model for the motor and present a Fourier analysis of the torque ripple, caused by the non-sinusoidal back electro-motive force (BEMF) and the cogging torque, where the main conclusion is that the frequencies present in the torque disturbance are integer multiples of six times the electric frequency. The resulting model is highly nonlinear. We propose an adaptive controller based on the internal model principle, where the resonant frequencies of the controller and the associated gains change according to the motor speed. This is achieved by replacing the time variable by the motor angle, which simplifies the nonlinear model. Our approach is passivity based and will work also for complex mechanical loads and several resonant frequencies. Simulation and experimental results are given to verify the new controller. We compare the performance of our adaptive algorithm with the well-known one from [Canudas de Wit, C., & Praly, L. (2000). Adaptive eccentricity compensation. IEEE Trans. Control Syst. Technol. 8, 757–766]. We find that it performs similarly in simple configurations, and it works also when the motor is part of a more complex system, for example, when the motor is connected to a load via a very flexible shaft.

Keywords: brushless DC; mutual torque; cogging torque; Park transformation; internal model; adaptive control

1. Introduction

A brushless DC motor (BLDCM) is a permanent-magnet synchronous machine (PMSM) connected to an inverter, usually operated by pulse width modulation and controlled by a processor that receives position measurements from a sensor (encoder) mounted on the motor shaft. BLDCMs offer high power density, reliability and efficiency. However, typical motors, driven by either rectangular or sine wave currents, exhibit torque pulsations or ripple.

At high speeds, torque ripple is mostly filtered out by the rotor inertia. However, at low speeds, the torque ripple produces undesirable speed variations, and inaccuracies in motion control. There are applications (e.g. computer tomography (Lee, Park, & Kwon, 2004), optics, machine tools (Jahns & Soong, 1996)) where very high precision is needed at low motor speeds.

In general, torque in a PMSM is developed by

- mutual torque,
- cogging torque and
- reluctance torque,

see Hanselman (1994) and Park, Park, Lee, and Harashima (2000). Imperfections of the motor geometry give rise to harmonics, which are usually multiples of six times the electric angular velocity, in the mutual torque (see also Degobert, Remy, Zeng, Barre, & Hautier, 2006; Hanselman, Hung, & Keshura, 1992). DC offset in the current sensors (and hence in the phase currents) leads to ripples in the mutual torque that are at the electric angular velocity, see the analysis in Gan and Qiu (2004).

There are two main control strategies for operating the inverter in a BLDCM: the "classical" one suits motors with trapezoidal back electro motive force (BEMF), and it produces rectangular current pulses lasting each for 60 electrical degrees. In this operation, only two phases are conducting at any time, but the torque will be close to constant, see Pillay and Krishnan (1989). Torque ripples will appear due to the imperfections of the current rectangles and of the BEMF – see Lu, Zhang, and Qu (2008) for methods to combat these ripples. The other strategy is by controlling the q component in order to control the electromagnetic torque. This leads to greater flexibility and it is better suited for PMSM with close to sinusoidal BEMF. In this paper, we consider only BLDCM operated under the latter (continuous) switching strategy.

Two general approaches have been proposed to reduce the torque ripple. One is to improve the motor's geometrical structure, see Hwang, Eom, Jung, Lee, and Kang (2001), Jahns and Soong (1996) and Ko and Kim (2004).

*Corresponding author. Email: mandel.yonatan@gmail.com

The second approach is to control the winding currents to overcome the disturbances, see for instance Ahn, Chen, and Dou (2005), Canudas de Wit and Praly (2000), Hung and Ding (1993), Le-Huy, Perret, and Feuillet (1986) and Malaize and Levine (2009). An interesting review of a wide range of design techniques for torque ripple minimization is in Jahns and Soong (1996). In Canudas de Wit and Praly (2000), an adaptive observer is designed to estimate the periodic torque disturbance. This estimate is then injected in order to cancel the disturbance. This algorithm needs a precise estimate of the friction in the system. Xu, Panda, Pan, Lee, and Lam (2004) use an iterative learning control (ILC) module in order to estimate the cyclic torque and record the reference current signals over one entire cycle. The ILC module then uses those signals to update the reference current for the next cycle. A gain-shaped sliding mode observer is used to estimate the torque ripple. In Gan and Qiu (2004), a gain scheduled robust two degree of freedom speed regulator, based on the internal model principle (IMP) and pole-zero placement, is developed to eliminate the torque ripples caused by DC offset (as discussed above). Their internal model changes its (single) resonant frequency according to the velocity reference. To prove local stability, the authors use linearization and the small gain theorem. Their proof uses the assumption that the disturbance frequency equals the reference electric velocity, which is debatable. In Hung and Ding (1993), a Fourier series decomposition is used to find a closed-form solution for the current harmonics that eliminate torque ripple and maximize efficiency simultaneously. In Park et al. (2000), the BEMF data, according to the rotor position, are measured and set up in a lookup table. Optimized reference current wave-forms are then obtained from the lookup table using the position and speed information from the shaft encoder. The motor currents are forced to track the reference currents. Petrović, Ortega, Stanković, and Tadmor (2000) design a passivity-based controller, relying on the principles of energy shaping and damping injection. The information about torque ripple harmonics used in the adaptation is extracted from the electrical subsystem, and a current controller is designed to achieve ripple minimization. Qian, Panda, and Xu (2004) implement an ILC scheme in the time domain to reduce periodic torque pulsations. A forgetting factor is introduced in this scheme to increase the robustness. However, this limits the extent to which torque pulsations can be suppressed. To eliminate this limitation, a modified ILC scheme is implemented in the frequency domain using Fourier series. Mattavelli, Tubiana, and Zigliotto (2005) propose the application of repetitive techniques to the current control in a field-oriented PMSM drive, where the q-axis current reference has been modified to achieve constant torque. Ferretti, Magnani, and Rocco (1998) present a compact model of the pulsating torque in a PMSM. An offline identification is proposed for the pulsating torque, to be used for suppressing the oscillations. In Ferretti, Magnani, and Rocco (1999), the technique of Ferretti et al. (1998) is extended to cope with variable motor speed using an adaptive compensator. Degobert et al. (2006), Ruderman, Ruderman, and Bertram (2013) and also Yepes et al. (2010) propose controllers that eliminate ripples of the mutual torque, using an IMP-based controller whose resonant frequencies are adjusted online according to the motor velocity.

In this paper, we propose a new type of controller to reduce the torque ripple caused by both the mutual torque and the cogging torque. *The idea is to use a resonant controller on the q-axis current reference but using the rotor angle in place of the time.* This controller may be called adaptive in the sense that the resonant frequencies (when regarded in the time domain) adjust themselves according to the motor speed.

In Section 2, we establish a model for the motor to cover all dynamics without any assumptions on the signals. In Section 3, we derive a formula for the electromagnetic torque and express the torque ripple caused by non-sinusoidal BEMF. In Section 4, the constant speed version of our controller is presented. The velocity loop, which uses the q-axis current, includes an internal model and a feedforward block. In Section 5, we introduce the adaptive version of the internal model, which works at variable speed. It is based on a transformation of linear differential equations, when time is replaced by motor angle. In Section 6, we give a short review of the Adaptive Eccentricity Algorithm from Canudas de Wit and Praly (2000) and Malaize and Levine (2009). In Section 7, we prove that, for the adaptive algorithm from Section 5, and under the restrictive assumption that the reference ω_{ref} is constant, the velocity error tends to zero. Simulation and experimental results are provided in Sections 8–10, with detailed comparisons with the controller of Canudas de Wit and Praly (2000). In particular, in Section 9 we consider a more complex system with a load connected to the motor via a flexible shaft.

2. Motor model – electrical part

We give the derivation of a mathematical model for a PMSM with one pair of poles per phase (similar to Zhong & Weiss, 2011).

Assume that the windings of a PMSM are connected in star, with each winding having the series resistance r. The basic voltage equations are

$$\tilde{e}_x = \frac{\mathrm{d}\Psi_x}{\mathrm{d}t} = -ri_x + v_x - v_{\mathrm{m}}, \tag{1}$$

where x can be one of a, b, c, $\tilde{e}_x$ are the induced voltages, Ψ_x are the phase winding flux linkages, i_x are the phase currents, and v_x are the phase voltages. The voltage v_m in the center of the star cannot be measured, but this is also not needed. Clearly,

$$i_a + i_b + i_c = 0. \tag{2}$$

We denote by θ the angle of the rotor such that $\theta = 0$ corresponds to the rotor, creating a flux parallel to the axis of winding a and in the direction of the flux created by $i_a > 0$.

Assuming a round (non-salient) rotor and a magnetically non-saturated machine, the flux linkage equations are

$$\Psi_a = Li_a + Mi_b + Mi_c + F(\theta),$$

$$\Psi_b = Mi_a + \mathrm{L}i_b + Mi_c + F\left(\frac{\theta - 2\pi}{3}\right),$$

$$\Psi_c = Mi_a + Mi_b + Li_c + F\left(\frac{\theta + 2\pi}{3}\right), \tag{3}$$

where L is the phase winding self-inductance, M is the mutual inductance between two windings (usually $M = -0.5\,\mathrm{L}$), $F(\theta)$ is the flux linkage through phase a due to the rotor, $F(\theta - 2\pi/3)$ is the flux linkage through phase b due to the rotor and similarly for phase c. Substituting Equation (2) in Equation (3), we get:

$$\Psi_a = L_s i_a + F(\theta),$$

$$\Psi_b = L_s i_b + F\left(\frac{\theta - 2\pi}{3}\right),$$

$$\Psi_c = L_s i_c + F\left(\frac{\theta + 2\pi}{3}\right), \tag{4}$$

where $L_s = L - M$ is the equivalent motor phase inductance. From electromagnetic field theory, we know that $\begin{bmatrix} LM \\ ML \end{bmatrix} > 0$. It follows that $L_s > 0$ (usually $L_s = 1.5\,\mathrm{L}$). In this model, we assume that L_s is independent of θ.

Rewriting Equation (4) in vector form, we get:

$$\underline{\Psi} = L_s \underline{i} + \underline{F}(\theta), \tag{5}$$

where $\underline{F}(\theta) = [F(\theta) F(\theta - 2\pi/3) F(\theta + 2\pi/3)]^{\mathrm{T}}$. Due to the motor structure, $F(\theta)$ is a periodic function with period 2π, so that we can write it as a Fourier series:

$$F(\theta) = \Psi_f \left[\sum_{n=0}^{\infty} \gamma_n \cos(n\theta) + \delta_n \sin(n\theta)\right],$$

where $\Psi_f > 0$ is a parameter that will be chosen later. If the rotor is symmetric, then $F(-\theta) = F(\theta)$ and hence the terms with $\sin(n\theta)$ disappear: $\delta_n = 0$.

We argue that $F(\theta)$ contains only odd harmonics. The symmetry of the rotor (the shape remains unchanged after a rotation of π) implies that $F(\theta + \pi) = -F(\theta)$. By subtracting $F(\theta + \pi)$ from $F(\theta)$, where

$$F(\theta) = \Psi_f \left[\sum_{k=0}^{\infty} \gamma_{2k} \cos(2k\theta) + \sum_{k=1}^{\infty} \gamma_{2k-1} \cos((2k-1)\theta)\right],$$

$$F(\theta + \pi) = \Psi_f \left[\sum_{k=0}^{\infty} \gamma_{2k} \cos(2k\theta) - \sum_{k=1}^{\infty} \gamma_{2k-1} \cos((2k-1)\theta)\right],$$

we get

$$2F(\theta) = F(\theta) - F(\theta + \pi) = 2\Psi_f \sum_{k=1}^{\infty} \gamma_{2k-1} \cos((2k-1)\theta).$$

Therefore,

$$F(\theta) = \Psi_f \sum_{n \mathrm{odd}} \gamma_n \cos(n\theta). \tag{6}$$

We normalize the coefficients such that $\sum_{n=1}^{\infty} \gamma_n = 1$. Then, we see that Ψ_f is the maximal flux linkage due to the rotor. We remark that for a "perfectly built" machine, we have no harmonics:

$$F(\theta) = \Psi_f \cos\theta. \tag{7}$$

By differentiating Equation (5), we get

$$\underline{\tilde{e}} = L_s \underline{\dot{i}} + \underline{e} \quad \text{where} \quad \underline{e} = \underline{\dot{F}}(\theta) = [e_a\, e_b\, e_c]^{\mathrm{T}}, \tag{8}$$

so that e_a, e_b, e_c are the BEMF on each phase due to the rotor movement. According to Equation (6), we have

$$e_a = -\omega \Psi_f \sum_{n\,\mathrm{odd}} \gamma_n n \sin(n\theta) \tag{9}$$

and similarly for e_b and e_c. Here, $\omega = \dot{\theta}$ is the mechanical angular speed. We remark that for a "perfectly built" motor,

$$e_a = -\omega\, \Psi_f \sin(\theta) \tag{10}$$

and similarly for e_b (with $\theta - 2\pi/3$ in place of θ) and for e_c. Comparing Equation (1) with Equation (8) and eliminating $\underline{\tilde{e}}$,we get:

$$L_s \underline{\dot{i}} = \underline{v} - r\underline{i} - \underline{e} - \underline{v}_m, \tag{11}$$

where $\underline{v} = [v_a\, v_b\, v_c]^{\mathrm{T}}$and $\underline{v}_m = [v_m\, v_m\, v_m]^{\mathrm{T}}$. Recall the Park transformation introduced in Park (1929), a unitary matrix U which transforms a vector from the a, b, c

coordinate system to the rotating d, q coordinate system:

$$\begin{bmatrix} f_d \\ f_q \\ f_0 \end{bmatrix} = U \cdot \begin{bmatrix} f_a \\ f_b \\ f_c \end{bmatrix},$$

where

$$U = \sqrt{\frac{2}{3}} \cdot \begin{bmatrix} \cos(\theta) & \cos\left(\theta - \frac{2\pi}{3}\right) & \cos\left(\theta + \frac{2\pi}{3}\right) \\ -\sin(\theta) & -\sin\left(\theta - \frac{2\pi}{3}\right) & -\sin\left(\theta + \frac{2\pi}{3}\right) \\ \frac{1}{\sqrt{2}} & \frac{1}{\sqrt{2}} & \frac{1}{\sqrt{2}} \end{bmatrix}. \quad (12)$$

Multiplying Equation (11) with the unitary matrix U from Equation (12), we get:

$$L_s U \cdot \begin{bmatrix} \dot{i}_a \\ \dot{i}_b \\ \dot{i}_c \end{bmatrix} = \begin{bmatrix} v_d \\ v_q \\ v_0 \end{bmatrix} - r \begin{bmatrix} i_d \\ i_q \\ i_0 \end{bmatrix} - \begin{bmatrix} e_d \\ e_q \\ e_0 \end{bmatrix} - \begin{bmatrix} 0 \\ 0 \\ \sqrt{3} v_m \end{bmatrix}.$$

Since, according to a short computation,

$$\begin{bmatrix} \dot{i}_d \\ \dot{i}_q \\ \dot{i}_o \end{bmatrix} = U \begin{bmatrix} \dot{i}_a \\ \dot{i}_b \\ \dot{i}_c \end{bmatrix} + \omega \begin{bmatrix} i_q \\ -i_d \\ 0 \end{bmatrix},$$

we get

$$L_s \cdot \begin{bmatrix} \dot{i}_d \\ \dot{i}_q \\ \dot{i}_o \end{bmatrix} - L_s \omega \begin{bmatrix} i_q \\ -i_d \\ 0 \end{bmatrix} = \begin{bmatrix} v_d \\ v_q \\ v_o \end{bmatrix} - r \begin{bmatrix} i_d \\ i_q \\ i_o \end{bmatrix} - \begin{bmatrix} e_d \\ e_q \\ e_o \end{bmatrix} - \begin{bmatrix} 0 \\ 0 \\ \sqrt{3} v_m \end{bmatrix}. \quad (13)$$

Notice that $i_o = 0$ (hence also $\dot{i}_o = 0$), due to Equation (2). Hence, according to the third equation (the third line) in (13), $v_o = e_o + \sqrt{3} v_m$.

Thus, the dynamic voltage equations in d, q coordinates are

$$L_s \dot{i}_d = v_d - r i_d - e_d + L_s \omega i_q,$$

$$L_s \dot{i}_q = v_q - r i_q - e_q - L_s \omega i_d.$$

For the case of a perfectly built motor as in Equation (7), using the Park transformation on Equation (10) we get, after some computation,

$$e_d = 0, \quad e_q = \sqrt{\frac{3}{2}} \cdot \omega \Psi_f. \quad (14)$$

3. The motor torque and BEMF

Torque production inside brushless PMSM is mainly due to the interaction between the permanent magnet field and the currents in the phase windings (mutual torque), and the interaction between the permanent magnets, located in the rotor, and the slotted iron structure of the stator (cogging torque). In this section, we derive formulas for these torques.

The total energy stored in a PMSM is

$$E_{\text{total}} = E_{\text{mag}} + E_{\text{kin}} = E_{\text{mag}} + \frac{1}{2} J \omega^2, \quad (15)$$

where E_{mag} is the energy stored in the magnetic field and E_{kin} is the kinetic energy of the rotating body with inertia J that contains the rotor. Differentiating Equation (15), we get:

$$\dot{E}_{\text{total}} = \dot{E}_{\text{mag}} + J \omega \dot{\omega}. \quad (16)$$

We denote by T_e the *electromagnetic torque* generated inside the motor, and by T_l the *mechanical load torque*. Substituting Newton's law, $J\dot{\omega} = T_e - T_l$, into Equation (16), we get:

$$\dot{E}_{\text{total}} = \dot{E}_{\text{mag}} + \omega (T_e - T_l). \quad (17)$$

On the other hand, $\dot{E}_{\text{total}}$ is the flow of power to the system.

This power is equal to the ingoing power minus the power lost in the resistors minus the power drawn by the load,

$$\dot{E}_{\text{total}} = \underline{v} \circ \underline{i} - r \underline{i} \circ \underline{i} - \omega T_l = (\underline{v} - r \underline{i}) \circ \underline{i} - \omega T_l, \quad (18)$$

where $\circ$ is the inner product in $\mathfrak{R}^3$. From Equations (17) and (18), we conclude:

$$\dot{E}_{\text{mag}} = (\underline{v} - r \underline{i}) \circ \underline{i} - \omega T_e.$$

Substituting Equation (11) into the above formula, we get:

$$\dot{E}_{\text{mag}} = (L_s \cdot \dot{\underline{i}} + \underline{e} + \underline{v}_m) \circ \underline{i} - \omega T_e. \quad (19)$$

Since $\underline{v}_m \circ \underline{i} = 0$, multiplying the above formula by dt, we get:

$$dE_{\text{mag}} = (L_s \cdot \dot{\underline{i}} + \underline{e}) \circ \underline{i} \cdot dt - T_e \, d\theta. \quad (20)$$

Integrating both sides of Equation (20), while assuming no motion of the rotor, to charge the magnetic field, we get (using that $d\theta = 0, \underline{e} = 0$)

$$E_{\text{mag}} = E_0(\theta) + \int_0^T L_s \cdot \dot{\underline{i}} \circ \underline{i} \cdot dt = E_0(\theta) + \int_0^T \frac{d}{dt} \left(\frac{1}{2} L_s \cdot \underline{i} \circ \underline{i} \right) dt = E_0(\theta) + \frac{1}{2} (L_s \cdot \underline{i}) \circ \underline{i}, \quad (21)$$

where $E_0(\theta)$ is the energy in the magnetic field due to the rotor magnet (while there are no currents) at the angle θ.

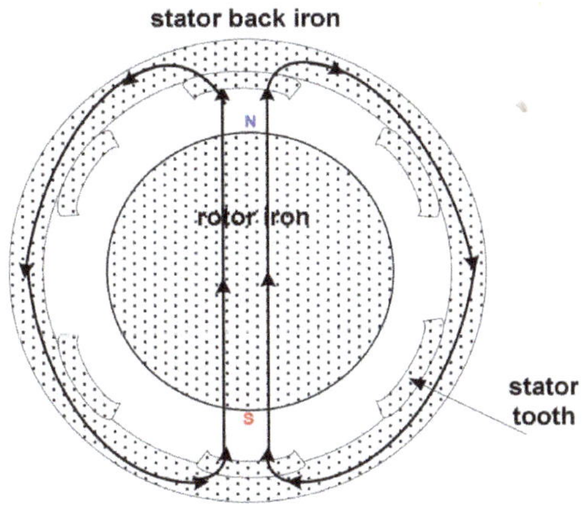

Figure 1. PMSM with stator slots.

In the sequel, we consider again the moving rotor. Differentiating Equation (21), we get

$$\dot{E}_{\text{mag}} = \frac{\mathrm{d}}{\mathrm{d}\theta} E_0(\theta) \cdot \omega + (L_s \cdot \underline{i}) \circ \dot{\underline{i}}.$$

From here and Equation (19), we conclude:

$$T_e = \frac{\underline{e} \circ \underline{i}}{\omega} - \frac{\mathrm{d}}{\mathrm{d}\theta} E_0(\theta). \qquad (22)$$

The term $T_{\text{em}} = e \circ i / \omega$ is called the *mutual torque*, while $T_{\text{ec}} = -(\mathrm{d}/\mathrm{d}\theta) E_0(\theta)$ is called the *cogging torque*.

The mutual torque may be calculated using phase currents and back EMF instant values, both in (a, b, c) and in (d, q) coordinates (we use that $i_0 = 0$):

$$T_{\text{em}} = \frac{\underline{e} \circ \underline{i}}{\omega} = \frac{(U \cdot \bar{e}) \circ (U \cdot \bar{i})}{\omega} = \frac{i_d e_d + i_q e_q}{\omega}. \qquad (23)$$

For a "perfectly built" motor, we have from Equation (14)

$$T_{\text{em}} = \frac{i_d \cdot 0 + i_q \cdot \sqrt{3/2} \Psi_f \omega}{\omega} = i_q \cdot \sqrt{3/2} \Psi_f. \qquad (24)$$

As mentioned earlier, cogging torque is created by the interaction between the permanent magnets in the rotor and the slotted iron structure of the stator. Usually, the stator has slots to hold the windings, see Figure 1. Due to these slots, the rotor has preferred directions and cogging is produced.

The following expression (25) for the electromagnetic cogging torque is taken from Hwang et al. (2001):

$$T_{ec} = \sum_{n=0}^{\infty} K_n n \sin(n N_L \theta), \qquad (25)$$

where K_n are constants determined by the machine's geometry and N_{L} is the least common multiple of the number of pairs rotor poles and the number of slots in the stator.

For the simple motor discussed earlier (one pair of stator poles per phase, one slot per stator pole and one pair of poles in the rotor), we have $N_{\text{L}} = 6$. More generally, N_{L} is a multiple of 6.

Recall that for a "non-perfectly built" motor, the BEMF due to the rotor movement can be described by Equation (9). Let us look at the third harmonic of the BEMF:

$$\begin{aligned}
e_{a_{\text{3rd harmonic}}} &= -\Psi_f \omega 3 \gamma_3 \sin(3\theta), \\
e_{b_{\text{3rd harmonic}}} &= -\Psi_f \omega 3 \gamma_3 \sin\left(3\left(\theta - \frac{2\pi}{3}\right)\right) \\
&= -\Psi_f \omega 3 \gamma_3 \sin(3\theta), \\
e_{c_{\text{3rd harmonic}}} &= -\Psi_f \omega 3 \gamma_3 \sin\left(3\left(\theta + \frac{2\pi}{3}\right)\right) \\
&= -\Psi_f \omega 3 \gamma_3 \sin(3\theta).
\end{aligned}$$

Since $i_a + i_b + i_c = 0$, we have from Equation (23) that the contribution of the third harmonic to the torque is $(\underline{e}_{\text{3rdharmonic}} \circ \underline{i}/\omega) = 0$. A similar result can be derived for all the odd multiples of 3.

Therefore, we conclude that the harmonics which contribute to the mutual torque, T_{em}, are multiples of $\sin(n\theta)$, where $n = 6p + 1$ for $p = 0, 1, 2, \ldots$ and $n = 6p - 1$ for $p = 1, 2, 3, \ldots$.

Recall from Equation (23) that in d, q coordinates, the electromagnetic torque is $T_{\text{em}} = (i_d e_d + i_q e_q)/\omega$. Since i_d and i_q are inputs to be chosen by the user, it is useful to look at the expressions for e_{d} and e_{q}. If we compute e_{d} and e_{q} from (9) and (12), we obtain

$$\begin{aligned}
e_d &= -\sqrt{\frac{3}{2}} \omega \Psi_f \left[\sum_{p=1}^{\infty} [\gamma_{6p+1}(6p+1) \right. \\
&\quad \left. + \gamma_{6p-1}(6p-1)] \sin(6p\theta) \right], \\
e_q &= \sqrt{\frac{3}{2}} \omega \Psi_f \left[\gamma_1 + \sum_{p=1}^{\infty} [\gamma_{6p+1}(6p+1) \right. \\
&\quad \left. - \gamma_{6p-1}(6p-1)] \cos(6p\theta) \right].
\end{aligned} \qquad (26)$$

Denoting for $p = 1, 2, 3, \ldots$

$$\begin{aligned}
\eta_{\text{d}p} &= -\sqrt{3/2} \Psi_f [\gamma_{6p+1}(6p+1) + \gamma_{6p-1}(6p-1)], \\
\eta_{\text{q}p} &= \sqrt{3/2} \Psi_f [\gamma_{6p+1}(6p+1) - \gamma_{6p-1}(6p-1)],
\end{aligned}$$

and $\eta_{\text{q}0} = \sqrt{3/2} \Psi_{\text{f}} \gamma_1$, we get from Equation (26)

$$\begin{aligned}
e_d &= \omega \left[\sum_{p=1}^{\infty} \eta_{\text{d}p} \sin(6p\theta) \right], \\
e_q &= \omega \left[\eta_{\text{q}0} + \sum_{p=1}^{\infty} \eta_{\text{q}p} \cos(6p\theta) \right].
\end{aligned} \qquad (27)$$

If the motor has Π pairs of poles per phase (instead of just one), then in Equations (25) and (27), we have to replace θ with $\Pi\theta$.

4. Velocity control (non-adaptive)

Figure 2 shows the current and velocity control loops of a BLDCM. Notice that $i_{d_{\text{ref}}} = 0$ and $i_{q_{\text{ref}}}$ is the sum of the velocity controller's output and a term from the feed-forward path with transfer function $\alpha = (1/\eta_{q0})(Js + D)$, which has the role of making the response faster. Here, D is the viscous friction coefficient of the motor, so that $T_l = T_L + D\omega$, where T_L is the external load torque. It is recommended to include a saturation block in the feedforward path.

In the sequel, we allow the motor to have Π pairs of poles per phase (until now we had $\Pi = 1$), so that we rely on the modified version of Equation (27). We will assume complete and instantaneous control of the currents, hence $\varepsilon_d = 0$ and $\varepsilon_q = 0$ in Figure 2. If we operate in the linear range of the converter and the motor (no saturation), then this is due to the fact that the current loop is much faster than the velocity loop (typical bandwidths are 1 kHzfor the current loop and 100 Hz for the velocity loop, see also Gan and Qiu, (2004). Substituting Equation (27) in Equation (23), while assuming $i_{d_{\text{ref}}} = 0$ and complete current control, we get

$$T_{\text{em}} = i_{q_{\text{ref}}}\left[\eta_{q0} + \sum_{p=1}^{\infty}\eta_{qp}\cos(6p\Pi\theta)\right]. \tag{28}$$

Using Equation (28), the block diagram in Figure 2 simplifies to the one in Figure 3.

In the sequel, we consider a control loop as in Figure 3. The special structure of this feedback system and the fact that $i_{q_{\text{ref}}}$ is a function of the feedback result in a steady-state output ω whose spectral contents contain harmonics

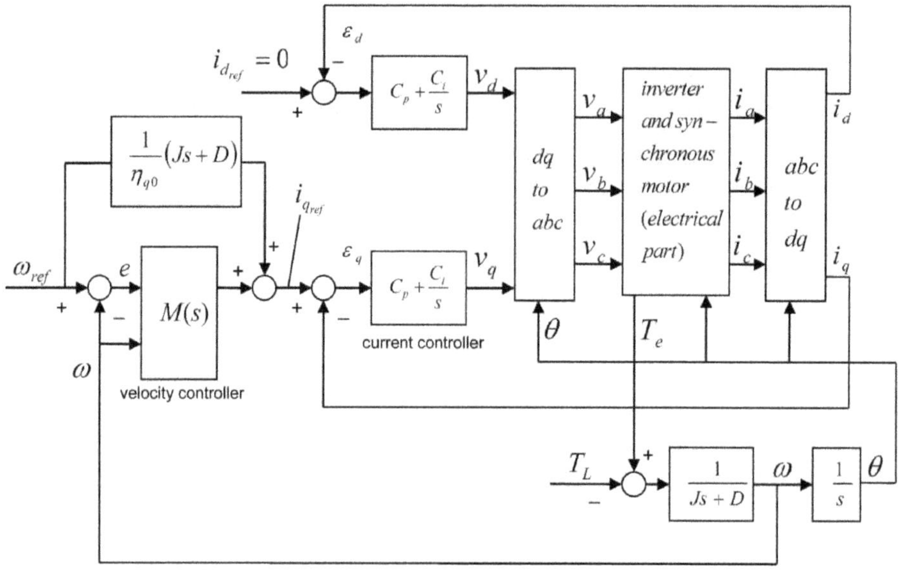

Figure 2. Full block diagram of the velocity control system. Here, T_L and T_e are the external load torque and the electromagnetic torque.

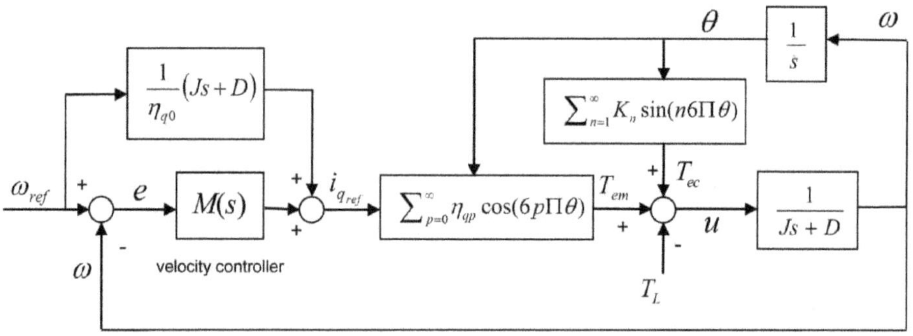

Figure 3. Block diagram of the simplified control system that corresponds to an ideal current control loop (i.e. no current tracking errors). The electromagnetic torque T_e from Figure 2 is $T_{\text{em}} + T_{\text{ec}}$, where T_{em} is the mutual torque and T_{ec} is the cogging torque.

whose frequency is a multiple of $6\Pi\omega_{\text{ref}}$. Thus, in steady state, ω will be periodic, with the fundamental frequency $6\Pi\omega_{\text{ref}}$. We will see this phenomenon in our simulations and experiments.

Based on the IMP, we propose for M a PI controller plus additional internal model terms:

$$M(s) = k_p + \frac{k_i}{s} + \sum_{m=1}^{N} \frac{k_m s}{s^2 + {\omega_m}^2}, \tag{29}$$

where $k_p, k_i, k_m > 0$ and $\omega_m = 6m\Pi\omega_{\text{ref}}$ is a resonant frequency of M. Such internal models have been used for instance in Jayawardhana and Weiss (2009), Knobloch, Isidori, and Flockerzi (1993) or Spitsa, Kuperman, Weiss, and Rabinovici (2006). The idea of the IMP is that M has infinite gain at the frequencies ω_m, resulting in zero tracking errors at these frequencies, if the closed-loop system is stable. Since there are a possibly large number of resonant frequencies, and they are multiples of a fundamental frequency, another way to build M would be using delay lines, as in repetitive control, see, for instance, Weiss (1997), but we have not explored this further.

We mention that in the terminology of van der Schaft (2000), the controller $M$is strictly input passive.

5. An adaptive controller

In this section, we modify the controller in Equation (29) to accommodate a variable motor speed. Internal models with variable resonant frequencies have appeared in Canudas de Wit and Praly (2000), Esbrook, Tan, and Khalil (2011), Lu et al. (2008) and Spitsa et al. (2006).

Recall from Equations (22), (25) and (28) that the electromagnetic torque generated inside the motor is

$$\begin{aligned} T_e &= T_{\text{em}} + T_{\text{ec}} \\ &= i_{q_{\text{ref}}} \left[\eta_{q0} + \sum_{p=1}^{\infty} \eta_{qp} \cos(6\Pi p\theta) \right] \\ &\quad + \sum_{n=0}^{\infty} K_n\, n \sin(nN_L\theta), \end{aligned}$$

where N_L is a multiple of 6Π.

From the above, it is obvious that the disturbance torque is periodic as a function of the motor positionθ, with frequencies which are usually multiples of 6Π. However, other integer frequencies may also appear, such as the frequency Π due to DC offsets (Gan & Qiu, 2004) (we mentioned this in the Introduction). Since the disturbance in the motor speed is a function of θ, we suggest using in the controller an internal model which is also a function of θ (instead of time). In order to do that, we visualize the input and the output of the controller as being functions of θ (instead of time). In this case, an internal model term to reduce the periodical disturbance, with frequency m, would be $k_m s/s^2 + m^2$. Overall, the controller would look as in Equation (29) and $\omega_{\text{ref}} = 1$. However, this is now a transfer function in an unconventional frequency domain, corresponding to θ as the time variable. To express our controller in the conventional time domain, first we revert from the Laplace to the θ domain where the term corresponding to the frequency m is described by the differential equation

$$p_{\theta\theta} + p \cdot m^2 = k_m e_\theta. \tag{30}$$

Here, e is the input to the IMP controller and p is the output.

Converting Equation (30) to the time domain, while using

$$\begin{aligned} p_\theta &= \frac{\mathrm{d}p}{\mathrm{d}t} \cdot \frac{\mathrm{d}t}{\mathrm{d}\theta} = \dot{p} \cdot \frac{1}{\omega}, \\ p_{\theta\theta} &= \left(\dot{p} \cdot \frac{1}{\omega} \right)_\theta = \frac{\mathrm{d}}{\mathrm{d}t} \left(\dot{p} \cdot \frac{1}{\omega} \right) \frac{\mathrm{d}t}{\mathrm{d}\theta} \\ &= \left[\ddot{p} \cdot \frac{1}{\omega} - \dot{p} \frac{\dot{\omega}}{\omega^2} \right] \cdot \frac{1}{\omega} = \frac{\ddot{p}}{\omega^2} - \dot{p} \frac{\dot{\omega}}{\omega^3}, \end{aligned}$$

and similarly $e_\theta = (\mathrm{d}e/\mathrm{d}t) \cdot (\mathrm{d}t/\mathrm{d}\theta) = \dot{e} \cdot (1/\omega)$, we get:

$$\ddot{p} - \dot{p}\frac{\dot{\omega}}{\omega} + p \cdot m^2 \cdot \omega^2 = k_m \dot{e} \omega. \tag{31}$$

This is a linear differential equation with variable coefficients (that depend on $\omega, \dot{\omega}$). A digital controller will be able to implement a good approximation of such a subsystem. Since in Equation (31) we have a division by ω, it is recommended to impose a lower bound on ω in this division: For some small $\varepsilon > 0$, if $\omega < \varepsilon$, then replace $\dot{\omega}/\omega$ with $\dot{\omega}/\varepsilon$.

The term k_i/s in Equation (29) has to be expressed in the normal time domain as well. This is much easier and we get:

$$\dot{p} = k_i e \omega. \tag{32}$$

The block diagram of the control system with the adaptive internal model controller (AIMC) is the same as in Figure 2 except that M is replaced with the AIMC block that contains a realization of both Equations (31) and (32).

6. The adaptive eccentricity compensation algorithm

In Canudas de Wit and Praly (2000), an adaptive observer, called *adaptive eccentricity compensator* (AEC), is proposed for the velocity control of a BLDCM that estimates the angle-dependent periodic torque disturbances acting on an electric motor. A proof for stability is given for a proportional controller acting on a very simplified motor model (an integrator from torque to angular velocity) using a Lyapunov function. A friction predictor is added to cancel the Coulomb friction, which is feedforward, so that it cannot react to changes in the load torque. In our implementation of the controller from Canudas de Wit and Praly (2000),

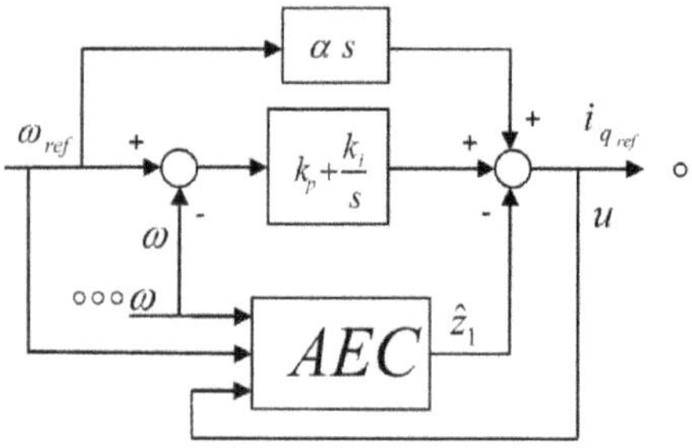

Figure 4. The velocity control loop with the AEC controller. The symbol ○ ○ ○ indicates a connection with the remaining part of the system, which is as in Figure 2.

used for comparison with our AIMC controller, we replace the friction predictor by an integral component in the controller, which can compensate any step load torque – this is in the PI block in the middle of Figure 4. Thus, we use a slightly better implementation of the AEC controller than the original one in Canudas de Wit and Praly (2000). The rest of the block diagram is as in Figure 2.

The controller of Canudas de Wit and Praly (2000) has been further developed in Malaize and Levine (2009) for the case of high precision positioning of a BLDCM (i.e. tracking of an angle reference signal). Thus, there is no need to filter out the noise originated in the discrete derivative of the encoder. On the other hand, tuning this new controller scheme is not that simple. The proof of stability in Malaize and Levine (2009) is more complex, and it too relies on a quadratic Lyapunov function. The adaptive observer structure, for the θ domain frequency m in Canudas de Wit and Praly (2000), is the following:

$$\begin{aligned}\alpha\dot{\hat{\omega}} &= u + \hat{z}_1 - h_0(\hat{\omega} - \omega),\\ \dot{\hat{z}}_1 &= \omega m^2 \hat{z}_2 - h_1(\hat{\omega} - \omega) - h_1(\omega_{\mathrm{ref}} - \omega),\\ \dot{\hat{z}}_2 &= -\omega \hat{z}_1.\end{aligned} \tag{33}$$

where h_0 and h_1 are positive constants to be chosen later. The control input, $u = i_{q_{\mathrm{ref}}}$, is given by the AEC block added to the output of a PI controller:

$$u = \alpha \cdot \dot{\omega}_{\mathrm{ref}} + k_p(\omega_{\mathrm{ref}} - \omega) + k_i \int_{t_0}^{t} (\omega_{\mathrm{ref}} - \omega)\, \mathrm{d}\sigma - \hat{z}_1 \tag{34}$$

7. A proof of convergence

We give a proof of convergence to zero of the speed tracking error e from Figure 3 under the restrictive assumption that the reference speed ω_{ref} is constant. We shall also assume that the system remains throughout in the reasonable operating region where the angular velocity ω and the total torque u acting on the motor are positive. We emphasize that this is not the "standard" situation from internal model-based control, because the disturbance is a superposition of sinusoidal signals in the variable θ, and not in the time variable.

LEMMA. *Consider the system* **P** *with transfer function*

$$P(s) = \frac{1}{\mathrm{J}s + D}$$

with input u and output ω and let $\omega_{\mathrm{ref}} > 0$ be a constant. Introduce the speed error $e = \omega_{\mathrm{ref}} - \omega$, the reference torque $u_{\mathrm{ref}} = D\omega_{\mathrm{ref}}$, the torque error $v = u_{\mathrm{ref}} - u$ and the Hamiltonian $H = Je^2\omega_{\mathrm{ref}}/2$. If we regard **P** *with θ (the integral of ω) as the time variable, with v as input and with e as output, then it becomes a nonlinear time-invariant system $\mathbf{P}_{\mathrm{NL}}$. In the region where $u, \omega \geq 0$, $\mathbf{P}_{\mathrm{NL}}$ is passive, i.e. $H_\theta \leq ve$.*

Proof Since $\dot{e} = (-De + v)/J$, we have

$$\begin{aligned}H_\theta &= \frac{H}{\omega} = \frac{Je\dot{e}\omega_{\mathrm{ref}}}{\omega} = e(-De + v)\frac{\omega_{\mathrm{ref}}}{\omega}\\ &= \left[-De^2 + ev\left(1 - \frac{\omega}{\omega_{\mathrm{ref}}}\right)\right]\frac{\omega_{\mathrm{ref}}}{\omega} + ev\\ &= -\frac{[D\omega_{\mathrm{ref}} + v]e^2}{\omega} + ev = \frac{-ue^2}{\omega} + ev.\end{aligned}$$

We see that if $u, \omega \geq 0$ then $H_\theta \leq ev$.

PROPOSITION. *Consider the feedback system from* Figure 3 *but with a perfectly built motor, meaning that $\eta_{qp} = 0$ for all $p > 0$, and with only finitely many terms in the sum defining the cogging torque T_{ec}. Assume that the adaptive internal model M is built according to Equations* (31) *and* (32), *with the resonant frequencies (in the θ domain) matching the frequencies of the cogging torque. Assume that $\omega_{\mathrm{ref}} > 0$ and T_{L} are constant and the system remains in the region where $u, \omega \geq 0$. Then,* $\lim_{t\to\infty} e(t) = 0$.

Proof. Using simple block diagram manipulations (moving the summation point for the feedforward signal further ahead until the right end of the diagram), we can see that Figure 3 is equivalent to the one shown in Figure 5.

Now, we regard this feedback system in the θ domain (i.e. we use θ in place of the time). Then, the feedback loop in the right upper corner of Figure 5 disappears, because θ is no longer a state variable, and instead $T_{\mathrm{ec}} - T_{\mathrm{L}}$ is now a disturbance generated by a finite-dimensional linear exosystem that has simple eigenvalues on the imaginary axis (T_{L} is generated by an eigenvalue at zero). The linear system with parameters J and D at the right end of Figure 5 becomes the nonlinear passive system $\mathbf{P}_{\mathrm{NL}}$ described in the lemma, with the state e (passivity holds as long as $u, \omega \geq 0$). The block diagram becomes very simple, and we leave it to the reader to redraw it. This new block diagram is the same as Figure 2 in Jayawardhana and Weiss (2009), with $\eta_{q0}M$ playing the role of the controller. It can be verified

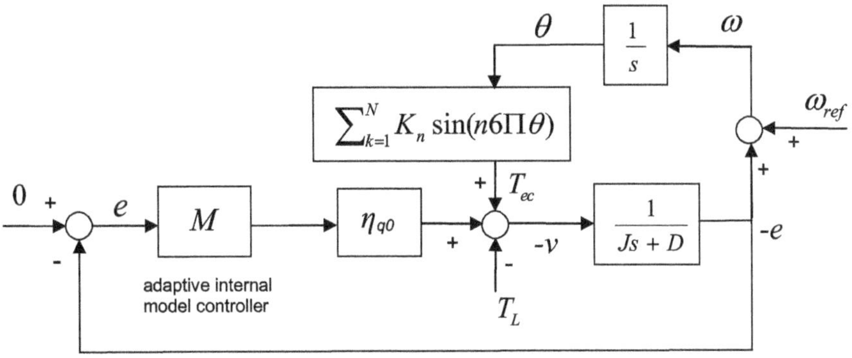

Figure 5. A block diagram equivalent to the one from Figure 3 when $\eta_{qp} = 0$ for all $p > 0$, and using an AIMC, as in the proposition.

that the assumptions in Theorem IV.3 from Jayawardhana and Weiss (2009) are satisfied, and this implies that indeed $e(t) \to 0$.

We remark that the conclusion of the proposition remains valid if we add an arbitrary L^2 component to the constant load torque T_L, and this follows from the same theorem in Jayawardhana and Weiss (2009).

8. Simulations without load

In this section (and also in the experiments), ω is measured by taking the discrete derivative from a high-resolution encoder with 20,000 counts per turn. Since the simulations and experiments are done at low motor speeds, a low-pass filter is used on these velocity measurements. We chose a second-order low-pass filter with natural frequency $\omega_n = 2\pi 200$ rad/sec and damping $\zeta = 0.7$. The angular acceleration (needed in Equation (31)) is computed by taking the discrete derivative of ω.

In order to facilitate the comparison of the AEC and AIMC controllers, the following assumptions are made:

- Since most of the ripple is due to the cogging, we assume $\eta_{q0} = \sqrt{3/2}\Psi_f$ and $\eta_{qp} = 0$ for $p \neq 0$.
- The motor has two pole pairs per phase, that is, $\Pi = 2$.
- D is assumed to be zero. Hence, comparing Figure 4 with Figure 2, we choose $\alpha = J/\eta_{q0}$.
- There is static friction on the rotor, 0.01 Nm, with sign depending on the direction of motion, of course.

We have obtained simulation results for the system shown in Figure 2, with: $\Psi_f = 0.0358$ V sec, $\eta_{q0} = 0.04384$ N m/A, $J = 332 \times 10^{-7}$ kg m^2, $\Pi = 2$, $K_1 = 0.006$ N m, $k_i = 10$ A, $k_p = 0.3125$ A sec, $L_s = 0.65 \times 10^{-3}$ H, $V_{bus} = 20$ V, $r = 0.5\Omega$. The coefficients of the PI part of the controller are chosen so that if we only use the PI part, we get a reasonable gain margin of 11 dB and a phase margin of 54°. For the first few simulations, the reference input is a step of size 2π rad/sec.

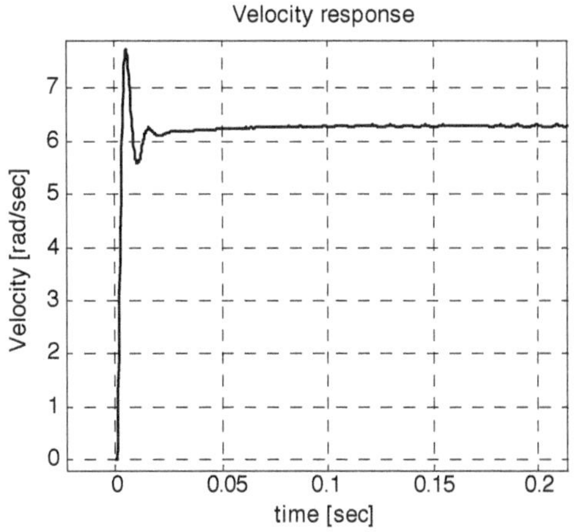

Figure 6. Velocity response for $T_{ec} = 0$, using a PI controller.

First, we simulate the velocity response for a "perfectly built" motor, that is, $T_{ec} = 0$. We see from Figure 6 that ω converges to ω_{ref}. The large overshoot is caused by the low-pass filter mentioned earlier. Next, we simulate the step response for a PMSM with cogging torque $T_{ec} = 0.006 \cdot \sin(12 \cdot \theta)$ (see Figure 7). We also show its steady-state frequency contents, that is, the absolute value of its fast Fourier transform (FFT) for the time interval [1,10] (sec). As we can see from Figure 7, the steady-state velocity is periodic, with the fundamental frequency $12\omega_{ref}$.

In the remaining simulations, we compare the two adaptive algorithms (AEC and AIMC). The PI part of the controllers is chosen as before. For the AEC controller, we choose $h_0, h_1 = 3$. For the AIMC controller, $k_1 = 1.8$. In both controllers, the gains are chosen such that if multiplied by 2 the system would become unstable.

In the next two figures, the reference input and the cogging torque are as in Figure 7. The velocity responses, while using the PI controller (alone) and the AIMC algorithm, are plotted in Figure 8. Figure 9 shows the

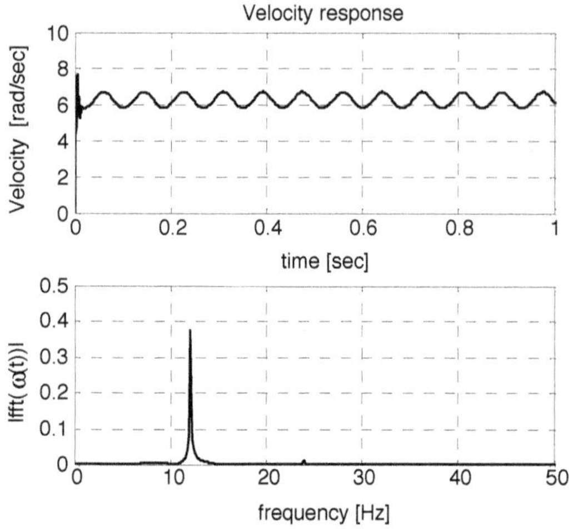

Figure 7. Velocity response and its steady-state frequency contents (FFT, absolute value) for a non-perfectly built motor with a PI controller.

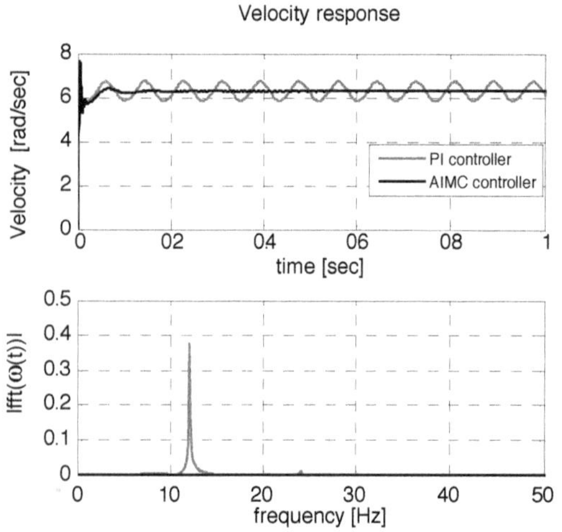

Figure 8. Velocity response and its steady-state frequency contents with a PI controller and with an AIMC controller, for the step velocity reference as in Figure 7.

velocity response with AEC algorithm compared with that of the AIMC algorithm.

To check the ability of both controllers to track a variable speed command, we changed the reference input. The new reference input and the velocity response using a simple PI controller are shown in Figure 10. The velocity responses, when using the adaptive algorithms, are shown in Figure 11. They are so close that we cannot distinguish them visually in Figure 11. A comparison between the velocity error signals of the two algorithms, after 3 s, is shown in Figure 12.

Next, we simulate a second harmonic for the cogging disturbance (though it does not exist in our experiment)

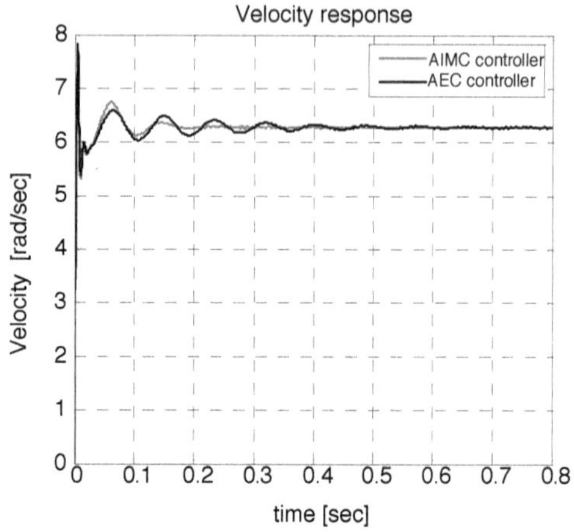

Figure 9. Comparison of the velocity response of both algorithms, for a step velocity reference, with cogging torque as in Figure 7.

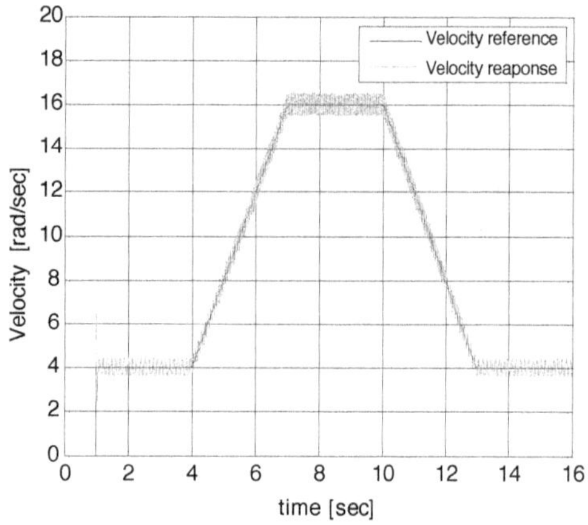

Figure 10. Velocity response for a variable reference velocity command using a simple PI controller, with cogging torque as in Figure 7.

and check both algorithms. We take $K_2 = 0.003$ N·m in Equation (25) (in addition to $K_1 = 0.006$ N m][· taken previously). To deal with the second harmonic as well, we add another AIMC block with $k_2 = 0.6$. The velocity response is shown in Figure 13. A comparison between the AIMC and the AEC for cogging with two harmonics is shown in Figure 14.

9. Simulation results with a flexible shaft and load

The AEC controller from Canudas de Wit and Praly (2000) was designed for a motor model that is just an integrator (from torque to angular velocity) and their proof of stability refers to this model. This model corresponds to a motor with a very simple mechanical load (just friction). The simulations in Section 8 use such a model and they show that in this case, the behavior of the control system

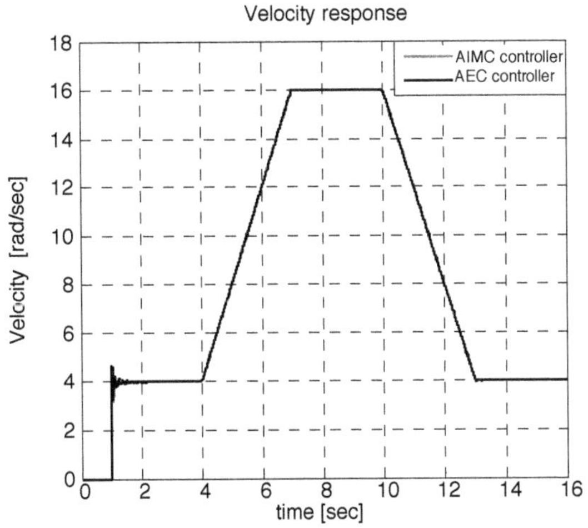

Figure 11. Velocity response for a variable speed command (as in Figure 10) for the two adaptive algorithms.

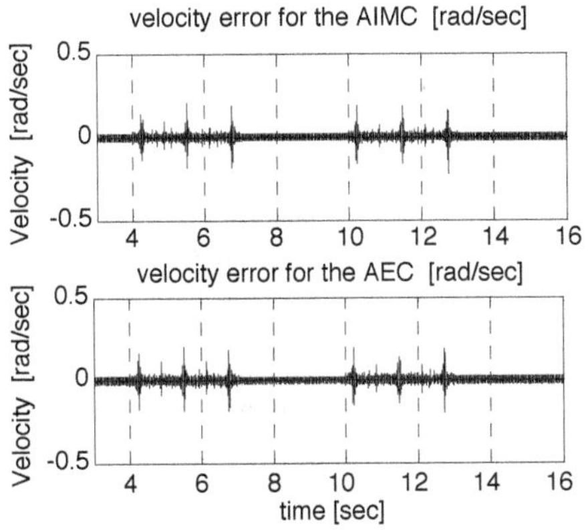

Figure 12. Velocity error for the same reference velocity and cogging torque as in Figure 10. The spikes are due to the resolution of the encoder.

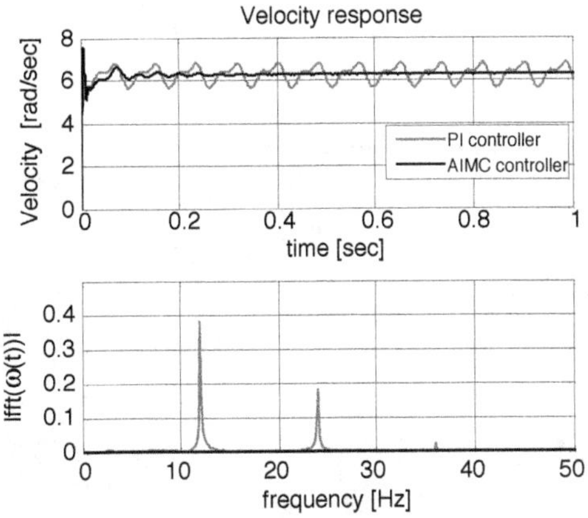

Figure 13. Velocity response and frequency contents in the presence of two cogging frequencies, using the PI controller and the AIMC algorithm, with $\omega_{\text{ref}} = 1$ Hz, leading to two resonant frequencies at 12 and 24 Hz.

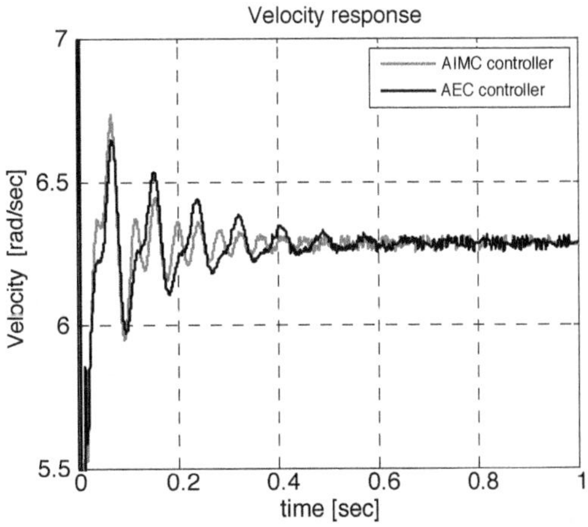

Figure 14. Velocity response in the presence of two cogging frequencies, using the AIMC and the AEC controllers, with a rotation speed of 1 Hz and two resonant frequencies at 12 and 24 Hz.

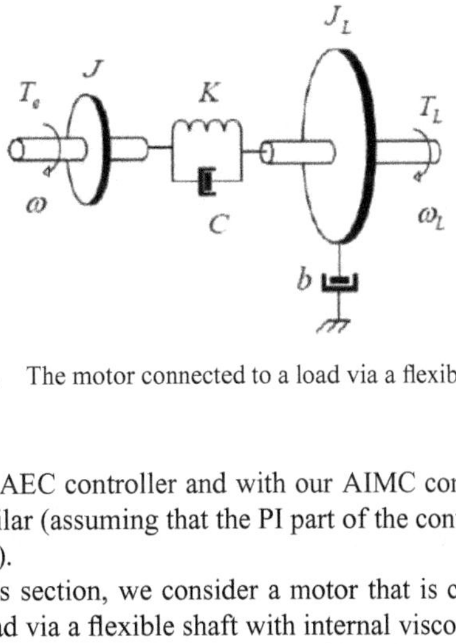

Figure 15. The motor connected to a load via a flexible shaft.

with the AEC controller and with our AIMC controller is very similar (assuming that the PI part of the controllers is the same).

In this section, we consider a motor that is connected to the load via a flexible shaft with internal viscous damping. The load has its own moment of inertia J_{L} and there is viscous friction between the load and the inertial reference system, with coefficient b. We denote by K and C the stiffness constant and the damping constant of the flexible shaft. The schematic representation of the mechanical system is shown in Figure 15.

We choose on purpose a very flexible (very soft) shaft, so that there will be a significant difference between the velocities at its two ends. The parameters are: $C = 0.005\,\text{Nm sec/rad}$, $K = 0.1\,\text{Nm/rad}$, $J_L = 332 \times 10^{-6}\,\text{kg m}^2$, $b = 0.01\,\text{Nm sec/rad}$.

For the simulations, we make the same assumptions as in Section 8 and we use the same motor parameters. The additions to the block diagram from Figure 2, in order to incorporate the flexible shaft and the load, are shown in Figure 16.

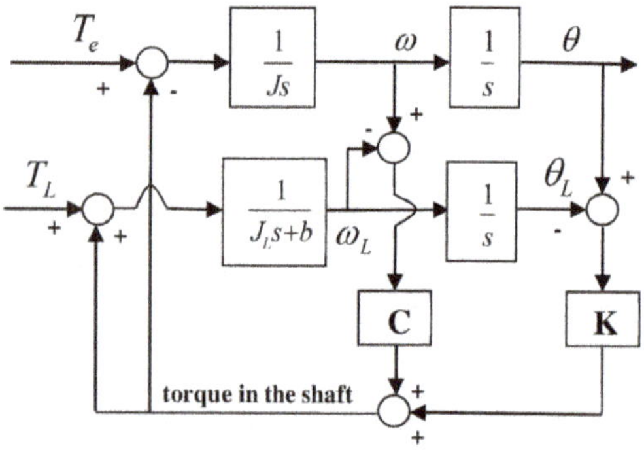

Figure 16. The block diagram of the mechanical system from Figure 15. This diagram should replace the two blocks in the right lower corner of Figure 2. The signal returning to the comparator on the left end of Figure 2 should be ω_L (if we want to control the velocity of the load). If the flexible shaft is sufficiently stiff, then, of course, $\omega_L \approx \omega$.

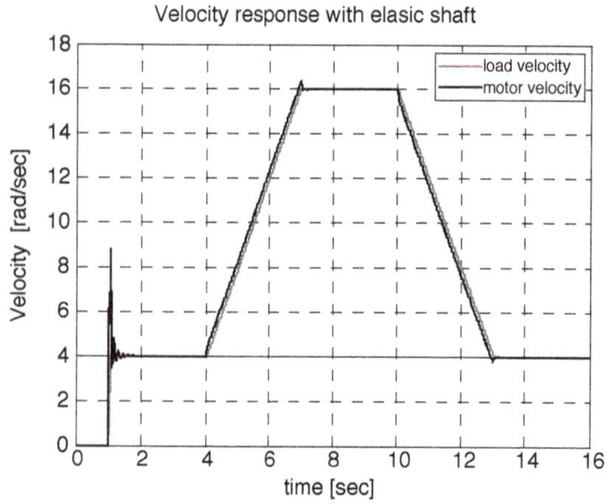

Figure 17. The motor and load velocities using a very flexible shaft and speed measurements from both the motor and the load. The adaptive controller is as described in Section 5, but it receives two different tracking errors e_L and e, as explained in the text.

One option is to close the velocity control feedback loop using the signalω, and to use exactly the same adaptive controller as in Section 5, with adjusted coefficients (we found that it performs well with $k_i$increased twice and k_1 decreased four times). When applying the reference signal from Figure 10, we observed that the cogging is not visible (like in Figure 11), but during the times when the reference signal increases or decreases between 4 and 16 rad/sec, there is an approximately constant velocity error of about 0.4 rad/sec (the load velocity ω_L lags behind ω). For an application where this velocity error is not acceptable, our solution is to mount another encoder on the load axis and use the tracking error $e_L = \omega_{ref} - \omega_L$ as the input of the PI part of the controller. The resonant part of the adaptive controller (described in (31)) should still receive the error signal $e = \omega_{ref} - \omega$, otherwise it becomes unstable.

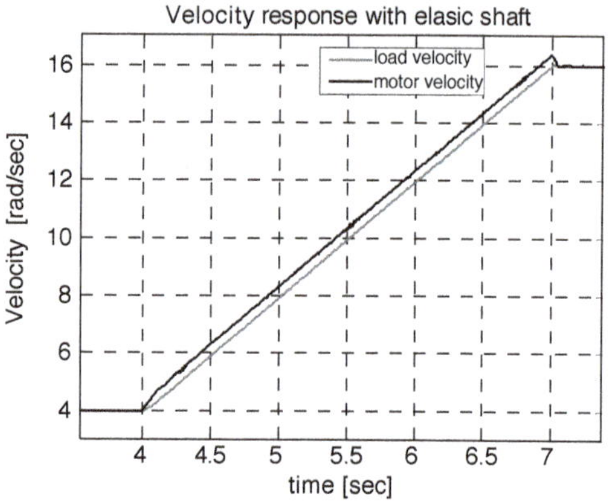

Figure 18. A zoom of Figure 17. Note that the motor velocity is ahead of the load velocity, which tracks the reference signal quite accurately.

With this control system, we obtain very good tracking for the reference signal from Figure 10, as shown in Figure 17. Figure 18 is a zoom of Figure 17.

10. Experimental results

For the experimental work, we used a RP23–54 24 V motor with an encoder of 20,000 counts/turn, a driver including the current sensors designed in the company Rafael and a dSpace setup, as shown in Figure 19. The sampling frequency for the current loop was 16 kHz and for the velocity loop, 4 kHz. The motor's nominal parameters are given in Section 8. The angular velocity was calculated inside the dSpace controller by taking the discrete derivative of the encoder output. We have closed the current loop in the dSpace controller.

10.1. A. Stability of the experimental system

The coefficients of the PI part of the controller have been chosen so that if we only use the PI part, we get a reasonable gain margin (8.5 dB) and phase margin (52°).

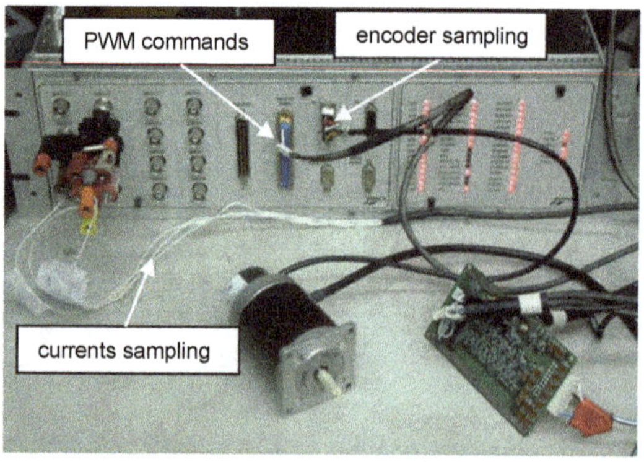

Figure 19. The equipment used for the experiments.

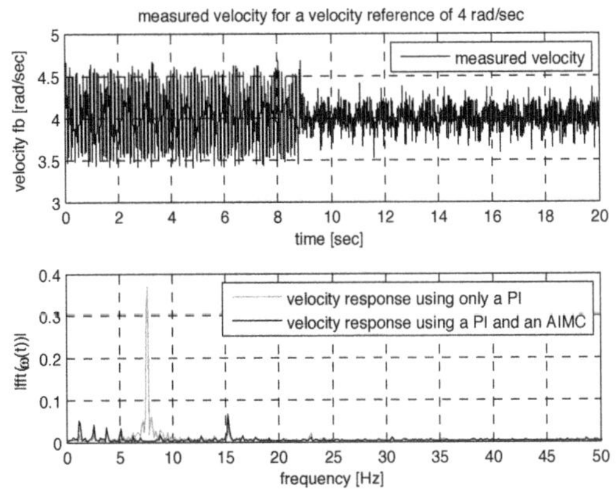

Figure 20. Velocity response and its steady-state frequency contents using a PI controller (0–9 s) and then an AIMC controller (from 9 to 20 s) for a velocity reference of 4 rad/sec. (The FFT was performed on the intervals 1–7 s for the PI, and 12–18 s for the AIMC).

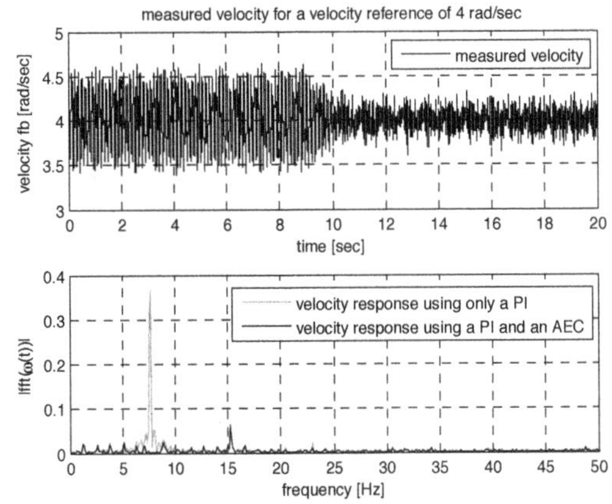

Figure 21. Velocity response and its steady-state frequency contents using a PI controller (0 up to 9 sec) and then an AEC controller (From 9 to 20 s) for a velocity reference of 4 rad/sec. (The FFT was performed on the intervals 1–7 s for the PI, and 12–18 s for the AEC).

10.2. B. Testing the adaptive algorithms

Figure 20 shows the motor velocity ω for $\omega_{\text{ref}} = 4$ rad/sec. We have used a PI controller up to 9 s, and then added an AIMC controller until 20 s. Figure 21 shows ω for the same ω_{ref} using a PI controller up to 9 s, and then switching to an AEC controller until 20 s. Since the AIMC and the AEC controllers give very similar results during the experiments (as can be seen by comparing Figures 20 and 21), in the next figures, we only give the results for the AIMC algorithm.

To check the adaptive controllers, the ω_{ref} command from Figure 10 was chosen. The velocity response using a PI controller versus an AIMC controller is shown in Figure 22. In order to see the reduction in the velocity error,

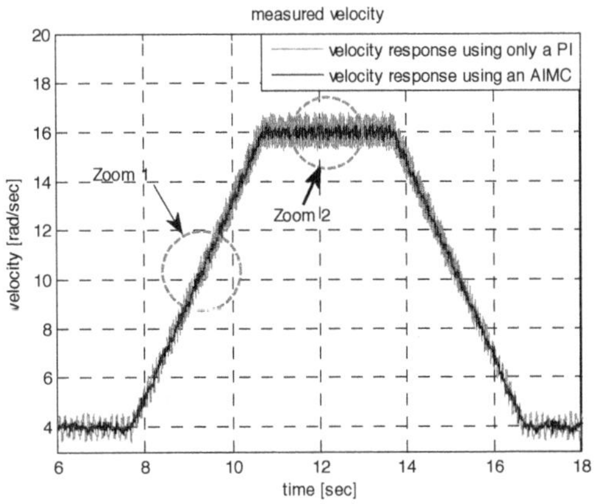

Figure 22. Velocity responses using a PI controller and an AIMC controller, for the reference command of Figure 10.

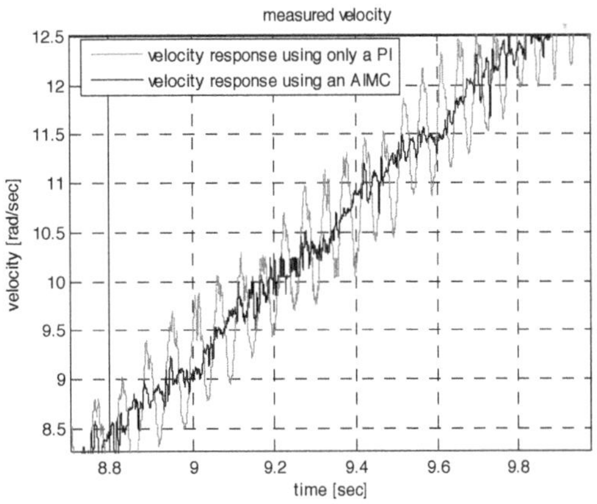

Figure 23. Velocity response using a PI versus an AIMC controller for a variable velocity reference, in the region marked "Zoom 1" in Figure 22.

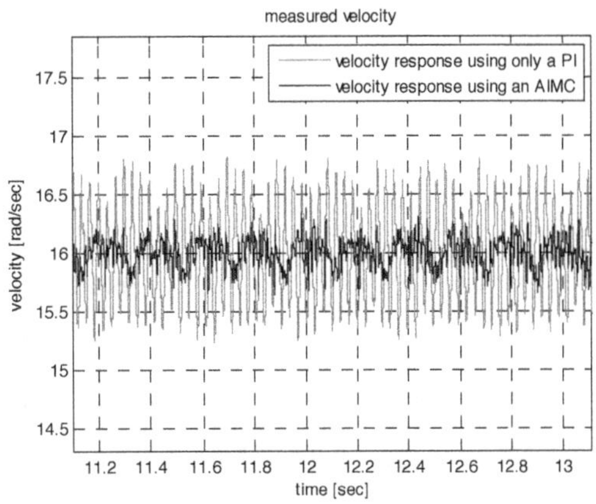

Figure 24. Velocity response using a PI versus an AIMC controller, for a constant reference command of 16 rad/sec, in the region marked as "Zoom 2" in Figure 22.

we view in Figures 23 and 24 two areas of the response, marked in Figure 22 as Zoom 1 and Zoom 2.

11. Conclusions

We have designed an internal model-based controller to suppress the torque ripple in a PMSM. The biggest challenge has been to get the controller to work at variable speed. To achieve this, we have used the rotor angle θ in place of the time variable in the formulation of the controller. We call this an adaptive internal model-based controller. Assuming a correct choice of resonant frequencies in the θ domain, we have proved convergence to zero of the tracking error under the restrictive assumption that ω_{ref} is constant. After giving a short review of the AEC from Canudas de Wit and Praly (2000), we have compared this controller with ours by simulations. The simulations show the ability of both controllers to track a variable ω_{ref} command while suppressing the effects of the cogging torque.

We have shown how to use our adaptive controller for a motor connected to a load via a flexible shaft and we have experimentally tested both controllers without load. The experiments confirm that both algorithms perform very well.

Acknowledgment

We are grateful to Dr Alex Ruderman from Nazarbayev University (Kazakhstan), formerly with the company Elmo Motion Control (Israel), for drawing our attention to this problem and providing background material.

Disclosure Statement

No potential conflict of interest was reported by the author(s).

Funding

This work was partly supported by the Israel Science Foundation grant 701/10.

References

Ahn, H. S., Chen, Y., & Dou, H. (2005). State periodic adaptive compensation of cogging and Coulomb friction in permanent-magnet linear motors. *IEEE Transaction on Magnetics*, *41*, 90–98.

Canudas de Wit, C., & Praly, L. (2000). Adaptive eccentricity compensation. *IEEE Transactions on Control Systems Technology*, *8*, 757–766.

Degobert, P., Remy, G., Zeng, J., Barre, P.-J., & Hautier, J.-P. (2006). *High performance control of the permanent magnet synchronous motor using self-tuning resonant controllers*. Proceedings of the South-eastern symposium on system theory, Cookeville, TN, USA.

Esbrook, A., Tan, X., & Khalil, H. K. (2011). *Tracking an unknown two-frequency reference using a frequency estimator-based servo compensator*. Proceedings of the 50th IEEE conference on decision and control, Orlando, FL, USA.

Ferretti, G., Magnani, G., & Rocco, P. (1998). Modeling, identification, and compensation of pulsating torque in permanent magnet AC motors. *IEEE Transaction on Industrial Electronics*, *45*, 912–920.

Ferretti, G., Magnani, G., & Rocco, P. (1999). *Torque ripple adaptive rejection in brushless motors*. Proceedings of the IEEE/ASME international conference on advanced intelligent mechatronics, Atlanta, USA.

Gan, W. C., & Qiu, L. (2004). Torque and velocity ripple elimination of AC permanent magnet motor control systems using the internal model principle. *IEEE Transaction on Mechatronics*, *9*, 436–447.

Hanselman, D. C. (1994). *Brushless permanent-magnet motor design*. New York: McGraw-Hill.

Hanselman, D., Hung, J. Y., & Keshura, M. (1992). *Torque ripple analysis in brushless permanent magnet motor drives*, Proceedings of the international conference on electrical machines, Manchester, UK, 823–827.

Hung, J. Y., & Ding, Z. (1993). Design of currents to reduce torque ripple in brushless permanent magnet motors. *IEE Proceedings B*, *140*, 260–266.

Hwang, S. M., Eom, J. B., Jung, Y. H., Lee, D. W., & Kang, B. S. (2001). Various design techniques to reduce cogging torque by controlling energy variation in permanent magnet motors. *IEEE Transaction on Magnetics*, *37*, 2806–2809.

Jahns, T. M., & Soong, W. L. (1996). Pulsating torque minimization techniques for permanent magnet AC motor drives – a review. *IEEE Transaction on Industrial Electronics*, *43*, 321–330.

Jayawardhana, B., & Weiss, G. (2009). State convergence of passive nonlinear systems with an L^2 input. *IEEE Transaction on Automatic Control*, *54*, 1723–1727.

Knobloch, H. W., Isidori, A., & Flockerzi, D. (1993). *Topics in control theory*. Basel: Birkhauser.

Ko, H. S., & Kim, K. J. (2004). Characterization of noise and vibration sources in interior permanent magnet brushless DC motors. *IEEE Transaction Magnetics*, *40*, 3482–3489.

Lee, K. W., Park, J. W., & Kwon, Y. A. (2004). *Research of speed ripple reduction method for computed tomography*. Proceedings of the annual conference on IEEE Industrial Electronics Society, Busan, Korea.

Le-Huy, H., Perret, R., & Feuillet, R. (1986). Minimization of torque ripple in brushless DC motor drives. *IEEE Transaction on Industrial Applications*, *22*, 748–755.

Lu, H., Zhang, L., & Qu, W. (2008). A new torque control method for torque ripple minimization of BLDC motors with unideal back EMF. *IEEE Transaction on Power Electronics*, *23*, 950–958.

Malaize, J., & Levine, J. (2009). *An observer based design for cogging forces cancellation in permanent magnet linear motors*. Proceeding of the 48th IEEE conference on. decision & control, Shanghai, China.

Mattavelli, P., Tubiana, L., & Zigliotto, M. (2005). Torque ripple reduction in PM synchronous motor drives using repetitive current control. *IEEE Transaction on Power Electronics*, *20*, 1423–1431.

Park, R. H. (1929). Two reactions theory of synchronous machines. *TAIEE*, *48*, 716–730. and Part II: *TAIEE*, 52, 338–350, 1933.

Park, S. J., Park, H. W., Lee, M. H., & Harashima, F. (2000). A new approach for minimum torque ripple maximum efficiency control of BLDC motor. *IEEE Transaction on Industrial Electronics*, *47*, 109–114.

Pillay, P., & Krishnan, R. (1989). Modeling, simulation and analysis of permanent-magnet motor drives, PART II: The brushless DC motor drive. *IEEE Transaction on Industry Applications*, *25*, 274–279.

Petrović, V., Ortega, R., Stanković, A., & Tadmor, G. (2000). Design and implementation of an adaptive controller for torque ripple minimization in PM synchronous motors. *IEEE Transaction on Power Electronics*, *15*, 871–880.

Qian, W., Panda, S. K., & Xu, J.-X. (2004). Torque ripple minimization in PM synchronous motors using iterative learning control. *IEEE Transaction on Power Electronics*, *19*, 272–279.

Ruderman, M., Ruderman, A., & Bertram, T. (2013). Observer-based compensation of additive periodic torque disturbances in permanent magnet motors. *IEEE Transaction on Industrial Informatics*, *9*, 1130–1138.

van der Schaft, A. (2000). *L_2-gain and passivity techniques in nonlinear control*. London: Springer-Verlag.

Spitsa, V., Kuperman, A., Weiss, G., & Rabinovici, R. (2006). *Design of a robust voltage controller for an induction generator in an autonomous power system using a genetic algorithm*. Proceeding of the ACC, Minneapolis, USA, 3475–3481.

Weiss, G. (1997). Repetitive control systems: Old and new ideas. In C. Byrneset al. (eds.), *System and control in the twenty-first century* (pp. 389–404). Basel: Birkhauser.

Xu, J. X., Panda, S. K., Pan, Y. J., Lee, T. H., & Lam, B. H. (2004). A modular control scheme for PMSM speed control with pulsating torque minimization. *IEEE Transaction on Industrial Electronics*, *51*, 526–536.

Yepes, A. G., Freijedo, F. D., Fernandez-Comesana, P., Malvar, J., Lopez, O., & Doval-Gandoy, J. (2010). *Torque ripple minimization in surface-mounted PM drives by means of PI + multi-resonant controller in synchronous reference frame*. Proceeding of the IECON, Glendale, AZ, USA.

Zhong, Q. C., & Weiss, G. (2011). Synchronverters: Inverters that mimic synchronous generators. *IEEE Transaction on Industrial Electronics*, *58*, 1259–1267.

12

On improving Popov's criterion for nonlinear feedback system stability

Y.V. Venkatesh[a,b]*

[a](formerly) Electrical Sciences Division, Indian Institute of Science, Bangalore, India;; [b]Department of ECE, National University of Singapore, Singapore

For the L_2-stability of a nonlinear single-input–single-output (SISO) feedback system, described by an integral equation and with the forward block transfer function $G(j\omega)$ and a first- and third-quadrant non-monotone nonlinearity $\varphi(\cdot) \in \mathcal{N}$ in the feedback path, we derive an interesting generalization of the celebrated criterion of Popov [(1962). Absolute stability of nonlinear systems of automatic control. *Automation and Remote Control*, *22*(8), 857–875]: $\Re(1 + j\alpha\omega)G(j\omega) > 0, 0 \leq \omega < \infty$, where $\alpha > 0$ is a constant. The generalization entails the addition of a general causal + anticausal O'Shea–Zames–Falb multiplier function whose time-domain L_1-norm is constrained by certain characteristic parameters (CPs) of the nonlinearity obtained from certain novel algebraic inequalities. If the nonlinearity is monotone or belongs to any prescribed subclass of $\mathcal{N}$, its CPs are reduced, thereby relaxing the time-domain constraint on the multiplier. An important special feature of the new stability results is a partial bridging of the significant gap between the Popov criterion and the stability results that appeared post-Popov in the form of considering monotone and other subclasses of nonlinearities in exchange for weakening the restrictions on the phase angle behaviour of $G(j\omega)$. Extensions to time-varying nonlinearities more general than those in the literature are also presented. Numerical examples are given to illustrate the theorems and to demonstrate their superiority over the existing literature.

Keywords: K–Y–P lemma; L_2-stability; Lur'e problem; Popov criterion; time-varying feedback systems

1. Introduction

In the analysis of problems arising in diverse areas of dynamical system design (such as satellite control, communication networks, and chemical plants), stability theory of nonlinear time-varying systems is an invaluable tool. Nonlinear differential and integral equations, which are typically used as mathematical models to describe such systems, are linearized (or perturbed) around, for instance, a periodic solution. The perturbed behaviour of a dynamical system is found to be more accurately modelled by nonlinear time-varying differential and integral equations, of which the latter are more general than the former, since they can describe infinite-dimensional systems and include the former as a special case. In this context, the feedback system of Figure 1(a) plays an important role in the analysis and synthesis of dynamical systems which are in practice an interconnection of subsystems subject to switching operations. The system of Figure 1(a) consists of a linear time-invariant part in the forward path and a nonlinear time-varying gain in the feedback path. A special case of Figure 1(a) is Figure 1(b) in which the feedback block is a time-invariant nonlinear gain. Primarily inspired by the classic papers of Nyquist and Bode on the frequency-domain analysis (and synthesis) of systems modelled by a special case of Figure 1(b), namely, a system with a constant linear gain in the feedback path, research workers have considered extension of similar ideas to the analysis of stability (and instability) of the system of Figure 1(a), the main subject of this paper, described by

$$v(t) = f(t) - k(t)\varphi(\sigma(t));$$

$$\sigma(t) = \sum_{m=0}^{\infty} g_m v(t - \tau_m) + \int_0^{\infty} g(\tau)v(t - \tau)\,\mathrm{d}\tau, \qquad (1)$$

where $\delta(t - \tau_m)$ is the Dirac delta function at instant $t = \tau_m$; $\sum_{m=0}^{\infty} g_m\delta(t - \tau_m) + g(t)$ is the impulse response of the time-invariant forward block with constant real sequences $\{g_m\}, \{\tau_m\}$, in which $\tau_m \geq 0, \forall m$; $g(\cdot)$ is a real-valued function in $[0, \infty)$; and $\sum_{m=0}^{\infty} |g_m| + \int_0^{\infty} |g(t)|\,\mathrm{d}t < \infty$, i.e. $\{g_m\} \in \ell_1$, and $g(\cdot) \in L_1$; real-valued gain $k(\cdot)$ assumes values in $[0, \infty)$; $f(\cdot), v(\cdot), \sigma(\cdot)$ are, respectively, the input, 'error' signal and output of the system; and $\varphi(\cdot)$, a real-valued function on $(-\infty, \infty)$, is a memoryless, first- and third-quadrant nonlinearity having the following basic properties: $\varphi(0) = 0$; there exist positive constants q_1 and q_2 with $q_1 < q_2$ such that $q_1\sigma^2 \leq \varphi(\sigma)\sigma \leq q_2\sigma^2$, $\sigma \neq 0$. We call the class of such nonlinearities $\mathcal{N}$, with its subclass of monotone nonlinearities denoted by $\mathcal{M}$. If the monotone nonlinearity also has the

*Email: yv.venkatesh@gmail.com

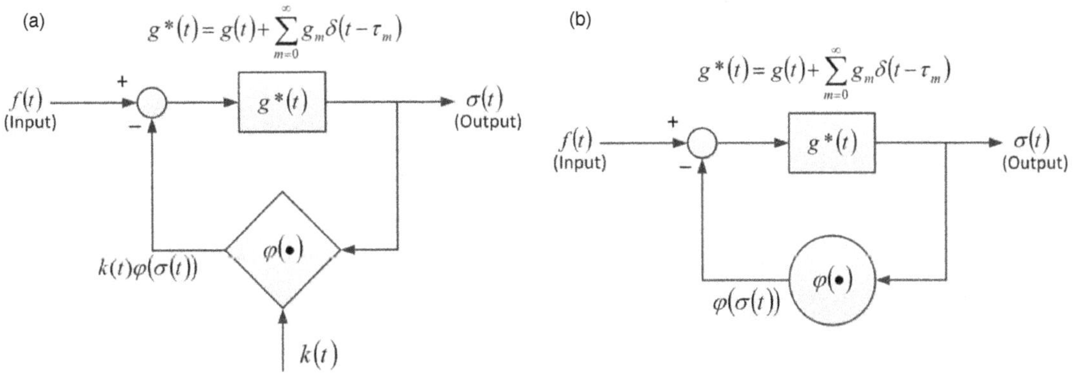

Figure 1. Block-schematics of systems under consideration. See text for details: (a) time-varying feedback system and (b) time-invariant feedback system.

'odd' property, i.e. $\varphi(-\sigma) = -\varphi(\sigma), \sigma \neq 0$, then its class is denoted by $\mathcal{M}_o$. The time-varying gain $k(\cdot)$ and the nonlinearity $\varphi(\cdot)$ in combination assume values in $[0, \infty)$. The transfer function of the forward block is given by $G(j\omega)$. For convenience in some manipulations, we denote the operator representing the time-invariant forward block by $\mathcal{G}(\cdot)$. Note that the class $\mathcal{N}$ of nonlinearities was originally proposed by Lur'e and Postnikov (1944) in their pioneering Lyapunov method-based work on a special case of Equation (1), namely, a differential equation with $k(t)$ replaced by a constant gain $K \in [0, \infty)$. In this case, the system is governed by the following time-invariant nonlinear integral equation:

$$v(t) = f(t) - K\varphi(\sigma(t));$$

$$\sigma(t) = \sum_{m=0}^{\infty} g_m v(t - \tau_m) + \int_0^{\infty} g(\tau) v(t - \tau)\, d\tau. \quad (2)$$

We consider the problems of L_2-stability of the systems governed by Equation (1) and by Equation (2). To this end, let $L_2[0, \infty)$ be the linear space of real-valued functions $x(\cdot)$ on $[0, \infty)$ with the property that $\int_0^{\infty} |x(t)|^2\, dt < \infty$, and equipped with the norm, $\|x(\cdot)\| = (\int_0^{\infty} |x(t)|^2\, dt)^{1/2}$. The nonlinear system described by Equation (1) is L_2-stable if $v \in L_2[0, \infty)$ for $f \in L_2[0, \infty)$, and an inequality of the type $\|v\| \leq C\|f\|$ holds where C is a constant.

1.1. *Brief survey of literature*

The class of (closed-loop) control systems *initiated* by Lur'e (1951) is governed, in the notation used by Kalman (1963), by the following equations:

$$\frac{d\underline{x}}{dt} = F\underline{x} - \underline{g}\varphi(\sigma); \quad \frac{d\xi}{dt} = -\underline{g}\varphi(\sigma); \quad \sigma = \underline{h}'\underline{x} + \rho\xi, \quad (3)$$

where the prime $'$ denotes transpose; σ, ξ, ρ are real scalars; $\underline{x}, \underline{g}, \underline{h}$ are real n-vectors; and F is real $n \times n$ matrix. The nonlinearity $\varphi(\sigma)$ is a real-valued continuous function which belongs to the class A_κ: $\varphi(0) = 0, 0 < \varphi(\sigma)\sigma < \kappa\sigma^2$. The problem then is to establish conditions for the global asymptotic stability – also called 'absolute stability' – of Equation (3) for any $\varphi \in A_\kappa$. It is well known that V. M. Popov established the following frequency-domain inequality for the absolute stability of Equation (3) in his celebrated paper (Popov, 1962):

$$\Re(2\alpha\rho + j\omega\beta)\left[\underline{h}'(j\omega I - F)^{-1}\underline{g} + \frac{\rho}{j\omega}\right] \geq 0$$

$$\text{for all real } \omega \quad (4)$$

holds for $2\alpha\rho = 1$ and some $\beta \geq 0$.

Earlier Lur'e (1951) had formulated the *algebraic* problem of finding necessary and sufficient conditions on ρ, g, h and F for the existence of a Lyapunov function comprising a quadratic form in $(\underline{x}, \sigma)$ and an integral of $\varphi(\sigma)$ to guarantee absolute stability of Equation (3) for class A_κ, where $\kappa = \infty$. It was left to Kalman (1963) to provide the complete answer in the form of a lemma, now known as the Kalman–Yakubovich–Popov (K–M–Y) lemma, containing certain matrix equations (and inequations), to the question of the existence of a Lyapunov function which guarantees absolute stability whenever Popov's criterion (4) is satisfied. See, in this context, Aizerman and Gantmacher (1964), LaSalle (1962), Lefschetz (1965) and Yakubovich (1962). For a generalization of this result (and of the circle criterion (Narendra and Goldwyn, 1964; Sandberg, 1964)) to input–output stability in an abstract setting, see the pioneering work of Zames (1966).

In the first 10 years or so after the publication of Popov's paper, a very large number of papers concentrated on the ramifications of Popov's criterion, as applied mainly to systems described by differential equations. In the course of analysing the challenge posed by the conjecture of Aizerman (1948) in light of Popov's criterion, Brockett and Willems (1965) relaxed the assumptions made on the phase angle behaviour of $G(j\omega)$ in exchange for (i) the choice of nonlinearity classes which are subsets of $\mathcal{N}$ and (ii) the use of multiplier functions more general than $(1 + j\alpha\omega)$ so that the 'real-part' conditions (in the frequency domain) are satisfied.

Brockett and Willems (1965) presented a list of third- and fourth-order $G(s)$ for which the null solutions of the linearized versions of the (time-invariant) system are asymptotically stable in the large for all positive gains K, but the Popov criterion cannot predict stability for $\varphi(\cdot) \in \mathcal{N}$, 'regardless of any additional assumptions placed on the coefficients of the system.' To deal with such problems, Brockett and Willems (1965) assumed monotone and odd monotone nonlinearities, and used RC-RL impedance functions. On the other hand, Lakshmi Thathachar, Srinath, and Ramapriyan (1966) explored the use of more general impedance (and hence positive real) functions with complex poles and zeros, and invoked the K–Y–P lemma. In this context, see also Narendra and Neuman (1966), Thathachar (1970) and Dewey and Jury (1966) for related results. However, the most significant result is due to O'Shea (1967) who proposed causal + anticausal functions with a time-domain L_1-norm constraint on the multiplier. See Zames and Falb (1968) for the infinite dimensional system counterpart based on the theory of positive operators. The multiplier used by O'Shea and Zames and Falb (hereafter called the OZF multiplier) is the most general multiplier function known in stability literature.

Remark 1 The system we consider is governed by the (infinite-dimensional) integral equation (1) which is more general than the (finite-dimensional) differential equation (3). When the following replacements are made, Equation (1) *effectively* reduces to the special case of Equation (3): (a) $k(t)$ is replaced by a constant gain $K \in [0, \infty)$; (b) $g(t)$ by the inverse Fourier transform $\mathfrak{g}(t) \doteq [\underline{h}'(j\omega I - F)^{-1}\underline{g} + \rho/j\omega]$; (c) $\sum_{m=0}^{\infty} g_m v(t - \tau_m)$ on the right-hand side of the second equation in Equation (1) by $\sigma(0)\mathfrak{g}(t)$ to represent the response to the initial condition $\sigma(0)$ of the finite dimensional system; (d) $f(t) \equiv 0, t \in [0, \infty)$. However, we consider the L_2-stability of Equation (1), for all $\varphi(\cdot) \in \mathcal{N}$, as against the global asymptotic stability of Equation (3). Furthermore, note that the stability results obtained can be made applicable to the case of finite feedback gain by appropriate transformations: if the feedback gain is confined to the sector $[0, \bar{K})$, then the transfer function $G(s)$ is replaced by $(G(s) + 1/\bar{K})$; and the time-varying gain $k(t)$ by $(k(t)/(1 + k(t)/\bar{K}))$. See Thathachar, Srinath, and Krishna (1966) and Huang, Venkatesh, Xiang, and Lee (2014a).

Remark 2 The impact of the K–Y–P lemma has been felt across the whole spectrum of system analysis and synthesis. Its applications to stability analysis are well known and too numerous to be listed here. In fact, elegant and extensive use has been made of the lemma in deriving stability conditions using multipliers having complex poles and zeros in the Lyapunov framework in Lakshmi Thathachar et al. (1966). For others, see the book Boyd, El Ghaoui, Feron, and Balakrishnan (1994) (pp. 123–128, 131–135). However, the K–Y–P lemma as such is not used in this paper because (i) we consider infinite-dimensional systems for which no counterparts of the K–Y–P lemma seem to exist; and, more importantly, (ii) for time-varying systems and for exploring anti-causal multiplier functions, it is not clear how to generate candidate Lyapunov functions, using the K–Y–P lemma or otherwise.

1.2. Motivation

A motivation to generalize the Popov criterion is the multiplier form of the Nyquist criterion which in the most general form reads: the system (2) with $\varphi(\sigma) \equiv \sigma$ is asymptotically stable for all constant gains $K \in [0, \bar{K})$, if there exists a frequency function, $Z(j\omega)$ such that $-\pi/2 \leq \arg\{(Z(j\omega)\} \leq \pi/2$ and $-\pi/2 < \arg\{Z(j\omega)(G(j\omega) + 1/\bar{K})\} < \pi/2$, where 'arg' denotes 'the phase angle of.' Alternatively, $\Re Z(j\omega) \geq 0$, and $\Re Z(j\omega)(G(j\omega) + 1/\bar{K}) > 0$, $\omega \in (-\infty, \infty)$, where '$\Re$' denotes 'the real part of.' The last two frequency-domain inequalities are called below (for convenience) as merely *real-part conditions*. See Brockett and Willems (1965) and Thathachar (1970). Note that neither a restriction of causality nor a time-domain L_1-norm (or otherwise) constraint has been imposed on $Z(j\omega)$. In particular, the chasm that exists between what we recognize as a very general frequency function $Z(j\omega)$ of the Nyquist criterion for linear time-invariant system (obtained from Equation (2) by setting $\varphi(\sigma) \equiv \sigma$) and the Popov multiplier function $(1 + j\alpha\omega)$ for the (absolute stability of the) nonlinear time-invariant system (2) is too big. In fact, a possible way of resolving Aizerman's conjecture Aizerman (1948) is to isolate classes of $G(j\omega)$ functions which obey Nyquist's criterion and for which $(1 + j\alpha\omega)$ for some $\alpha > 0$ is also a multiplier function. In this context, it is to be noted that the *transition* from the class $\mathcal{N}$ of nonlinearities to the class of linear gains is too abrupt.

In the present paper, we adopt a converse approach to deal with Aizerman's conjecture. More, specifically, in order to derive L_2-stability criteria for the system (2) with $\varphi(\cdot) \in \mathcal{N}$ we explore the use of the OZF multiplier itself. The results can specialized to be applicable to systems governed by differential equations of the type considered by Popov and others. It is known that under certain general conditions, L_2-stability of systems governed by differential equations can be shown to be equivalent to their absolute stability.

Another motivation (to generalize the Popov criterion) arose from some recent stability results for multi-input–muliti-output (MIMO) time-varying continuous- and discrete-time nonlinear systems, which constitute a generalizations of the system (1). See Huang, Venkatesh, Xiang, and Lee (2014b) and Venkatesh (2014). It was found that for continuous-time MIMO L_2-stability as also for discrete-time MIMO ℓ_2-stability, certain algebraic inequalities involving multi-variable nonlinearities in generalized two-variable polynomial forms (extending quadratic and bi-quadratic forms) can be exploited to modify (and partially dispense with) the existing constraints on multiplier functions commonly used in the literature. Characteristic parameters (CPs) of the nonlinearity obtained from the extremal values of the ratios of polynomial forms in two variables govern the time-domain constraints on the multiplier.

1.3. Main contributions

We establish two stability theorems: Theorem 1 for the time-invariant system (2) with $\varphi(\cdot) \in \mathcal{N}$ and Theorem 2 for the time-varying system (1) for the same class $\mathcal{N}$ of nonlinearities. In Theorem 1, the stability conditions are expressed in terms of a causal + anticausal multiplier function $Z(j\omega)$ subject to the following requirements: [C1-1] - $\Re Z(j\omega)G(j\omega) > 0$, $\omega \in (-\infty, \infty)$; and [C1-2] – a time-domain constraint on the inverse Fourier transform of $Z(j\omega)$, depending on certain CPs of the nonlinearity to be defined later below. On the other hand, in Theorem 2, the stability requirements are [C2-1] – same as [C1-1] of Theorem 1; [C2-2] – a global bound on positive and negative lobes of the normalized rate of variation $\theta(t) \doteq dk(t)/dt/k(t)$; and [C2-3] – this is a modified version of [C1-2] of Theorem 1 that depends on the choice of the global bounds in [C2-2].

1.4. Organization of the paper

The next section (Section 2) is concerned with the main assumptions, problem formulation, and mathematical preliminaries, including the definitions of the CPs of two classes $\mathcal{N}$ and $\mathcal{M}$ of nonlinearities needed in the proof of the stability theorems. Section 3 presents the first main stability result (Theorem 1) of the paper (meant for the L_2-stability of the time-invariant nonlinear system (1)) with $\varphi(\cdot) \in \mathcal{N}$. In Section 4, the second main result of the paper (Theorem 2) deals with the L_2-stability of the nonlinear time-varying system (1) with $\varphi(\cdot) \in \mathcal{N}$. In a subsection, the L_2-stability of the same system is analysed when there is a periodic switching of nonlinearities. Examples are given in Section 5 with a view to illustrate the application of the theorems and to exhibit the superiority of the stability conditions over those in the literature. In Section 6, the new stability results are critically examined for a *theoretical* comparison with the literature, and some open problems are presented. Section 7 concludes the paper, followed by appendices which contain proofs of the various lemmas used in the proofs of the theorems.

2. Assumptions, problem formulation, and preliminaries

For the system (1), the impulse response of the linear block is assumed to be in $L_1 \cap L_2$, and when the nonlinearity $\varphi(\sigma)$ is replaced by $q_2\sigma$ (with constant $q_2 > 0$) and time-varying gain $k(t)$ by the constant gain $K \in [0, \infty)$, its solutions are in $L_1 \cap L_2$, which implies that the zeros of $(1 + KG(s))$ for $K \in [0, \infty)$ lie strictly in the left-half ($\Re s < -\delta \leq 0$) of the complex plane.

Problem formulation: Given $G(s)$ satisfying the above assumptions, find conditions (i) on $\varphi(\cdot) \in \mathcal{N}$ for the L_2-stability of the system (2); and (ii) on $\varphi(\cdot) \in \mathcal{N}$ and $k(t)$ for the L_2-stability of the system (1).

Preliminaries: For any real-valued function $x(\cdot)$ on $[0, \infty)$ and any $T \geq 0$, we define the *truncated function* $x_T(\cdot)$ by: $x_T(t) = x(t)$ for $0 \leq t \leq T$; and $x_T(t) = 0$ for $t < 0$ and $t > T$. Let L_{2e} be the space of those real-valued functions $x(\cdot)$ on $[0, \infty)$ whose truncations $x_T(\cdot)$ belong to $L_2[0, \infty)$ for all $T \geq 0$. In order to establish stability of the system under consideration, we first assume infinite *escape time* for the solution of the system with $f \in L_2$ and the solution belongs to L_{2e}. We then show that, under certain conditions on $\varphi(\cdot)$, $k(t)$ and $G(j\omega)$, the solution actually belongs to $L_2[0, \infty)$.

Consider the class of operators $\mathcal{Z} \in L_{2e} \to L_{2e}$, satisfying an equation of the type

$$\mathcal{Z}\{\sigma(t)\} = \sigma(t) + \alpha\frac{d\sigma}{dt} + \sum_{m=1}^{\infty}\{z_m\sigma(t-\tau_m) + z'_m\sigma(t+\tau'_m)\} + \int_{-\infty}^{\infty} z(\tau)\sigma(t-\tau)\,d\tau, \tag{5}$$

where (real) constant $\alpha > 0$; the real sequences $\{z_m\}$ and $\{z'_m\}$ are in ℓ_1, i.e. $\sum_{m=1}^{\infty}(|z_m| + |z'_m|) < \infty$; sequences $\{\tau_m\}$ and $\{\tau'_m\}$ are in $[0, \infty)$; $z(\cdot)$ is a real-valued function on $(-\infty, \infty)$, and is in $L_1(-\infty, \infty)$, i.e. $\int_{-\infty}^{\infty} |z(t)|\,dt < \infty$. Its Fourier transform is given by

$$Z(j\omega) = 1 + j\alpha\omega + \sum_{m=1}^{\infty}\{z_m(e^{-j\tau_m\omega}) + z'_m(e^{j\tau'_m\omega})\} + \int_{-\infty}^{\infty} z(t)e^{-j\omega t}\,dt. \tag{6}$$

The multiplier function used by O'Shea (1967) and Zames and Falb (1968) for dealing with the class $\mathcal{M}$ of monotone nonlinearities is Equation (6), and called earlier (above) as the OZF multiplier. For convenience in later manipulations, let

(1) z_m^+ and $-z_m^-$ denote, respectively, the positive and negative coefficients from the set $\{z_m\}$, i.e. $z_m = z_m^+ - z_m^-$, $m = 1, 2, \ldots$; $z_m'^+$ and $-z_m'^-$ denote, respectively, the positive and negative coefficients from the set $\{z'_m\}$. i.e. $z'_m = z_m'^+ - z_m'^-$, $m = 1, 2, \ldots$;
(2) $z_c(t) = z(t)$ for $t \geq 0$; and $z_a(t) = z(t)$ for $t < 0$, so that $z(t) = z_c(t) + z_a(t)$ for $t \in (-\infty, \infty)$;
(3) $z_c^+(t) = z_c(t)$ if $z_c(t) > 0, t \in [0, \infty)$; else $z_c^+(t) = 0, t \in [0, \infty)$; $-z_c^-(t) = z_c(t)$ if $z_c(t) < 0, t \in [0, \infty)$;

else $z_c^-(t) = 0, t \in [0, \infty)$, i.e. $z_c(t) = z_c^+(t) - z_c^-(t)$, $t \in [0, \infty)$; and

(4) $z_a^+(t) = z_a(t)$ if $z_a(t) > 0, t \in (-\infty, 0]$; else $z_a^+(t) = 0, t \in (-\infty, 0]$. Similarly, $-z_a^-(t) = z_a(t)$ if $z_a(t) < 0, t \in (-\infty, 0]$; else $z_a^-(t) = 0, t \in (-\infty, 0]$, i.e. $z_a(t) = z_a^+(t) - z_a^-(t), t \in (-\infty, 0]$.

Theorems 1 and 2 (and their proofs) are based on new algebraic inequalities, involving the nonlinearity $\varphi(\cdot)$, which are motivated by the well-known algebraic inequality $-\frac{1}{2}(a^2 + b^2) \leq ab \leq \frac{1}{2}(a^2 + b^2)$ for arbitrary, real scalars a and b. Based on these inequalities, the characteristic parameters (CPs) of $\varphi(\cdot) \in \mathcal{N}, \mathcal{M}$ are defined as follows. Let $\Phi(\sigma) \doteq \int_0^\sigma \varphi(\xi)\,\mathrm{d}\xi > 0$ for $\sigma \neq 0$; and for (real) $x, y \in (-\infty, \infty)$, let $\Psi(x, y) \doteq \{\varphi(x)x + \varphi(y)y\}$. Then (i) for $\varphi(.) \in \mathcal{N}$, $\nu_i \varphi(\sigma)\sigma \leq \Phi(\sigma) \leq \nu_s \varphi(\sigma)\sigma$ and $-\mu_i \Psi(x, y) \leq \varphi(x)y \leq \mu_s \Psi(x, y)$, where the CPs $\nu_i \geq 0, \nu_s > 0, \mu_i > 0$, and $\mu_s > 0$ are defined by

$$\nu_i \doteq \inf_{\substack{\varphi(\cdot)\in\mathcal{N} \\ x\neq 0}} \frac{\Phi(x)}{\varphi(x)x}; \quad \nu_s \doteq \sup_{\substack{\varphi(\cdot)\in\mathcal{N} \\ x\neq 0}} \frac{\Phi(x)}{\varphi(x)x};$$

$$-\mu_i \doteq \inf_{\substack{\varphi(\cdot)\in\mathcal{N} \\ x,y\neq 0}} \frac{\varphi(x)y}{\Psi(x,y)}; \quad \mu_s \doteq \sup_{\substack{\varphi(\cdot)\in\mathcal{N} \\ x,y\neq 0}} \frac{\varphi(x)y}{\Psi(x,y)}; \quad \text{and} \quad (7)$$

(ii) for $\varphi(.) \in \mathcal{M}$, $\eta_i \varphi(\sigma)\sigma \leq \Phi(\sigma) \leq \eta_s \varphi(\sigma)\sigma$; and $-\gamma_i \Psi(x, y) \leq \varphi(x)y \leq \gamma_s \Psi(x, y)$, where the CPs $\eta_i \geq 0, \eta_s > 0, \gamma_i > 0$, and $\gamma_s > 0$ are defined by

$$\eta_i \doteq \inf_{\substack{\varphi(\cdot)\in\mathcal{M} \\ x\neq 0}} \frac{\Phi(x)}{\varphi(x)x}; \quad \eta_s \doteq \sup_{\substack{\varphi(\cdot)\in\mathcal{M} \\ x\neq 0}} \frac{\Phi(x)}{\varphi(x)x};$$

$$-\gamma_i \doteq \inf_{\substack{\varphi(\cdot)\in\mathcal{M} \\ x,y\neq 0}} \frac{\varphi(x)y}{\Psi(x,y)}; \quad \gamma_s \doteq \sup_{\substack{\varphi(.)\in\mathcal{M} \\ x,y\neq 0}} \frac{\varphi(x)y}{\Psi(x,y)}. \quad (8)$$

Note that $\eta_s \leq 1$ for $\varphi(\cdot) \in \mathcal{M}$. See Table 1 for typical values of μ_i and μ_s for $\varphi(\cdot) \in \mathcal{N}$ and $\mathcal{M}$. In the last item of Table 1, the slope of $\varphi(\sigma)$ is negative for $x \in [1.214, 2.1275]$. It is conjectured that, for an arbitrary $\varphi(\cdot) \in \mathcal{N}$, the upper limit of $\mu_i = \mu_s = 1$.

3. Main results-1

With the preliminaries settled, we now present the first main result of the paper.

Table 1. Typical values of μ_i and μ_s for $\varphi(\cdot) \in \mathcal{N}$ and $\mathcal{M}$.

No.	Nonlinearity $\varphi(\cdot)$	Class of $\varphi(\cdot)$	μ_i	μ_s
1	$(x - 0.3x^3 + 0.03x^5)$	$\mathcal{N}$	0.7892	0.7892
2	$(x + 0.3x^3 + 0.03x^5)$	$\mathcal{M}$	0.636	0.636
3	$(x - 0.3x^3 + 0.1x^5)$ for $x < 0$ $(x - 0.3x^3 + 0.03x^5)$ for $x \geq 0$	$\mathcal{N}$	0.9389	0.7892

THEOREM 1 *The nonlinear system* (1) *with* $\varphi(\cdot) \in \mathcal{N}$ *and* $k(t)$ *replaced by a constant gain* $K \in [0, \infty)$ *is* L_2*-stable, if there exists a multiplier function* $Z(\mathrm{j}\omega)$ *of the form Equation* (6) *such that [H-1] for some positive constant* δ, $\Re Z(\mathrm{j}\omega)G(\mathrm{j}\omega) \geq \delta > 0,\ \omega \in (-\infty, \infty)$; *and [H-2]* $\sum_{m=1}^{\infty}\{(\mu_i z_m^+ + \mu_s z_m^-) + (\mu_i z_m'^+ + \mu_s z_m'^-)\} + \int_0^\infty (\mu_i z_c^+(\tau) + \mu_s z_c^-(\tau))\,\mathrm{d}\tau + \int_{-\infty}^0 (\mu_i z_a^+(\tau) + \mu_s z_a^-(\tau))\,\mathrm{d}\tau) < \frac{1}{2}$, *where* μ_i *and* μ_s *are defined by Equation* (7).

COROLLARY T1-1 *If* $\varphi(\cdot) \in \mathcal{M}$, *Theorem* 1 *holds if, in condition [H-2]* μ_i *and* μ_s, *are replaced respectively by* γ_i *and* γ_s *as defined by Equation* (8).

The proof of Theorem 1, which follows the strategy developed in Venkatesh (1978), requires the following lemma which is proved in Appendix 2.

LEMMA 1 *With the multiplier function* $Z(j\,\omega)$ *defined by Equation* (6), *the integral*

$$\lambda_1(T) \doteq \int_0^T \varphi(\sigma_T(t))\mathcal{Z}\{\sigma_T(t)\}\,\mathrm{d}t \quad (9)$$

satisfies the inequality $\lambda_1(T) \geq \alpha(\Phi(\sigma_T(T)) - \Phi(\sigma_T(0)))$, *where* $\Phi(\sigma) = \int_0^\sigma \varphi(\xi)\,\mathrm{d}\xi$, *for all* $\sigma_T(t)$ *in the domain of* $\mathcal{Z}$ *and for all* $T \geq 0$, *if condition [H-2] of Theorem* 1 *is satisfied. Note that* $\Phi(\sigma) > 0$ *for* $\sigma \neq 0$.

Proof of Theorem 1. Consider the integral, for any $T > 0$,

$$\rho_1(T) \doteq \int_0^T f_T(t)\mathcal{Z}\{\mathcal{G}\{v_T(t)\}\}\,\mathrm{d}t, \quad (10)$$

where $\mathcal{G}\{v_T(t)\} \doteq \int_0^t g(\tau)v_T(t-\tau)\,\mathrm{d}\tau$. It follows from $f_T(t) = v_T(t) + K\varphi(\sigma_T(t))$ in Equation (1) that

$$\rho_1(T) = \int_0^T v_T(t)\mathcal{Z}\{\mathcal{G}\{v_T(t)\}\}\,\mathrm{d}t + \int_0^T K\varphi(\sigma_T(t))\mathcal{Z}\{\sigma_T(t)\}\,\mathrm{d}t. \quad (11)$$

Let $V_T(\mathrm{j}\omega)$ denote the Fourier transform of $v_T(t)$. (Note that the subscript T in $V_T(\mathrm{j}\omega)$ refers to the fact that the original time-function is truncated. There is no truncation of the Fourier transform.) Applying the Parseval theorem (see Appendix 1) to the first integral on the right-hand side of Equation (11), we obtain

$$\int_0^T v_T(t)\mathcal{Z}\{\mathcal{G}\{v_T(t)\}\}\,\mathrm{d}t = \frac{1}{2\pi}\int_{-\infty}^{\infty} V_T(-\mathrm{j}\omega)Z(\mathrm{j}\omega)G(\mathrm{j}\omega)V_T(\mathrm{j}\omega)\,\mathrm{d}\omega. \quad (12)$$

Invoking the condition [H-1] of Theorem 1, $\Re Z(\mathrm{j}\omega) G(\mathrm{j}\omega) \geq \delta > 0$ for some $\delta > 0$, we obtain the following inequality:

$$\int_0^T v_T(t)\mathcal{Z}\{\mathcal{G}\{v_T(t)\}\}\,\mathrm{d}t = \frac{1}{2\pi}\int_{-\infty}^{\infty} Z(\mathrm{j}\omega)G(\mathrm{j}\omega)|V_T(\mathrm{j}\omega)|^2\,\mathrm{d}\omega \geq \frac{\delta}{2\pi}\|v_T\|^2. \tag{13}$$

By virtue of Lemma 1, the second integral right-hand side of Equation (11),

$$\int_0^T K\varphi(\sigma_T(t))\mathcal{Z}\{\sigma_T(t)\}\,\mathrm{d}t \geq \alpha(\Phi(\sigma_T(T)) - \Phi(\sigma_T(0))), \tag{14}$$

where, to recall, $\Phi(\sigma) = \int_0^\sigma \varphi(\xi)\,\mathrm{d}\xi$. By applying the Parseval theorem to Equation (10), combining the result with Equations (13) and (14), and noting that $\Phi(\sigma) \geq 0, \forall\sigma$, we obtain

$$\delta\|v_T\|^2 \leq \alpha\Phi(\sigma_T(0)) + 2\pi\int_0^T f_T(t)\mathcal{Z}\{\mathcal{G}\{v_T(t)\}\}\,\mathrm{d}t = \int_{-\infty}^{\infty} F_T(-\mathrm{j}\omega)Z(\mathrm{j}\omega)G(\mathrm{j}\omega)V_T(\mathrm{j}\omega)\,\mathrm{d}\omega. \tag{15}$$

Now using the Cauchy–Schwarz inequality in Equation (15) leads to

$$\int_{-\infty}^{\infty} F_T(-\mathrm{j}\omega)Z(\mathrm{j}\omega)G(\mathrm{j}\omega)V_T(\mathrm{j}\omega)\,\mathrm{d}\omega \leq \sup_{-\infty<\omega<\infty} |Z(\mathrm{j}\omega)G(\mathrm{j}\omega)|\|f_T\|\|v_T\|. \tag{16}$$

Note that $\sup_{-\infty<\omega<\infty}|Z(\mathrm{j}\omega)G(\mathrm{j}\omega)|$ is finite by virtue of the assumptions on $Z(\cdot)$ and $G(\cdot)$. Let $C = \sup_{-\infty<\omega<\infty} | Z(\mathrm{j}\omega)G(\mathrm{j}\omega) |$ and $\alpha_0 \doteq \alpha\Phi(\sigma_T(0))$, which is independent of T and can be assumed bounded. Then, from Equations (15) and (16), we obtain the inequality $\delta\|v_T\|^2 \leq \alpha_0 + C\|f_T\|\|v_T\|$ which is valid for all $T > 0$. Since δ, C and α_0 are independent of T, we conclude that $\|v\| \leq C'\|f\| + \sqrt{\alpha_0' + C'^2/4\|f\|^2}$, where $\alpha_0' = \alpha_0/\delta$ and $C' = C/\delta$. The theorem is proved.

4. Main results-2

We now derive L_2-stability conditions for the system (1). To this end, let $h(t)$ be a nonnegative, integrable and bounded function on $t \in [0,\infty)$, and let $\varpi(t) \doteq \mathrm{e}^{-\int_{t_0}^t h(\tau)\,\mathrm{d}\tau}$. Assume that the integral $\int_{t_0}^t h(\tau)\,\mathrm{d}\tau \leq M < \infty, t \in [0,\infty)$ and $0 < \epsilon \leq \lim_{t\in\infty}\int_{t_0}^t h(\tau)\,\mathrm{d}\tau \leq M < \infty$. Then, $\varpi(t)$ is a bounded positive function. Note that $(\mathrm{d}\varpi/\mathrm{d}t)/\varpi(t) = -h(t)$, which is non-positive. We assume that (i) $k(t) \in [\epsilon,\infty)$ for $t \geq 0$, where (constant) $\epsilon > 0$, is a piecewise-continuous function of bounded variation with first-order (i.e. jump-)discontinuities in ℓ_1; and (ii) $k(t)$ is made up of the continuous part, $k_c(t)$, and the discontinuous part $k_d(t)$: discontinuities at instants t_{m+} correspond to positive jumps, α_m^+; and instants t_{m-} corresponding to negative jumps, α_{m-}^-.

The derivative of $k(t)$ is then given by

$$\frac{\mathrm{d}k}{\mathrm{d}t} = \frac{\mathrm{d}k_c}{\mathrm{d}t} + \sum_m\{\alpha_m^+\delta(t-t_{m+}) + \alpha_m^-\delta(t-t_{m-})\}. \tag{17}$$

Furthermore, let $\theta(t) = \mathrm{d}k/\mathrm{d}t/k(t)$. At the positive discontinuities, t_{m+}, of $k(t)$, the value of $k(t)$ is, by convention, taken as $k(t_{m+}^-)$ where t_{m+}^- is the instant just to the left of t_{m+}. Similarly, at the negative discontinuities, t_{m-}, of $k(t)$, the value of $k(t)$ is taken as $k(t_{m-}^-)$ where t_{m-}^- is the instant just to the left of t_{m-}. Note that, based on the assumptions on $k(t)$, $k(t_{m-}^-) \neq 0$, and $k(t_{m+}^-) \neq 0$, $t \geq 0$. Furthermore, let $\theta^+(t) = \theta(t)$, for $\theta(t) > 0$; $\theta^+(t) = 0$, for $\theta(t) \leq 0$; and $\theta^-(t) = \theta(t)$, for $\theta(t) < 0$; $\theta^-(t) = 0$, for $\theta(t) \geq 0$. Also, let $\theta_c(t) = \mathrm{d}k_c/\mathrm{d}t/k_c(t)$. Note that $\theta(t) = \theta^+(t) + \theta^-(t)$, where

$$\theta^+(t) = \theta_c^+(t) + \sum_m \psi_{m+}\delta(t-t_{m+}); \quad \text{and}$$
$$\theta^-(t) = \theta_c^-(t) + \sum_m \psi_{m-}\,\delta(t-t_{m-}). \tag{18}$$

where $\psi_{m+} \doteq \alpha_m^+/k(t_{m+}^-)$ and $\psi_{m-} \doteq \alpha_m^-/k(t_{m-}^-)$. We need some additional preliminaries. By the statement that $\varpi(t)\mathrm{e}^{-\xi t}k(t)$ for $t \in [0,\infty)$ is nonincreasing, we mean that

$$\frac{\mathrm{d}(\varpi(t)k(t)\mathrm{e}^{-\xi t})}{\mathrm{d}t} = \mathrm{e}^{-\xi t}\left\{\frac{\mathrm{d}\varpi}{\mathrm{d}t} - \xi\varpi(t)\right\}k(t) + \mathrm{e}^{-\xi t}\varpi(t)\frac{\mathrm{d}k(t)}{\mathrm{d}t} \leq 0, \quad t \in [0,\infty). \tag{19}$$

And, similarly, by the statement that $\varpi(t)\mathrm{e}^{\zeta t}k(t)$ for $t \in [0,\infty)$ is nondecreasing, we mean that

$$\frac{\mathrm{d}(\varpi(t)k(t)\mathrm{e}^{\zeta t})}{\mathrm{d}t} = \mathrm{e}^{\zeta t}\left\{\frac{\mathrm{d}\varpi}{\mathrm{d}t} + \zeta\varpi(t)\right\}k(t) + \mathrm{e}^{\zeta t}\varpi(t)\frac{\mathrm{d}k(t)}{\mathrm{d}t} \geq 0, \quad t \in [0,\infty). \tag{20}$$

In view of the assumptions on $\varpi(t)$ and $k(t)$, it can be shown that (19) and (20) together reduce to the following inequality:

$$-\zeta \leq \left(\frac{\mathrm{d}\varpi/\mathrm{d}t}{\varpi(t)}\right) + \theta(t) \leq \xi. \tag{21}$$

THEOREM 2 *The system* (1) *with* $\varphi(\cdot) \in \mathcal{N}$; $k(t) \in [\epsilon,\infty)$ *for* $t \geq 0$, *where (constant)* $\epsilon > 0$; *and* $\|v_\varepsilon\|^2 \doteq \int_0^\infty \mathrm{e}^{2\varepsilon t}(v(t))^2\,\mathrm{d}t$, *where* $\varepsilon > 0$ *is an arbitrarily small number, is exponentially* L_2*-stable in the sense that* $\|v_\varepsilon\| \leq C_1\|f\| + \sqrt{C_0 + (C_1^2/4)\|f\|^2}$, *where* C_1, C_2 *are constants, if there exist a multiplier function* $Z(\mathrm{j}\omega)$ *defined* *by Equation* (6); *a bounded positive function* $\varpi(\cdot)$ *as defined above; and nonnegative constants,* ξ, ζ, *such that [H-1]: (for the* $\ddot{I}_{\mathfrak{k}}$ *defined above)* $\sup_{-\infty<\omega<\infty} \|Z(j\omega-\varepsilon)G(j\omega-\varepsilon)\| < \infty$; *and* $\Re Z((j\omega-\varepsilon)G(j\omega-\varepsilon)) > 0, \omega \in (-\infty, \infty)$; $[H-2]: \varpi(t)e^{-\xi t}k(t)$ *is nonincreasing and* $\varpi(t)e^{\zeta t}k(t)$ *is nondecreasing for all* $t \in [0, \infty)$; $[H-3]: \alpha\nu_s\xi + \sum_{m=1}^{\infty}\{(1+e^{\xi\tau_m})(\mu_i z_m^+ + \mu_s z_m^-) + (1+e^{\zeta\tau'_m})(\mu_i z'^+_m + \mu_s z'^-_m)\} + \int_0^{\infty}\{(1+e^{\xi\tau})(\mu_i z_c^+(\tau) + \mu_s z_c^-(\tau)\}\,d\tau + \int_{-\infty}^{0}\{(1+e^{-\zeta\tau})(\mu_i z_a^+(\tau) + \mu_s z_a^-(\tau)\}\,d\tau < 1$, *where* ν_s, μ_i, *and* μ_s *are defined in Equation* (7).

The proof of Theorem 2 depends on Lemma 2 given below. The proofs of both are given in Appendix 3. (Also see in this context Venkatesh, 1978, pp. 572–573.)

LEMMA 2 *With the operator* $\mathcal{Z}$ *defined by Equation* (5), *the integral*

$$\lambda_2(T) \doteq \int_0^T \varpi(t)k(t)\varphi(\sigma_T(t))\mathcal{Z}\sigma_T(t)\,dt > -\alpha\varpi(0)k(0)\Phi(\sigma_T(0))), \tag{22}$$

for all σ_T *in the domain of* $\mathcal{Z}$ *and for all* $T \geq 0$, *if conditions [H-2] and [H-3] of Theorem 2 are satisfied.*

Condition [H-2] of Theorem 2 is equivalent to an appropriate bound on the *normalized* rate of variation of $k(t)$, as made explicit by the following corollary.

COROLLARY L2-1 *With* $\theta^+(t)$ *and* $\theta^-(t)$ *as defined above, if* $\varpi(t)e^{-\xi t}k(t)$ *is nonincreasing and* $\varpi(t)e^{\zeta t}k(t)$ *is nondecreasing for* $t \in [0, \infty)$, *then, for some positive constants* N_1 *and* N_2, *for all finite* $T > 0$ *and for all* $t_0 \geq 0$,

$$\frac{1}{T}\int_{t_0}^{T+t_0} \theta^+(t)\,dt \leq N_1, \quad -N_2 \leq \frac{1}{T}\int_{t_0}^{T+t_0} \theta^-(t)\,dt; \tag{23}$$

$$\text{for } \xi \neq \zeta, \quad \lim_{T\to\infty} \frac{1}{T}\int_{t_0}^{T+t_0} \theta^+(t)\,dt \leq \xi,$$

$$-\zeta \leq \lim_{T\to\infty} \frac{1}{T}\int_{t_0}^{T+t_0} \theta^-(t)\,dt, \tag{24}$$

but, for $\xi = \zeta, \theta^+(t)$ *and* $\theta^-(t)$ *are unrestricted.*

If $k(t)$ is piecewise constant, then $k_c(t)$ is identically zero. Let M denote the number of discontinuities in a finite interval $T > 0$. Then, by invoking Equations (18), (24) becomes,

$$\frac{1}{T}\sum_{m_0}^{m_0+M} \psi_{m+} \leq N_1, -N_2 \leq \frac{1}{T}\sum_{m_0}^{m_0+M} \psi_{m-} \quad \text{and}$$

$$\lim_{T\to\infty} \frac{1}{T}\sum_m \psi_{m+} \leq \xi, -\zeta \leq \lim_{T\to\infty} \frac{1}{T}\sum_m \psi_{m-}, \tag{25}$$

but, for $\xi = \zeta$, $\sum_m \psi_{m+}$ and $\sum_m \psi_{m-}$ are unrestricted. The proof of Lemma 2 is similar to that of Lemma 2 of Venkatesh (1978) (pp. 572, 577–579), and is hence omitted. When $k(t)$ is periodic with period $\mathfrak{p}$, then Equations (23)–(25) reduce respectively to

$$\frac{1}{\mathfrak{p}}\int_0^{\mathfrak{p}} \theta^-(t)\,dt \geq -\zeta, \quad \frac{1}{\mathfrak{p}}\int_0^{\mathfrak{p}} \theta^+(t)\,dt \leq \xi; \tag{26}$$

$$\left(\frac{1}{\mathfrak{p}}\right)\sum_{m, t_{m-}\in\mathfrak{D}} \psi_m(t_{m-}) \geq -\zeta, \quad \left(\frac{1}{\mathfrak{p}}\right)\sum_{m, t_{m+}\in\mathfrak{D}} \psi_m(t_{m+}) \leq \xi, \tag{27}$$

where $\mathfrak{D}$ denotes the semi-closed interval $(0, \mathfrak{p}]$.

4.1. *Switching nonlinearities*

We now consider the case of switching between nonlinearities in Equation (1). To this end, let the nonlinear gain consists of two nonlinearities $\varphi_1(\cdot)$ and $\varphi_2(\cdot)$, both belonging to class $\mathcal{N}$, which switch from one to the other periodically, with the *apparent* dwell times of $\varphi_1(\cdot)$ and $\varphi_2(\cdot)$ given, respectively, by $\mathfrak{d}_1$ and $\mathfrak{d}_2$. The fundamental period of switching is $\mathfrak{p} \doteq \mathfrak{d}_1 + \mathfrak{d}_2$, Let $k_1(t), k_2(t) \in [\epsilon, \infty)$ for $t \geq 0$ and for some $\epsilon > 0$, be periodic with the same period $\mathfrak{p}$ and active, respectively, during the regimes of $\varphi_1(\cdot)$, and $\varphi_2(\cdot)$. In Equation (1), with $u(t)$ denoting the step function, let

$$k(t)\varphi(\sigma) = k_1(t)\varphi_1(\sigma)\sum_{m=0}^{\infty}(u(t-m\mathfrak{p}) - u(t-m\mathfrak{p}-\mathfrak{d}_1)) + k_2(t)\varphi_2(\sigma)\sum_{m=0}^{\infty} \times (u(t-m\mathfrak{p}-\mathfrak{d}_1) - u(t-(m+1)\mathfrak{p})). \tag{28}$$

For simplicity in later manipulations, let $U_1(t) \doteq \sum_{m=0}^{\infty}(u(t-m\mathfrak{p}) - u(t-m\mathfrak{p}-\mathfrak{d}_1))$ and $U_2(t) \doteq \sum_{m=0}^{\infty}(u(t-m\mathfrak{p}-\mathfrak{d}_1) - u(t-(m+1)\mathfrak{p}))$, so that Equation (28) becomes $k(t)\varphi(\sigma) = k_1(t)U_1(t)\varphi_1(\sigma) + k_2(t)U_2(t)\varphi_2(\sigma)$. Note that $U_1(t)$ and $U_2(t)$ are periodic with period $\mathfrak{p}$. We now establish a corollary to Theorem 2 using a special case of the multiplier operator $\mathcal{Z}$ defined by Equation (5). Let

$$\mathcal{Z}_p\{\sigma(t)\} \doteq \sigma(t) + \sum_{m=1}^{\infty}\{z_m\sigma(t-m\mathfrak{p}) + z'_m\sigma(t+m\mathfrak{p})\}, \tag{29}$$

where, as before, the real sequences $\{z_m\}$ and $\{z'_m\}$ are in ℓ_1, i.e. $\sum_{m=1}^{\infty}(|z_m| + |z'_m|) < \infty$; sequences $\{\tau_m\}$ and $\{\tau'_m\}$ are in $[0, \infty)$. Its Fourier transform is given by

$$Z_p(j\omega) = 1 + \sum_{m=1}^{\infty}\{z_m e^{-jm\mathfrak{p}\omega} + z'_m e^{jm\mathfrak{p}\omega}\}. \tag{30}$$

For (real) $x, y \in (-\infty, \infty)$ and with $r = 1, 2$, let $\Psi_r(x, y) \doteq \{\varphi_r(x)x + \varphi_r(y)y\}$. Then (i) for $\varphi_r(\cdot) \in \mathcal{N}$, $-\mu_{i,r}\Psi_r(x, y) \leq$

$\varphi_r(x)y \le \mu_{s,r}\Psi_r(x,y)$, where the CPs $\mu_{i,r} > 0$, and $\mu_{s,r} > 0$ are defined by

$$-\mu_{i,r} \doteq \inf_{\substack{\varphi_r(\cdot)\in\mathcal{N}\\ x,y\neq 0}} \frac{\varphi_r(x)y}{\Psi_r(x,y)}, \quad \mu_{s,r} \doteq \sup_{\substack{\varphi_r(\cdot)\in\mathcal{N}\\ x,y\neq 0}} \frac{\varphi_r(x)y}{\Psi_r(x,y)},$$

$$r = 1,2, \quad \mu_{i,*} \doteq \max(\mu_{i,1}, \mu_{i,2}),;$$

$$\mu_{s,*} \doteq \max(\mu_{s,1}, \mu_{s,2}); \tag{31}$$

(ii) for $\varphi_r(\cdot) \in \mathcal{M}$, $-\gamma_{i,r}\Psi_r(x,y) \le \varphi_r(x)y \le \gamma_{s,r}\Psi_r(x,y)$, where the CPs $\gamma_{i,r} > 0$, and $\gamma_{s,r} > 0$ are defined by

$$-\gamma_{i,r} \doteq \inf_{\substack{\varphi_r(\cdot)\in\mathcal{M}\\ x,y\neq 0}} \frac{\varphi_r(x)y}{\Psi_r(x,y)}, \quad \gamma_{s,r} \doteq \sup_{\substack{\varphi_r(\cdot)\in\mathcal{M}\\ x,y\neq 0}} \frac{\varphi_r(x)y}{\Psi_r(x,y)},$$

$$r = 1,2, \quad \gamma_{i,*} \doteq \max(\gamma_{i,1}, \gamma_{i,2}),$$

$$\gamma_{s,*} \doteq \max(\gamma_{s,1}, \gamma_{s,2}). \tag{32}$$

COROLLARY T2-1 *The system* (1) *with* $\varphi_1(\cdot), \varphi_2(\cdot) \in \mathcal{N}$, *and* $k_1(t), k_2(t)$ *as defined above, is* L_2*-stable, if there exist a multiplier operator* $\mathcal{Z}_p$ *defined by Equation* (29) *such that [H-1]:* $\sup_{-\infty<\omega<\infty} \|Z_p(j\omega)G(j\omega)\| < \infty$; *and* $\Re Z_p((j\omega)G(j\omega)) > 0, \omega \in (-\infty,\infty)$; *and [H-2]:* $\sum_{m=1}^{\infty}\{\mu_{i,*}(z_m^+ + z_m'^+) + \mu_{s,*}(z_m^- + z_m'^-)\} < \frac{1}{2}$, *where* $\mu_{i,*}$ *and* $\mu_{s,*}$ *are as defined in Equation* (31).

Its proof is similar to the proof of Theorem 1 and depends on the following corollary which is proved in Appendix 4.

COROLLARY L2-2 *With the operator multiplier* $\mathcal{Z}_p$ *defined by* (29), *the integral*

$$\lambda_{2p}(T) \doteq \int_0^T k(t)\varphi(\sigma_T(t))\mathcal{Z}_p\{\sigma_T(t)\}\,dt \tag{33}$$

satisfies the inequality $\lambda_{2p}(T) \ge 0$, *for all* $\sigma_T(t)$ *in the domain of* $\mathcal{Z}_p$ *and for all* $T \ge 0$ *if condition [H-2] of Corollary T2-1 is satisfied.*

5. Examples

The difference (in form) between the Popov criterion and the circle criterion is that the former uses the (frequency domain) multiplier function $(1 + jq\omega)$, where the real constant $q > 0$, but the latter (i.e. the circle criterion) uses none. The starting point, then, for the illustrations to follow is that the phase angle behaviour of $G(j\omega)$ is such that L_2-stability *cannot* be established by either the circle criterion as applied to the nonlinear time-varying systems with $\varphi(\cdot) \in \mathcal{N}$, i.e. $\Re G(j\omega) \not> 0$, $\omega \in (-\infty,\infty)$, or the Popov criterion as applied to the nonlinear time-invariant system, i.e. there does not exist a (real) constant $q > 0$ such that $\Re(1 + jq\omega)G(j\omega) \not> 0$, $\omega \in (-\infty,\infty)$. For an application of the new theorems to the examples, we recall that the CPs ν_s, μ_i and μ_s of the nonlinearity $\varphi(\cdot) \in \mathcal{N}$ are defined by Equation (7); the CPs γ_i and γ_s for $\varphi(\cdot) \in \mathcal{M}$ are defined by Equation (8); and $\bar{K}$ denotes the upper limit of the time-varying gain $k(t)$ in the case of Equation (1), and of the constant gain K in Equation (2). Here, we implicitly assume, without loss of generality, that $0 < \varphi(\sigma)\sigma \le \sigma^2$ for all $\sigma \neq 0$.

Example 1 In the sixth-order system of O'Shea (1967) (pp. 726–727), the k_2 (in O'Shea's notation) corresponds to the sector limit of the nonlinearity. Since we assume gain-transformed system in our theorems, the function

$$G_1(s) + \left(\frac{1}{k_2}\right) = \left(\frac{(s+0.005)(s+0.1)(s+1000)}{(s+0.0001)(s+2)(s+50)}\right)^2 \tag{34}$$

(where we have used subscript 1 for the transfer function of the forward block to distinguish it from our use of the notation) corresponds to our $G(s)$. We choose a multiplier function of the form

$$Z(j\omega) = 1 + j\alpha_0\omega + \frac{\alpha_1}{(j\omega + \beta_1)} + \frac{\alpha_2}{(j\omega + \beta_2)}, \tag{35}$$

where the parameters $\alpha_0, \alpha_1, \alpha_2, \beta_1$ and β_2 are to be computed subject to the constraint that the real part condition [H-1] of Theorems 1 and 2 is to be satisfied. O'Shea's multiplier function corresponds to $\alpha_0 = 0.01927; \alpha_1 = 1.8992; \beta_1 = -2.0; \alpha_2 = -0.000247; \beta_2 = 0.005$. This set can be slightly improved – by way of (subsequently) relaxing the constraints on the CPs of $\varphi(\cdot)$ or/and on the normalized rate of variation of $k(t)$ when applied to system (1), but without attempting an explicit cancellation of poles and zeros – to $\alpha_0 = 0.01927; \alpha_1 = 1.684; \beta_1 = -2.0; \alpha_2 = -0.00025; \beta_2 = 0.005$. It is found that an entirely different and improved set of parameters is $\alpha_0 = 0.0049; \alpha_1 = 1.576; \beta_1 = -1.9; \alpha_2 = -0.0; \beta_2 = 0.0$.

Application of Theorem 1: If we use O'Shea's multiplier function, we can conclude that the system (2) with $\varphi(\cdot) \in \mathcal{N}$ is L_2-stable if, from condition [H-2], the inequality $\mu_s < 0.5005$ is satisfied. The conclusions of using the slightly improved O'Shea's multiplier are almost the same. In contrast, if we use the new parameters, condition [H-2] leads to the inequality $\mu_s < 0.6028$ which is an improvement over the earlier inequality. A contribution of Theorem 1 is that the nonlinear time-invariant system (2), while being absolutely stable, according to OShea, for $\varphi(\cdot) \in \mathcal{M}$, is, *in fact*, stable for $\varphi(\cdot) \in \mathcal{N}$, if the CP μ_s of the nonlinearity obeys the inequality given above.

Application of Theorem 2: From the same O'Shea's multiplier function, we can conclude that the system (1) with $\varphi(\cdot) \in \mathcal{N}$ is (exponentially) L_2-stable if there exist nonnegative constants ξ and ζ such that condition [H-2] is satisfied; and with CPs ν_s and μ_s defined by Equation (7) and $0 \le \xi < 0.005$ and $0 \le \zeta < 2$, the inequality $\{0.01927\nu_s\xi + 0.999\mu_s + \mu_s(0.0002473\int_0^{\infty} e^{(\xi-0.005)t}\,dt + 1.899\int_{-\infty}^{0} e^{(2-\zeta)t}\,dt) < 1\}$, obtained from condition [H-3], is satisfied. Evidently, we can strike a trade-off among ξ, ζ, ν_s and μ_s to satisfy the last inequality. For instance, suppose we set $\xi = 0.003$ and $\zeta = 1.5$, then the inequality becomes $(0.00006\nu_s + 4.87943\mu_s) < 1$, In other words, the system (1) is (exponentially) L_2-stable if (i) the nonlinearity with $\varphi(\cdot) \in \mathcal{N}$ has CPs ν_s and μ_s which together satisfy the last inequality; and, with $\xi = 0.003$ and $\zeta = 1.5$, (ii) the positive lobes $\theta^+(t)$ and negative lobes $\theta^-(t)$ of the normalized rate of variation of the time-varying gain $k(t)$ satisfy the inequalities (23), (24), and/or (25), depending on the nature of $k(t)$. The optimized parameters of O'Shea's multiplier function improve the above conclusions marginally.

In contrast, the consequences of using the new set of parameters are as follows. With $\xi \geq 0$ and $0 \leq \zeta < 1.9$, the inequality obtained from condition [H-3] is $\{0.0049\nu_s\xi + \mu_s(0.8295 + 1.576\int_{-\infty}^{0} e^{(1.9-\zeta)t}\,dt) < 1\}$, Suppose we set $\xi = 0.003$ and $\zeta = 1.5$ as before, then the last inequality becomes $(0.000015\nu_s + 4.7695\mu_s) < 1$ which allows a larger sub-class of nonlinearities in $\mathcal{N}$ than the inequality obtained above for O'Shea's multiplier function. Note that the constraints on $\theta^+(t)$ and $\theta^-(t)$ of the normalized rate of variation of $k(t)$ are the same as with the use of the O'Shea multiplier function (because the values of ξ and ζ have been retained).

Note that Corollary T2-1 (meant for the case of a periodic switching of nonlinearities) of Theorem 2 cannot be applied to the present problem because the multiplier function does not have the periodic structure of Equation (30).

Example 2 O'Shea (1967) also considers the example of Dewey and Jury (1966) with the transfer function $G_1(s) = 40/(s(s+1)(s^2+0.8s+16))$. The time-invariant system (2) with $\varphi(\sigma) \equiv \sigma$ with the above $G_1(s)$ is asymptotically stable for $k \in [\epsilon, 1.76)$ where constant $\epsilon > 0$. For an application of our theorems, we deal with the gain-transformed

$$G(s) \doteq \left(G_1(s) + \left(\frac{1}{1.76}\right)\right). \tag{36}$$

According to O'Shea (1967), the multiplier function

$$Z(j\omega) = 1 + j10^{-14}\omega + j0.999\sin(1.1118\omega) \tag{37}$$

satisfies the real-part condition as well as the time-domain constraint on it as required by his Theorem 2 (on p. 725 of the quoted reference) for the absolute stability of the system (2) with odd-monotone (i.e. class $\mathcal{M}_o$ of) nonlinearities having slopes in the sector $(\epsilon, 1.76)$, where $\epsilon > 0$.

In contrast, from condition [H-2] of our Theorem 1, the system is L_2-stable, in fact, for $\varphi(\cdot) \in \mathcal{N}$, if its CPs μ_i and μ_s satisfy the inequality $(\mu_i + \mu_s) < 1.001$. When we use this multiplier in our Theorem 2, meant for the time-varying system (1), we need to satisfy the following inequality: $10^{-14}\nu_s\xi + 0.4995(\mu_i e^{\xi} + \mu_s e^{\zeta}) < 0.001$, on the basis of which any desired trade-off can be struck among (i) the CPs ν_s, μ_i, μ_s of $\varphi(\cdot) \in \mathcal{N}$; and (ii) the (global) upper bound ξ on the positive lobes $\theta^+(t)$ and (global) lower bound ζ on the negative lobes $\theta^-(t)$ of the normalized rate of variation of $k(t)$. In the application of our Theorems 1 and 2, note that there is no explicit bound on the slope of the nonlinearity. Note further that Corollary T2-1 of Theorem 2 cannot be used because the term $j10^{-14}\omega$ in the multiplier function affects the periodic structure of the multiplier function as required by Equation (30). Hence, nonlinear switching cannot be handled by this multiplier.

Interestingly, the following multiplier functions without the $(j\omega)$ term also satisfy not only the real-part condition [H-1] of Theorem 2, but also the periodicity requirement of its Corollary T2-1:

$$Z_1(j\omega) = 1 + j1.109\sin(1.1\omega); \tag{38}$$

$$Z_2(j\omega) = 1 + j1.62\sin(1.082\omega); \quad \text{and} \tag{39}$$

$$Z_3(j\omega) = 1 + j2.6\sin(1.067\omega). \tag{40}$$

We can use these three multipliers separately in applying Theorem 2 to the system (1) with $\varphi(\cdot) \in \mathcal{N}$ and $k(t)$ periodic with the fundamental period $\mathfrak{p}$. Note that such an application corresponds to a special case of Corollary T2-1 (meant for the case of a periodic switching of nonlinearities), since we consider only one of each $k(t)$ and $\varphi(\cdot)$. As a consequence, we find that the system (1) is L_2-stable if (i) $\mathfrak{p} = 1.1$, and $(\mu_i + \mu_s) < 0.9091$; or (ii) $\mathfrak{p} = 1.082$, and $(\mu_i + \mu_s) < 0.6713$; or (iii) $\mathfrak{p} = 1.067$, and $(\mu_i + \mu_s) < 0.3846$. A by-product of the results is that smaller the period of $k(t)$, the more severe is the constraint on the CPs of the nonlinearity. There are *no* restrictions on the rate of variation of $k(t)$. To apply Corollary T2-1 in full, let us now assume that, in the system (1), two nonlinearities $\varphi_1(\cdot)$ and $\varphi_2(\cdot)$, both belonging to class $\mathcal{N}$, switch from one to the other periodically with period $\mathfrak{p}$, and the associated time-varying gains $k_1(t)$ and $k_2(t)$ also have the same period, then the system (1) is L_2-stable, if, in the last three inequalities (above in this paragraph), μ_i and μ_s are replaced, respectively, by $\mu_{i,*}$ and $\mu_{s,*}$, defined by Equation (31). It may be further noted that, in the applications of Theorem 2 and Corollary T2-1, no constraints are imposed on the dwell-time characteristics of $k(t)$ or of switching nonlinearities.

Example 3 In the course of illustrating integral quadratic constraint-based system analysis, which includes absolute stability, Megretski and Rantzer (1997) (pp. 822–824) consider a third-order system with

$$G(s) = \frac{s^2}{(s^3 + 2s^2 + 2s + 1)}, \tag{41}$$

a saturation nonlinearity and a unit-gain element characterized by delay parameter θ_0, as a result of which the transfer function of the forward block becomes $G(s)e^{-\theta_0 s}$. The authors analyse the absolute stability of the system without and with delay.

The linear system obtained from Equation (2) by setting $\varphi(\sigma) \equiv \sigma$ is asymptotically stable for $k \in [0, \infty)$. For nonlinear time-varying systems, the circle criterion gives the upper bound ($\bar{K}$) on the gain as approximately 8.13; and for a successful application of the Popov criterion to the system (2), this gain is approximately 8.9.

In this example of Megretski and Rantzer (1997), the aspect relevant to the application of our theorems is the reference to an odd monotone (i.e. class $\mathcal{M}_o$ of) nonlinearities (of which saturation nonlinearity is an example). In this case, $\bar{K}$ is allowed to be arbitrarily large. To establish the absolute stability of the system *without delay*, they suggest the use of an OZF multiplier function of the form

$$Z(j\omega) = 1 + \frac{\alpha}{(\beta - j\omega)}. \tag{42}$$

It is found (from our experimental work) that $\alpha = -1.3704$ and $\beta = 1.4$. The assumption of an odd monotone nonlinearity is crucial to an application of either Theorem 2 of O'Shea (1967) (p. 725) (or, equivalently, the corresponding theorem in Zames and Falb (1968)). Furthermore, when there is non-zero delay, there is a need to arrive at the Routh–Hurwitz limit for the feedback gain of the system (2) after setting $\varphi(\sigma) \equiv \sigma$.

Our stability Theorems 1 and 2 do not restrict the sign of the impulse response function of the multiplier; and the same multiplier is applicable to the nonlinearity class $\mathcal{N}$ for both the systems (2) and (1). In this particular case, the consequences of applying our theorems are as follows. From Theorem 1, the system (2) with $\varphi(\cdot) \in \mathcal{N}$ is L_2-stable, if $\mu_s < 0.5108$. From Theorem 2, the system (1) with $\varphi(\cdot) \in \mathcal{N}$ is L_2-stable, if, for $0 \le \zeta < 1.4$, the inequality $(0.7143 + 1/(1.4 - \zeta))\mu_s < 1$ is satisfied. In the latter case, we can trade-off between μ_s and ζ. Suppose we set $\zeta = 1$, then the last inequality reduces to $\mu_s < 0.3111$. Since ξ can be allowed to be arbitrarily large, there is no upper bound on the positive lobes $\theta^+(t)$ of the normalized rate of variation of the time-varying gain $k(t)$. On the other hand, since $\zeta = 1$, the (global) lower bound on the negative lobes $\theta^-(t)$ of the normalized rate of variation of the time-varying gain $k(t)$ must satisfy the inequalities (23), (24), and/or (25), depending on the nature of $k(t)$.

For the same problem, typically and without any attempt to optimize the parameters of the multiplier function, we find the following: (i) by setting $\bar{K} = 11.4$, the multiplier function

$$Z_1(j\omega) = (1 - j0.199 \sin 3.57\omega); \quad \text{and} \tag{43}$$

(ii) by setting $\bar{K} = 16.0$, the multiplier function

$$Z_2(j\omega) = (1 - j0.46 \sin 2.7\omega), \tag{44}$$

satisfy separately the condition [H-1] of our Theorem 2 (as also of Theorem 1) with $G(j\omega)$ replaced by $(G(j\omega) + 1/\bar{K})$. The corresponding stability results are as follows. With $\varphi(\cdot) \in \mathcal{N}$ and $k(t)$ periodic with the fundamental period $\mathfrak{p}$, the system (1) is L_2-stable if i) for $\bar{K} = 11.4, \mathfrak{p} = 3.57$, and $(\mu_i + \mu_s) < 5.0251$; and (ii) for $\bar{K} = 16.0$, $\mathfrak{p} = 2.7$, and $(\mu_i + \mu_s) < 2.1739$. From these results, it is evident that, for the L_2-stability of Equation (1), the larger the upper bound $\bar{K}$ is, the smaller will be the period of $k(t)$ and the more severe the constraint on the CPs of the nonlinearity.

It is found that, when there is delay in the system, $\bar{K}$ is to be restricted. Even with such a restriction, designing a suitable multiplier is quite complicated. A typical result is as follows. With $\theta_0 = 0.2$ and $\bar{K} = 8.8$, the Popov-multiplier function $Z(j\omega) = 1 + j0.0375\omega$ satisfies condition [H-1] of Theorem 2, in which we replace $G(j\omega)$ by $e^{-\theta_0\omega}G(j\omega)$. Therefore, the system (1) with $G(s)$ replaced by $e^{-\theta_0 s}G(s)$ and with $\varphi(\cdot) \in \mathcal{N}$ is (exponentially) L_2-stable if (from condition H-3]) the inequality $0.0375(\nu_s\xi) < 1$, or $\nu_s\xi < 26.6667$, is satisfied, in which we can strike a trade-off between ξ and ν_s. (Note that ζ can assume arbitrarily large values.) Suppose we set $\xi = 50$, then the CP $\nu_s < 0.5333$. The (global) upper bound on the positive lobes $\theta^+(t)$ of the normalized rate of variation of the time-varying gain $k(t)$ must satisfy the inequalities (23), (24), and/or (25), depending on the nature of $k(t)$, with $\xi = 50$. Note that Corollary T2-1 of Theorem 2 cannot be considered for the system (1) with switching nonlinearities, because the multiplier does not have the structure of Equation (29).

Example 4 The fourth-order system considered by Brockett and Willems (1965) (Part 2, p. 410) has the transfer function

$$G(s) = \frac{(10s + 1)(2s + 1)}{(s^2 + 20s + 400)(2s^2 + 5s + 4)}, \tag{45}$$

(which is Equation (22) in the same reference). Brockett and Willems (1965) conclude that the time-invariant nonlinear feedback system (2) with $\varphi(\cdot) \in \mathcal{M}_o$ is absolutely stable. However, it is found that there does exist the Popov-multiplier function $(1 + 0.0401s)$ using which the Popov criterion $\Re((1 + j0.0401\omega)G(j\omega)) > 0$, $-\infty < \omega < \infty$ is verified. Therefore, the time-invariant nonlinear feedback system is absolutely stable for $\varphi(\cdot) \in \mathcal{N}$. In other words, there is *no need* to use a multiplier function meant for monotone nonlinearities as found in Brockett and Willems (1965) (Part 2).

With the above choice of the Popov-multiplier function, we apply our Theorem 2 to the time-varying nonlinear feedback system with $\varphi(\cdot) \in \mathcal{N}$. Note that (i) the condition [H-1] of Theorem 2 is satisfied for an arbitrarily small $\varepsilon > 0$; (ii) and the multiplier function $Z(j\omega)$ defined by Equation (6) now has only one term, namely $j\alpha\omega$. As a

Table 2. Application of Theorem 1. See text for details.

Example No.	G (s)	Multiplier function	Multiplier parameters	CPs of $\varphi(\cdot)$	Constraints on CPs
1	(34)	(35)	$(\alpha_0, \alpha_1, \alpha_2, \beta_1, \beta_2)$	μ_s	–
			(0.01927, 1.8992, −2.0, −0.000247, 0.005) (O'Shea)	μ_s	$\mu_s < 0.5005$
			(0.01927, 1.684, −2.0, −0.00025, 0.005) (Improved)	μ_s	$\mu_s < 0.5006$
			(0.0049, 1.576, −1.9, −0.0, 0.0) (New)	μ_s	$\mu_s < 0.6028$
2	(36)	(37)	See (37)	μ_i, μ_s	$(\mu_i + \mu_s) < 1.001$
3	(41)	(42)	(α, β)	μ_s	
			(−1.3704, 1.4)		$\mu_s < 0.5108$

Table 3. Application of Theorem 2. See Table 2 for the $G(s)$ and corresponding multiplier functions, and the text for details.

Example no.	A typical choice of (ξ, ζ)	Constraint on CPs of $\varphi(\cdot)$	Constraints[a] on $(\theta^+(t), \theta^-(t))$
1	$\xi = 0.003, \zeta = 1.5$	$(0.00006\nu_s + 4.87943\mu_s) < 1$	(23) and (24), or (25), or (26) and (27)
	Same as above	$(0.000052\nu_s + 4.69\mu_s) < 1$	Same as above
	Same as above	$(0.000015\nu_s + 4.7695\mu_s) < 1$	Same as above
2	(ξ, ζ)	$10^{-14}\nu_s\xi + 0.4995(\mu_i e^{\xi} + \mu_s e^{\zeta}) < 1.001$	(23) and (24), or (25)
3	$\zeta = 1; 0 \leq \xi < \infty$	$\mu_s < 0.3111$	$\theta^+(t) < \infty$; $\theta^-(t)$ obeys (23) and (24), or (25) with $\zeta = 1$.

[a] See text for details.

consequence, in the condition [H-3] of Theorem 2, the summation involving $z_c(\tau)$ and $z_a(\tau)$ vanishes, and ζ can be allowed to be arbitrarily large. Therefore, the system (1) is (exponentially) L_2-stable if (from condition H-3]) the inequality $0.0401(\nu_s\xi) < 1$, or $\nu_s\xi < 24.9376$, is satisfied. Now invoking Corollary A17 of Theorem 2, we can strike a trade-off between the (global) bound ξ on the positive lobes $\theta^+(t)$ of the normalized rate of variation of $k(t)$ and the CP ν_s of the nonlinearity $\varphi(\cdot) \in \mathcal{N}$ for the (exponential) L_2-stability of the system (1). Suppose we set $\xi = 25$, then the CP $\nu_s < 0.9975$. Note that Corollary T2-1 of Theorem 2 cannot be considered for the system (1) with switching nonlinearities, because the multiplier does not have the structure of Equation (29).

Tables 2–4 contain a summary of the computational results obtained from the above Examples 5.1–5.3 for the L_2-stability of:

(i) system (2), using Theorem 1 and with $\varphi(\cdot) \in \mathcal{N}$ whose CPs are defined by Equation (7); and
(ii) system (1), using Theorem 2 with the same $\varphi(\cdot)$ as for system (2), and with $k(t)$ aperiodic if the multiplier function is aperiodic (in the frequency domain), and $k(t)$ periodic if the multiplier function is periodic (in the frequency domain).

For Example 5.4 and also for Example 5.3 with delay, the Popov multiplier can be used to satisfy the real-part condition of Theorems 1 and 2, i.e. there is no need for extra terms in the multiplier function. Therefore, these two cases are not included in Tables 2 and 3. Table 4 is meant for an application of Theorem 2 to the special case of a periodic $k(t)$ with period $\mathfrak{p}$. In this table, the following details are omitted to avoid repetition of the contents of Table 2: $G(s)$ and the multiplier function parameters corresponding to Examples 5.1–5.3. Note that, with a multiplier function which is periodic in the frequency domain, Corollary T2-1 contains no restrictions on the rate of variation of the time-varying gain(s).

Table 4. Application of Theorem 2 to Examples 5.2 and 5.3 for periodic $k(t)$ with period $\mathfrak{p}$. See text for details.

Example no.	Multiplier function	$\mathfrak{p}$	Constraint on CPs of $\varphi(\cdot)$
2	(38)	1.1	$(\mu_i + \mu_s) < 0.9091$
	(39)	1.082	$(\mu_i + \mu_s) < 0.6713$
	(40)	1.067	$(\mu_i + \mu_s) < 0.3846$
$3, \bar{K} = 11.4$	(43)	3.57	$(\mu_i + \mu_s) < 5.0251$
$3, \bar{K} = 16.0$	(44)	2.7	$(\mu_i + \mu_s) < 2.1739$

6. Comparisons and critique

(1) There seem to be *no* results in the literature on the stability of nonlinear time-invariant/time-varying feedback systems with $\varphi(\cdot) \in \mathcal{N}$ in which a multiplier distinct from the Popov multiplier $(1 + j\alpha\omega)$ has been used.

(2) The L_2-stability conditions of Venkatesh (1978) (Theorem 1, p. 571) are a special case of Theorem 2. In fact, as applied to the class $\mathcal{M}$ of monotone nonlinearities, Theorem 2 of the present paper can be considered as an alternative form of Theorem 1 of Venkatesh (1978), in the

sense that the CPs γ_i and γ_s obtained from Equation (8) are distinct from the CPs δ_i and δ_s that appear in Theorem 1 of Venkatesh (1978). The problem of establishing which CPs of the class $\mathcal{M}$ of nonlinearities lead to more relaxed constraints on the (normalized) rate of variation of $k(t)$ seems to be open.

(3) In Huang et al. (2014b), the only theorem that has some tangible relationship with Theorem 2 of the present paper is Theorem 4B on page 25. However, the multiplier function in Theorem 4B of that reference is the counterpart of the Popov multiplier, i.e. it is a special case of Equation (6). A generalization of that theorem, namely, Theorem 4B, to correspond to Theorem 2 of the present paper is an open problem. The other theorems of Huang et al. (2014b) meant for nonlinear MIMO systems apply to nonlinearities which are monotone (and its variations). Even in these cases, the multiplier function is a special case of (the matrix counterpart of) Equation (6). More specifically, the former does not contain the matrix counterparts of the causal and anti-causal functions $z_c(t)$ and $z_a(t)$ of Equation (6).

(4) As far as Example 5.2 above is concerned, if we were to use the matrix counterpart of the O'Shea multiplier in Theorem 4B of Huang et al. (2014b), then the stability theorems are applicable only to those systems in which the vector nonlinearities have the property of path independence for line integrals involving them, and also possess the property of monotonicity (or its variations). However, if we were to use the (matrix) counterparts of the new multipliers, Equations (38)–(40), of Example 5.2 in the theorems of Huang et al. (2014a), then the vector nonlinearities in Huang et al. (2014b) need to belong to the monotone class (or its variations). In other words, the stability conditions of Huang et al. (2014a) cannot be applied to the case when the vector nonlinearities belong to (the vector counterpart of) class $\mathcal{N}$. A generalization of the theorems of Huang et al. (2014a) to hold for the case of vector nonlinearities belonging to class $\mathcal{N}$ is an open problem.

(5) While we can include any type of restriction on nonlinearities in the definition of their CPs, it would be highly desirable to improve Theorems 1 and 2 by invoking more effective CPs of the class $\mathcal{N}$ of nonlinearities such that, when we specialize the theorems to apply to a linear time-invariant system, the bound on the time-domain L_1-norm of the multiplier tends to an arbitrarily large value. In effect, the goal is to arrive at the Nyquist criterion as a limiting case of a stability result for nonlinear systems having a nonlinearity belonging to class $\mathcal{N}$. In this context, it is interesting to compare this with the introduction of the class of power-law (monotone) nonlinearities in Brockett and Willems (1965) and its exploitation in Thathachar (1970). Note that the problem posed above is of a different nature.

(6) In the framework used in the paper, we cannot estimate the domain of attraction of the origin. In fact, the relationship between L_2-stability and the domain of attraction as studied in the literature (using Lyapunov functions) does not seem to be known. See Hu, Huang, and Lin (2004) for an application of linear matrix inequalities (in a Lyapunov framework) to compute domains of attraction. From another point of view, the implications of Theorem 2 for robust absolute stability need further study. In this context, see, for instance, Liu and Molchanov (2002) in which a Lyapunov framework has been used to derive some robust absolute stability criteria for certain types of time-varying uncertainties and multiple time-varying nonlinearities. Similarly, it would be interesting to study the relationship between Theorems 1 and 2 and the results in, for instance, Wada, Ikeda, Ohta, and Siljak (1998) on parametric absolute stability to deal with parametric uncertainties and input reference values.

(7) To recall, Section 2 and its subsection on switching nonlinearities deal with the L_2-stability of the system (1) under the general assumption that the system is asymptotically stable when $\varphi(\sigma) \equiv \sigma$ and $k(t) \equiv K \in [0, \infty)$. It is not known how to modify the framework adopted here to deal with the possibility of stabilizing an unstable time-invariant system by switching operations on either the time-varying gain or the nonlinearity (or both).

(8) As mentioned in the introduction, the KYP-lemma proves that Popov's stability criterion for the system (2) with $\varphi(\cdot) \in \mathcal{N}$ is equivalent to the existence of a Lyapunov function comprising a quadratic form and an integral of the nonlinearity. There exist various generalization of this lemma. See, for instance, Iwasaki and Hara (2000). An interesting open problem is to find possible Lyapunov function candidates for Theorem 1, as also their generalization for Theorem 2. Such candidates, if they exist, facilitate computation of finite domains of attraction on the basis of Theorems 1 and 2.

(9) One of the goals of researchers is to find a graphical interpretation (in the frequency domain) of the real-part condition of Theorems 1 and 2. In contrast with the simple, elegant and completely graphical version of Popov's theorem for time-invariant nonlinear systems, the variations on that theorem (using more general multiplier functions) not only lack simplicity in the frequency domain, but also require satisfaction of time-domain integral inequalities involving the inverse Fourier transforms of the multiplier functions. See Venkatesh (1978) (pp. 573–575) for one of the few attempts in this genre for time-varying nonlinear feedback systems with $\varphi(\cdot) \in \mathcal{M}$. A similar procedure can be adopted for $\varphi(\cdot) \in \mathcal{N}$, but is more complex. It is possible, however, to convert the real-part condition and the time-domain constraints on the multiplier function of Theorems 1 and 2 to a non-convex optimization problem. Details are omitted due to lack of space.

(10) Since we deal with periodically switching nonlinearities (with periodic gains having the same period) as one of the applications of the main results, the very interesting survey paper of Shorten, Wirth, Mason, Wulff, and King (2007) on the stability of switching and hybrid systems is relevant here. Note, however, that Shorten et al. (2007) do not consider generalization of the Popov theorem to systems with time-varying nonlinearities with $\varphi(\cdot) \in \mathcal{N}$, and described by integral equations. Zevin and Pinsky (2005) present frequency-domain absolute (asymptotic) stability (and instability) conditions for a system, which is described by a Volterra equation and controlled by a nonlinear sector-restricted feedback having a time-varying delay, to be absolutely (asymptotically) stable (and unstable). The stability conditions are independent of the delay. Interestingly, these authors provide examples of systems satisfying the Aizerman conjecture. However, they do not consider time-varying feedback systems of the type (1). On the other hand, Dehghan and Ong (2012) introduce the concepts of *dwell-time invariance* and *maximal constraint admissible dwell-time-invariant set* for discrete-time switching systems under dwell-time switching, and derive a necessary and sufficient condition for asymptotic stability of the origin of the switching systems under dwell-time switching. In contrast, we consider continuous-time systems and dispense with dwell-time considerations for the L_2-stability of time-varying nonlinear systems described by Equation (1). See Venkatesh (2014) for more general results (dispensing with dwell-time considerations) for the ℓ_2-stability of discrete-time MIMO systems.

(11) Based on the computational experiments that gave typical values of μ_i and μ_s for specific nonlinearities, as listed in Table 1, it is conjectured that, when nothing is known about the precise structure of $\varphi(\cdot) \in \mathcal{N}$, the upper limit of both μ_i and μ_s is 1. A consequence of this is the following conjecture:

Generalized Popov Theorem: The nonlinear system (1) with $\varphi(\cdot) \in \mathcal{N}$ and $k(t)$ replaced by a constant gain $K \in [0, \infty)$ is L_2-stable, if there exists a multiplier function $Z(\mathrm{j}\omega)$ of the form Equation (6) such that [H-1] for some positive constant δ, $\Re Z(\mathrm{j}\omega)G(\mathrm{j}\omega) \geq \delta > 0$, $\omega \in (-\infty, \infty)$; and [H-2] $\sum_{m=1}^{\infty}(|z_m| + |z'_m|) + \int_0^{\infty}|z_c(\tau)|\mathrm{d}\tau + \int_{-\infty}^{0}|z_a(\tau)|\mathrm{d}\tau < \frac{1}{2}$.

Conjectured Generalization of Theorem 1 of Venkatesh (1978) (p. 571): The system (1) with $\varphi(\cdot) \in \mathcal{N}$; $k(t) \in [\epsilon, \infty)$ for $t \geq 0$, where (constant) $\epsilon > 0$; and $\|v_\varepsilon\|^2 \doteq \int_0^{\infty} \mathrm{e}^{2\varepsilon t}(v(t))^2\,\mathrm{d}t$, where $\varepsilon > 0$ is an arbitrarily small number, is *exponentially* L_2-stable in the sense that $\|v_\varepsilon\| \leq C_1\|f\| + \sqrt{C_0 + (C_1^2/4)\|f\|^2}$, where C_1, C_2 are constants, if there exist a multiplier function $Z(\mathrm{j}\omega)$ defined by Equation (6); a bounded positive function $\varpi(\cdot)$ as defined above; and nonnegative constants, ξ, ζ, such that [H-1]: (for the $\ddot{I}$ţ defined above) $\sup_{-\infty<\omega<\infty}\|Z(\mathrm{j}\omega{-}\varepsilon)G(\mathrm{j}\omega{-}\varepsilon)\|<\infty$; and $\Re Z((\mathrm{j}\omega - \varepsilon)G(\mathrm{j}\omega{-}\varepsilon))>0, \omega \in (-\infty, \infty)$; [H-2]: $\varpi(t)\mathrm{e}^{-\xi t}k(t)$ is nonincreasing and $\varpi(t)\mathrm{e}^{\zeta t}k(t)$ is nondecreasing for all $t \in [0, \infty)$; and [H-3]: $\alpha\nu_s\xi + \sum_{m=1}^{\infty}\{(1 + \mathrm{e}^{\xi\tau_m})|z_m| + (1 + \mathrm{e}^{\zeta\tau'_m})|z'_m|\} + \int_0^{\infty}(1 + \mathrm{e}^{\xi\tau})|z_c(\tau)|\,\mathrm{d}\tau + \int_{-\infty}^{0}(1 + \mathrm{e}^{-\zeta\tau})$ $|z_a(\tau)|\,\mathrm{d}\tau < 1$, where ν_s is defined by (7).

7. Conclusions

For the L_2-stability of time-invariant and time-varying single-input–single-output feedback systems with *non-monotone* nonlinearities, we have derived new frequency-domain criteria in terms of the transfer function of the linear time-invariant part and a general multiplier function originally employed for *monotone nonlinearities*. The results provide a preliminary bridge between the Popov criterion (for first- and third-quadrant nonlinearities) and the results of the literature on monotone and other nonlinearities for time-invariant nonlinear systems. This bridge is established via certain CP of the nonlinearities obtained from new algebraic inequalities. In some sense, the results can be treated as a quantified (and somewhat *baroque*) improvement of Popov's criterion whose necessity or otherwise *cannot* be established from the presented results. Without doubt, Popov's criterion has an everlasting and apparently impregnable beauty. Examples are given not only to illustrate the theorems, but also to demonstrate their superiority over the existing stability conditions of the literature.

In common with many of the results in the literature, a limitation of the framework used in the paper is that it appears to be impossible to find L_2-stabilization conditions in the frequency domain for an unstable transfer function of the forward block or for a feedback system in which the gain is in the unstable Routh–Hurwitz sector. On the other hand, the same framework in an extended form has been used to derive new frequency-domain L_2-stability conditions for continuous-time MIMO systems in Huang et al. (2014b), and ℓ_2-stability conditions for discrete-time MIMO systems in Venkatesh (2014) for (in both the cases) aperiodic/periodic time-varying and nonlinear feedback gains.

Acknowledgments

The author wishes to express grateful thanks to the expert reviewers for their critical comments and valuable suggestions which have led to the present, significantly improved version of the paper.

Disclosure statement

No potential conflict of interest was reported by the authors.

References

Aizerman, M. A. (1948). On one problem concerning the stability in the 'large' of dynamic systems. *Uspexi Matemacheskii Nauk*, *4*(4), 186–188.

Aizerman, M. A., & Gantmacher, F. R. (1964). *Absolute stability of regulator systems*. San Francisco, CA: Holden-Day.
Boyd, S., El Ghaoui, L., Feron, E., & Balakrishnan, V. (1994). *Linear matrix inequalities in systems and control theory*. SIAM studies in applied mathematics. Philadelphia: SIAM.
Brockett, R. W., & Willems, J. L. (1965). Frequency-domain stability criteria – Parts 1 and 2. *IEEE Transaction on Automatic Control*, *10*(3, 4), 407–413, 255–261.
Dehghan, M., & Ong, C. J. (2012). Discrete-time switching linear system with constraints: Characterization and computation of invariant sets under dwell-time consideration. *Automatica*, *48*(5), 964–969.
Dewey, A., & Jury, E. (1966). A stability inequality for a class of nonlinear feedback systems. *IEEE Transaction on Automatic Control*, *11*(1), 54–62.
Hu, T., Huang, B., & Lin, Z. (2004). Absolute stability with a generalized sector condition. *IEEE Transaction on Automatic Control*, *49*(4), 535–548.
Huang, Z. H., Venkatesh, Y. V., Xiang, C., & Lee, T. H. (2014). Frequency-domain L_2-stability conditions for switched linear and nonlinear SISO systems. *International Journal of Systems Science*, *45*(3), 682–703.
Huang, Z., Venkatesh, Y. V., Xiang, C., & Lee, T. H. (2014). Frequency-Domain L_2-stability conditions for time-varying linear and nonlinear MIMO systems. *Control Theory and Technology*, *12*(1), 1–22.
Iwasaki, T., & Hara, S. (2000). Generalized KYP lemma: Unified frequency domain inequalities with design applications. *IEEE Trans. on Automatic Control*, *50*(1), 41–59.
Kalman, R. E. (1963). Lyapunov functions for the problem of Lure in automatic control. *Proceedings of the National Academy of Sciences*, *49*(2), 201–205.
Lakshmi Thathachar, M. A., Srinath, M. D., & Ramapriyan, H. K. (1966). On a modified Lur'e problem. *IEEE Transaction on Automatic Control*, *11*(1), 62–68.
LaSalle, J. P. (1962). Complete stability of a nonlinear control system. *Proceedings of the National Academy of Sciences*, *48*(3), 600–603.
Lefschetz, S. (1965). Liapunov stability and controls. *SIAM Journal on Control*, *3*(1), 1–6.
Liu, D., & Molchanov, A. (2002). Criteria for robust absolute stability of time-varying nonlinear continuous-time systems. *Automatica*, *38*(4), 627–637.
Lur'e, A. I. (1951). *Neikotorlye Nelineinye Zadachi Teorii Avtom'aticheskogo Regulirovaniya*. Gostekhizdat: Moscow. Book in Russian.
Lur'e, A. I., & Postnikov, V. N. (1944). On the theory of stability of control systems. *Prikladnaya Matematica i Mekhanika*, *8*(3), 246–248.
Megretski, A., & Rantzer, A. (1997). System analysis via integral quadratic constraints. *IEEE Transaction on Automatic Control*, *42*(5), 819–830.
Narendra, K., & Goldwyn, R. (1964). A geometrical criterion for the stability of certain nonlinear non-autonomous systems. *IEEE Transaction on Circuit Theory*, *11*(3), 406–407.
Narendra, K. S., & Neuman, C. P. (1966). Stability of a class of differential equations with a single monotone nonlinearity. *SIAM Journal on Control*, *4*(2), 295–308.
O'Shea, R. P. (1967). An improved frequency time domain stability criterion for autonomous continuous systems. *IEEE Transaction on Automatic Control*, *12*(6), 719–724.
Popov, V. M. (1962). Absolute stability of nonlinear systems of automatic control. *Automation and Remote Control*, *22*(8), 857–875.
Sandberg, I. W. (1964). A frequency domain condition for the stability of feedback systems containing a single time varying non-linear element. *Bell System Technical Journal*, *43*, 1601–1638.
Shorten, R., Wirth, F., Mason, O., Wulff, K., & King, C. (2007). Stability criteria for switched and hybrid systems. *SIAM Review*, *49*(4), 545–592.
Thathachar, M. A. L. (1970). Stability of systems with power-law nonlinearities. *Automatica*, *6*, 721–730.
Thathachar, M. A. L., Srinath, M. D., & Krishna, G. (1966). Stability with a nonlinearity in a sector. *IEEE Transaction on Automatic Control*, *11*(4), 311–312.
Venkatesh, Y. V. (1978). Global variation criteria for the L_2-stability of nonlinear time varying systems. *SIAM Journal on Mathematical Analysis*, *19*(3), 568–581.
Venkatesh, Y. V. (2014). On the ℓ_2-stability of time-varying linear and nonlinear discrete-time MIMO systems. *Control Theory and Technology*, *12*(3), 250–274.
Wada, T., Ikeda, M., Ohta, Y., & Siljak, D. D. (1998). Parametric absolute stability of Lur'e systems. *IEEE Transaction on Automatic Control*, *43*(11), 1649–1653.
Yakubovich, V. A. (1962). The solution of certain matrix inequalities in automatic control theory. *Doklady Akademii Nauk SSSR*, *143*, 1304–1307.
Zames, G. (1966). On the input–output stability of time-varying nonlinear feedback systems, Parts 1 and 2. *IEEE Transaction on Automatic Control*, *11*(2, 3), 465–476, 228–238.
Zames, G., & Falb, P. L. (1968). Stability conditions for systems with monotone and slope restricted nonlinearities. *SIAM Journal on Control*, *6*, 89–108.
Zevin, A. A., & Pinsky, M. A. (2005). Absolute stability criteria for a generalized Lur'e problem with delay in the feedback. *SIAM Journal on Control and Optimization*, *43*(6), 2000–2008.

Appendix 1. Parseval's theorem

Suppose $f_1(\cdot)$ and $f_2(\cdot)$ are real-valued functions defined on $[0, \infty)$, and belong to the class of $L_1 \cap L_2$ functions. Then

$$\int_0^{\infty} f_1(t) f_2(t)\,\mathrm{d}t = \frac{1}{2\pi}\int_{-\infty}^{\infty} F_1(\mathrm{j}\omega)F_2(-\mathrm{j}\omega)\,\mathrm{d}\omega,$$

where F_1, F_2 are Fourier transforms of $f_1(t)$ and $f_2(t)$.

Appendix 2. Proof of Lemma 1

The integral of Equation (9) can be rewritten as

$$\lambda_1(T) = \int_0^T \varpi(\sigma_T(t))\left\{\sigma_T(t) + \alpha\frac{\mathrm{d}\sigma_T}{\mathrm{d}t} + \sum_{m=1}^{\infty} z_m\sigma_T(t-\tau_m) + z_m'\sigma_T(t+\tau_m') + \int_{-\infty}^{\infty} z(\tau)\sigma_T(t-\tau)\,\mathrm{d}t\right\}\mathrm{d}t. \quad \text{(A1)}$$

We split Equation (A1) into its components and simplify wherever required as follows. Let

$$\lambda_{1-1}(T) \doteq \alpha\int_0^T\left\{\frac{\mathrm{d}\sigma_T}{\mathrm{d}t}\right\}\varphi(\sigma_T(t))\,\mathrm{d}t = \alpha\int_{\sigma_T(0)}^{\sigma_T(T)}\varphi(\sigma_T)\,\mathrm{d}\sigma_T = \alpha(\Phi(\sigma_T(T)) - \Phi(\sigma_T(0))), \quad \text{(A2)}$$

where $\Phi(\sigma) = \int_0^{\sigma}\varphi(\xi)\,\mathrm{d}\xi > 0$ for $\sigma \neq 0$.

We assume that interchanges of the operations of summation and integration and of two integrals (one with respect to τ and the other with respect to t) in Equation (A1) are valid. Let

$$\lambda_{1-2}(T) \doteq \int_0^T \sum_{m=1}^{\infty} z_m \sigma_T(t-\tau_m)\varphi(\sigma_T(t))\,dt$$
$$= \sum_{m=1}^{\infty}(z_m^+ - z_m^-)\int_0^T \sigma_T(t-\tau_m)\varphi(\sigma_T(t))\,dt, \quad \text{(A3)}$$

$$\lambda_{1-3}(T) \doteq \int_0^T \sum_{m=1}^{\infty} z'_m \sigma_T(t+\tau'_m)\varphi(\sigma_T(t))\,dt$$
$$= \sum_{m=1}^{\infty}(z_m'^+ - z_m'^-)\int_0^T \sigma_T(t+\tau'_m)\varphi(\sigma_T(t))\,dt, \quad \text{(A4)}$$

$$\lambda_{1-4}(T) \doteq \int_0^T \varphi(\sigma_T(t))\left\{\int_0^{\infty} z_c(\tau)\sigma_T(t-\tau)\,d\tau\right\}dt$$
$$= \int_0^{\infty}(z_c^+(\tau) - z_c^-(\tau))\left\{\int_0^T \varphi(\sigma_T(t))\sigma_T(t-\tau)\,dt\right\}d\tau, \quad \text{(A5)}$$

$$\lambda_{1-5}(T) \doteq \int_0^T \varphi(\sigma_T(t))\left\{\int_{-\infty}^{0} z_a(\tau)\sigma_T(t-\tau)\,d\tau\right\}dt$$
$$= \int_{-\infty}^{0}(z_a^+(\tau) - z_a^-(\tau))\left\{\int_0^T \varphi(\sigma_T(t))\sigma_T(t-\tau)\,dt\right\}d\tau. \quad \text{(A6)}$$

We invoke Equation (7) defining the CPs of $\varphi(\cdot)$ to reduce Equation (A3) to the following inequality:

$$\lambda_{1-2}(T) \geq -\sum_{m=1}^{\infty}(\mu_i z_m^+ + \mu_s z_m^-) \times \int_0^T \{\varphi(\sigma_T(t))\sigma_T(t) + \varphi(\sigma_T(t-\tau_m))\sigma_T(t-\tau_m)\}\,dt, \quad \text{(A7)}$$

in which the integral involving the integrand $\varphi(\sigma_T(t-\tau_m))\sigma_T(t-\tau_m)$ can be simplified as follows by changing the variable of integration to $\eta \doteq (t-\tau_m)$:

$$\int_0^T \varphi(\sigma_T(t-\tau_m))\sigma_T(t-\tau_m)\,dt = \int_{-\tau_m}^{T-\tau_m}\varphi(\sigma_T(\eta))\sigma_T(\eta)\,d\eta, \quad \text{(A8)}$$

which, on noting that $\sigma_T(\eta) = 0$ for $\eta < 0$ and changing the (dummy) variable of integration back to t, becomes

$$\int_0^T \varphi(\sigma_T(t-\tau_m))\sigma_T(t-\tau_m)\,dt = \int_0^{T-\tau_m}\varphi(\sigma_T(t))\sigma_T(t)\,dt. \quad \text{(A9)}$$

We now recall the property of the nonlinearity that $\varphi(\sigma)\sigma > 0, \forall \sigma \neq 0$ to reduce Equation (A9) to the inequality

$$\int_0^T \varphi(\sigma_T(t-\tau_m))\sigma_T(t-\tau_m)\,dt \leq \int_0^T \varphi(\sigma_T(t))\sigma_T(t)\,dt, \quad \text{(A10)}$$

which in combination with Equation (A7) gives

$$\lambda_{1-2}(T) \geq -2\sum_{m=1}^{\infty}(\mu_i z_m^+ + \mu_s z_m^-)\int_0^T \varphi(\sigma_T(t))\sigma_T(t)\,dt. \quad \text{(A11)}$$

Along similar lines, we can show that

$$\lambda_{1-3}(T) \geq -2\sum_{m=1}^{\infty}(\mu_i z_m'^+ + \mu_s z_m'^-)\int_0^T \varphi(\sigma_T(t))\sigma_T(t)\,dt. \quad \text{(A12)}$$

We now consider $\lambda_{1-4}(T)$ as defined in Equation (A5), and adopt (with minor obvious modifications) the procedure employed above for $\lambda_{1-2}(T)$. With respect to the integral with the integrand $\varphi(\sigma_T(t-\tau))\sigma_T(t-\tau)\,dt$, we change the variable of integration to $\eta \doteq (t-\tau)$, and simplify as before (by invoking the property of truncated functions) to arrive at the following inequality:

$$\lambda_{1-4}(T) \geq -2\left(\int_0^{\infty}(\mu_i z_c^+(\tau) + \mu_s z_c^-(\tau))\,d\tau\right) \times \int_0^T \varphi(\sigma_T(t))\sigma_T(t)\,dt. \quad \text{(A13)}$$

Similarly, we can show that

$$\lambda_{1-5}(T) \geq -2\left(\int_{-\infty}^{0}(\mu_i z_a^+(\tau) + \mu_s z_a^-(\tau))\,d\tau\right) \times \int_0^T \varphi(\sigma_T(t))\sigma_T(t)\,dt. \quad \text{(A14)}$$

Using the inequalities (A11)–(A14) along with Equation (A2) in Equation (A1), we obtain

$$\lambda_1(T) \geq \int_0^T \left\{1 - \left(2\sum_{m=1}^{\infty}(\mu_i z_m^+ + \mu_s z_m^-) + (\mu_i z_m'^+ + \mu_s z_m'^-) + \int_0^{\infty}(\mu_i z_c^+(\tau) + \mu_s z_c^-(\tau))\,d\tau + \int_{-\infty}^{0}(\mu_i z_a^+(\tau) + \mu_s z_a^-(\tau))\,d\tau\right)\varphi(\sigma_T(t))\sigma_T(t)\,dt\right\}$$
$$+ \alpha(\Phi(\sigma_T(T)) - \Phi(\sigma_T(0))), \quad \text{(A15)}$$

from which we conclude that $\lambda_1(T) \geq \alpha(\Phi(\sigma_T(T)) - \Phi(\sigma_T(0)))$ if condition [H-2] of Theorem 1 is satisfied. The lemma is proved.

Appendix 3. Proof of Lemma 2

The integral of Equation (22) can be rewritten as

$$\lambda_2(T) = \int_0^T \varphi(t)k(t)\varphi(\sigma_T(t)) \times \left\{\sigma_T(t) + \alpha\frac{d\sigma_T}{dt} + \sum_{m=1}^{\infty}(z_m\sigma_T(t-\tau_m) + z'_m\sigma_T(t+\tau'_m)) + \int_{-\infty}^{\infty} z(\tau)\sigma_T(t-\tau)\,d\tau\right\}dt. \quad \text{(A16)}$$

Let $\lambda_2(T) \doteq \lambda_{2-1}(T) + \lambda_{2-2}(T)$, where

$$\lambda_{2-1}(T) = \int_0^T \varpi(t)k(t)\varphi(\sigma_T(t)) \times \left\{\beta_1\sigma_T(t) + \alpha\frac{d\sigma_T}{dt} + \sum_{m=1}^{\infty}(z_m\sigma_T(t-\tau_m) \times + \int_0^{\infty} z_c(\tau)\sigma_T(t-\tau)\,d\tau\right\}dt; \quad \text{and} \quad \text{(A17)}$$

$$\lambda_{2-2}(T) = \int_0^T \varpi(t)k(t)\varphi(\sigma_T(t)) \times \left\{\beta_2\sigma_T(t) + \sum_{m=1}^{\infty}(z'_m\sigma_T(t+\tau'_m) + \int_{-\infty}^{0} z_a(\tau)\sigma_T(t-\tau)\,\mathrm{d}\tau\right\}\mathrm{d}t, \quad \text{(A18)}$$

where positive constants β_1 and β_2 are chosen such that $(\beta_1 + \beta_2) = 1$ a As in the proof of Lemma 1, we assume that interchanges of summation and integration operations; and of two integrals (one with respect to τ and the other with respect to t) are valid. We split Equation (A17) into its components and simplify wherever required as follows:

$$\lambda_{2-1a}(T) \doteq \beta_1 \int_0^T \varpi(t)k(t)\varphi(\sigma_T(t))\mathrm{e}^{-\xi t}\{\mathrm{e}^{\xi t}\sigma_T(t)\}\,\mathrm{d}t;\ \lambda_{2-2a}(T) \doteq \alpha \int_0^T \varpi(t)k(t)\varphi(\sigma_T(t))\mathrm{e}^{-\xi t}\left\{\mathrm{e}^{\xi t}\frac{\mathrm{d}\sigma_T}{\mathrm{d}t}\right\}\mathrm{d}t; \quad \text{(A19)}$$

$$\lambda_{2-3a}(T) \doteq \sum_{m=1}^{\infty} z_m \left\{\int_0^T \varpi(t)k(t)\varphi(\sigma_T(t))\mathrm{e}^{-\xi t}(\mathrm{e}^{\xi t}\sigma_T(t-\tau_m))\,\mathrm{d}t\right\}; \quad \text{(A20)}$$

$$\lambda_{2-4a}(T) \doteq \int_0^{\infty} z_c(\tau) \times \left\{\int_0^T \varpi(t)k(t)\varphi(\sigma_T(t))\mathrm{e}^{-\xi t}(\mathrm{e}^{\xi t}\sigma_T(t-\tau))\,\mathrm{d}t\right\}\mathrm{d}\tau; \quad \text{(A21)}$$

$$\lambda_{2-1b}(T) \doteq \beta_2 \int_0^T \varpi(t)k(t)\varphi(\sigma_T(t))\mathrm{e}^{\zeta t}\{-\mathrm{e}^{\zeta t}\sigma_T(t)\}\,\mathrm{d}t; \quad \text{(A22)}$$

$$\lambda_{2-3b}(T) \doteq \sum_{m=1}^{\infty} z'_m \left\{\int_0^T \varpi(t)k(t)\varphi(\sigma_T(t))\mathrm{e}^{\zeta t}(\mathrm{e}^{-\zeta t}\sigma_T(t+\tau'_m))\,\mathrm{d}t\right\};$$

and (A23)

$$\lambda_{2-4b}(T) \doteq \int_{-\infty}^{0} z_a(\tau) \times \left\{\int_0^T \varpi(t)k(t)\varphi(\sigma_T(t))\mathrm{e}^{\zeta t}(\mathrm{e}^{-\zeta t}\sigma_T(t-\tau))\,\mathrm{d}t\right\}\mathrm{d}\tau. \quad \text{(A24)}$$

Since $\varpi(t)k(t)\mathrm{e}^{-\xi t}$ is nonincreasing, by the second mean value theorem, there is a point T' in [0, T] for which the integrals of Equations(A19)–(A21) become

$$\lambda_{2-1a}(T) = \varpi(0)k(0)\beta_1 \int_0^{T'} \mathrm{e}^{\xi t}\varphi(\sigma_T(t))\sigma_T(t)\,\mathrm{d}t;\ \lambda_{2-2a}(T) = \varpi(0)k(0)\alpha \int_0^{T'} \mathrm{e}^{\xi t}\varphi(\sigma_T(t))\frac{\mathrm{d}\sigma_T}{\mathrm{d}t}\,\mathrm{d}t; \quad \text{(A25)}$$

$$\lambda_{2-3a}(T) = \varpi(0)k(0)\sum_{m=1}^{\infty} z_m \times \left\{\int_0^{T'} \mathrm{e}^{\xi t}\varphi(\sigma_T(t))\sigma_T(t-\tau_m)\,\mathrm{d}t\right\}; \quad \text{and (A26)}$$

$$\lambda_{2-4a}(T) = \varpi(0)k(0)\int_0^{\infty} z_c(\tau) \times \left\{\int_0^{T'} \mathrm{e}^{\xi t}\varphi(\sigma_T(t))\sigma_T(t-\tau)\,\mathrm{d}t\right\}\mathrm{d}\tau. \quad \text{(A27)}$$

The second integral in Equation (A25) can be simplified by integrating by parts and by invoking the definition of the CP ν_s in Equation (7) to obtain

$$\int_0^{T'} \mathrm{e}^{\xi t}\varphi(\sigma_T(t))\frac{\mathrm{d}\sigma_T}{\mathrm{d}t}\,\mathrm{d}t = \mathrm{e}^{\xi T'}\Phi(\sigma_T(T')) - \Phi(\sigma_T(0)) - \xi\int_0^{T'} \mathrm{e}^{\xi t}\Phi(\sigma_T(t))\,\mathrm{d}t \geq \mathrm{e}^{\xi T'}\Phi(\sigma_T(T')) - \Phi(\sigma_T(0)) - \nu_s\xi\int_0^{T'} \mathrm{e}^{\xi t}\varphi(\sigma_T(t))\sigma_T(t)\,\mathrm{d}t, \quad \text{(A28)}$$

where $\Phi(\sigma) = \int_0^{\sigma}\varphi(\xi)\,\mathrm{d}\xi > 0$ for $\sigma \neq 0$. Inequality (A28) in combination with the definition of $\lambda_{2-2a}(T)$ in Equation (A25), leads to the inequality

$$\lambda_{2-2a}(T) \geq \varpi(0)k(0)\left\{\mathrm{e}^{\xi T'}\Phi(\sigma_T(T')) - \Phi(\sigma_T(0)) - \nu_s\xi\int_0^{T'} \mathrm{e}^{\xi t}\varphi(\sigma_T(t))\sigma_T(t)\,\mathrm{d}t\right\}, \quad \text{(A29)}$$

As far as integrals (A26) and (A27) are concerned, by changing the variables of integration in them respectively to $u \doteq (t - \tau_m)$ and $u \doteq (t - \tau)$, exploiting the property of (time-)truncated functions, invoking the definition of CPs μ_i and μ_s in Equation (7), and recalling the nonnegativity of $\varphi(\sigma)\sigma$, as was done in the proof of Lemma 1, we get the following inequalities:

$$\lambda_{2-3a}(T) \geq -\varpi(0)k(0)\left\{\sum_{m=1}^{\infty}((1+\mathrm{e}^{\xi\tau_m})(\mu_i z_m^+ + \mu_s z_m^-))\right\} \times \left\{\int_0^{T'} \mathrm{e}^{\xi t}\varphi(\sigma_T(t))\sigma_T(t)\,\mathrm{d}t\right\}, \quad \text{(A30)}$$

$$\lambda_{2-4a}(T) \geq -\varpi(0)k(0)\left\{\int_0^{\infty}(1+\mathrm{e}^{\xi\tau})(\mu_i z_c^+(\tau) + \mu_s z_c^-(\tau))\,\mathrm{d}\tau\right\} \times \left\{\int_0^{T'} \mathrm{e}^{\xi t}\varphi(\sigma_T(t))\sigma_T(t)\,\mathrm{d}t\right\}, \quad \text{(A31)}$$

We use Equations (A25) and (A29)–(A31) in Equation (A17) to obtain the inequality

$$\lambda_{2-1}(T) \geq \alpha\varpi(0)k(0)\mathrm{e}^{\xi T'}(\Phi(\sigma_T(T')) - \Phi(\sigma_T(0))) + \varpi(0)k(0) \times \left\{\beta_1 - \alpha\nu_s\xi - \sum_{m=1}^{\infty}((1+\mathrm{e}^{\xi\tau_m})(\mu_i z_m^+ + \mu_s z_m^-)) - \int_0^{\infty}(1+\mathrm{e}^{\xi\tau})(\mu_i z_c^+(\tau) + \mu_s z_c^-(\tau))\,\mathrm{d}\tau\right\} \times \int_0^{T'} \mathrm{e}^{\xi t}\varphi(\sigma_T(t))\sigma_T(t)\,\mathrm{d}t, \quad \text{(A32)}$$

from which we conclude that $\lambda_{2-1}(T) > -\alpha\varpi(0)k(0)\Phi(\sigma_T(0))$ if

$$\alpha v_s\xi + \sum_{m=1}^{\infty}((1 + e^{\xi\tau_m})(\mu_i z_m^+ + \mu_s z_m^-)) + \int_0^{\infty}(1 + e^{\xi\tau})(\mu_i z_c^+(\tau) + \mu_s z_c^-(\tau))\,d\tau < \beta_1. \quad (A33)$$

We now consider $\lambda_{2-2}(T)$ defined by Equation (A18). Since $\varpi(t)k(t)e^{\zeta t}$ is nondecreasing, by the (extended) second mean value theorem, there is a point T'' in [0, T] for which the integrals of Equations (A1)–(A3) respectively become

$$\lambda_{2-1b}(T) = \beta_2\varpi(T)k(T)e^{\zeta T}\int_{T''}^{T} e^{-\zeta t}\varphi(\sigma_T(t))\sigma_T(t)\,dt; \quad (A34)$$

$$\lambda_{2-3b}(T) = \varpi(T)k(T)e^{\zeta T}\sum_{m=1}^{\infty} z'_m \times \left\{\int_{T''}^{T} e^{-\zeta t}\varphi(\sigma_T(t))\sigma_T(t+\tau_m)\,dt\right\}; \quad \text{and} \quad (A35)$$

$$\lambda_{2-4b}(T) = \varpi(T)k(T)e^{\zeta T}\int_{-\infty}^{0} z_a(\tau) \times \left\{\int_{T''}^{T} e^{-\zeta t}\varphi(\sigma_T(t))\sigma_T(t-\tau)\,dt\right\}d\tau. \quad (A36)$$

In the integral of Equation (A35), let $u \doteq (t + \tau'_m)$, invoke the property of (time-)truncated functions, recall the definitions of the CPs μ_i and μ_s in Equation (7), and follow the line of simplification adopted above for Equation (A26) to get

$$\lambda_{2-3b}(T) \geq -\varpi(T)k(T)e^{\zeta T}\left\{\sum_{m=1}^{\infty}(1 + e^{\zeta\tau'_m})(\mu_i z'^+_m + \mu_s z'^-_m)\right\} \times \int_{T''}^{T} e^{-\zeta t}\varphi(\sigma_T(t))\sigma_T(t)\,dt. \quad (A37)$$

Similarly, from Equation (A36), by changing the variable of integration in its integral to $u \doteq (t - \tau)$, and following a procedure similar to the above, we obtain

$$\lambda_{2-4b}(T) \geq -\varpi(T)k(T)e^{\zeta T} \times \left\{\int_{-\infty}^{0}(1 + e^{\zeta\tau})(\mu_i z_a^+(\tau) + \mu_s z_a^-(\tau))\,d\tau\right\} \times \int_{T''}^{T} e^{-\zeta t}\varphi(\sigma_T(t))\sigma_T(t)\,dt. \quad (A38)$$

We combine Equations (A34), (A37) and (A38) with Equation (A18) to get the inequality

$$\lambda_{2-2}(T) \geq \varpi(T)k(T)e^{\zeta T}\left\{\beta_2 - \sum_{m=1}^{\infty}((1 + e^{\zeta\tau'_m})(\mu_i z'^+_m + \mu_s z'^-_m)) - \int_{-\infty}^{0}(1 + e^{-\zeta\tau})(\mu_i z_a^+(\tau) + \mu_s z_a^-(\tau))\,d\tau\right\} \times \int_{T''}^{T} e^{-\zeta t}\varphi(\sigma_T(t))\sigma_T(t)\,dt, \quad (A39)$$

from which we conclude that $\lambda_{2-2}(T) > 0$ if

$$\sum_{m=1}^{\infty}((1 + e^{\zeta\tau'_m})(\mu_i z'^+_m + \mu_s z'^-_m)) + \int_{-\infty}^{0}(1 + e^{-\zeta\tau})(\mu_i z_a^+(\tau) + \mu_s z_a^-(\tau))\,d\tau \times \int_{T''}^{T} e^{-\zeta t}\varphi(\sigma_T(t))\sigma_T(t)\,dt < \beta_2. \quad (A40)$$

Since $\lambda_2(T) = \lambda_{2-1}(T) + \lambda_{2-2}(T)$, from Equations (A33) and (A40) we conclude that

$$\lambda_2(T) > -\alpha\varpi(0)k(0)\Phi(\sigma_T(0)), \quad (A41)$$

if conditions [H-2] and [H-3] of Theorem 2 are satisfied. Lemma 2 is proved.

Proof Consider the integral, for any $T > 0$,

$$\rho_2(T) \doteq \int_0^T \varpi(t)f_T(t)\mathcal{Z}\mathcal{G}\{v_T(t)\}\,dt, \quad (A42)$$

where $\mathcal{G}\{v_T(t)\} = \int_0^t g(\tau)v_T(t-\tau)\,d\tau$. It follows from $f_T(t) = v_T(t) + k(t)\varphi(\sigma_T(t))$ in Equation (1) that

$$\rho_2(T) = \int_0^T \varpi(t)v_T(t)\mathcal{Z}\{\mathcal{G}\{v_T(t)\}\}\,dt + \int_0^T \varpi(t)k(t)\varphi(\sigma_T(t))\mathcal{Z}\{\sigma_T(t)\}\,dt. \quad (A43)$$

From condition [H-1], there exists an $\varepsilon > 0$, however small, such that $\Re Z(j\omega - \varepsilon)G(j\omega - \varepsilon) > \delta > 0$ for $\omega \in (-\infty, \infty)$. There exists a $\varepsilon > 0$ from condition [H-1] of the theorem on the basis of which the first integral on the right-hand side of Equation (A43) is rewritten as follows:

$$\int_0^T \varpi(t)v_T(t)\mathcal{Z}\{\mathcal{G}\{v_T(t)\}\}\,dt = \int_0^T \varpi(t)e^{-2\varepsilon t}(v_T(t)e^{\varepsilon t})(e^{\varepsilon t}\mathcal{Z}\mathcal{G}\{v_T(t)\})\,dt. \quad (A44)$$

Now, from the assumptions made on $\varpi(t)$, $e^{-2\varepsilon t}\varpi(t)$ is non-increasing. From the second mean value theorem as applied to Equation (A44), there exists a $T' \in [0, T]$ such that

$$\int_0^T \varpi(t)v_T(t)\mathcal{Z}\{\mathcal{G}\{v_T(t)\}\}\,dt = \varpi(0)\int_0^{T'}(v_T(t)e^{-\varepsilon t})(e^{-\varepsilon t}\mathcal{Z}\mathcal{G}\{v_T(t)\}\,dt. \quad (A45)$$

We let $V_T(j\omega)$ denote, as before, the Fourier transform of (the time-truncated function) $v_T(t)$, and apply the Parseval theorem to the integral on the right-hand side of Equation (A45). For this process, we note that there is no loss of generality in assuming that the upper limit T' of the integral can be replaced by T itself, and $v_T(t)$ can be set to zero in the interval $(T', T]$. We get

$$\int_0^T \varpi(t)v_T(t)\mathcal{Z}\{\mathcal{G}\{v_T(t)\}\}\,dt = \frac{\varpi(0)}{2\pi}\int_{-\infty}^{\infty} V_T(-(j\omega - \varepsilon))Z(j\omega - \varepsilon) \times G(j\omega - \varepsilon)V_T(j\omega - \varepsilon)\,d\omega. \quad (A46)$$

Invoking the condition [H-1] of the theorem, namely, for some $\delta > 0$, $\Re Z(j\omega - \varepsilon)G(j\omega - \varepsilon) \geq \delta > 0, \omega \in (-\infty, \infty)$, the following inequality holds:

$$\int_0^T \varpi(t)v_T(t)\mathcal{Z}\{\mathcal{G}\{v_T(t)\}\}\,dt$$
$$= \frac{\varpi(0)}{2\pi}\int_{-\infty}^{\infty} Z(j\omega-\varepsilon)G(j\omega-\varepsilon)|V_T(j\omega-\varepsilon)|^2 d\omega$$
$$\geq \frac{\varpi(0)\delta}{2\pi}\int_0^T e^{2\varepsilon t}(v_T(t))^2\,dt. \quad \text{(A47)}$$

By virtue of Lemma 2, the second integral right-hand side of Equation (A43),

$$\int_0^T \varpi(t)k(t)\varphi(\sigma_T(t))\mathcal{Z}\{\sigma_T(t)\}\,dt \geq -\alpha\varpi(0)k(0)\Phi(\sigma_T(0))), \quad \text{(A48)}$$

where, to recall, $\Phi(\sigma) = \int_0^\sigma \varphi(\xi)\,d\xi$. With (an arbitrarily small) $\varepsilon > 0$, we rewrite Equation (A42) as follows:

$$\rho_2(T) = \int_0^T \varpi(t)e^{-\varepsilon t}f_T(t)(e^{\varepsilon t}\mathcal{Z}\mathcal{G}\{v_T(t)\})\,dt. \quad \text{(A49)}$$

Based on the assumed property of $\varpi(t)$, we infer that $e^{-\varepsilon t}\varpi(t)$ is nonincreasing. Invoking the second mean value theorem in Equation (A48), there exists a $T' \in [0, T]$ such that

$$\rho_2(T) = \varpi(0)\int_0^{T'} f_T(t)(e^{\varepsilon t}\mathcal{Z}\mathcal{G}\{v_T(t)\})\,dt. \quad \text{(A50)}$$

Noting that there is no loss of generality, as before, in assuming that $f_T(t) = 0$ and $v_T(t) = 0$ for $t \in [T', T]$, we apply the Parseval theorem to Equation (A49) to obtain

$$\rho_2(T) = \frac{\varpi(0)}{2\pi}\int_{-\infty}^{\infty} F_T(-j\omega)Z(j\omega-\varepsilon)G(j\omega-\varepsilon)V_T(j\omega-\varepsilon)\,d\omega, \quad \text{(A51)}$$

which can be reduced to an inequality (in steps) as follows:

$$\rho_2(T) \leq \frac{\varpi(0)}{2\pi}\sup_{-\infty<\omega<\infty} Z(j\omega-\varepsilon)G(j\omega-\varepsilon) \times \int_0^\infty f_T(t)(e^{\varepsilon t}v_T(t))\,dt$$
$$\leq \frac{\varpi(0)}{2\pi}\sup_{-\infty<\omega<\infty} Z(j\omega-\varepsilon)G(j\omega-\varepsilon)\|f_T\| \times \sqrt{\int_0^\infty e^{2\varepsilon t}(v_T(t))^2\,dt} \quad \text{(A52)}$$

the last step being based on an application of the Cauchy–Schwartz inequality. With the relationship (A43) in mind, we now combine the inequalities (A50), (A47) and (A46) to get

$$\frac{\varpi(0)\delta}{2\pi}\int_0^T e^{2\varepsilon t}(v_T(t))^2\,dt$$
$$\leq \alpha\varpi(0)k(0)\Phi(\sigma_T(0))) + \frac{\varpi(0)}{2\pi}\sup_{-\infty<\omega<\infty} \times Z(j\omega-\varepsilon)G(j\omega-\varepsilon)\|f_T\|\sqrt{\int_0^T e^{2\varepsilon t}(v_T(t))^2\,dt}. \quad \text{(A53)}$$

Noting that $\sup_{-\infty<\omega<\infty}|Z(j\omega-\varepsilon)G(j\omega-\varepsilon)| \doteq C$ is finite by virtue of the assumptions on $Z(\cdot)$ and $G(\cdot)$, Equation (A51) can be simplified to give

$$\int_0^T e^{2\varepsilon t}(v_T(t))^2\,dt$$
$$\leq \frac{2\pi\alpha}{\delta}k(0)\Phi(\sigma_T(0)) + \frac{C}{\delta}\|f_T\|\left(\int_0^T e^{2\varepsilon t}(v_T(t))\,dt\right)^{1/2}$$
$$\leq C_0 + C_1\|f_T\|\left(\int_0^T e^{2\varepsilon t}(v_T(t))^2\,dt\right)^{1/2}, \quad \text{(A54)}$$

where $C_0 \doteq (2\pi\alpha/\delta)k(0)\Phi(\sigma_T(0))$ and $C_1 \doteq (C/\delta)$ are finite. From Equation (A52), we get the following inequality:

$$\int_0^T e^{2\varepsilon t}(v_T(t))^2\,dt \leq C_0 + C_1\|f_T\|\left(\int_0^T e^{2\varepsilon t}(v_T(t))^2\,dt\right)^{1/2}, \quad \text{(A55)}$$

which is valid for all $T > 0$. Since C_0 and C_1 are independent of T, we conclude that $\|v_\varepsilon\| \leq C_1\|f\| + \sqrt{C_0 + C_1^2/4\|f\|^2}$, where $\|v_\varepsilon\|^2 \doteq \int_0^\infty e^{2\varepsilon t}(v(t))^2\,dt$. The theorem is proved.

Appendix 4. Proof of Corollary L2-1

The integral of Equation (33) can be rewritten as

$$\lambda_{2p}(T) = \int_0^T k(t)\varphi(\sigma_T(t)) \times \left\{\sigma_T(t) + \sum_{m=1}^{\infty}(z_m\sigma_T(t-m\mathfrak{p})) + z'_m\sigma_T(t+m\mathfrak{p})\right\}dt. \quad \text{(A56)}$$

As before, we assume that an interchange of the operations of summation and integration in Equation (A55) is valid. In Equation (101), we replace $k(t)\varphi(\sigma)$ by the right-hand side of Equation (28), and split $\lambda_{2p}(T)$ in terms of its components and simplify as follows.

Let

$$\lambda_{2p-1}(T) \doteq \int_0^T \sum_{m=1}^{\infty} z_m\sigma_T(t-m\mathfrak{p})k(t)\varphi(\sigma_T(t))\,dt$$
$$= \sum_{m=1}^{\infty}(z_m^+ - z_m^-)\int_0^T \sigma_T(t-m\mathfrak{p})k(t)\varphi(\sigma_T(t))\,dt$$
$$= \sum_{m=1}^{\infty}(z_m^+ - z_m^-)\int_0^T \sigma_T(t-m\mathfrak{p})k_1(t)U_1(t)\varphi_1(\sigma_T(t)) + k_2(t)U_2(t)\varphi_2(\sigma_T(t))\,dt. \quad \text{(A57)}$$

In the last integral of the right-hand side of Equation (A56), we change the variable of integration from t to $\xi \doteq (t - m\mathfrak{p})$, invoke Equation (31) defining the CPs of $\varphi_1(\cdot)$ and $\varphi_2(\cdot)$, use the periodicity property of $U_1(t)$ and $U_2(t)$ along with the properties of the time-truncated function – the last step being similar to what was done in the proof of Lemma 1 – and recall that $k(t)\varphi(\sigma_T(t)) = \{k_1(t)\varphi_1(\sigma_T(t)) + k_2(t)\varphi_2(\sigma_T(t))\}$ to reduce Equation (A56) to the following inequality:

$$\lambda_{2p-1}(T) \geq -\sum_{m=1}^{\infty}(\mu_{i,1}z_m^+ + \mu_{s,1}z_m^-) \times \int_0^T [k_1(t)U_1(t)\{\varphi_1(\sigma_T(t))\sigma_T(t) + \varphi_1(\sigma_T(t-m\mathfrak{p}))\sigma_T(t-m\mathfrak{p})\}]\,\mathrm{d}t + (\mu_{i,2}z_m^+ + \mu_{s,2}z_m^-) \times \int_0^T [k_2(t)U_2(t)\{\varphi_2(\sigma_T(t))\sigma_T(t) + \varphi_2(\sigma_T(t-m\mathfrak{p})) \times \sigma_T(t-m\mathfrak{p})\}]\,\mathrm{d}t$$

$$\geq -2\sum_{m=1}^{\infty}(\mu_{i,*}z_m^+ + \mu_{s,*}z_m^-)\int_0^T k(t)\varphi(\sigma_T(t))\sigma_T(t)\,\mathrm{d}t. \qquad (A58)$$

Similarly, let

$$\lambda_{2p-2}(T) \doteq \int_0^T \sum_{m=1}^{\infty} z'_m\sigma_T(t+\tau'_m)k(t)\varphi(\sigma_T(t))\,\mathrm{d}t$$

$$= \sum_{m=1}^{\infty}(z_m'^+ - z_m'^-)\int_0^T \sigma_T(t+\tau'_m)k(t)\varphi(\sigma_T(t))\,\mathrm{d}t$$

$$= \sum_{m=1}^{\infty}(z_m'^+ - z_m'^-)\int_0^T \sigma_T(t+m\mathfrak{p})k_1(t)U_1(t)\varphi_1(\sigma_T(t)) + k_2(t)U_2(t)\varphi_2(\sigma_T(t))\,\mathrm{d}t. \qquad (A59)$$

Following the line of simplification and reduction of $\lambda_{2p-1}(T)$ above, we can show that $\lambda_{2p-2}(T)$ obeys the following inequality:

$$\lambda_{2p-2}(T) \geq -2\sum_{m=1}^{\infty}(\mu_{i,*}z_m'^+ + \mu_{s,*}z_m'^-) \times \int_0^T k(t)\varphi(\sigma_T(t))\sigma_T(t)\,\mathrm{d}t. \qquad (A60)$$

Using the inequalities (A58) and (A59) in (A55), we obtain

$$\lambda_{2p}(T) \geq \left\{1 - 2\sum_{m=1}^{\infty}(\mu_{i,*}(z_m^+ + z_m'^+) + \mu_{s,*}(z_m^- + z_m'^-))\right\} \times \int_0^T k(t)\varphi(\sigma_T(t))\sigma_T(t)\,\mathrm{d}t \qquad (A61)$$

from which we conclude that $\lambda_{2p}(T) \geq 0$, if condition [H-2] of Corollary T2-1 is satisfied. The lemma is proved.

13

Robust output feedback stabilization for discrete-time systems with time-varying input delay

Shoulin Hao[a], Tao Liu[a]*, Jie Zhang[b], Ximing Sun[a] and Chongquan Zhong[a]

[a]*Institute of Advanced Control Technology, Dalian University of Technology, Dalian 116024, People's Republic of China;* [b]*School of Chemical Engineering and Advanced Materials, Newcastle University, Newcastle upon Tyne NE1 7RU, UK*

In this paper, a robust output feedback stabilization method is proposed for discrete-time systems subject to time-varying input delay. By introducing an augmented state, the robust stabilization problem is converted into that of a delay-free system. Using Artstein's reduction method and the scaled-bounded real lemma, sufficient conditions for robust stability are established via matrix inequalities, by which the output feedback controller can be derived. A tuning parameter is introduced to efficiently solve these matrix inequalities, therefore facilitating the improvement of control performance. An illustrative example is used to show the effectiveness and merit of the proposed method.

Keywords: time-varying delay; scaled-bounded real lemma; dynamic output feedback; matrix inequality

1. Introduction

Time delay is usually associated with industrial applications (Liu & Gao, 2012; Seborg, Edgar, & Mellichamp, 2004). Advanced control design for industrial processes with time delay has drawn a lot of attention in the past decades. For the convenience of control design, various delays in the system response have been mostly merged into the state delay or input delay for consideration in the literature. For the case of state delay, Xia, Liu, Shi, Rees, and Thomas (2007) investigated the stability of discrete-time systems with a constant delay by using a lifting method. Regarding time-varying state delay, a few stability criteria were proposed in the literature, for example, Wu (2003) and Wu and Grigoriadis (2001), for which the main concern was focused on reducing the control conservatism. By using a free-weighting matrix, a flexible output feedback control design for a discrete-time system with a time-varying state delay was presented by He, Wu, Liu, and She (2008). Based on the scaled small gain theorem, new stability criteria were proposed in the recent paper (Li & Gao, 2011) in terms of linear matrix inequalities in combination with an approximation on the state delay. For industrial batch processes with time-varying state delay, two-dimensional stability conditions were established for robust closed-loop iterative learning control (ILC) design (Liu & Gao, 2010). For the case of input delay, Yue (2004) addressed the problem of robust feedback stabilization for uncertain continuous-time input-delayed systems. For stable/unstable or minimum/non-minimum phase systems with long time delay, Albertos and Garcia (2009) proposed a robust control method based on a predicted undelayed output in frequency domain, and later improved the adjusting capability between the output performance and robust stability by introducing a tuning parameter (Garcia & Albertos, 2013). Robust stability analysis of the filtered Smith predictor control structure was addressed by Normey-Rico, Garcia, and Gonzalez (2012) for stable or unstable processes with time-varying input delay. An internal model control-based ILC method was proposed to cope with uncertain input delay for batch process operation (Liu, Gao, & Wang, 2010). By comparison, a predictor-based controller design was given by Gonzalez, Sala, and Sanchis (2013) based on the analysis of the Lyapunov–Krasovskii stability condition developed by Manitius and Olbrot (1979), which was then extended in the recent paper (Gonzalez, Sala, & Albertos, 2012) to allow for larger delay variation.

Recently, Najafi, Hosseinnia, Sheikholeslam, and Karimadini (2013) addressed the problem of state feedback control for continuous-time systems with known or time-invariant input delay by sequential sub-predictors, which was further extended to output feedback in the recent paper (Najafi, Sheikholeslam, Wang, & Hosseinnia, 2014). Based on using an interval observation technique, an output feedback stabilization method for time-varying input delay systems without model uncertainties was proposed by Polyakov, Efimov, Perruquetti, and Richard (2013). In discrete-time domain, robust stabilization of linear discrete-time systems with time-varying input delay was studied by Gonzalez (2013) using Artstein's reduction

*Corresponding author. Emails: liurouter@ieee.org; tliu@dlut.edu.cn

method (Artstein, 1982) and the scaled-bounded real lemma (Apkarian & Gahinet, 1995), leading to superior control performance in comparison with previous methods.

Considering that output feedback is widely used in engineering practice and state measurement is not available in many industrial applications, this paper proposes an dynamic output feedback control method to stabilize discrete-time systems with input delay, based on the output measurement and past input information. Moreover, the input delay variation and plant uncertainties are also taken into account for robust control. By introducing an augmented state, such a system description is transformed into a delay-free model involved with uncertainties for robust stabilization. Sufficient conditions for robust stability are established in terms of bilinear matrix inequalities which can be solved efficiently by configuring a tuning parameter. For clarity, the paper is organized as follows. In Section 2, the control problem and some preliminary knowledge for analysis are presented. By using the scaled-bounded real lemma, stability analysis of the transformed systems is given in Section 3. In Section 4, the design of dynamic output feedback controller is detailed. An illustrative example is provided in Section 5 to demonstrate the effectiveness of the proposed method. Finally, some conclusions are given in Section 6.

Throughout this paper, the following notations are used: $\Re^{n\times m}$ denotes an $n \times m$ real matrix space. For any matrix $P \in \Re^{m\times m}$, $P > 0$ (or $P < 0$) means P is a positive- (or negative-) definite symmetric matrix, in which the symmetric elements are indicated as '*'. Denote by P^{T} the transpose of P, and by P^{-1} the inverse of P. The identity vector/matrix with appropriate dimension is denoted by I.

2. Problem description and preliminary knowledge

Consider a discrete-time system with time-varying input delay described by

$$\begin{aligned} x(t+1) &= (A+\Delta A(t))x(t) + (B+\Delta B(t))u(t-d(t)),\\ y(t) &= Cx(t),\\ x(t) &= \varphi(t), \quad t=-h_2,\ldots,0.\ h_1 \le d(t) \le h_2, \end{aligned} \tag{1}$$

where $A \in \Re^{n\times n}$, $B \in \Re^{n\times m}$, and $h_{12} = h_2 - h_1$. $x(t)$ is the state, $u(t)$ is the control input, and $\varphi(t)$ $(t = -h_2, \ldots, 0)$ is a given initial condition sequence. Denote by $\Delta A(t)$ and $\Delta B(t)$ the model uncertainties which may be described by the following form:

$$(\Delta A(t)\ \Delta B(t)) = \delta F \Delta(t)\,(E_A\ E_B), \tag{2}$$

where F, E_A, and E_B are known matrices of appropriate dimensions; $\Delta(t)$ is a time-varying matrix with unknown elements, satisfying $\Delta(t)^{\mathrm{T}}\Delta(t) \le I$, and δ is a positive scalar reflecting the bound of uncertainties. Note that only the measured output and the past input information will be used to design the controller for the convenience of implementation.

Define the following state transformation in terms of Artstein's reduction method (Artstein, 1982),

$$z(t) = x(t) + \phi_t^u(h_1) + \phi_t^u(h_2), \tag{3}$$

where $\phi_t^u(h) : L_2([0,\infty], \Re^m) \to \Re^n$ is an operator with respect to $u(t)$ defined by

$$\phi_t^u(h) \triangleq \sum_{i=0}^{h-1} A^{-i-1}\frac{B}{2}u(t-h+i), \quad h \in \mathbb{N}^+. \tag{4}$$

For the ease of comprehension, the following lemma from the proof of Proposition 1 in Gonzalez (2013) is briefly presented as below, which will be used in the later analysis.

Lemma 1 *The system described by Equation (1) can be transformed into the following form using the state transformation in Equation (3),*

$$\begin{aligned} z(t+1) &= Az(t) + \theta_B u(t) + \delta F\omega_{\Sigma_t} + \frac{h_{12}B}{2}\omega_{d_t},\\ \hat{y}(t) &= Cz(t), \end{aligned} \tag{5}$$

$$\begin{aligned} \sigma_{\Sigma_t} &= E_A z(t) + \theta_E u(t-1) + E_B u(t) + \sum_{i=1}^{2}\theta_i\omega_{it}\\ &\quad + \frac{h_{12}E_B}{2}\omega_{d_t}, \end{aligned} \tag{6}$$

where

$$\theta_B = \sum_{r=1}^{2} A^{-h_r}\frac{B}{2}, \quad \theta_E = -\sum_{r=1}^{2}\sum_{j=0}^{h_r-1} E_A A^{-j-1}\frac{B}{2},$$

$$\omega_{\Sigma_t} = \Delta(t)\sigma_{\Sigma_t},$$

$$m_1 = \left\| \sum_{r=1}^{2}\sum_{f=0}^{h_r-1} z^{-f} \right\|_\infty,$$

$$m_2 = \left\| \sum_{r=1}^{2}\sum_{j=0}^{h_r-1}\sum_{f=1}^{h_r-j-1} A^{-j-1}\frac{B}{2}z^{-f} \right\|_\infty,$$

$$\theta_1 = -m_1\frac{E_B}{2}, \quad \theta_2 = m_2 E_A, \quad \upsilon(t) = u(t) - u(t-1),$$

$$\omega_{d_t} = \frac{2}{h_{12}}\left(u(t-d(t)) - \frac{1}{2}(u(t-h_1) + u(t-h_2))\right).$$

Note that ω_{d_t} can be expressed as $\omega_{d_t} = \Delta_d \upsilon(t)$, $\Delta_d : \upsilon \to \omega_d$ is a time-varying delay operator satisfying $||\Delta_d||_\infty \le 1$. $\omega_{it} = \Delta_i\upsilon(t), i = 1,2.$, and $\Delta_i : \upsilon \to \omega_i$ is an operator satisfying $||\Delta_i||_\infty = 1$ and

$\sum_{r=1}^{2}\Omega_{ir}^{\upsilon_t}=m_i\Delta_i, i=1,2,$ *where*

$$\Omega_{1r}^{\upsilon_t}=\sum_{f=0}^{h_r-1}\upsilon(t-f),\quad \Omega_{2r}^{\upsilon_t}=\sum_{j=0}^{h_r-1}\sum_{f=1}^{h_r-j-1}A^{-j-1}\frac{B}{2}\upsilon(t-f).$$

For the convenience of analysis, by introducing an augmented state $\eta(t)=[z^{\mathrm{T}}(t)\ u^{\mathrm{T}}(t-1)]^{\mathrm{T}}$, *a reformulation of Equations (5) and (6) can be obtained (Jungers, Castelan, Moraes, & Moreno, 2013),*

$$\begin{aligned}\eta(t+1)&=\overline{A}\eta(t)+\overline{\theta}_B u(t)+\delta\overline{F}\omega_{\sum_t}+\overline{B}_d\omega_{d_t},\\ \widehat{y}(t)&=\overline{C}\eta(t),\end{aligned}\tag{7}$$

$$\sigma_{\sum_t}=\overline{E}\eta(t)+E_B u(t)+\sum_{i=1}^{2}\theta_i\omega_{it}+\frac{h_{12}E_B}{2}\omega_{d_t},\tag{8}$$

where

$$\overline{A}=\begin{bmatrix}A&0\\0&0\end{bmatrix},\quad \overline{\theta}_B=\begin{bmatrix}\theta_B\\I\end{bmatrix},\quad \overline{F}=\begin{bmatrix}F\\0\end{bmatrix},\tag{9}$$

$$\overline{B}_d=\begin{bmatrix}\frac{h_{12}B}{2}\\0\end{bmatrix},\quad \overline{E}=\begin{bmatrix}E_A^{\mathrm{T}}\\\theta_E^{\mathrm{T}}\end{bmatrix}^{\mathrm{T}},\quad \overline{C}=\begin{bmatrix}C^{\mathrm{T}}\\0\end{bmatrix}^{\mathrm{T}}.\tag{10}$$

Concerning the augmented system shown in Equation (7), a full-order dynamic output feedback controller is proposed as

$$\begin{aligned}x_c(t+1)&=A_c x_c(t)+B_c\widehat{y}(t),\\ u(t)&=C_c x_c(t)+D_c\widehat{y}(t),\end{aligned}\tag{11}$$

where $x_c(t)$ *is the controller state,* A_c, B_c, C_c, D_c *are the controller matrices to be designed later.*

Remark 1 For the convenience of control design, the measured output is combined with the past input information to construct a delay-free predicted output shown in Equation (5).

3. Stability analysis

Based on the augmented system description in Equations (7) and (8), applying the controller shown in (11) results in the following closed-loop system,

$$x_g(t+1)=\widehat{A}x_g(t)+\delta\widehat{F}\omega_{\sum_t}+\widehat{B}\omega_{d_t},\tag{12}$$

$$\sigma_{\sum_t}=\widehat{C}x_g(t)+\sum_{i=1}^{2}\theta_i\omega_{it}+\frac{h_{12}E_B}{2}\omega_{d_t},\tag{13}$$

where

$$x_g(t)=\begin{bmatrix}\eta(t)\\x_c(t)\end{bmatrix},\tag{14}$$

and

$$\widehat{A}=\begin{bmatrix}\overline{A}+\overline{\theta}_B D_c\overline{C}&\overline{\theta}_B C_c\\B_c\overline{C}&A_c\end{bmatrix},\quad \widehat{F}=\begin{bmatrix}\overline{F}\\0\end{bmatrix},\tag{15}$$

$$\widehat{B}=\begin{bmatrix}\overline{B}_d\\0\end{bmatrix},\quad \widehat{C}=\begin{bmatrix}\overline{E}+E_B D_c\overline{C}&E_B C_c\end{bmatrix}.\tag{16}$$

For stability analysis, the following theorem is given.

THEOREM 1 *The system described by Equation (1) using the controller shown in Equation (11) is asymptotically stable if there exist matrices* $P>0$, $S>0$ *and scalars* $\varepsilon_1>0$, $\varepsilon_2>0$ *such that the following matrix inequality holds:*

$$\begin{bmatrix}-P&A_s&B_sW_1&0\\ *&-P^{-1}&0&C_s^{\mathrm{T}}\\ *&*&-W_1&W_1D_s^{\mathrm{T}}\\ *&*&*&-W_2\end{bmatrix}<0,\tag{17}$$

where $A_s=\widehat{A}$, $B_s=\begin{bmatrix}\delta\widehat{F}&\widehat{B}&0&0\end{bmatrix}$, $\rho=\delta^{-1}$, $\widehat{L}=\begin{bmatrix}L+D_c\overline{C}&C_c\end{bmatrix}$, $L=\begin{bmatrix}0&-I\end{bmatrix}$, $W_1=\mathrm{diag}\{\rho I_{l_1},S,\varepsilon_1 I_m,\varepsilon_2 I_n\}$, $W_2=\mathrm{diag}\{\rho I_{l_1},S,\varepsilon_1 I_m,\varepsilon_2 I_m\}$.

$$C_s=\begin{bmatrix}\widehat{C}\\\widehat{L}\\\widehat{L}\\\widehat{L}\end{bmatrix},\quad D_s=\begin{bmatrix}0&\frac{h_{12}E_B}{2}&\theta_1&\theta_2\\0&0&0&0\\0&0&0&0\\0&0&0&0\end{bmatrix}.$$

Proof Using Equations (12), (13), and Lemma 1, we have

$$\begin{bmatrix}x_g(t+1)\\\overline{\sigma}_{\sum_t}\end{bmatrix}=\begin{bmatrix}A_s&B_s\\C_s&D_s\end{bmatrix}\begin{bmatrix}x_g(t)\\\overline{\omega}_t\end{bmatrix},\quad \overline{\omega}_t=\bar{\Delta}\overline{\sigma}_{\sum_t},\tag{18}$$

where

$$\bar{\Delta}=\mathrm{diag}\{\Delta(t),\Delta_d,\Delta_1,\Delta_2\},\quad ||\bar{\Delta}||_\infty\le 1,\tag{19}$$

$$\overline{\omega}_t=\left(\omega_{\sum_t}^{\mathrm{T}}\quad\omega_{d_t}^{\mathrm{T}}\quad\omega_{1t}^{\mathrm{T}}\quad\omega_{2t}^{\mathrm{T}}\right)^{\mathrm{T}},$$

$$\overline{\sigma}_{\sum_t}=\left(\sigma_{\sum_t}^{\mathrm{T}}\quad\upsilon^{\mathrm{T}}(t)\quad\upsilon^{\mathrm{T}}(t)\quad\upsilon^{\mathrm{T}}(t)\right)^{\mathrm{T}}.\tag{20}$$

Using the scaled-bounded real lemma (Apkarian & Gahinet, 1995) and the scaling matrices W_1 and W_2 that satisfy $W_1\bar{\Delta}=\bar{\Delta}W_2$, we are sure the system in Equation (18) is stable if there exist matrices $P>0$ and $S>0$ such that

$$\begin{pmatrix}-P&A_s&B_s&0\\ *&-P^{-1}&0&C_s^{\mathrm{T}}\\ *&*&-W_1{}^{-1}&D_s^{\mathrm{T}}\\ *&*&*&-W_2\end{pmatrix}<0.\tag{21}$$

Moreover, the inequality in (21) implies that

$$\begin{pmatrix}-P&A_s\\ *&-P^{-1}\end{pmatrix}<0,\tag{22}$$

which guarantees the internal stability of the system in (18). Pre- and post-multiplying (21) by $\mathrm{diag}\{I,I,W_1,I\}$, we obtain the matrix inequality in (17). The proof is completed.

It should be noted that the dimension of the identity matrices in W_1 and W_2 may be different if Δ is not square. In fact, $\mathrm{diag}\{I,I,W_1,I\}$ is symmetric diagonal matrix due to W_1 is symmetric.

4. Design of dynamic output feedback controller

Based on the above stability analysis, the output feedback controller can be derived as stated in the following theorem.

THEOREM 2 *Given a time-varying delay $d(t)$ satisfying $h_1 \le d(t) \le h_2$, there exists a dynamic output feedback controller in the form of Equation (11), such that the closed-loop system (12) and (13) is asymptotically stable if there exist matrices $S > 0, \overline{P} > 0$, $\widehat{A}_c, \widehat{B}_c, \widehat{C}_c, \widehat{D}_c$ and scalars $\varepsilon_1 > 0$, $\varepsilon_2 > 0$, ρ satisfying the following matrix inequality:*

$$\begin{bmatrix} -\overline{P} & J_1 & J_2 & 0 \\ * & J_3 & 0 & J_4 \\ * & * & -\overline{W}_1 & J_5 \\ * & * & * & -W_2 \end{bmatrix} < 0, \tag{23}$$

where

$$\overline{P} = \begin{bmatrix} P_{11} & P_{12} \\ * & P_{22} \end{bmatrix}, \quad J_5 = \begin{bmatrix} 0 & 0 & 0 & 0 \\ \frac{h_{12} E_B^{\mathrm{T}}}{2} & 0 & 0 & 0 \\ \varepsilon_1 \theta_1^{\mathrm{T}} & 0 & 0 & 0 \\ \varepsilon_2 \theta_2^{\mathrm{T}} & 0 & 0 & 0 \end{bmatrix},$$

$$J_1 = \begin{bmatrix} U_1^{\mathrm{T}} \overline{A} + \widehat{B}_c \overline{C} & \widehat{A}_c \\ \overline{A} + \overline{\theta}_B \widehat{D}_c \overline{C} & \overline{A} V_1 + \overline{\theta}_B \widehat{C}_c \end{bmatrix},$$

$$J_2 = \begin{bmatrix} U_1^{\mathrm{T}} \overline{F} & U_1^{\mathrm{T}} \overline{B}_d & 0 & 0 \\ \overline{F} & \overline{B}_d & 0 & 0 \end{bmatrix},$$

$$J_3 = \begin{bmatrix} P_{11} - U_1^{\mathrm{T}} - U_1 & P_{12} - I - M^{\mathrm{T}} \\ * & P_{22} - V_1^{\mathrm{T}} - V_1 \end{bmatrix},$$

$$J_4 = \begin{bmatrix} \overline{E}^{\mathrm{T}} + \overline{C}^{\mathrm{T}} \widehat{D}_c^{\mathrm{T}} E_B^{\mathrm{T}} & \Pi_1 & \Pi_1 & \Pi_1 \\ V_1^{\mathrm{T}} \overline{E}^{\mathrm{T}} + \widehat{C}_c^{\mathrm{T}} E_B^{\mathrm{T}} & \Pi_2 & \Pi_2 & \Pi_2 \end{bmatrix},$$

$$\Pi_1 = L^{\mathrm{T}} + \overline{C}^{\mathrm{T}} \widehat{D}_c^{\mathrm{T}}, \quad \Pi_2 = V_1^{\mathrm{T}} L^{\mathrm{T}} + \widehat{C}_c^{\mathrm{T}},$$

$$M = V_1^{\mathrm{T}} U_1 + V_2^{\mathrm{T}} U_2, \quad \overline{W}_1 = \mathrm{diag}\{\rho I_{l_1}, S^{-1}, \varepsilon_1 I_m, \varepsilon_2 I_n\}.$$

V_2 and U_2 can be obtained by using a singular value decomposition on $M - V_1^{\mathrm{T}} U_1$, corresponding to the controller parameters given by

$$D_c = \widehat{D_c},$$

$$C_c = (\widehat{C}_c - D_c \overline{C} V_1) V_2^{-1},$$

$$B_c = U_2^{-\mathrm{T}} (\widehat{B}_c - U_1^{\mathrm{T}} \overline{\theta}_B D_c),$$

$$A_c = U_2^{-\mathrm{T}} (\widehat{A}_c - U_1^{\mathrm{T}} \overline{A} V_1 - U_1^{\mathrm{T}} \overline{\theta}_B D_c \overline{C} V_1 - U_2^{\mathrm{T}} B_c \overline{C} V_1 - U_1^{\mathrm{T}} \overline{\theta}_B C_c V_2) V_2^{-1}.$$

Proof The proof is based on a suitable congruence transformation and changes of matrix variables. We introduce a slack variable G which is a nonsingular matrix.

Firstly, we assume the inequality (23) holds. Define

$$G = \begin{bmatrix} V_1 & \blacklozenge \\ V_2 & \blacklozenge \end{bmatrix}, \quad G^{-1} = \begin{bmatrix} U_1 & \blacklozenge \\ U_2 & \blacklozenge \end{bmatrix} y, \quad \Omega = \begin{bmatrix} U_1 & I \\ U_2 & 0 \end{bmatrix},$$

where '$\blacklozenge$' denotes the elements that are uniquely determined from equations $GG^{-1} = G^{-1}G = I$. Note that the matrix U_2 included in G^{-1} is assumed to be invertible.

It is easy to verify

$$J_1 = \Omega^{\mathrm{T}} A_s G \Omega, \quad J_2 = \Omega^{\mathrm{T}} B_s W_1,$$

$$J_3 = \Omega^{\mathrm{T}} (P - G^{\mathrm{T}} - G) \Omega, \quad J_4 = \Omega^{\mathrm{T}} G^{\mathrm{T}} C_s^{\mathrm{T}},$$

where the following changes of variables are used:

$$\overline{P} = \Omega^{\mathrm{T}} P \Omega, \quad \widehat{D}_c = D_c,$$

$$\widehat{C}_c = D_c \overline{C} V_1 + C_c V_2, \quad \widehat{B}_c = U_1^{\mathrm{T}} \overline{\theta}_B D_c + U_2^{\mathrm{T}} B_c,$$

$$\widehat{A}_c = U_1^{\mathrm{T}} \overline{A} V_1 + U_1^{\mathrm{T}} \overline{\theta}_B D_c \overline{C} V_1 + U_2^{\mathrm{T}} B_c \overline{C} V_1 + U_1^{\mathrm{T}} \overline{\theta}_B C_c V_1 + U_2^{\mathrm{T}} A_c V_2.$$

Thus, inequality (23) can be rewritten as

$$\begin{bmatrix} -\Omega^{\mathrm{T}} P \Omega & \Omega^{\mathrm{T}} A_s G \Omega & \Omega^{\mathrm{T}} B_s W_1 & 0 \\ * & \Omega^{\mathrm{T}} (P - G^{\mathrm{T}} - G) \Omega & 0 & \Omega^{\mathrm{T}} G^{\mathrm{T}} C_s^{\mathrm{T}} \\ * & * & -W_1 & W_1 D_s^{\mathrm{T}} \\ * & * & * & -W_2 \end{bmatrix} < 0. \tag{24}$$

Since U_2 is invertible, matrix Ω should be invertible. Performing a congruent transformation to Equation (24) by pre- and post-multiplying $\mathrm{diag}\{\Omega^{-\mathrm{T}}, \Omega^{-\mathrm{T}}, I, I\}$ and $\mathrm{diag}\{\Omega^{-1}, \Omega^{-1}, I, I\}$, respectively, we obtain

$$\begin{bmatrix} -P & A_s G & B_s W_1 & 0 \\ * & P - G^{\mathrm{T}} - G & 0 & G^{\mathrm{T}} C_s^{\mathrm{T}} \\ * & * & -W_1 & W_1 D_s^{\mathrm{T}} \\ * & * & * & -W_2 \end{bmatrix} < 0. \tag{25}$$

Using the well-known inequality,

$$(G^{\mathrm{T}} - P) P^{-1} (G - P) \ge 0. \tag{26}$$

we have

$$- G^{\mathrm{T}} P^{-1} G \le P - G^{\mathrm{T}} - G. \tag{27}$$

Combining Equations (25) and (27) leads to the inequality (17) in Theorem 1. This completes the proof.

Remark 2 Note that the sufficient condition in (23) is not a strict linear matrix inequality due to that S^{-1} enters in $\overline{W}_1$ in a nonlinear manner. It is therefore suggested to let the matrix variable S be an identity matrix when using the LMI toolbox to find a feasible solution, but in exchange for conservativeness. For a single-input system, it can be easily verified that the positive-definite matrix S is exactly a positive scalar. Hence, we can introduce a tuning parameter

$\beta > 0$ instead of S and monotonically decrease or increase it to solve inequality (23), owing to that β enters inequality (23) linearly. For a multiple-input system, we can solve inequality (23) by letting $S = \beta I$ and adjusting β to obtain a feasible solution in the same way, or using a cone complementarity linearization algorithm proposed by Ghaoui, Oustry, and AitRami (1997).

It should be noted that the robust performance against model uncertainties with a bounded delay interval h_{12} can be evaluated by using the following optimization procedure,

$$\min \quad \rho \quad \text{s.t.}(23)$$

where the robust performance level against model uncertainties is indicated by $\bar{\delta} = \rho^{-1}$.

5. Illustration

Consider the example studied by Zhang, Xu, and Zou (2008),

$$x(t+1) = \begin{bmatrix} 1.0078 & 0.0301 \\ 0.5202 & 1.0078 \end{bmatrix} x(t) + \begin{bmatrix} -0.0001 \\ -0.0053 \end{bmatrix} u(t). \tag{28}$$

The state response of the system with no input ($u(t) = 0$) and initial condition $x_0 = (0.1, 0.1)^{\mathrm{T}}$ is shown in Figure 1, indicating that the system is unstable. For illustration, here we assume the system output $y(t) = \begin{bmatrix} 10 & 1 \end{bmatrix} x(t)$, which is used for feedback control. The corresponding full-order controller parameters can be obtained by solving the stability condition in inequality (23) with a fixed tuning parameter $\beta = 500$ for the system without model uncertainties,

$$A_c = 10^3 \times \begin{bmatrix} -0.0007 & 0.0000 & 0.0000 \\ -4.3312 & 0.0020 & 0.0002 \\ -0.0245 & 0.0000 & -0.0000 \end{bmatrix},$$

$$B_c = 10^3 \times \begin{bmatrix} -0.0006 \\ -1.5446 \\ -0.0087 \end{bmatrix},$$

$$C_c = \begin{bmatrix} 33.3430 & -0.0155 & -0.0014 \end{bmatrix}, \quad D_c = 11.8905.$$

The following state feedback control law was given in Zhang et al. (2008),

$$u(t) = \begin{bmatrix} 110.6827 & 34.6980 \end{bmatrix} x(t - d(t)), \tag{29}$$

which could stabilize the system in Equation (28), where $d(t)$ is a time-varying delay and satisfies $1 \le d(t) \le 4$. The simulation results for the system without model uncertainties are shown in Figures 2 and 3.

It is seen from Figure 2 that the state response of the closed-loop system by the proposed method recovers to zero obviously faster than that of Zhang et al. (2008). Figure 3 shows that an apparently smaller amount of control effort is needed by the proposed method in comparison with that of Zhang et al. (2008).

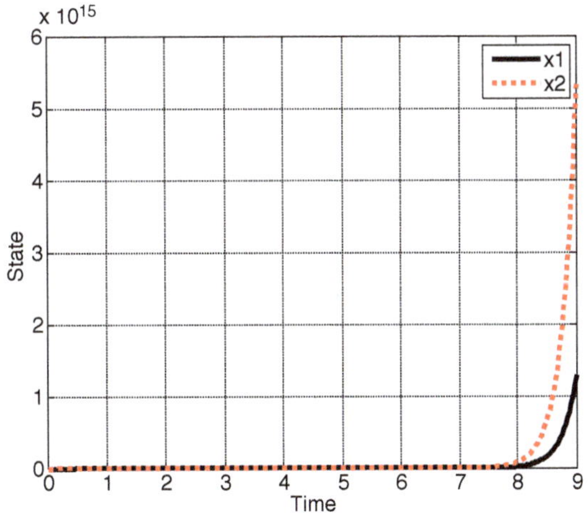

Figure 1. State response of the open-loop system with $x_0 = (0.1, 0.1)^{\mathrm{T}}$.

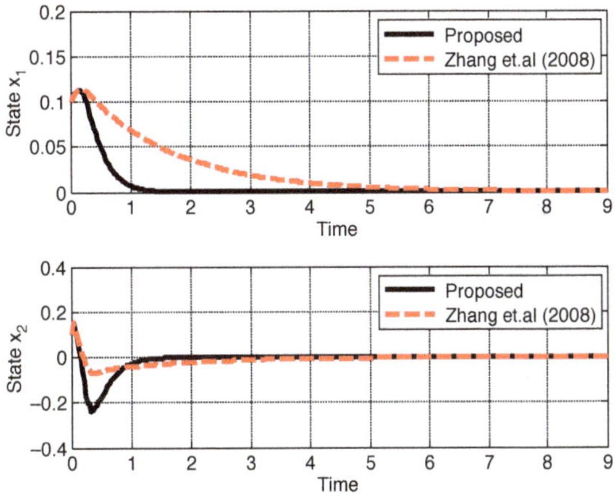

Figure 2. State response of the closed-loop system without model uncertainties for an initial condition $x_0 = (0.1, 0.1)^{\mathrm{T}}$.

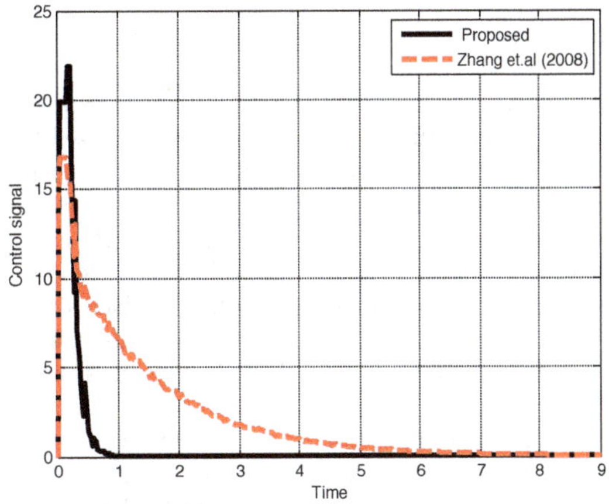

Figure 3. Control signal of the closed-loop system without model uncertainties for an initial condition $x_0 = (0.1, 0.1)^{\mathrm{T}}$.

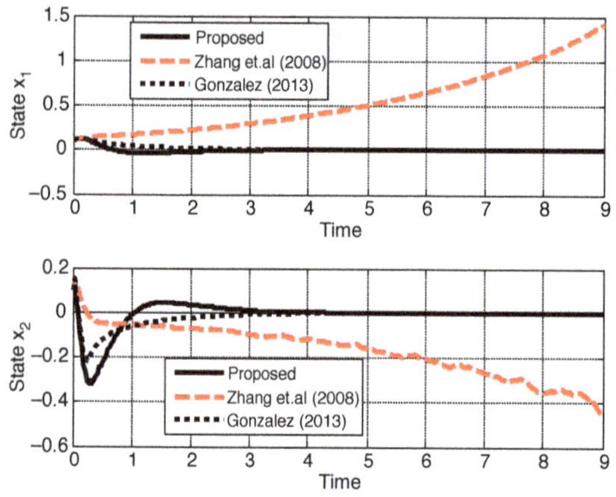

Figure 4. State response of the closed-loop system with model uncertainties.

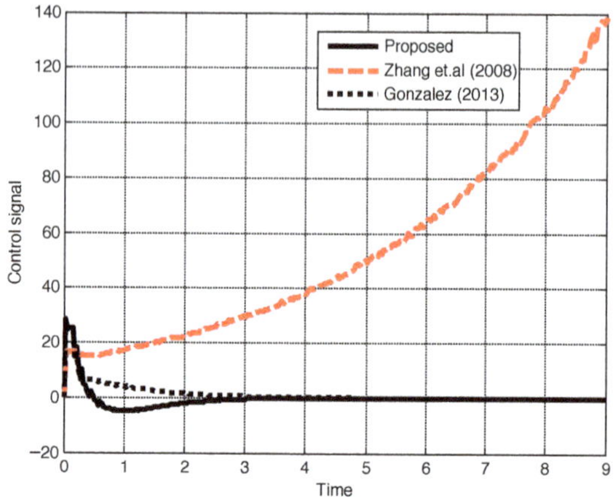

Figure 5. Control action of the closed-loop system with model uncertainties.

Then, assume that besides the delay variation, there exist model uncertainties with $F = (0.1, 0)^{\mathrm{T}}$, $E_A = (0.1, 0.2)$, $E_B = 0.01$ as assumed by Gonzalez (2013). A robust state feedback control law can be obtained as $K = (204.5466, 52.5988, 0.0046)$ by solving the LMI conditions in the cited reference, which allows for a model uncertainty bound of $\delta = 0.454$.

Using the proposed method yields the following full-order output controller matrices by fixing $\beta = 450$ in solving the stability condition (23),

$$A_c = 10^3 \times \begin{bmatrix} -0.0026 & 5.1686 & 0.0013 \\ -0.0000 & 0.0034 & 0.0000 \\ -0.0000 & 0.0000 & -0.0000 \end{bmatrix},$$

$$B_c = 10^3 \times \begin{bmatrix} -5.4921 \\ -0.0027 \\ -0.0000 \end{bmatrix},$$

$$C_c = \begin{bmatrix} 0.0123 & -24.2497 & -0.0062 \end{bmatrix}, \quad D_c = 25.7674.$$

which allows for the maximum model uncertainty bound of $\delta = 0.3735$.

With $\Delta(t) = 0.5$ and the initial state condition of $x_0 = (0.1, 0.1)^{\mathrm{T}}$, the resulting state response of the closed-loop system is shown in Figure 4, and the control signal is plotted in Figure 5. It is seen that the proposed method holds the control system robust stability well against model uncertainties and, in contrast, the closed-loop system becomes unstable by using the state feedback control method given by Zhang et al. (2008). Note that the proposed output feedback control method gives a similar control performance with that of Gonzalez (2013) which was based on using state feedback. Moreover, it can be verified that the maximum input delay variation of $1 \leq d(t) \leq 9$ is allowed by the proposed method as well as that of Gonzalez (2013), based on only using the output measurement.

6. Conclusion

Robust stabilization of discrete-time systems with time-varying input delay and model uncertainties has been addressed in this paper. Correspondingly, an output feedback control method has been proposed in contrast with a recently developed method (Gonzalez, 2013) depending on state measurement. By transforming such a process model into a delay-free one via introducing an augmented state, the scaled-bounded real lemma is adopted to establish sufficient stabilization conditions in terms of matrix inequality, which can be simply solved by configuring a tuning parameter. An illustrative example from the literature has been used to show the effectiveness and superior performance of the proposed output feedback control method.

Disclosure statement

No potential conflict of interest was reported by the authors.

Funding

This work is supported in part by the National Thousand Talents Program of China, NSF China [grant number 61473054], and the Fundamental Research Funds for the Central Universities of China.

References

Albertos, P., & Garcia, P. (2009). Robust control design for long time-delay systems. *Journal of Process Control*, *19*(10), 1640–1648.

Apkarian, P., & Gahinet, P. (1995). A convex characterization of gain-scheduled H_∞ controllers. *IEEE Transactions on Automatic Control*, *40*(5), 853–864.

Artstein, Z. (1982). Linear systems with delayed control: A reduction. *IEEE Transactions on Automatic Control*, *27*(4), 869–879.

Garcia, P., & Albertos, P. (2013). Robust tuning of a generalized predictor-based controller for integrating and unstable systems with long time-delay. *Journal of Process Control, 23*(8), 1205–1216.

Ghaoui, L., Oustry, F., & AitRami, M. (1997). A cone complementarity linearization algorithm for static output feedback and related problems. *IEEE Transactions on Automatic Control, 42*(8), 1171–1176.

Gonzalez, A. (2013). Robust stabilization of linear discrete-time systems with time-varying input delay. *Automatica, 49*(9), 2919–2922.

Gonzalez, A., Sala, A., & Albertos, P. (2012). Predictor-based stabilization of discrete time varying input delay systems. *Automatica, 48*(2), 454–457.

Gonzalez, A., Sala, A., & Sanchis, R. (2013). LK stability analysis of predictor-based controllers for discrete-time systems with time-varying actuator delay. *Systems & Control Letters, 62*(9), 764–769.

He, Y., Wu, M., Liu, G., & She, J. (2008). Output feedback stabilization for a discrete-time system with a time-varying delay. *IEEE Transactions on Automatic Control, 53*(10), 2372–2377.

Jungers, M., Castelan, E., Moraes, V., & Moreno, U. (2013). A dynamic output feedback controller for NCS based on delay estimates. *Automatica, 49*(3), 788–792.

Li, X., & Gao, H. (2011). A new model transformation of discrete-time systems with time-varying delay and its application to stability analysis. *IEEE Transactions on Automatic Control, 56*(9), 2172–2178.

Liu, T., & Gao, F. (2010). Robust two-dimensional iterative learning control for batch processes with state delay and time-varying uncertainties. *Chemical Engineering Science, 65*(23), 6134–6144.

Liu, T., & Gao, F. (2012). *Industrial process identification and control design: Step-test and relay-experiment-based methods*. London: Springer.

Liu, T., Gao, F., & Wang, Y. (2010). IMC-based iterative learning control for batch processes with uncertain time delay. *Journal of Process Control, 20*(2), 173–180.

Manitius, A., & Olbrot, A. (1979). Finite spectrum assignment problem for systems with delays. *IEEE Transactions on Automatic Control, 24*(4), 541–553.

Najafi, M., Sheikholeslam, F., Wang, Q. & Hosseinnia, S. (2014). Robust H_∞ control of single input-delay systems based on sequential sub-predictors. *IET Control Theory and Applications, 8*(13), 1175–1184.

Najafi, M., Hosseinnia, S., Sheikholeslam, F., & Karimadini, M. (2013). Closed-loop control of dead time systems via sequential sub-predictors. *International Journal of Control, 86*(4), 599–609.

Normey-Rico, J., Garcia, P., & Gonzalez, A. (2012). Robust stability analysis of filtered Smith predictor for time-varying delay processes. *Journal of Process Control, 22*(10), 1975–1984.

Polyakov, A., Efimov, D., Perruquetti, W., & Richard, J. (2013). Output stabilization of time-varying input delay systems using interval observation technique. *Automatica, 49*(11), 3402–3410.

Seborg, D., Edgar, T., & Mellichamp, D. (2004). *Process dynamic and control* (2nd ed.). New Jersey: John Wiley & Sons.

Wu, F. (2003). Robust quadratic performance for time-delayed uncertain linear systems. *International Journal of Robust Nonlinear Control, 13*(2), 153–172.

Wu, F., & Grigoriadis, K. (2001). LPV systems with parameter-varying time delays: Analysis and control. *Automatica, 37*(2), 221–229.

Xia, Y., Liu, G., Shi, P., Rees, D., & Thomas, E. (2007). New stability and stabilization conditions for systems with time-delay. *International Journal of Systems Science, 38*(1), 17–24.

Yue, D. (2004). Robust stabilization of uncertain systems with unknown input delay. *Automatica, 40*(2), 331–336.

Zhang, B., Xu, S., & Zou, Y. (2008). Improved stability criterion and its applications in delayed controller design for discrete-time systems. *Automatica, 44*(11), 2963–2967.

14

Effective fault-tolerant control paradigm for path tracking in autonomous vehicles

Afef Fekih* and Shankar Seelem

Department of Electrical and Computer Engineering, University of Louisiana at Lafayette, Lafayette, LA, USA

A novel fault-tolerant control paradigm that integrates fault detection (FD) with optimal control for path tracking is designed to ensure accurate path tracking in the presence of faults. The proposed approach is designed to maintain vehicle stability, dynamics, and maneuverability in the event of a faulty steering system. A sensor fusion-based fault detection and identification approach is proposed to accurately detect and identify sensor faults when they occur. A weight adjustment algorithm is considered to ensure accurate detection while providing robustness against parameter variations and uncertainties. Following FD and using the estimated fault vector, a fault-tolerant controller is designed to guarantee the stability of the closed loop system. The proposed controller incorporates a linear quadratic regulator (LQR)-based algorithm with a feed-forward gain. The LQR-based controller is designed to maintain system stability under faulty conditions while operating the dynamic system at minimum cost. The proposed approach is validated using a ground vehicle required to track various paths while being subject to multiple fault scenarios. For accurate performance analysis, vehicle handling and dynamics were implemented using CarSim, a high-fidelity vehicle simulator. Effective path tracking capabilities, vehicle handling, and stabilization under both fault-free and faulty conditions are the main positive features of the proposed approach.

Keywords: fault detection and identification (FDI); fault-tolerant control (FTC); linear quadratic regulator (LQR); sensor fusion; observer

1. Introduction

As modern day autonomous vehicles become more and more complex and highly integrated, the vulnerability of their components to faults/failures increases (Cheng, 2011; Fekih, 2014). Defects in sensors, actuators, or the system itself can degrade overall system performance. Undetected, faults can develop into failures which probably increase with the increased complexity of the system. Moreover, mitigating unsatisfactory performances or even instability caused by the unpredictable faults in actuators, sensors, or other components is of foremost priority, especially in safety-critical systems such as ground vehicles.

According to the International Federation of Automatic Control SAFEPROCESS technical committee (Gustaffson, 2000), a fault is defined as any unpermitted deviation of at least one characteristic property or parameter of the system from the acceptable/usual/standard condition, while a failure is a permanent interruption of a system's ability to perform its function under specific operating conditions. In order to maintain high levels of performance and guarantee proper system behavior, it is important that faults be promptly detected and identified and appropriate remedies be applied to prevent system malfunctions. Diagnosis is the primary stage of active fault-tolerant control (FTC) systems. Its goal is to perform two main decision tasks: fault detection (FD), consisting of deciding whether or not a fault has occurred, and fault isolation, consisting of deciding which element of the system has failed (Fekih, 2014).

A FTC system is a control system specifically designed to automatically accommodate faults among system components while maintaining system stability along with a desired level of overall performance (Blanke & Staroswiecki, 2006; Noura, Theilliol, Ponsart, & Noura, 2009). The key issue of a FTC system is to prevent local faults from developing into system failures that can end the mission of the system and cause safety hazards for man and environment. Existing efforts in FTC design can be classified into two main approaches: the passive and active approaches (Jiang & Yu, 2012). In the passive approach, robust control techniques are used to ensure that the control loop system remains insensitive to certain faults. The effectiveness of this strategy, which usually assumes very restrictive repertory of faults, depends upon the robustness of the nominal closed-loop system. In the active approach, a new control system is redesigned according to the estimation of the fault performed by the fault detection and identification (FDI) filter and according to the specifications to be met by the faulty system. In contrast to passive approaches that are mostly conservative, active approaches

*Corresponding author. Email: afef.fekih@louisiana.edu

are able to deal with a large number of fault scenarios and can handle a certain number of unforeseen faults that were not considered at the design stage.

A growing body of research in this area has resulted in a number of FD and FTC schemes for ground vehicles (Arogeti, Wang, Low, & Yu, 2008; Chen, Song, & Li, 2011; Dong, Verhaegen, & Holweg, 2008; Fekih & Seelem, 2012; Herpin, Fekih, Golconda, & Lakhotia, 2007; Laureiro, Benmoussa, Touati, Merzouki, & OuldBouamama, 2014; Morteza & Fekih, 2014a, 2014b; Tabbache, Benbouzid, Kheloui, & Bourgeot, 2011; Wang & Wang, 2013; Yang, Cocquempot, & Jiang, 2008). A passive actuator FTC was proposed for Four-Wheel Independently Actuated electric ground vehicle in Wang and Wang (2013). The approach exploits the redundancy of the system and groups actuators with similar faults in one subsystem and applies control allocation to distribute the control effort. Actuator grouping was attempted to reduce the significant computational cost typically associated with control allocation. In Laureiro et al. (2014) a bond graph model-based FD approach and an FTC were designed for an over-actuated heavy size autonomous vehicle. The approach relied heavily on analytical redundancy relations derived from the bond graph model. A robust and adaptive FTC tracking approach was proposed in Chen et al. (2011). An FTC strategy which considers a maximum-likelihood voting algorithm was proposed for sensor faults in Tabbache et al. (2011). Note that most of these works exploited system redundancy and required high computational costs, drawbacks that might prevent their real-time implementation.

In this paper, an FTC framework that implicitly integrates FDI with FTC is designed for the automatic steering of an autonomous ground vehicle subject to sensor faults. The proposed controller is based on a linear quadratic regulator (LQR) augmented with a feed-forward gain. The LQR-based controller is designed to place the system's eigenvalues in the stable region while operating the dynamic system at minimum cost function. An observer-based FDI approach is proposed to detect and identify sensor faults when they occur. Using the estimated fault vector, the fault-tolerant controller is designed to maintain system stability when faults occur. The proposed framework is implemented on a ground vehicle required to follow a given path, while being subject to sensor faults. The steering controller is designed to maintain vehicle stability, dynamics, and maneuvrability in the event of a faulty steering system.

Compared with the existing work already reported in the literature (Arogeti et al., 2008; Chen et al., 2011; Laureiro et al., 2014; Morteza & Fekih, 2014b; Tabbache et al., 2011; Wang & Wang, 2013), the contributions of this paper are in the following aspects:

- It presents a complete FTC design with the FDI algorithm as integral part of the framework and applies it to the automatic steering of an autonomous vehicle.
- The FDI algorithm incorporates a weight adjustment algorithm to ensure accurate detection, while providing robustness against parameter variations and uncertainties.
- It integrates the optimal properties of the LQR framework with an observer-based fault detection scheme to achieve effective fault tolerance.
- It provides an easy to implement algorithm which achieves fault-tolerance with optimum computational costs. This is crucial in autonomous vehicles, which often work under tight real-time deadlines and cannot tolerate prolonged delays in control reconfiguration.

The rest of this paper is organized as follows. Section 2 presents the dynamic model of the vehicle and discusses the design specifications. The proposed control paradigm is detailed in Section 3. Section 4 is dedicated to the performance analysis of the proposed algorithm. Finally, some concluding remarks end this paper in Section 5.

2. Vehicle dynamic model and problem formulation

A dynamic model of the vehicle, with the front and rear wheels lumped together into a pair of single wheels at the center of gravity (CG) (Fekih & Deveriste, 2013), is considered as shown in Figure 1.

Assuming constant longitudinal velocity and combining the lateral forces with the available slip angles, the vehicle's dynamic model is defined as follows (Fekih & Deveriste, 2013):

$$\begin{bmatrix} \dot{v}_y \\ \dot{r} \end{bmatrix} = \begin{bmatrix} \dfrac{-(c_f + c_r)}{mv_x} & \dfrac{l_r c_r - l_f c_f}{mv_x} - v_x \\ \dfrac{l_r c_r - l_f c_f}{I_z v_x} & \dfrac{-(l_f^2 c_f + l_r^2)}{I_z v_x} \end{bmatrix} \begin{bmatrix} v_y \\ r \end{bmatrix} + \begin{bmatrix} \dfrac{c_f}{m} \\ \dfrac{l_f c_f}{m} \end{bmatrix} \delta, \tag{1}$$

where $\dot{v}_y$ is the rate of change of lateral velocity, $\dot{r}$ is the yaw rate of the vehicle, δ is the steering angle, θ is the yaw

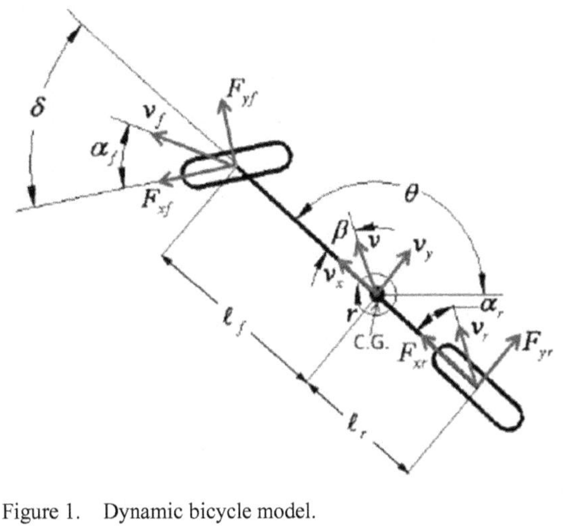

Figure 1. Dynamic bicycle model.

angle (orientation angle of the vehicle with respect to the X axis), and v_y, v_x are the lateral and longitudinal velocity, respectively. c_f and c_r are the cornering stiffness of the front and rear tires, respectively. ℓ_f is the distance from the CG to the front axle and ℓ_r is the distance from the CG to the rear axle. I_z is the vehicle yaw moment of inertia. The remaining variables and parameters are defined in the Appendix (Table A1). Since our objective is to develop a steering control system for automatic lane keeping, the state variables are being expressed in terms of position and orientation error. If we consider a vehicle traveling at a constant velocity on a road of a large radius with curvature k and assume a constant longitudinal velocity, the rate of change of the desired orientation of the vehicle is given by

$$\dot{r}_{\text{des}} = v_x k, \tag{2}$$

where $\dot{r}_{\text{des}}$ is the desired yaw rate and k is the curvature of the road. The desired path lateral acceleration of the vehicle can be written as

$$\dot{v}_y = k v_x^2. \tag{3}$$

Define e as the distance of the CG of the vehicle from the center line of the path and e_1 as the yaw angle error of the vehicle with respect to the path; then, we have

$$\begin{aligned} \ddot{e} &= \dot{v}_y + v_x(\dot{r} - \dot{r}_{\text{des}}), \\ \dot{e} &= \dot{v}_y + v_x(r - r_{\text{des}}), \\ e_1 &= r - r_{\text{des}}. \end{aligned} \tag{4}$$

The state-space model in tracking error variables is therefore given by

$$\dot{x} = Ax + B_1\delta + B_2\dot{r}_{\text{des}}, \tag{5}$$

with $x = [e \quad \dot{e} \quad e_1 \quad \dot{e}_1]^{\text{T}}$.

$$A = \begin{bmatrix} 0 & 1 & 0 & 0 \\ 0 & -\dfrac{(c_f + c_r)}{m v_x} & \dfrac{c_f + c_r}{m} & \dfrac{l_f c_f - l_r c_r}{m v_x} \\ 0 & 0 & 0 & 1 \\ 1 & \dfrac{l_r c_r - l_f c_f}{I_z v_x} & \dfrac{l_f c_f - l_r c_r}{I_z} & -\dfrac{(l_f^2 c_f + l_r^2 c_r)}{I_z v_x} \end{bmatrix}, \ldots,$$

$$B_1 = \begin{bmatrix} 0 \\ \dfrac{c_f}{m} \\ 0 \\ \dfrac{l_f c_f}{m} \end{bmatrix}, \ldots B_2 = \begin{bmatrix} 0 \\ \dfrac{l_r c_r - l_f c_f}{m v_x} - v_x \\ 0 \\ \dfrac{-(l_f^2 c_f + l_r^2 c_r)}{I_z v_x} \end{bmatrix}.$$

The output vector of the system consists of measurements from the two sets of sensors. Sensor failures are modeled as additive signals to the sensor output as follows:

$$y = [y_1 \quad y_2]^{\text{T}} = Cx + Ff, \tag{6}$$

where y_1 and y_2 are the measurements from the two sensors which measure the lateral deviation of the vehicle and f is the fault signal, which is a function of time and state x, and C is the output matrix, defined as $C = \begin{bmatrix} C_1 \\ C_2 \end{bmatrix} = \begin{bmatrix} 1 & 0 & d_1 & 0 \\ 1 & 0 & -d_2 & 0 \end{bmatrix}$, where d_1 and d_2 are, respectively, the distances from the CG of the vehicle to the front and rear bumpers where the sensors are located. $F = [1 \quad 0]^{\text{T}}$ in the event of failure of sensor 1 and $F = [0 \quad 1]^{\text{T}}$ in the event of failure of sensor 2, respectively. Here, we consider that the lateral sensing system consists of two sets of sensors which provide the information of the lateral deviation.

3. FTC paradigm

The faults under investigation are sensor faults with varying severities and types. The FTC objectives are to maintain vehicle stability, dynamics, and maneuverability in the event of faulty sensors.

3.1. *Observer-based FDI algorithm*

The following are some important features of the vehicle's dynamic model that can aid in designing an easy-to-implement FDI algorithm: (1) it has two zero eigenvalues and (2) (A, C_1) and (A, C_2) are observable. This implies that we can estimate the state through either y_1 or y_2. This makes FDI easy to implement with minimum computational cost.

The observability properties of the vehicle imply that we can build two observers, each of which is being driven by a single sensor output. Furthermore, in order to ensure that no erroneous estimates of the state are obtained under sensor failures, we fuse the sensor output and the estimated output from one observer prior to their use by the other observer. Fusion blocks play the role of switches, which select the healthy signal. Then post-filters are designed such that the transfer functions from fault signals to residuals have consistent behavior in order to facilitate fault identification.

Output fusion is a convex combination of the sensor output and the estimated output from the observer. The fused output y_{fi} is given by

$$y_{fi} = (1 - \lambda_i)y_i + \lambda_i \hat{y}_i^j, \tag{7}$$

where $\hat{y}_i^j$ is the estimate of the ith output from the jth observer, and $i, j = 1, 2$, and λ_i is a weight function. The weights $\lambda_i \in [0, 1]$ are adjusted using a weight adjustment algorithm. When there are no faults, that is, $\lambda_i = 0$, then the fused output y_{fi} is identical to the sensor output y_i. When faults occur, the corresponding values of λ_i will increase toward one. When $\lambda_i = 1$, the sensor output is incorrect and therefore is not being taken into account at all. The observers will switch between two configurations according to the relative size of the weights as follows:

If $\lambda_1 < \lambda_2$, then the observers are defined by

$$\dot{\hat{x}}_1 = A\hat{x}_1 + B_1\delta + L_1(y_{f1} - \hat{y}_1^1) + \lambda_1 L_1 C_1(\hat{x}_1 - \hat{x}_2),$$
$$\dot{\hat{x}}_2 = A\hat{x}_2 + B_1\delta + L_2(y_{f2} - \hat{y}_2^2). \tag{8}$$

If $\lambda_1 > \lambda_2$, then

$$\dot{\hat{x}}_1 = A\hat{x}_1 + B_1\delta + L_1(y_{f1} - \hat{y}_1^1)$$
$$\dot{\hat{x}}_2 = A\hat{x}_2 + B_1\delta + L_2(y_{f2} - \hat{y}_2^2) + \lambda_2 L_2 C_2(\hat{x}_2 - \hat{x}_1), \tag{9}$$

where L_1 and L_2 are the solutions of the characteristic equation defined by det(sI-A + L_1C_1) and det(sI-A + L_2C_2), respectively. Observers (8)–(9) are variations of the Luenberger observer where the fused outputs replace the sensor outputs.

3.2. *Threshold generation logic*

For accurate FD, define the following fault indicator function or threshold logic:

$$\|r_i(t)\| < T_h \Rightarrow \text{fault-free conditions,}$$
$$\|r_i(t)\| > T_h \Rightarrow \text{faulty conditions,} \tag{10}$$

where T_h is a predefined threshold typically chosen based on the application at hand. Note that setting low thresholds results in high false-positive rates (alarms are issued under no fault conditions), and setting high thresholds increases the false-negative rates (alarms are missed when faults occur). Clearly, the selection of the thresholds is closely related to robustness and sensitivity of the residual generator. Different analysis procedures are used depending on the techniques employed to generate the residual signal. The most widely used approaches to analyze the residual signal generated by observers are threshold logic and limit monitoring. Threshold level selection methods are generally problem specific and are not useful for a general case (Hsiao & Tomizuka, 2005). To avoid improper FD, threshold level selection is often done on the basis of the designer's experience and in response to problem requirements.

3.3. *Post-filters*

Consider the error vector:

$$e_{yi} = \begin{bmatrix} y_1 - \hat{y}_1^1 & y_1 - \hat{y}_1^2 & y_2 - \hat{y}_2^1 & y_2 - \hat{y}_2^2 \end{bmatrix}^T. \tag{11}$$

Residuals are generated by filtering e_y through post-filters $M_i(s)$, that is,

$$r_i = M_i e_{yi} \quad i = 1,2,3,4. \tag{12}$$

$M_i(s)$ define the transfer functions from the faulty signals to the residuals such that the residuals from the two observers are comparable in magnitude. Note that r_1 and r_2 are related to sensor 1 and r_3 and r_4 are related to sensor 2, respectively. Notice that observers (8)–(9) are coupled, that is, faults in either of the two sensors affect all residuals. The problem of identifying the exact fault sensor can be solved by using properly designed post-filters (Zhang, Ding, Lam, & Wang, 2003).

Define the state-space transfer function from fault f to e_y by

$$V_i(s) = C(sI - A)^{-1}B + F, \tag{13}$$

where F is a matrix which represents the sensor failure as follows:

$$F = \begin{bmatrix} 1 & 0 \end{bmatrix}^T \quad \text{If sensor 1 fails,}$$
$$F = \begin{bmatrix} 0 & 1 \end{bmatrix}^T \quad \text{If sensor 2 fails.}$$

Consider the scenarios when sensor 1 has failed and $\lambda_1 < \lambda_2$, then from Equation (12) we have

$$V_1(s) = -(1-\lambda_1)C_1(sI - A + (1-\lambda_1)L_1C_1)^{-1}L_1 + 1 = \frac{n_1(s)}{d(s)}, \tag{14}$$

$$V_3(s) = -(1-\lambda_1)C_2(sI - A + (1-\lambda_1)L_1C_1)^{-1}L_1 = \frac{(1-\lambda_1)n_3(s)}{d(s)}, \tag{15}$$

where $((n_1(s), d(s))$ and $(n_3(s), d(s))$ are the co-prime pairs of the polynomials defined as follows:

$$n_1(s) = \det\left(\begin{bmatrix} sI - A & L_1 \\ 0 & 1 \end{bmatrix} \begin{bmatrix} I & 0 \\ (1-\lambda_1)C_1 & 1 \end{bmatrix}\right), \tag{16}$$

$$n_3(s) = \det\left(\begin{bmatrix} sI - A & L_1 \\ 0 & 1 \end{bmatrix} \begin{bmatrix} I & 0 \\ (1-\lambda_1 C_2 & 1 \end{bmatrix}\right), \tag{17}$$

where $n_1(s)$ and $n_3(s)$ are also independent of λ_1.

Factorizing $n_1(s) = n_1^+(s)n_1^-(s)$ and $n_3(s) = n_3^+(s)n_3^-(s)$, where $n_i^+(s)$ and $n_i^-(s)$, i = 1,3, are the factors of $n_i(s)$ which have their roots in the left half plane.

Choosing:

$$M_1(s) = \frac{n_1{}^+(s)}{n_3^-(s)k(s)}, \tag{18}$$

$$M_3(s) = \frac{n_3^+(s)}{n_1^-(s)k(s)}, \tag{19}$$

where $k(s)$ is a Hurwitz polynomial such that $M_1(s)$ and $M_3(s)$ are proper and stable, yields

$$(1-\lambda_1)M_1(s)V_1(s) = M_3(s)V_3(s). \tag{20}$$

This implies that if we choose post-filters M_i's such that $a_1M_1V_1 = M_3V_3$ and $a_2M_4V_4 = M_2V_2$ for some real numbers $a_1 > 0$ and $a_2 < 1$, we can define the following identification rules:

If $\lambda_1 < \lambda_2$, and a fault was detected, $|r_1| > |r_3|$ indicates that sensor 1 has failed, while $|r_1| < |r_3|$ implies that sensor 2 has failed.

Similarly, if $\lambda_1 > \lambda_2$, and a fault has been detected, $|r_2| > |r_4|$ suggests that sensor 1 has failed, while $|r_2| < |r_4|$ implies that sensor 2 has failed. In order to accommodate faults, when these latter happen, the controller considers the fused outputs, that is, y_{f1} and y_{f2} instead of the sensor outputs y_1 and y_2.

If the fault occurs, the faulty sensor output is replaced by the observer output.

The next step after residual generation is the analysis of the residual signal for FD. The residual generator takes the sensor measurements as inputs and generates residuals. The latter are small, ideally zero, when there are no fault; but when a fault occurs, the residuals are significantly large. Due to the effect of disturbances, model uncertainties, and measurement noise, the residuals are different from zero even when there are no faults. A robust residual generator is proposed next to alleviate these effects while remaining sensitive to faults.

3.4. *Weight adjustment algorithm*

If any fault has been detected and identified, weights λ_i, $i = 1, 2$ in the fusion blocks will be adjusted. Suppose sensor 1 fails, then we adjust the weights according to the following 1st order differential equations:

$$\dot{\lambda}_1 = -\alpha(\lambda_1 - g(|[r_1 \quad r_2]^{\mathrm{T}}|) \tag{21}$$

$$\dot{\lambda}_2 = -\alpha\lambda_2. \tag{22}$$

If sensor 2 fails, then the adaption rule becomes

$$\dot{\lambda}_1 = -\alpha\lambda_1 \tag{23}$$

$$\dot{\lambda}_2 = -\alpha(\lambda_2 - g(|[r_3 \quad r_4]^{\mathrm{T}}|), \tag{24}$$

where g is a sigmoid function defined as follows: g: R $\rightarrow$[0, 1] $g(x) = 1/(1 + e^{-ax})$ $a > 0$ with $a \in [-10, 10]$. The value of a is determined based of the fault magnitude. For instance, complete failure of sensor 1 results in $a = -10$, hence the use of information from residuals r_1 and r_2. If no failure is reported, then $a = +10$, resulting in $g = 0$, hence residual signals not being considered in the adjustment algorithm. Values between -10 and 10 represent faults with various magnitudes. The sufficient conditions for convergence of the estimated state are $|\dot{\lambda}_1| < \alpha$ and $\lambda_1 + \lambda_2 \leq 1$. The parameter $\alpha > 0$ and is a trade-off between stability and FDI performance. Large α increases the response of the FDI unit to faults, while small α results in a slowly varying condition which guarantees system stability.

3.5. *FTC strategy*

When a fault is detected, the state variables are reconstructed accordingly and the feedback controller is redesigned as follows:

$$\delta(t) = -K_f\hat{x}(t), \tag{25}$$

where K_f is the feedback controller gain when the lateral control system enters into a degraded mode, that is, when the information is lost in one of the lateral deviation sensors. A fault in any of the sensors results in a change in the output measurements and the state variables are defined as follows:

$$\dot{x}(t) = Ax(t) + B_1\delta(t) + B_2\dot{r}_{\mathrm{des}}, \tag{26}$$

$$y(t) = \bar{C}x(t) + Ff, \tag{27}$$

where A is the state matrix, B_1 is the control matrix, and $\bar{C}$ is the output matrix of the faulty system; $\dot{r}_{\mathrm{des}}$ is the desired yaw rate, f is the additive fault signal, and F represents sensor faults. Now an optimal estimator can easily be designed by solving the relevant Riccati equation associated with the system given by Equations (25) and (26). Assuming the system is observable, the state estimates $\hat{x}$ are defined as follows:

$$\dot{\hat{x}}(t) = (A - L\bar{C})\hat{x}(t) + B_1\delta(t) + Ly(t), \tag{28}$$

where L is the observer gain which is defined by $L = Y\bar{C}B_1^{-1}$ and Y is the positive semi-definite stabilizing solution of the following algebraic Riccati equation:

$$(A - B_1\bar{C})Y + Y(A - \bar{C})^{\mathrm{T}} - Y\bar{C}Y + B_1B_1{}^{\mathrm{T}} = 0. \tag{29}$$

In order to guarantee the stability of the proposed observers, define the estimation error $\varepsilon(t)$ as follows:

$$\varepsilon(t) = x(t) - \hat{x}(t). \tag{30}$$

The error dynamics are stable if and only if the matrix $U = \left[\begin{smallmatrix} A-LC & K \\ -C & 1 \end{smallmatrix}\right]$ is Hurwitz stable.

Note that the matrix U can be written as follows:

$$U = \begin{bmatrix} A & B \\ 0 & 1 \end{bmatrix} - \begin{bmatrix} L \\ 1 \end{bmatrix}[C \quad 0]. \tag{31}$$

Therefore, the poles of the matrix U can be arbitrarily assigned, provided that $\left(\left[\begin{smallmatrix} A & B \\ 0 & 1 \end{smallmatrix}\right], [C \quad 0]\right)$ is observable. Hence, stability of the proposed observer is guaranteed by the proper choice of observer gain L, which is selected in a way such that the matrix U is Hurwitz stable.

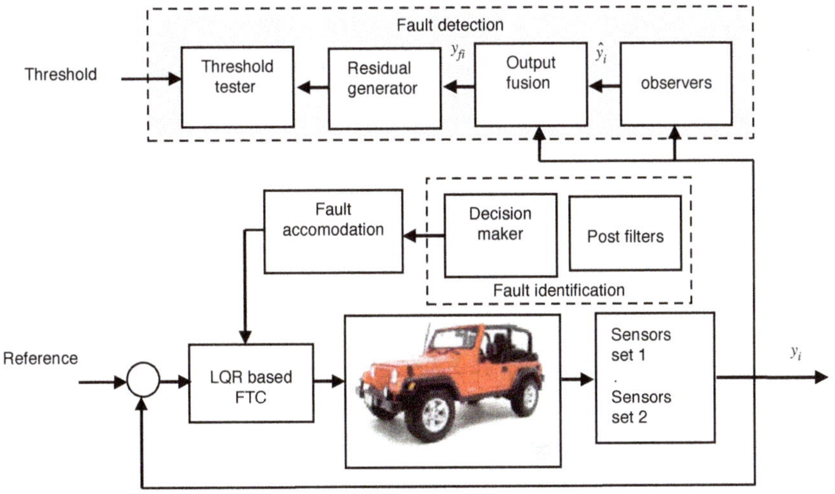

Figure 2. Schematic diagram of the proposed FTC framework.

Define the objective cost function to be minimized by the controller as follows:

$$J = \int_0^\infty \hat{x}^{\mathrm{T}}(t)Q\hat{x}(t) + \delta^{\mathrm{T}}(t)R\delta(t), \qquad (32)$$

where P satisfies the following Riccati equation:

$$P = A^{\mathrm{T}}PA - A^{\mathrm{T}}PB_1(R + B_1^{\mathrm{T}}PB_1)^{-1}B_1^{\mathrm{T}}PA + Q. \qquad (33)$$

Here, Q is a diagonal weighting matrix with an entry for each state corresponding to the performance aspects contributing to the cost function and R is the weighting value corresponding to the control effort contributing to the cost function.

The proposed FTC framework is illustrated in Figure 2.

The approach uses two observers driven by sensor outputs. Following FD, faulty sensors are identified. The state variables are then constructed from the output of the healthy sensor and the controller is updated accordingly to ensure proper steering and maintain system stability when faults occur.

4. Application of the FTC framework to an automotive steering system

To validate the performances of the proposed control paradigm, we provide a series of computer experiments using various paths and fault scenarios. For accurate evaluation, the proposed controller is implemented using CarSim (Mechanical Corporation). The latter is used in the automotive industry as the standard by which vehicle handling and dynamics are tested. It provides a high-fidelity and complete model of the vehicle and its environment. The performance of the proposed control paradigm is compared to that of the CarSim steering controller, which details are illustrated in the Appendix. Two driving maneuvers are chosen to perform the various experiments as detailed in the following.

4.1. *Lane change maneuver*

Lane change maneuver is a common test for vehicle handling as it represents an essential collision avoidance maneuver. A lane change path is chosen to demonstrate the tracking capability on a straight path as well as the response to a quick, yet continuous transient section (position and curvature). Experiments on this path are performed at a constant longitudinal speed of 30 m/s (108 km/h) and considering a road adhesion factor of one.

A double lane change path is selected to illustrate the tracking capability as well as the steering control of the vehicle on a straight path. Figures 3 and 4 show the vehicle following a lane change maneuver. Computer experiments were first carried out when 90% of the information from the sensor is lost and the front sensor has failed at $t = 4$ sec. For comparison purposes, simulations were carried out with and without FTC. The fault considered here is an abrupt change in the front sensor.

Figures 5 and 6 depict the responses of the vehicle at a longitudinal speed of 40 m/s (144 km/h) and considering a road adhesion factor of 0.75 when 90% of the sensor information is lost ($a = 3$). The observer gains in this case were $L_1 = 10$ and $L_2 = 20$, while the controller gain was $K = 80$. Note a reduction in the friction between the road and the tire of the vehicle in this case. We can observe from the figures that when the FTC approach is considered, the vehicle is able to recover from the fault fairly quickly and follow the prescribed path without much delay. This is important since lane change control operation is high bandwidth in nature and cannot tolerate significant delays in the

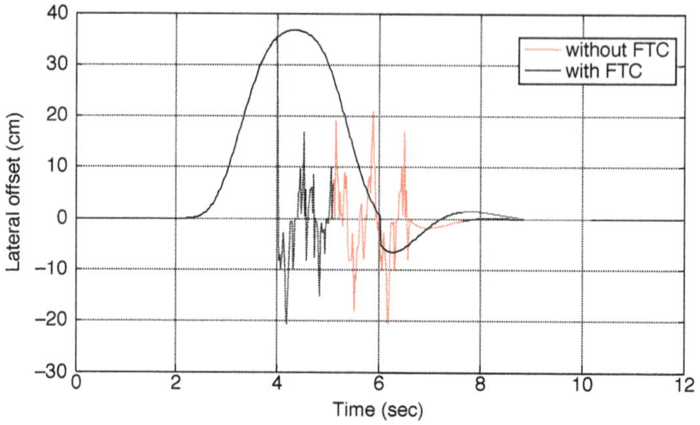

Figure 3. Lateral offset with 90% loss at $v = 30$ m/s and $u = 1$.

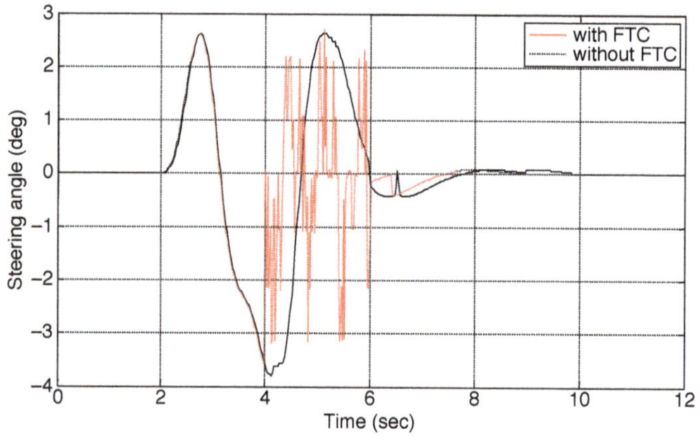

Figure 4. Steering angle with 90% loss at $v = 30$ m/s and $u = 1$.

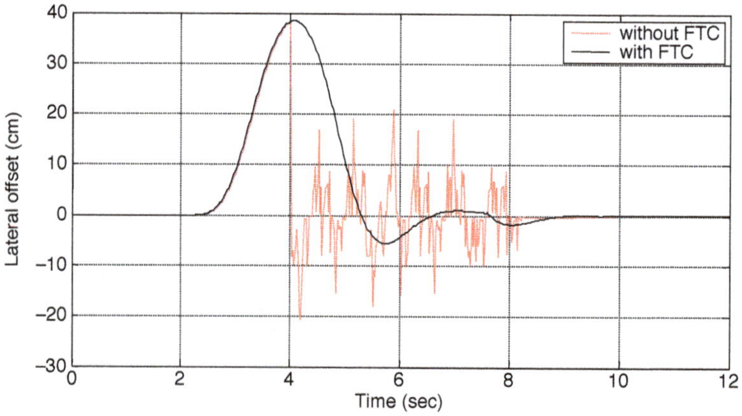

Figure 5. Steering angle with 90% loss at $v = 40$ m/s and $u = 0.75$.

control loop. Therefore, quick accommodation of faults in the lane-keeping control system is an important issue. Note though that due to the increase in speed and decrease in the road adhesion factor, the vehicle takes more time to follow the path. In contrast, the vehicle is not able to track the set path after the fault occurrence in the case without FTC.

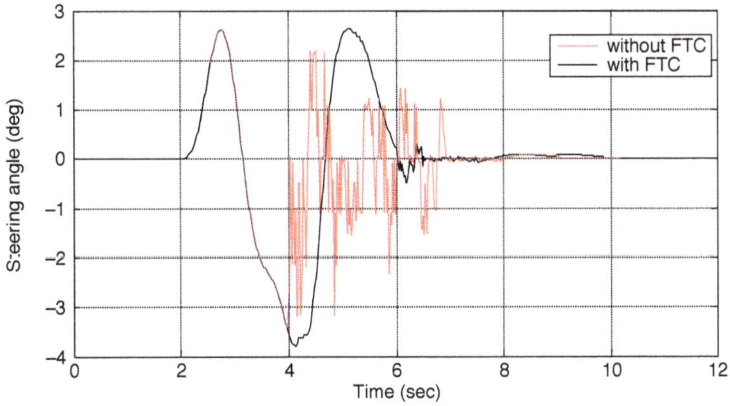

Figure 6. Steering angle with 90% loss at $v = 40$ m/s and $u = 0.75$.

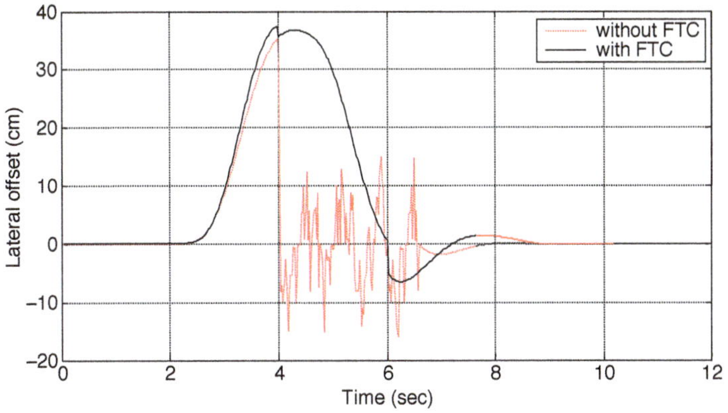

Figure 7. Lateral offset with 70% sensor information loss at $v = 40$ m/s and $u = 0.75$.

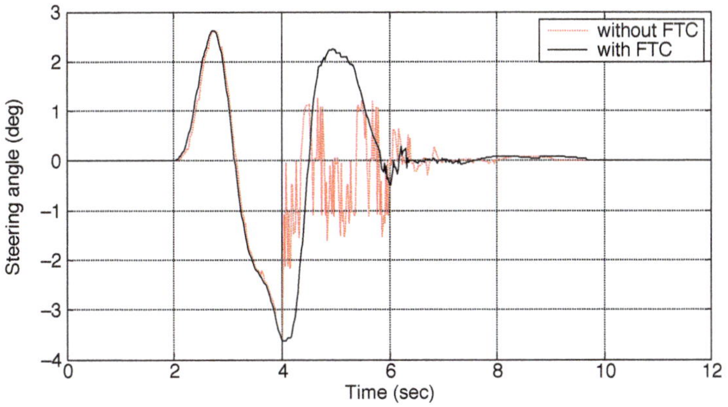

Figure 8. Steering angle at 70% sensor information loss at $v = 40$ m/s and $u = 0.75$.

Next, a 70% sensor information loss is considered ($a = 1$). The observer gains in this case were $L_1 = 8$ and $L_2 = 10$, while the controller gain was $K = 50$. Figures 7 and 8 illustrate the response of the vehicle with a longitudinal speed of 40 m/s (108 km/h) and considering a road adhesion factor of one.

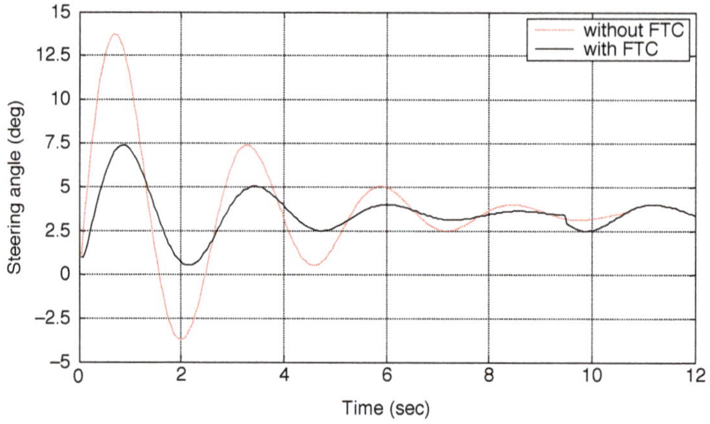

Figure 9. Lateral offset on a circular track with $v = 30$ m/s and $u = 1$.

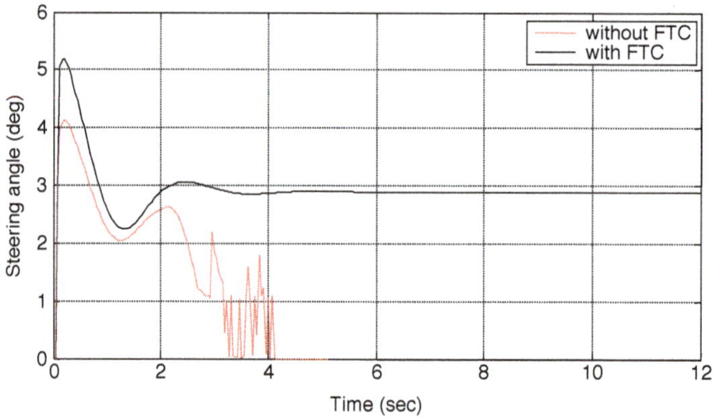

Figure 10. Steering angle on a circular track with $v = 30$ m/s and $u = 1$.

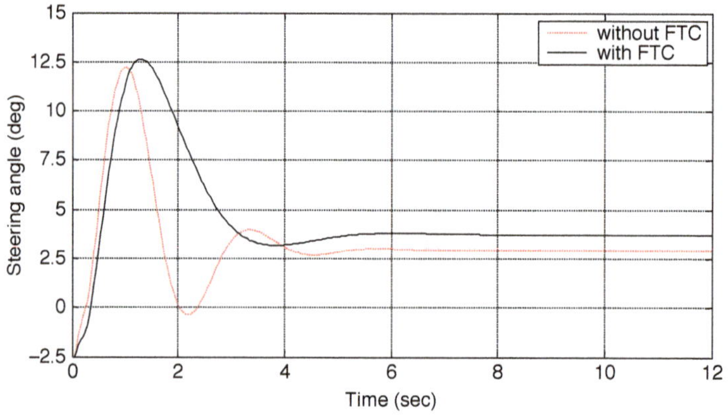

Figure 11. Lateral offset at 70% sensor information loss with $v = 30$ m/s and $u = 1$.

In this case, as expected fewer disturbances can be observed in the presence of faults when compared to the case with 90% information loss.

4.2. *Circular track*

In this section, we consider a circular track of 500 ft in order to show the performance of the observer-based FTC

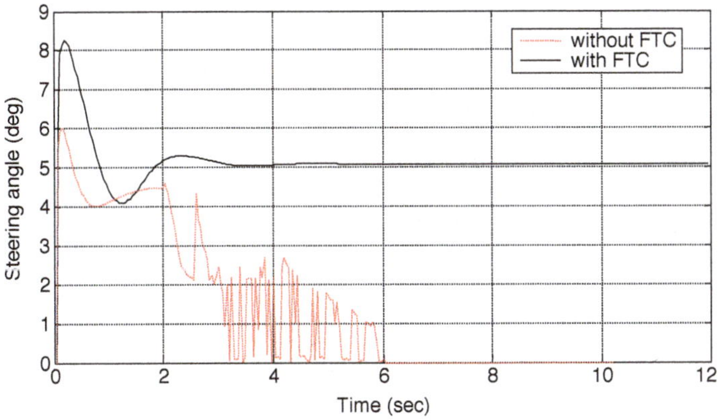

Figure 12. Steering angle at 70% sensor information loss with $v = 30$ m/s and $u = 1$.

algorithm. This path can provide valuable insight into the handling of a vehicle as well as some important characteristics of the proposed controller. Hence, this path was chosen to illustrate the steady-state characteristics of the controller while executing a constant nonzero curvature path. Experiments on this path are performed at a constant longitudinal speed of 30 m/s (108 km/h) and considering a road adhesion factor of one. Figures 9 and 10 illustrate the response of the vehicle when considering 90% of sensor information loss.

Note that information from the sensor is lost after 2 s and the vehicle requires more time to follow the path, as shown in Figure 9. The steering controller is stabilized after 4 s and system stability is maintained as shown in Figure 10. Next, the simulation was carried out with 70% sensor information loss. Figures 11 and 12 illustrate the lateral offset and steering angle in such case.

The simulation results show the effectiveness of the proposed FTC algorithm in improving the vehicle response under different paths and using various driving scenarios under several faulty conditions. Note that compared to the case without an FTC algorithm, the system is not able to recover from the sensor fault and the controller loses its steering capabilities shortly after the occurrence of the sensor fault. This is more prominent when the vehicle is following a circular path.

5. Conclusion

This paper presented an effective FTC paradigm that integrates the optimal properties of the LQR framework with an observer-based FD scheme to maintain vehicle stability and ensure handling in the presence of faults. A weight adjustment algorithm is incorporated in the FDI unit to ensure robust performance in the presence of parameter variations and disturbances. For accurate validation, the control routines were implemented in MATLAB/Simulink environment and tested using CarSim, a high-fidelity vehicle simulator. The results confirmed the ability of the proposed FTC framework to effectively monitor the system and ensure correct tracking performance under faulty conditions. Future work will focus on the integration of the proposed methodology with environment information devices such as radars and vision systems.

Funding

This work is partially supported by the Louisiana Board of Regents Support Fund contract numbers LEQSF (2012-15)-RD-A-26 and LEQSF-EPS (2015)-PFUND-421 and by LaSPACE/NASA [grant number NNX10AI40H] sub awards No. 84415 and No. 89632.

Disclosure statement

No potential conflict of interest was reported by the authors.

References

Arogeti, S., Wang, D., Low, C., & Yu, M. (2008, January). Fault detection isolation and estimation in a vehicle steering system. *IEEE Transaction in Industrial Electronics*, *59*(12), 1–2.

Blanke, M., & Staroswiecki, M. (2006). *Diagnosis and fault-tolerant control* (2nd ed.). Berlin: Springer-Verlag.

Chen, H., Song, Y., & Li, D. (2011). *Fault-tolerant tracking control of FW-steering autonomous vehicles*. Proceedings of the Chinese Control and Decision Conference CCDC, Mianyang, June 25–27, 2008, pp. 92–97.

Cheng, H. (2011). *Autonomous intelligent vehicles. Theory, algorithms and implementation*. London: Springer-Verlag.

Dong, J., Verhaegen, M., & Holweg, E. (2008, July 6–11). *Closed-loop subspace predictive control for fault tolerant MPS design*. Proceedings of the 17th IFAC World Congress, Seoul, pp. 3216–3221.

Fekih, A. (2014, June). *Fault diagnosis and fault tolerant control design for aerospace systems: A bibliographical review*. Proceedings of the 2014 American Control Conference, Portland, OR, pp. 1286–1291.

Fekih, A., & Deveriste, D. (2013, June 17–19). *A Fault-Tolerant Steering Control Design for Automatic Path Tracking in Autonomous Vehicles*. Proceedings of the American Control Conference, Washington, DC.

Fekih, A., & Seelem, S. (2012, October 3–5). *A fault tolerant control design for automatic steering control of ground vehicles*. Proceedings of IEEE Multi-Conference on Systems and Controls, Croatia, Dubrovnic, pp. 1491–1496.

Gustaffson, F. (2000). *Adaptive filtering and change detection*. Chichester: John Wiley & Sons.

Herpin, J., Fekih, A., Golconda, S., & Lakhotia, A. (2007, December). Steering control of the autonomous vehicle: CajunBot. *AIAA Journal of Aerospace Computing, Information, and Communication, 4*, 1134–1142.

Hsiao, T., & Tomizuka, M. (2005, June 8–10). *Threshold selection for timely fault detection in feedback control systems*. Proc. of the American Control Conference, Vol. 5, Portland, pp. 3303–3308.

Jiang, J., & Yu, X. (2012, April). Fault tolerant control systems: A comparative study between active and passive approaches. *Annual Reviews in Control, 36*(1), 60–72.

Laureiro, R., Benmoussa, S., Touati, Y., Merzouki, R., & Ould-Bouamama, B. (2014, January). Integration of fault diagnosis and fault-tolerant control for health monitoring of a class of MIMO intelligent autonomous vehicles. *IEEE Transaction on Vehicular Technology, 63*(1), 30–39.

Mechanical Corporation. Retrieved from www.carsim.com

Morteza, M., & Fekih, A. (2014a). A stability guaranteed robust fault tolerant control design for vehicle suspension systems subject to actuator faults and disturbances. *IEEE Transactions on Control Systems Technology, PP*(99), 1–10.

Morteza, M., & Fekih, A. (2014b, March). Adaptive PID-sliding mode fault tolerant control approach for vehicle suspension systems subject to actuator faults. *IEEE Transactions on Vehicular Technology, 63*(3), 1041–1054.

Noura, H., Theilliol, D., Ponsart, J.-C., & Noura, A. (2009). *Fault-tolerant control systems. Design and practical applications*. London: Springer-Verlag.

Tabbache, B., Benbouzid, M., Kheloui, A., & Bourgeot, J. (2011, June 27–30). *DSP-based sensor fault-tolerant control of electric vehicle power trains*. Proceedings of IEEE International Symposium on Industrial Electronics, Gdansk, pp. 2085–2090.

Wang, R., & Wang, J. (2013, March). Passive actuator fault-tolerant control for a class of over-actuated nonlinear systems and applications to electric vehicles. *IEEE Transaction on Vehicular Technology, 62*(3), 972–985.

Yang, H., Cocquempot, V., & Jiang, B. (2008, June 25–27). *Hybrid fault tolerant tracking control design for electric vehicles*. Proceedings of the Mediterranean Conference on Control and Automation, Ajaccio, pp. 1210–1215.

Zhang, M., Ding, S., Lam, J., & Wang, H. (2003, March). An LMI approach to design robust fault detection filter for uncertain LTI systems. *Automatica, 39*(3), 543–550.

Appendix

A.1. CarSim steering controller

The theory and the algorithm used to steer a road vehicle in CarSim environment are described in this section. The algorithm is intended to provide optimal control for a continuous linear system:

$$\dot{x} = Ax + Bu + Hv \tag{A1}$$

$$y = Cx + Du + Ev. \tag{A2}$$

where x is an array of n state variable, u is control input, y is an output, and A, B, C, D, E, and H are matrices with constant coefficients. The control objective is to determine the value of u to predict output to $y(t)$ to match a target $y_{\text{target}}(t)$ over some preview time t.

The system has initial conditions x_0 at time $t = 0$, a constant input u, and a constant disturbance v, then the time response is defined by

$$x(t) = e^{At}x_0 + \int_0^t e^{A\eta}Bud\eta + \int_0^t e^{A\eta}Hvd\eta. \tag{A3}$$

The term $e^{A\eta}$ is an $(n \times n)$ matrix called the state transition matrix. Each coefficient in the matrix is the portion of state variable i at time t that is linearly related to the state variable j at time zero. The two integrals in Equation (A3) define the forced responses to each state variable due to constant control u and disturbance v over the time interval.

Combining Equations (A2) and (A3), we obtain the following output response:

$$y(t) = Cx = Ce^{At}x_0 + C\left[\int_0^t e^{A\eta}\mathrm{d}\eta\right][Bu + Hv]. \tag{A4}$$

A control response scalar $g(t)$ and a disturbance response scalar $h(t)$ are defined to relate the responses over the time t. A free response array F is defined and relates the state variables at time 0 to the resulting output variable y at time t.

$$F(t) = \mathrm{Ce}^{\mathrm{At}} \tag{A5}$$

The response equation is defined by

$$y(t) = F(t)x_0 + g(t)u + h(t)v \tag{A6}$$

To determine the optimal control, a quadratic performance index J is defined by

$$J = \frac{1}{T}\int_0^T \{y_{\text{target}}(t) - y(t)\}^2 W(t)\mathrm{d}t. \tag{A7}$$

Here, $W(t)$ is an arbitrary weighting function. A optimal control law is designed by minimizing the cost function J, representing the squared deviation of response variable $y(t)$ relative to the target function $y_{\text{target}}(t)$. The control function u minimizing J can be found by substituting Equation (A6) into (A7) and taking a partial derivative of J with respect to u.

$$J = \frac{1}{T}\int_0^T (F(t)x_0 + g(t)u + h(t)v - y_{\text{target}}(t))^2 W(t)\mathrm{d}t. \tag{A8}$$

Solving for u results in

$$u = \frac{\int_0^T \{y_{\text{target}}(t) - F(t)x_0 - h(t)v\}g(t)W(t)\mathrm{d}t}{\int_0^T g(t)^2 W(t)\mathrm{d}t}. \tag{A9}$$

In practice, the integrals over T can be replaced with finite summations

$$u = \frac{\sum_{i=1}^{m} (y_{\text{target}}(t) - F_i x_0 - h_i v)g_i W_i}{\sum_{i=1}^{m} g_i^2 W_i}. \tag{A10}$$

The algorithm is programmed to generate a steering wheel angle in the vehicle solver program for a given target path. The algorithm synthesizes the target path over the preview time and calculates the optimal front steering effort u to minimize deviations from the path. It also delays the driver steering control by a constant time. The geometry of the road is given as a sequence of X and Y coordinates that define a reference line. Station S is defined as the distance along the reference line, typically a road centerline. For each pair of X–Y coordinates, a corresponding increment of S is computed by

$$S_i = S_{i-1} + \sqrt{(X_i - X_{i-1})^2 + (Y_i - Y_{i-1})^2}. \quad \text{(A11)}$$

To calculate the optimal steering control using Equation (A10), the target position is needed at each point considered in the summation. The station target location is

$$S_{\text{targ},i} = S + \frac{iV_x T}{m}, \quad \text{(A12)}$$

where V_x is the vehicle forward speed.

The controller calculations are performed using an axis system where the vehicle is located such that the center of the vehicle front axle is at $X = 0$ and $Y = 0$ and the X and Y axes are aligned with the longitudinal and lateral axes of the vehicle. The target lateral translation is calculated by first getting the inertial X and Y coordinates of the path as a function of the station at the target location (S_{targ}):

$$Y_{\text{target}} = [Y(S_{\text{targ}}) - Y_V]\cos(\psi) - [X(S_{\text{targ}}) - X_V]\sin(\psi). \quad \text{(A13)}$$

A.2. *Vehicle parameters*

Table A1. Vehicle parameters.

M	Vehicle mass, 1573 kg
c_f, c_r	Cornering stiffness of front/rear wheels 2*60,000 N/rad
l_f	Distance between the front wheels and the center of gravity, 1.137 m
l_r	Distance between the rear wheels and the center of gravity, 1.530 m
I_z	Yaw moment of inertia, 2753 kg m^2

15

Maneuver-based control of the 2-degrees of freedom underactuated manipulator in normal form coordinates

Carsten Knoll* and Klaus Röbenack

Faculty of Electrical and Computer Engineering, Institute of Control Theory, Technische Universität Dresden, Dresden, Germany

In this contribution, we provide a constructive way to transform a generic Lagrangian mechanical control system into the well-known Byrnes–Isidori normal form. Then, we restrict ourselves to the underactuated two-joint-manipulator in the horizontal plane. That system fails Brocketts condition and thus achieving point-to-point control is a challenging task. The system is analyzed in normal form coordinates, which allows to design a subordinate sliding mode controller and identify three discrete symmetries. Using these preliminary steps, finally a maneuver-based control scheme is proposed for equilibrium transition. Thereby each maneuver corresponds to a suitable sliding surface defined in normal form coordinates.

Keywords: underactuated manipulator; equilibrium transition; Lagrangian Byrnes–Isidori normalform; discrete symmetry; sliding mode control; maneuver-based control

1. Introduction

Underactuated mechanical systems are an interesting and illustrative field of control theory (Fantoni & Lozano, 2001; Spong, 2000). Many different control approaches have been applied to such systems, see, for example, Graichen, Treuer, and Zeitz (2007), Glück, Eder, and Kugi (2013), Riachy, Orlov, Floquet, Santiesteban, and Richard (2008), and Knoll and Röbenack (2013). Among them, underactuated manipulators (possessing some unactuated joints) are an important subclass, see De Luca, Iannitti, Mattone, and Oriolo (2002) and the references therein. Possible applications include areas where the reduction in weight and/or cost is desired while tasks can still be fulfilled with fewer actuators, like in space robotics (Yesiloglu & Temeltas, 2010; Yoshida, 1997) or when considering simple pick-and-place mechanisms (Oriolo & Nakamura, 1991).

Depending on how many joints lack an actuator, where they are located and which further assumptions are made, for example, regarding friction and potential forces, the control of underactuated manipulators is more or less difficult. In some cases, for example, the one discussed here, the system is not controllable, while in other cases even the property of linearizability via static feedback is fulfilled (Franch, Reyes, & Agrawal, 2013), allowing relatively easy controller design.

In this paper, we consider a 2-degrees of freedom (DOF) manipulator in the horizontal plane with an active joint at the base and a passive joint between the links. Motivated by pick-and-place tasks, our objective is the equilibrium transition of that system.

Due to Oriolo and Nakamura (1991) it is well-known that this manipulator does not fulfill the so-called Brockett condition (Brockett, 1983), which is necessary for the existence of a continuous differentiable feedback that stabilizes an isolated equilibrium point.

Because of the combination of an easy model with challenging control properties this system was subject to many studies before. Some of them, e.g. Arai and Tachi (1991) and Mareczek, Buss, and Schmidt (1999), rely on the presence of a holding brake which, however, might be seen as some kind of actuator. Other approaches assume the presence of considerable static friction, either explicitly (Mahindrakar, Rao, & Banavar, 2006), or implicitly (Scherm & Heimann, 2000) by means of the numerical stability of the algorithm. Like in the present contribution, the model investigated in Nakamura, Suzuki, and Koinuma (1997) and De Luca, Mattone, and Oriolo (2000) is assumed to be frictionless. The two approaches both use periodic input signals leading to high actuator activity, a longer duration and path length of the transition in comparison to the presence of a brake or static friction.

This paper contains two main contributions: Firstly, basing on the partial feedback linearization of a general underactuated system, we introduce its transformation to the "Lagrangian Byrnes–Isidori normal form (LBINF)" and give a simple proof of its global existence in terms of a closed-form formula. This representation allows for

*Corresponding author. Email: carsten.knoll@tu-dresden.de

any underactuated system to separate the effect of the input from system-inherent dynamics ("drift"), such that the corresponding vector fields are always orthogonal. In Olfati-Saber (2001, Section 3.7) a related but different approach has been proposed earlier, to transform an underactuated system to Byrnes–Isidori normal form, see also Choukchou-Braham, Cherki, Djemai, and Busawon (2014, Section 4.2). However, in that approach the existence of the transformation depends on an involutivity condition and the calculation depends on the solution of partial differential equations. Therefore, in general it is not possible to express that transformation in closed form, even if it exists. Furthermore, due to the simple structure of our transformation, some of the new state variables can be interpreted as quasi-velocities, see, for example, Cameron and Book (1997). This facilitates the controller design in the following.

The second contribution is, as an extension to Knoll and Röbenack (2010, 2011a), the presentation of a novel approach for the point-to-point control of the underactuated planar 2-DOF manipulator. In contrast to the existing approaches, it does neither rely on a brake or static friction nor on the application of periodic inputs.

This paper is organized as follows: In Section 2, we establish the transformation of a general underactuated mechanical system to the LBINF. Next, in Section 3, the model of the 2-DOF manipulator is stated and transformed into that representation. Section 4 is devoted to the analysis of the manipulator system in the new coordinates. Thereby, a generic subordinate sliding mode control is constructed and three discrete symmetry-mappings are identified, which substantially simplify the control design. Finally, in Section 5, the equilibrium transition is performed via successive execution of suitable maneuvers, each one corresponding to a special sliding surface defined in coordinates of the normal form.

The application of the control strategy proposed in Section 5 depends on a number of details not worthwhile to be included in the written text. Each single detail might be considered as a straightforward consequence of the outlined ideas. However, to facilitate the ability to reproduce and critically examine our results we provide the full source code (using the Python programming language) for the related simulations, see Knoll (2014a). This practice follows the argumentation of Ince, Hatton, and Graham-Cumming (2012).

2. Input–output linearization and Byrnes–Isidori normal form

2.1. Input–output linearization of underactuated systems

In this section, we consider a holonomic[1] mechanical system with n DOF. Its equations of motion, typically obtained from the Lagrangian formalism, read

$$\mathbf{M}(\mathbf{q})\ddot{\mathbf{q}} + \mathbf{C}(\mathbf{q},\dot{\mathbf{q}}) + \mathbf{K}(\mathbf{q},\dot{\mathbf{q}}) = \mathbf{B}(\mathbf{q})\tau. \tag{1}$$

Thereby, $\mathbf{q}$, $\dot{\mathbf{q}}$ and $\ddot{\mathbf{q}}$ denote the generalized coordinates, velocities and accelerations, respectively, $\mathbf{M}(\mathbf{q})$ is the (positive definite) mass matrix, $\mathbf{C}(\mathbf{q},\dot{\mathbf{q}})$ describes the action of centrifugal and Coriolis forces and $\mathbf{K}$ collects the conservative and dissipative forces resulting from potential energy changes and friction. The m-dimensional vector τ contains the generalized forces from the actuators which are assigned to the scalar differential equations of (1) via the $n \times m$-matrix $\mathbf{B}(\mathbf{q})$. Obviously, the quantities $\mathbf{q}$, $\dot{\mathbf{q}}$ and $\ddot{\mathbf{q}}$ are time-dependent elements of $\mathbb{R}^n$. However for better readability the time argument is not explicitly written.

We assume that the system is underactuated, that is, $m < n$, and that the actuators are not redundant. Then, $\mathbf{B}(\mathbf{q})$ has full column rank for all $\mathbf{q} \in \mathbb{R}^m$ which implies the existence of $\mathbf{T}(\mathbf{q}) \in \mathbb{R}^{n\times n}$, such that $\mathbf{T}(\mathbf{q})\mathbf{B}(\mathbf{q}) = (\mathbf{I}_m, \mathbf{0})^{\mathrm{T}}$. By left-multiplying Equation (1) with $\mathbf{T}(\mathbf{q})$ and performing a suitable transformation of the coordinates, the system (1) can be rewritten as

$$\begin{pmatrix} \mathbf{M}_{11}(\mathbf{q}) & \mathbf{M}_{12}(\mathbf{q}) \\ \mathbf{M}_{12}^{\mathrm{T}}(\mathbf{q}) & \mathbf{M}_{22}(\mathbf{q}) \end{pmatrix} \begin{pmatrix} \ddot{\mathbf{q}}_1 \\ \ddot{\mathbf{q}}_2 \end{pmatrix} + \begin{pmatrix} \mathbf{C}_1(\mathbf{q},\dot{\mathbf{q}}) \\ \mathbf{C}_2(\mathbf{q},\dot{\mathbf{q}}) \end{pmatrix} + \begin{pmatrix} \mathbf{K}_1(\mathbf{q},\dot{\mathbf{q}}) \\ \mathbf{K}_2(\mathbf{q},\dot{\mathbf{q}}) \end{pmatrix} = \begin{pmatrix} \tau_1 \\ \mathbf{0} \end{pmatrix}. \tag{2}$$

For convenience, we also denote the new generalized coordinates $\mathbf{q} = (\mathbf{q}_1^{\mathrm{T}}, \mathbf{q}_2^{\mathrm{T}})^{\mathrm{T}}$. In this representation, the whole system splits up into an m-dimensional fully actuated subsystem with joint coordinates $\mathbf{q}_1$ and an $(n-m)$-dimensional subsystem without direct actuation.

For facilitation of the analysis and the controller design often an input–output linearization (also called "partial feedback linearization") of Equation (2) is performed (see, e.g. De Luca et al., 2002; Sastry, 1999) by applying the nonlinear static feedback

$$\begin{aligned} \tau_1 &= [\mathbf{M}_{11}(\mathbf{q}) - \mathbf{M}_{12}(\mathbf{q})\mathbf{M}_{22}^{-1}(\mathbf{q})\mathbf{M}_{12}^{\mathrm{T}}(\mathbf{q})]\mathbf{a} \\ &\quad - \mathbf{M}_{12}(\mathbf{q})\mathbf{M}_{22}^{-1}(\mathbf{q})(\mathbf{C}_2(\mathbf{q},\dot{\mathbf{q}}) + \mathbf{K}_2(\mathbf{q},\dot{\mathbf{q}})) \\ &\quad + \mathbf{C}_1(\mathbf{q},\dot{\mathbf{q}}) + \mathbf{K}_1(\mathbf{q},\dot{\mathbf{q}}). \end{aligned} \tag{3}$$

The resulting partial linearized system reads

$$\ddot{\mathbf{q}}_1 = \mathbf{a}, \tag{4a}$$

$$\ddot{\mathbf{q}}_2 = -\mathbf{M}_{22}^{-1}(\mathbf{q})(\mathbf{C}_2(\mathbf{q},\dot{\mathbf{q}}) + \mathbf{K}_2(\mathbf{q},\dot{\mathbf{q}}) + \mathbf{M}_{12}^{\mathrm{T}}(\mathbf{q})\mathbf{a}). \tag{4b}$$

According to Equation (4a), the new input $\boldsymbol{a}$ corresponds to the generalized accelerations of the actuated coordinates. In other words, the feedback (3) can be interpreted as inner control loop for the actuated joints.

In control theory, the input affine state-space representation of a system

$$\dot{\mathbf{x}} = \mathbf{f}(\mathbf{x}) + \mathbf{g}(\mathbf{x})\mathbf{u} \tag{5}$$

is very common. In this representation, $\mathbf{x}$ denotes the time-dependent state taking values in $\mathbb{R}^N$ (or an open subset thereof), $\mathbf{f}$ is the so-called drift vector field and $\mathbf{g}$ is a state-dependent $N \times m$ matrix assigning the components of the m-dimensional input $\mathbf{u}$ to the N differential equations in Equation (5). Clearly, by choosing the state components

$$\mathbf{x} = (\mathbf{q}_1^{\mathrm{T}}, \dot{\mathbf{q}}_1^{\mathrm{T}}, \mathbf{q}_2^{\mathrm{T}}, \dot{\mathbf{q}}_2^{\mathrm{T}})^{\mathrm{T}} =: (\mathbf{x}_1^{\mathrm{T}}, \mathbf{x}_2^{\mathrm{T}}, \mathbf{x}_3^{\mathrm{T}}, \mathbf{x}_4^{\mathrm{T}})^{\mathrm{T}} \tag{6}$$

and introducing n definitional equations of the form $\dot{x}_j = x_k$ the system of n second-order ODEs (4) can be rewritten in the form Equation (5) with $N = 2n$ and $\mathbf{u} = \boldsymbol{a}$:

$$\dot{\mathbf{x}}_1 = \mathbf{x}_2, \tag{7a}$$

$$\dot{\mathbf{x}}_2 = \mathbf{a}, \tag{7b}$$

$$\dot{\mathbf{x}}_3 = \mathbf{x}_4, \tag{7c}$$

$$\dot{\mathbf{x}}_4 = -\bar{\mathbf{M}}_{22}^{-1}(\mathbf{x}_1, \mathbf{x}_3)(\bar{\mathbf{C}}_2(\mathbf{x}) + \bar{\mathbf{K}}_2(\mathbf{x}) + \bar{\mathbf{M}}_{12}^{\mathrm{T}}(\mathbf{x}_1, \mathbf{x}_3)\mathbf{a}). \tag{7d}$$

The bar over the quantities indicates the reordering of the arguments to take the definition of the state (6) into account.

2.2. *Byrnes–Isidori normal form*

Equation (7d) shows that except in the special case where $\bar{\mathbf{M}}_{12}(\mathbf{x}_1, \mathbf{x}_3) \equiv 0$ the accelerations of the non-actuated joints are also influenced by $\boldsymbol{a}$. In terms of the input–output-normal form Equations (7c), (7d) represent the internal dynamics. As we will see in Sections 4.1 and 5 system analysis and control design are facilitated by the introduction of new coordinates, say $\mathbf{z} := (\mathbf{z}_1, \mathbf{z}_2, \mathbf{z}_3, \mathbf{z}_4)$, along with a state transformation which alters the internal dynamics such that it is not directly influenced by the input. This corresponds to a transformation to the Byrnes–Isidori normal form (Isidori, 1995), which in general refers to a decomposition of the system into mere integrator chains and an internal dynamics which does not depend on the input. In the scope of underactuated mechanical systems, we can restrict the generality and formulate an adapted definition:

DEFINITION 1 *A dynamical system of the form* (5) *is said to be in LBINF, if* $\mathbf{f}$ *and* $\mathbf{g}$ *are compatible with the representation*

$$\begin{pmatrix} \dot{\mathbf{z}}_1 \\ \dot{\mathbf{z}}_2 \\ \dot{\mathbf{z}}_3 \\ \dot{\mathbf{z}}_4 \end{pmatrix} = \begin{pmatrix} \mathbf{z}_2 \\ \mathbf{0} \\ \mathbf{f}_3(\mathbf{z}) \\ \mathbf{f}_4(\mathbf{z}) \end{pmatrix} + \begin{pmatrix} \mathbf{0} \\ \boldsymbol{I}_m \\ \mathbf{0} \\ \mathbf{0} \end{pmatrix} \mathbf{a}. \tag{8}$$

THEOREM 1 *Every mechanical system given in the form* (1) *can be transformed to the form* (8) *by a globally invertible change of state coordinates.*

Proof The equivalence between Equations (1) and (7) is straightforward.[2] Now, the LBINF coordinates are given by

$$\mathbf{z}_i := \mathbf{x}_i, \quad i = 1, 2, 3 \tag{9a}$$

and

$$\mathbf{z}_4 := \mathbf{x}_4 + \bar{\mathbf{M}}_{22}^{-1}(\mathbf{x}_1, \mathbf{x}_3)\bar{\mathbf{M}}_{12}^{\mathrm{T}}(\mathbf{x}_1, \mathbf{x}_3)\mathbf{x}_2. \tag{9b}$$

The first three components of the transformation are identical and the last one, given by Equation (9b) can be inverted globally as well as

$$\mathbf{x}_4 = \mathbf{z}_4 - \bar{\mathbf{M}}_{22}^{-1}(\mathbf{z}_1, \mathbf{z}_3)\bar{\mathbf{M}}_{12}^{\mathrm{T}}(\mathbf{z}_1, \mathbf{z}_3)\mathbf{z}_2. \tag{10}$$

To complete the proof, we notice that in the expression for the time derivative of $\mathbf{z}_4$ the coefficients of $\boldsymbol{a}$ nullify each other:

$$\begin{aligned} \dot{\mathbf{z}}_4 &= \underbrace{-\bar{\mathbf{M}}_{22}^{-1}(\bar{\mathbf{C}}_2 + \bar{\mathbf{K}}_2) - \bar{\mathbf{M}}_{22}^{-1}\bar{\mathbf{M}}_{12}^{\mathrm{T}}\mathbf{a}}_{\dot{\mathbf{x}}_4} \\ &\quad + \bar{\mathbf{M}}_{22}^{-1}\bar{\mathbf{M}}_{12}^{\mathrm{T}}\underbrace{\mathbf{a}}_{\dot{\mathbf{x}}_2} + \left(\frac{\mathrm{d}}{\mathrm{d}t}\left(\bar{\mathbf{M}}_{22}^{-1}\bar{\mathbf{M}}_{12}^{\mathrm{T}}\right)\right)\mathbf{z}_2 \\ &= -\bar{\mathbf{M}}_{22}^{-1}(\bar{\mathbf{C}}_2 + \bar{\mathbf{K}}_2) + \left(\frac{\mathrm{d}}{\mathrm{d}t}\left(\bar{\mathbf{M}}_{22}^{-1}\bar{\mathbf{M}}_{12}^{\mathrm{T}}\right)\right)\mathbf{z}_2. \end{aligned} \tag{11}$$

Remark 1 The vector $\mathbf{z}_4$ is a weighted sum of the actuated and non-actuated velocities and can be interpreted as a quasi-velocity (Cameron & Book, 1997). It obviously has the same physical dimension like $\mathbf{x}_4$ $(= \dot{\mathbf{q}}_2)$. In states with $\mathbf{x}_2 = \mathbf{0}$, we have $\mathbf{z}_4 = \mathbf{x}_4$, which is interesting for equilibrium transitions (see Section 4.1, Definition 2).

Remark 2 In contrast to the special Byrnes–Isidori normal form proposed earlier in Olfati-Saber (2001, Section 3.7), here no additional constraints regarding involutivity have to be fulfilled. The transformation to the normal form always exists and is given in closed form by Equation (9). Note that, for a general multi-input system (5) the Byrnes–Isidori normal form exists only if the distribution spanned by the input vector fields is involutive, cf. (Isidori, 1995, Proposition 5.1.2). However, for the mechanical system (7), the input vector fields depend only on $\mathbf{x}_1, \mathbf{x}_3$ but are zero in the first and third block (7a) and (7c), respectively. This structure implies that all Lie brackets between the input vector fields of Equation (7) are zero vector fields, that is, the associated distribution is involutive.

The LBINF representation has two main advantages: The system input $\boldsymbol{a}$ only occurs in the acceleration-equations of the actuated joints and can therefore be eliminated by interpreting the velocities as inputs themselves. This might be reasonable where a subordinate speed control is implemented in order to handle backslash and friction. Additionally, the systems motion through the state space is easier to understand and thus to control which will be illustrated in Section 4.1 and the following ones.

3. The underactuated manipulator

3.1. *System description*

From now on, we consider the underactuated two-degree-of-freedom horizontal manipulator with a friction-free non-actuated second joint, cf. Figure 1.

The configuration space for this system is considered to be $\mathbb{S}_1 \times \mathbb{S}_1$, that is, for the two angles the value of 2π is identified with the value 0. For the actual calculations this assumption does not matter, but it simplifies the transition scheme because any desired joint angle can be reached in both rotating directions.

By means of the Lagrangian formalism the equations of motions are obtained in the form of Equation (2). A subsequent partial linearization leads to the model

$$\dot{x}_1 = x_2, \tag{12a}$$

$$\dot{x}_2 = a, \tag{12b}$$

$$\dot{x}_3 = x_4, \tag{12c}$$

$$\dot{x}_4 = -\kappa x_2^2 \sin x_3 - (1 + \kappa \cos x_3)a, \tag{12d}$$

where the only remaining parameter κ is dimensionless and describes the distribution of mass on the second arm. Like many other mechanical systems the manipulator model (12) possesses time reversal symmetry (Knoll & Röbenack, 2011b) which will be used later for simplifying the trajectory planning, cf. Section 4.3.

The Jacobian linearization of this system is not controllable and, moreover, the system violates Brocketts necessary conditions (Brockett, 1983; Oriolo & Nakamura, 1991). In other words, there exists no continuous differentiable feedback of the state, that is, no continuous differentiable and static control algorithm, which stabilizes an isolated equilibrium point of the system.[3] Therefore, despite the simple model (12), the control of this system and especially equilibrium transitions are a challenging problem.

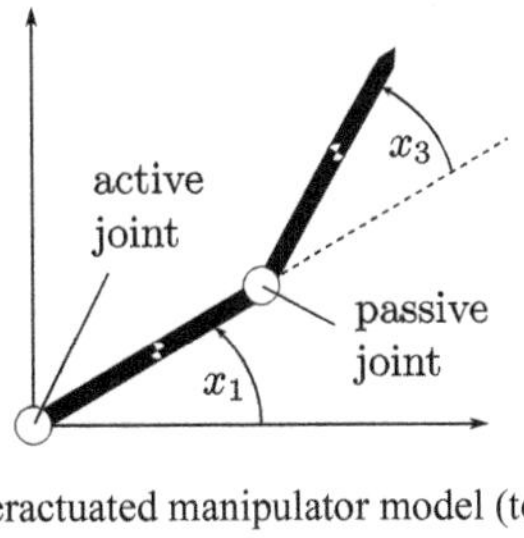

Figure 1. Underactuated manipulator model (top view).

As in the general case, discussed in the previous section, the input a affects both the double integrator subsystem for the active joint and the second subsystem for the passive joint. The transformation to the LBINF and its inverse are given by

$$z_i := x_i \quad \forall i \in \{1, 2, 3\}, \tag{13a}$$

$$z_4 := x_4 + (1 + \kappa \cos x_3)x_2 \tag{13b}$$

and

$$x_4 = z_4 - (1 + \kappa \cos z_3)z_2. \tag{13c}$$

This implies

$$\dot{z}_4 = -\kappa x_2 \sin x_3 (x_2 + x_4) \tag{14a}$$

$$= -\kappa z_2 \sin z_3 (z_4 - \kappa \cos z_3 z_2) \tag{14b}$$

and thus, the dynamics in LBINF read

$$\underbrace{\begin{pmatrix} \dot{z}_1 \\ \dot{z}_2 \\ \dot{z}_3 \\ \dot{z}_4 \end{pmatrix}}_{\dot{\mathbf{z}}} = \underbrace{\begin{pmatrix} z_2 \\ 0 \\ z_4 - (1 + \kappa \cos z_3)z_2 \\ -\kappa z_2 \sin z_3 (z_4 - \kappa \cos z_3 z_2) \end{pmatrix}}_{\mathbf{f}(\mathbf{z})} + \underbrace{\begin{pmatrix} 0 \\ 1 \\ 0 \\ 0 \end{pmatrix}}_{\mathbf{g}(\mathbf{z})} a. \tag{15}$$

4. System analysis in normal form

4.1. *Projection to the z_2–z_4-plane*

The model representation (15) has the advantage to separate the influence of the input a and the drift. In other words, the two vector fields $\mathbf{f}$ and $\mathbf{g}$ are always orthogonal to each other. Especially interesting is the projection into the two-dimensional subspace spanned by the second and fourth unit coordinate vector, that is, the z_2–z_4-plane:

$$\underbrace{\begin{pmatrix} \dot{z}_2 \\ \dot{z}_4 \end{pmatrix}}_{\dot{\bar{\mathbf{z}}}} = \underbrace{\begin{pmatrix} 0 \\ -\kappa z_2 \sin z_3 (z_4 - \kappa \cos z_3 z_2) \end{pmatrix}}_{\bar{\mathbf{f}}(\mathbf{z})} + \underbrace{\begin{pmatrix} 1 \\ 0 \end{pmatrix}}_{\bar{\mathbf{g}}(\mathbf{z})} a, \tag{16}$$

where z_3 can be interpreted as a parameter rather than a component of the state. Obviously, $\bar{\mathbf{g}}$ and $\bar{\mathbf{f}}$ are parallel to the z_2 and z_4 axes, respectively. From a topological point of view, only considering the direction of the drift vector field $\bar{\mathbf{f}}$, the projected system is similar to a double integrator, see Knoll and Röbenack (2011a) for details. In other words, except for some singular z_3-values, in one of the four quadrants the drift always points towards the abscissa (dark gray) and in another one it points away from it (light gray), cf. Figure 2. The remaining two quadrants are crossed by the straight line

$$z_4 = \kappa \cos z_3 z_2 \tag{17}$$

which, together with the z_4-axis, indicates where the sign of the drift changes.

This situation makes the system dynamics much more predictable and hence allows the construction of a maneuver-based equilibrium transition.

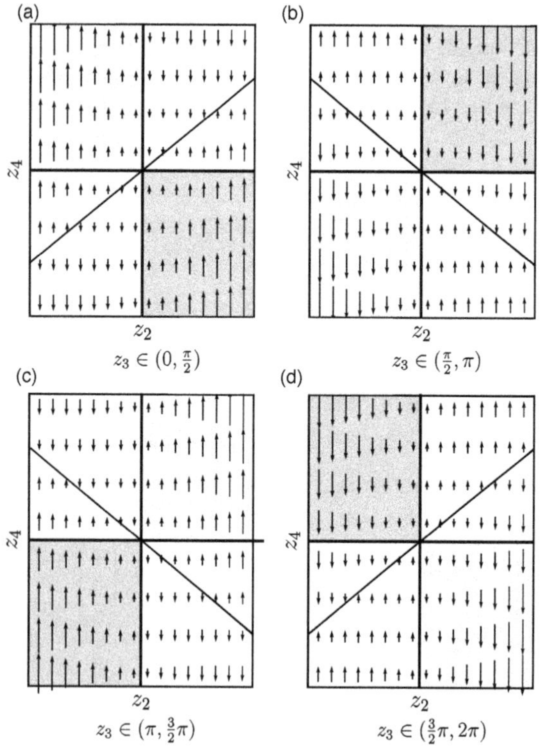

Figure 2. Drift vector field of the projected manipulator dynamics (16). For (almost) any value of z_3 in one quadrant the drift is topologically equivalent to the (negative) double integrator. In other words: the drift points towards (away from) the abcissa (dark and light shading, respectively). The diagonal line indicates where the drift changes its sign, see Equation (17).

4.2. *Sliding mode control in the z_2–z_4-plane*

These maneuvers will be performed by stabilizing the system to appropriate sliding surfaces which are constructed as follows. We consider the curve

$$z_2 = \varphi(z_4) \tag{18}$$

in the z_2–z_4-plane, where $\varphi(\cdot)$ is a differentiable function (up to a finite number of exception points where its derivative might be discontinuous). Associated to this "switching curve" is the "switching function"

$$\Phi(\mathbf{z}) := z_2 - \varphi(z_4). \tag{19}$$

Clearly, the condition $\Phi(\mathbf{z}) = 0$ constitutes a three-dimensional submanifold of the state space of system (15).

THEOREM 2 *The sliding surface given by $\Phi(\mathbf{z}) = 0$ is asymptotically stabilized by the feedback law*

$$v = \varphi'(z_4)f_4(\mathbf{z}) - \gamma_1 \operatorname{sign}(\Phi(\mathbf{z})) \mid \Phi(\mathbf{z})|^{\gamma_2} \tag{20}$$

with $\gamma_1 > 0$ and $\gamma_2 \in (0, 1)$.

Proof Attractiveness and invariance of the set $\{\mathbf{z} \in \mathbb{R}^4 \mid \Phi(\mathbf{z}) = 0\}$ is shown by the Lyapunov argument

$$\begin{aligned} \frac{1}{2}\frac{\mathrm{d}}{\mathrm{d}t}\Phi^2(\mathbf{z}) &= \Phi(\mathbf{z})\dot{\Phi}(\mathbf{z}) \\ &= \Phi(\mathbf{z})(\dot{z}_2 - \varphi'(z_4)\dot{z}_4) \\ &= \Phi(\mathbf{z})(v - \varphi'(z_4)f_4(\mathbf{z})) \\ &= -\gamma_1|\Phi(\mathbf{z})|^{1+\gamma_2} \leq 0, \end{aligned} \tag{21}$$

cf. (Slotine & Li, 1991, Chapter 7).

Remark 3 If the system is in sliding regime, that is, if $\Phi(\mathbf{z}) = 0$, then it is governed by the so-called reduced dynamics

$$\dot{z}_3 = z_4 - (1 + \kappa \cos z_3)\varphi(z_4) =: \bar{f}_3(z_3, z_4) \tag{22a}$$

$$\dot{z}_4 = \kappa\varphi(z_4) \sin z_3(\kappa\varphi(z_4) \cos z_3 - z_4) =: \bar{f}_4(z_3, z_4), \tag{22b}$$

and therefore behaves like an autonomous system. On the other hand, the particular choice of $\varphi(\cdot)$ is a remaining degree of freedom to reach the control objective of the respective maneuver.

4.3. *Symmetry properties*

In this section, we identify some properties of the system which allow the direct construction of new solution trajectories from given solution trajectories. As mentioned above, the manipulator possesses time reversal symmetry because it is a conservative mechanical system (see Knoll & Röbenack, 2011b for further explanation). Using the reduced dynamics this is expressed by the observation

$$\bar{f}_3(-z_3, z_4) = \bar{f}_3(z_3, z_4), \tag{23a}$$

$$\bar{f}_4(-z_3, z_4) = -\bar{f}_4(z_3, z_4). \tag{23b}$$

Symmetry Property 1 (time reversal symmetry). Suppose $t \mapsto (z_3(t), z_4(t))$ is a solution of Equation (22) for $t \in [0, \tau]$. Then, another solution for this interval is given by

$$\hat{z}_3(t) := -z_3(\tau - t), \tag{24a}$$

$$\hat{z}_4(t) := z_4(\tau - t). \tag{24b}$$

Proof We differentiate $(\hat{z}_3, \hat{z}_4)$ w.r.t. time, apply the symmetry (23) to change the sign of the respective first arguments and then substitute using Equation (24):

$$\dot{\hat{z}}_3(t) = \bar{f}_3(\underbrace{-z_3(\tau - t)}_{\hat{z}_3(t)}, \underbrace{z_4(\tau - t)}_{\hat{z}_4(t)}), \tag{25a}$$

$$\dot{\hat{z}}_4(t) = \bar{f}_4(\underbrace{-z_3(\tau - t)}_{\hat{z}_3(t)}, \underbrace{z_4(\tau - t)}_{\hat{z}_4(t)}). \tag{25b}$$

The result is the system dynamics of Equation (22) but with the newly constructed solution.

This property allows to consider the submaneuvers of leaving and reaching an equilibrium point as essentially the same problem. The only difference is the sign of the time variable. Furthermore, there is a related property associated with Equation (23).

Symmetry Property 2 (recurrence property). Let $t \mapsto (z_3(t), z_4(t))$ be a solution of Equation (22) for $t \in [0, \tau]$ and suppose that $z_3(\tau) = k\pi$ with $k \in \mathbb{Z}$. Then, another solution for the interval $[\tau, 2\tau]$ is given by

$$\hat{z}_3(t) := 2z_3(\tau) - z_3(2\tau - t), \tag{26a}$$

$$\hat{z}_4(t) := z_4(2\tau - t). \tag{26b}$$

The proof of this property follows the same lines as the previous one and is therefore omitted. As a consequence of the recurrence property, in general there are periodic solutions during the sliding regime. The only reasons for non-periodic solutions are: (1) the system reaches an equilibrium, (2) the sliding regime is interrupted, for example, by a change of the switching curve, or (3) the solution ceases to exist due to final escape time. However, the latter case can be excluded if $|\varphi(\cdot)|$ is bounded.[4]

Now, we additionally assume point symmetry for the switching curve, that is, $\varphi(-z_4) = -\varphi(z_4)$, from which immediately follows

$$\bar{f}_3(-z_3, -z_4) = -\bar{f}_3(z_3, z_4), \tag{27a}$$

$$\bar{f}_4(-z_3, -z_4) = -\bar{f}_4(z_3, z_4). \tag{27b}$$

Symmetry Property 3 (quadrant equivalence). Let $t \mapsto (z_3(t), z_4(t))$ be a solution of Equation (22) for $t \in [0, \tau]$ and suppose $\varphi(\cdot)$ to be point symmetric. Then, another solution for $t \in [0, \tau]$ is given by

$$\hat{z}_3(t) := -z_3(t), \tag{28a}$$

$$\hat{z}_4(t) := -z_4(t). \tag{28b}$$

Again, the proof is similar to the argument used in the case of the time reversal symmetry and therefore omitted. The obvious consequence from the quadrant equivalence property is, that submaneuvers (or switching curves) only have to be planned for the first and second quadrant of the z_2–z_4-plane and then can be trivially adapted for the third and fourth quadrant, respectively.

4.4. *Parking regime*

Clearly, all equilibrium points of Equation (15) are located at the origin of the z_2–z_4-plane. The projected transition-trajectory must thus leave this point and reach this point somehow. Due to the loss of controllability near the equilibria, actually reaching such a state is quite hard. Considering the drift, this difficulty is reflected by the vanishing of the drift on the z_4-axis.

DEFINITION 2 *For system* (15) *all states with* $z_2 = 0$ *are called* parking regime.

This denomination results from the fact that whenever the active joint stands still (for finite time), the only non-vanishing time derivative in Equation (15) is $\dot{z}_3 = \text{const}$. In the z_2–z_4-projection the state then remains unchanged, that is, the projected system is "parking". However, the drift conditions in the plane of course do change with z_3. For the motion planning this property allows, simply to wait in the parking regime until suitable drift conditions for the subsequent maneuver are reached.

Another consequence of the investigation of the projected drift vector field is that the tangent of any solution trajectory of Equation (15) containing a parking regime must be parallel to the z_2-axis, where $z_2 = 0$ holds. Since equilibria also fulfill Definition 2, it is clear that trajectories leaving or reaching the origin must have the tangent line $z_4 = 0$, that is, the z_2-axis.

5. Equilibrium transition via maneuver-based control

5.1. *Maneuver overview*

We consider a desired transition between the initial state $\mathbf{z}^*$ and the final state $\mathbf{z}^\dagger$, both of which are assumed to be equilibrium points. The transition between these states can be split up into several maneuvers which are connected by time periods where the system is in the parking regime, that is, we have $z_2(t) = 0$ and thus $\dot{z}_3(t) = z_4(t) = \pm z_4^{\mathrm{p}}$. Therefore, any desired drift condition can be achieved simply by waiting. The maximal waiting time is given by $2\pi/z_4^{\mathrm{p}}$ and therefore $|z_4^{\mathrm{p}}|$ should be chosen sufficiently large. For the sake of simplicity, we consider $|z_4^{\mathrm{p}}|$ to be fixed for all occurring parking regimes.

To any of the maneuvers we can associate a curve in the z_2–z_4-plane, which can be stabilized by sliding mode control. Obviously, the first maneuver ("A") has to start in the initial equilibrium $\mathbf{z}^*$ and it ends on the z_4-axis in the parking regime. Conversely, the last maneuver ("D") has to reach the final equilibrium $\mathbf{z}^\dagger$ and it starts on the z_4-axis. As will be shown in Section 4.3, maneuvers A and D can be regarded as equivalent due to time reversal symmetry.

For the simplest combinations of initial and final states, these two maneuvers already suffice for equilibrium transition. In general, however, two more maneuvers are necessary. Maneuver B serves to change the sign of $\dot{z}_3$. In other words, it transfers the system from a parking regime with $z_4 = z_4^{\mathrm{p}}$ to another parking regime with $z_4 = -z_4^{\mathrm{p}}$, or vice versa. The objective of maneuver C is to adapt z_1 such that, after the final maneuver D, the active joint is in its desired position. Thereby we exploit the recurrence property, and,

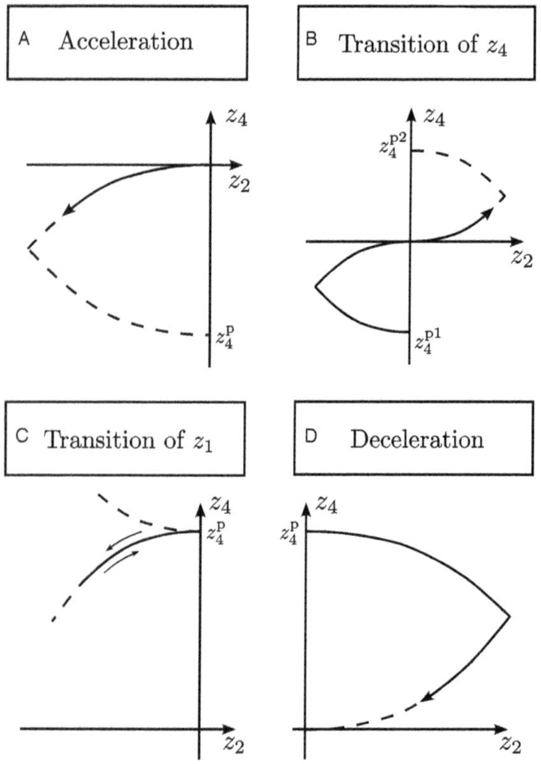

Figure 3. Principle succession of the maneuvers to perform an equilibrium point transition.

additionally, the fact that the coordinate z_1 is cyclic, that is, the dynamics of the system is independent of its value. Figure 3 shows how the switching curves of these four maneuvers might look like.

The further sections require addressing the components of the state at different stages of the transition. To this end, we introduce the following notation: The letters "A", "B", "C", and "D" used as head-index denominate the corresponding maneuver, the letter "p" indicates the parking regime and the symbols $*$ and $\dagger$ stand for the beginning and the end of a maneuver, respectively. For example, z_3^{*A} is the angle of the second joint in the start equilibrium, z_4^{pA} is the z_4 value in the parking regime associated with maneuver A and $z_1^{\dagger C}$ is the angle of the first joint after the completion of maneuver C. The head index "eq" indicates the equilibrium value of a quantity – either at the start or end of the transition.

5.2. *Leaving and reaching equilibrium points*

In this section, we consider the transfer of the system from parking regime to an equilibrium (maneuver D), and its time reversal counterpart (maneuver A). Thereby, depending on the z_3 value of the concerned equilibrium, two cases can occur. Firstly, we discuss the case $z_3^{eq} \in (\pi/2, \pi) \cup (\pi, \frac{3}{2}\pi)$, with the maneuvers D1 and A1. After that, we extend the results to $z_3^{eq} \in (0, \pi/2) \cup (\frac{3}{2}\pi, 2\pi)$ for the construction of maneuvers A2 and D2.

From the investigation of the drift in Section 4.1, it becomes obvious that any projected trajectory containing a point with $z_2 = 0$ must have a tangent in this point which is parallel to the z_2-axis. Clearly, this also holds true for the switching curve (19). In other words: $\varphi(\cdot)$ must have infinite slope at its roots. This property is complied by power functions with exponents smaller then one. From Knoll and Röbenack (2011a), we know that for $z_2, z_4 > 0$ and $z_3 \in (\pi/2, \pi)$.

$$\varphi(z_4) := \begin{cases} (\mu\beta|z_4|)^{1/\beta} & \text{if } \dfrac{z_4^p - z_4}{z_4^p} > \dfrac{1}{2}, \\ (\mu\beta|z_4^p - z_4|)^{1/\beta} & \text{else} \end{cases} \tag{29}$$

with $2 < \beta < 3$ is a suitable choice for maneuver D1 in the first quadrant.[5]

Of course, the state z_3 changes during the maneuver which causes two issues: Firstly, the drift conditions change during the execution and secondly, the final value of z_3 depends on the z_3-start value (of maneuver D1) and this dependency cannot be expressed in closed form. However, for the motion planning a closed-form expression is not necessary. Given the desired final value $z_3^\dagger$, the start value for maneuver D1 can be obtained by backward integration, that is, by stabilizing the system $\dot{\mathbf{z}} = -\mathbf{f}(\mathbf{z}) - \mathbf{g}(\mathbf{z})a$ on the same sliding surface.[6] The z_3-value where the system reaches the parking regime in backward running time, obviously is the start value z_3^* for maneuver D1 in forward time.

Regarding the first issue, it is clear that a mere *quantitative* change of the drift vector field is not a problem for the execution of maneuver D1. As long as the sign of the drift does not change, the reduced dynamics lead the system towards the equilibrium. If, however, z_3 leaves the interval $(\pi/2, \pi)$ the z_2–z_4-projection of the system dynamics are no longer topologically equivalent to the vector field depicted in Figure 2(b). In other words, if the drift changes its sign, the origin cannot be reached. From Equation (15) we extract

$$\dot{z}_3 = z_4 - (1 + \kappa \cos z_3) z_2 \tag{30}$$

and hence, all states with $\dot{z}_3 = 0$ must lay on the straight line

$$z_4 = (1 + \kappa \cos z_3) z_2. \tag{31}$$

For the sake of simplicity, we assume $\kappa < 1$. This so-called "strong inertial coupling" (Spong, 1998) can be reached for any combination of arm lengths by means of suitable constructional measures (Knoll, Leist, & Röbenack, 2011). Then, the straight line always passes from the third to the first quadrant of the z_2–z_4-plane. This means that, the switching curve (29) crosses this line somewhere in the first quadrant, and thus z_3 increases at the beginning of the

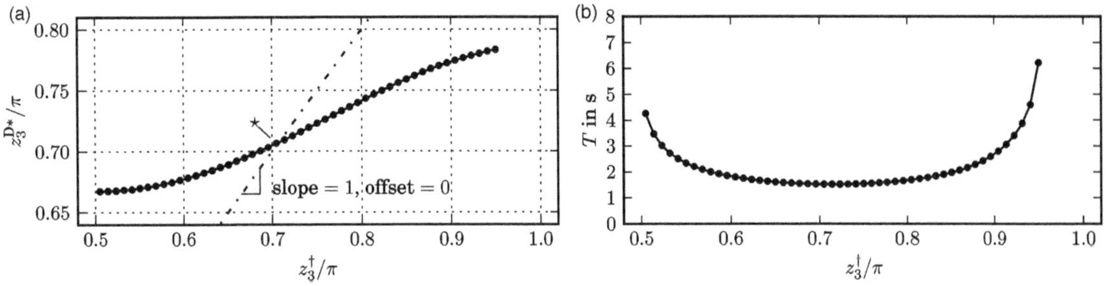

Figure 4. (a): The z_3-starting-value for maneuver D1 in dependency of the desired final passive joint angle $z_3^\dagger$. The bisecting line (dash-dotted) and the intersection point will be used for the construction of maneuver B (Section 5.3). (b): Duration of maneuver D1 for each final value. Each dot in both diagrams corresponds to a single simulation run.

maneuver and decreases at its end. From this qualitative consideration, we can suspect that the overall change of z_3 during maneuver D1 is "relatively small" and the construction (29) hence is a good choice for the sliding surface. This conjecture is confirmed by simulation experiments,[7] see Figure 4.

By applying the quadrant equivalence (Symmetry Property 3) and identifying $z_3 \in (-\pi, -\pi/2)$ with $z_3 \in (\pi, \frac{3}{2}\pi)$, this approach for maneuver D1 can be easily adapted for desired final conditions with $z_3^\dagger \in (\pi, \frac{3}{2}\pi)$. From Figure 2(c) follows that the switching line then lies in the third quadrant.

Due to time reversal symmetry it is clear, that (with slight adaption) the results for the deceleration maneuver also hold true for maneuver A1, that is, to leave the start equilibrium and reach a parking regime. In particular, the switching line through the first quadrant ($z_4^{\mathrm{P}} > 0$), which is used to reach equilibrium states with $z_3^\dagger \in (\pi/2, \pi)$, also serves to leave equilibria with $z_3^* \in (\pi, \frac{3}{2}\pi)$, while the switching line in the third quadrant (same curve mirrored along both coordinate axes, hence $z_4^{\mathrm{P}} < 0$) can either be used for breaking with $z_3^\dagger \in (\pi, \frac{3}{2}\pi)$ or accelerating with $z_3^* \in (\pi/2, \pi)$.

However, for the remaining two intervals $(0, \pi/2)$ and $(\frac{3}{2}\pi, 2\pi)$ the situation is different. In these cases, the double-integrator-like drift conditions lie in the second and fourth quadrant, see Figure 2(a) and 2(d). From the positive slope of the line (31) implicitly follows that in these quadrants the absolute value of $\dot{z}_3$ is relatively high which causes the drift to change its sign shortly after the maneuver start and therefore inhibits the same kind of maneuver as in the cases above.

The construction of a maneuver-pattern, which incorporates this sign change of the drift, is illustrated best by means of an acceleration-maneuver ("A2") for $z_3^* \in (\frac{3}{2}\pi, 2\pi)$ (fourth quadrant). The generalizations to the opposite time direction and the other quadrant can then be performed by the Symmetry properties 1 and 3.

The maneuver A2 is divided into two phases, each one associated with one value[8] of $\mathrm{sign}(\dot{z}_4)$. In each phase, the

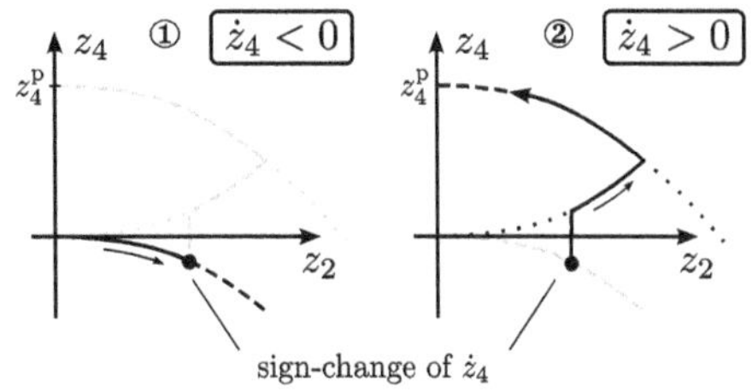

Figure 5. Schematic illustration of the two phases of maneuver A2. Phase 1: The drift points downwards until z_3 reaches a certain value. The sliding surface is defined by an adapted single branch version of Equation (29). Phase 2: Drift points upwards. A sliding surface similar to the maneuver A1 can be used, augmented with a lower bound on the value of $\varphi(\cdot)$ in the lower branch. Maneuver D2 uses the same switching curves. Then, due to time inversion, phase 1 and 2 switch its rolls and the direction of the arrows is reverted.

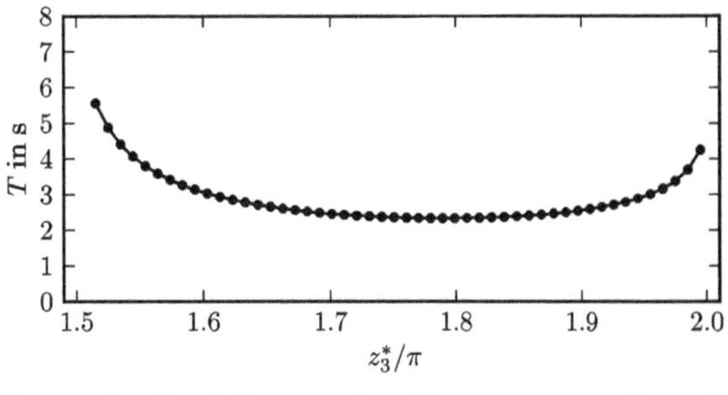

Figure 6. Simulation results for maneuver A2 with $z_3^* \in (\frac{3}{2}\pi, 2\pi)$. Obviously, the maneuver can be executed with acceptable duration everywhere, except near the interval boundary.

system is stabilized onto an suitable sliding surface, see Figures 5 and 6 for an overview and the simulation source code (Knoll, 2014a) for details.

Together the maneuvers A1, A2 and D1, D2 can be used to leave or reach, respectively, almost any equilibrium point. However, the values $z_3 = (k/2)\pi$, $k \in \mathbb{Z}$ must be excluded due to the singular drift conditions in these cases. Additionally, z_3^{eq} values close to these singularities might cause too long maneuver durations, see Figures 4 and 6.

5.3. *Transition of z_4 (Maneuver B)*

When leaving or reaching an equilibrium, due to the drift conditions (see Figure 2) the equilibrium-angle of the second joint determines the sign of z_4 in the associated parking regime. If the desired initial and final states are given such that the z_4^{p}-values, associated to the appropriate maneuvers A and D, do not match, an additional maneuver ("B") has to perform the transfer $z_4^{\mathrm{pA}} = \pm|z_4^{\mathrm{p}}| \to z_4^{\mathrm{pD}} = \mp|z_4^{\mathrm{p}}|$. As stated above, during parking regime z_4 is equivalent to $\dot{z}_3$, that is, to the angular velocity of the second joint. Thus, the task of maneuver B is the inversion of the rotating direction of that joint.

Although there naturally exists an infinite variety of possible motion patterns for this goal, in the present context it is easily achieved by the combination of maneuvers D1 and A1, described above. With this choice and the assumptions $z_4^{\mathrm{pA}} < 0$ and $z_4^{\mathrm{pD}} > 0$, we have the following situation: The system is in parking regime before maneuver B begins, which makes it possible to wait until $z_3 = z_3^{*\mathrm{B}} \in (\pi, \frac{3}{2}\pi)$. From Figure 2 it is obvious, that the third quadrant can be used for D1-breaking. Furthermore, we can choose a suitable value for $z_3^{*\mathrm{B}}$ such that the equilibrium reached by this (intermediate) breaking maneuver has the same z_3-value. This is expressed in Figure 4(a) by the intersection (marked with a star) of the simulated curve with the bisecting line of the first quadrant (dash-dotted). After reaching the z_2–z_4-origin, the manipulator is immediately re-accelerated by maneuver A1 (as second part of maneuver B) in the first quadrant. Due to time reversal symmetry this motion is the exact inversion of the first part. Consequently, the z_3 value at the beginning and the end are identical as well. Finally, the desired parking regime with z_4^{pD} is reached with $z_3(t^{\mathrm{B}\dagger}) = z_3^{*\mathrm{B}}$.

5.4. *Transition of z_1 (Maneuver C)*

The last missing design step for a transition from almost any desired equilibrium to almost any other is the proper adaption of the actuated joint angle z_1, which is the task of maneuver C. As for the previously treated maneuvers we initially assume a certain situation (here: $z_4^{\mathrm{pD}} > 0$), and later extend the results by applying the symmetry properties.

For $z_4^{\mathrm{pD}} > 0$ Figure 2 implies that the subsequent maneuver D takes place in the first quadrant, thus we have

$$\Delta z_1^{\mathrm{D}} := \int_{t^{\mathrm{D}*}}^{t^{\mathrm{D}\dagger}} z_2(t)\,\mathrm{d}t > 0. \tag{32}$$

That value is known a priory and only depends on $z_3^{\dagger}$ because z_1 is cyclic.

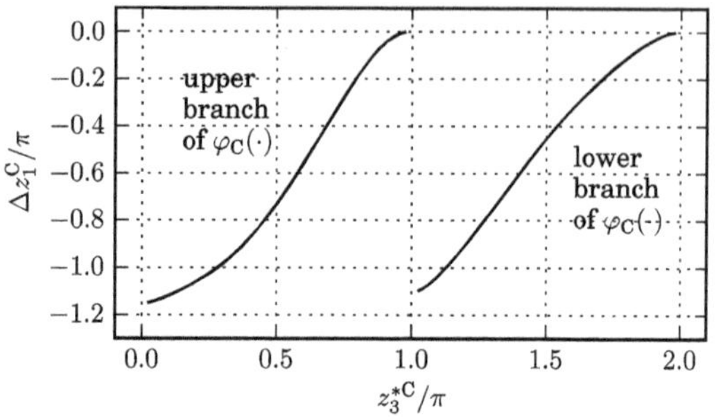

Figure 8. Value of Δz_1^{C} (displacement of active joint) in dependency of the z_3-start value of maneuver C (obtained by numerical simulation). If $z_3^{*\mathrm{C}} < \pi$ according to Figure 2 the drift in the second quadrant points upwards, therefore the upper branch of the sliding surface (cf. (35) and Figure 7(a)) is active. For $z_3^{*\mathrm{C}} > \pi$, the drift points downwards and thus the lower branch is used. For motion planning this mapping can be inverted piecewise to obtain $z_3^{*\mathrm{C}}$ for a given Δz_1^{C}.

To reach the desired final equilibrium $(z_1^{\dagger}, 0, z_3^{\dagger}, 0)$, it has to be ensured that at the beginning of maneuver D

$$z_1^{\mathrm{D}*} = z_1^{\dagger} - \Delta z_1^{\mathrm{D}} \tag{33}$$

holds. During parking regime, in which the system is between maneuvers B and C as well as between C and D, respectively, $z_1(\cdot)$ is constant. Therefore, the necessary displacement Δz_1^{C} of the active joint during maneuver C is easily determined by

$$\Delta z_1^{\mathrm{C}} = z_1^{\mathrm{C}\dagger} - z_1^{\mathrm{C}*} = z_1^{\mathrm{D}*} - z_1^{\mathrm{B}\dagger} \overset{33}{=} z_1^{\dagger} - \Delta z_1^{\mathrm{D}} - z_1^{\mathrm{B}\dagger}. \tag{34}$$

To achieve this displacement the recurrence property can be exploited (see Section 4.3). For a suitable sliding surface (i.e. $\varphi(\cdot)$ is bounded and does not pass through the origin) the resulting motion in z_2–z_4-plane is recurrent: every point on the sliding surface is reached twice. Hence, starting the maneuver in parking regime implies ending it in parking regime. Obviously, the actual shape of the sliding surface for maneuver C is a (function-valued and thus infinite dimensional) degree of freedom. For convenience, we again use a power function similar to Equation (29), namely

$$\varphi_{\mathrm{C}}(z_4) = -(\mu\beta|z_4^{\mathrm{p}} - z_4|)^{1/\beta}. \tag{35}$$

Note that, due to the absolute value $|\cdot|$ the graph of this function has two branches.[9] The negative sign indicates that the curve is located in the left half of the z_2–z_4-plane, see Figure 7. Consequently, during the motion through the left half-plane we have $z_2 < 0$ and thus $\Delta z_1^{C} < 0$. The actual value of Δz_1^{C} depends on the starting conditions of maneuver C, that is, on $z_3^{*\mathrm{C}}$. While this dependency cannot be obtained in closed form, it is straightforward to calculate it numerically, see Figure 8.

For any given value of $\Delta z_1^{\mathrm{C}} \in [-\pi, 0)$ the displacement can be achieved with a single instance of maneuver C. In

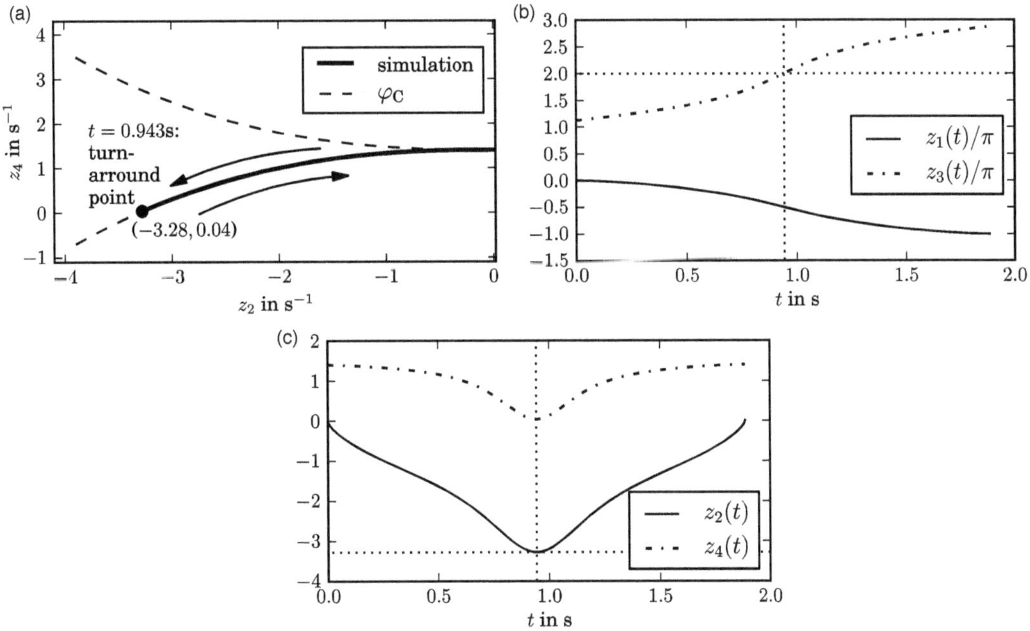

Figure 7. Exemplary simulation result for maneuver C (with $\Delta z_1^C \stackrel{!}{=} \pi$). Note that the sliding surface according to (35) has two branches(dashed lines in subfigure (a)). During one run of maneuver C only one of these branches is used, depending on the sign of the drift at the beginning of the maneuver. In this example the lower branch is used.

fact, there are two possible solutions for z_3^{*C} corresponding to the two branches of the sliding surface (35). From these, the "nearest" one (considering the $\mathbb{S}_1$-Topology) should be chosen, to minimize the waiting time in parking regime. If a bigger displacement is desired, that is, $\Delta z_1^C \in (-2\pi, -\pi)$ is needed, then the simplest approach is to perform two instances of maneuver C, each one for achieving a displacement of $\Delta z_1^C/2$. Finally, note that every value from outside the interval $(-2\pi, 0]$ can be mapped to its interior by

$$\tilde{\Delta z_1} := (\Delta z_1 \bmod 2\pi) - 2\pi. \quad (36)$$

Because z_1 is the angle of a revolute joint, this mapping does not affect the actual configuration of the manipulator, cf. Section 3.1.

Due to the quadrant equivalence (Symmetry property 1) the extension of maneuver C to $z_4^p < 0$ with a sliding surface in the right half-plane is straightforward.

5.5. *Maneuver summary*

All four maneuvers (A, B, C, D) enable us to perform almost arbitrary equilibrium transitions. Due to adverse drift conditions equilibria with z_3-values close to integral multiples of $\pi/2$ must be excluded. The actual sizes of the excluded z_3 intervals depend on the maneuver duration which can be tolerated, cf. for example, Figure 4(b). In Figure 9 a complete equilibrium transition with all maneuvers and parking regimes is depicted. See (Knoll, 2014b) for the respective animations.

Remark 4 Similar to the simple ODE $\dot{x} = \sqrt{x}$ there is an issue with the uniqueness of the solution of the reduced dynamics which is related with the "infinite slope" dz_2/dz_4 of the used sliding surfaces for $z_2 = 0$ (see Figure 3 and Equation (29)). The respective intersection point of the sliding surface with the z_4-axis (either an equilibrium or an parking regime), cannot be left by sliding mode control. However, if once $|z_2| > 0$ then the solution of the reduced dynamics is unique due to $\varphi(\cdot)$ being locally Lipschitz. For the (numerical) realization of the maneuvers this suggests to choose $|z_2(t = 0)| = \varepsilon > 0$. For ε small enough, its actual value becomes irrelevant to the solution of the dynamics and the input trajectories obtained by this "trick" nevertheless lead the unspoilt system to the desired state.

Remark 5 Of course the proposed motion schemes are by far not unique. In fact, for every single part of the motion (i.e. for every maneuver) there exists an infinite degree of freedom for the shape of the sliding surface, which might be used to optimize the motion w.r.t. overall duration or other suitable criteria. However, the aim of the presented solution is to show the principle applicability of the concept.

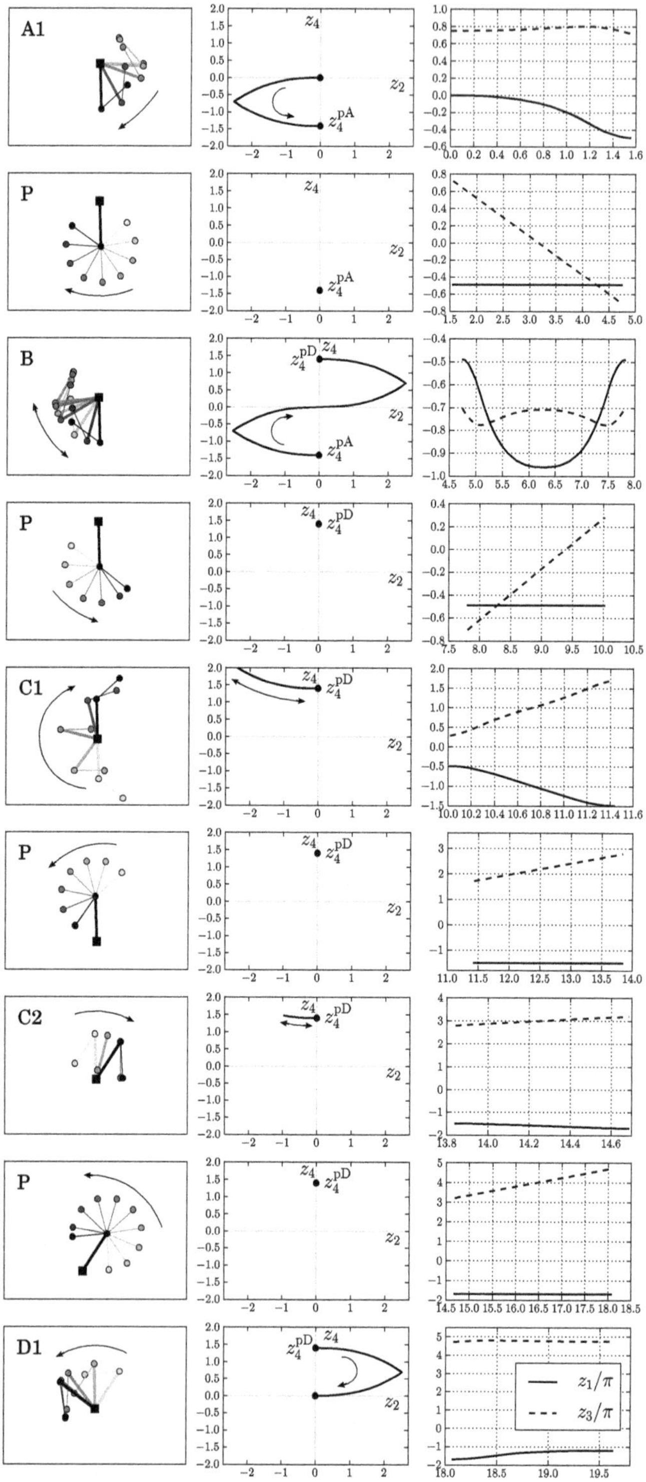

Figure 9. Simulation results for a complete equilibrium transition. Left column: Motion in working space. Middle column: Projection of the trajectory (and hence the switching functions) to the z_2-z_4-subspace. Right column: Time evolution of the joint angles $z_1(t), z_3(t)$. For a video of these results see (Knoll, 2014b).

6. Conclusion and outlook

In this contribution, we defined the LBINF as a special case of the Byrnes–Isidori normal form for generic Lagrangian mechanical systems. Furthermore, we constructively proved its existence for systems given by Equation (1). Then, we analyzed the frictionless underactuated 2-DOF manipulator in that normal form, especially the drift conditions in the z_2–z_4-subspace. On that basis, the controller (20) was defined which restricts the system dynamics to a submanifold (sliding surface) of the state space, determined by $z_2 = \varphi(z_4)$. Additionally, useful symmetries and the parking regime were identified. Finally, the equilibrium transition was decomposed into several subtasks, each of which is complied by a suitable maneuver, that is, by the suitable choice of the sliding function and appropriate initial conditions (obtained during parking regime).

The feasibility of the proposed controller is demonstrated by detailed simulation studies (Knoll, 2014a, 2014b) and first experimental results (Knoll et al., 2011). Currently, the experimental setup is reworked in order to implement the full equilibrium transition.

Besides the experimental realization there are some more questions for further investigation. One research direction concerns the application of the LBINF to other underactuated control systems and its relation to important control properties like linearizability by static or dynamic feedback, differential flatness and configuration flatness (see Knoll & Röbenack, 2014; Rathinam & Murray, 1998 and the references therein) or accessibility.

Another direction deals with the consequences of the Brockett condition, mentioned in Section 3.1. This result precludes the existence of a continuous differentiable feedback of the state of the original system to stabilize an isolated equilibrium. However, it does not preclude a nonsmooth state feedback or a dynamic extension of the system.

Acknowledgments

The authors would like to thank Matthias Franke, Chenzi Huang and Christian Mächler for valuable discussions and the anonymous reviewers for their useful comments and suggestions.

Funding

This work has been supported by Deutsche Forschungsgemeinschaft (DFG) under research [grant number RO 2427/2-1].

Notes

1. Although underactuated systems are sometimes referenced as "second-order nonholonomic", for example, in Oriolo and Nakamura (1991), many of them are holonomic in a physical sense.
2. Note the redefinition of $\mathbf{q}$ along with the elimination of $\boldsymbol{B}(\mathbf{q})$.
3. In such a case (Oriolo & Nakamura, 1991) suggest either nonsmooth feedback or a different control objective. In particular, the authors propose a smooth controller, which stabilizes a 1-dimensional submanifold of all equilibrium points. The case of a dynamic extension of the state, which might be another possibility to cope with the violation of Brocketts condition, is not discussed.
4. Then $\bar{f}_3(\cdot,\cdot)$ and $\bar{f}_4(\cdot,\cdot)$ are globally Lipschitz and hence finite escape time cannot occur.
5. This choice for φ is made for the sake of simplicity. In regions which are not near to the z_4-axis the curve $\varphi(\cdot)$ could have an arbitrary shape as long as it is continuous, (piecewise) differentiable and takes finite positive values.
6. Note that, due to time reversal symmetry, this also could be interpreted as an execution of maneuver A1 with appropriate adaption of the state.
7. As numerical values $\kappa = 0.9$ and $|z_4|^{\mathrm{P}} = 1.4\mathrm{s}^{-1}$ are used for all simulations.
8. For convenience, we define $\mathrm{sign}(0) := 1$.
9. They are called "upper" and "lower" branch because in the z_2–z_4-plane the argument of φ_{C} runs vertically.

References

Arai, H., & Tachi, S. (1991). Position control of a two degree of freedom manipulator with passive joint. *IEEE Transaction on Industrial Electronics*, *38*(1), 15–20.

Brockett, R. W. (1983). Asymptotic stability and feedback stabilization. In R. Brockett, R. S. Millmann, & H. J. Sussmann (Eds.), *Differential geometric control theory* (pp. 181–191). Boston, MA: Birkhäuser.

Cameron, J. M., & Book, W. J. (1997). Modeling mechanisms with nonholonomic joints using the Boltzmann–Hamel equations. *The International Journal of Robotics Research*, *16*(1), 47–59.

Choukchou-Braham, A., Cherki, B., Djemai, M., & Busawon, K. (2014). *Analysis and control of underactuated mechanical systems*. Cham: Springer International Publishing Switzerland.

De Luca, A., Iannitti, S., Mattone, R., & Oriolo, G. (2002). Underactuated manuipulators: Control properties and techniques. *Machine Intelligence and Robotic Control*, *4*, 113–125.

De Luca, A., Mattone, R., & Oriolo, G. (2000). Stabilization of an underactuated planar 2R manipulator. *International Journal of Robust and Nonlinear Control*, *10*, 181–198.

Fantoni, I., & Lozano, R. (2001). *Non-linear control for underactuated mechanical systems*. London: Springer.

Franch, J., Reyes, A., & Agrawal, S. K. (2013). *Differential flatness of a class of n – DOF planar manipulators driven by an arbitrary number of actuators*. Proceedings of the European Control Conference (ECC), 2013, Zurich (pp. 161–166).

Glück, T., Eder, A., & Kugi, A. (2013). Swing-up control of a triple pendulum on a cart with experimental validation. *Automatica*, *49*(3), 801–808.

Graichen, K., Treuer, M., & Zeitz, M. (2007). Swing-up of the double pendulum on a cart by feedforward and feedback control with experimental validation. *Automatica*, *43*(1), 63–71.

Ince, D. C., Hatton, L., & Graham-Cumming, J. (2012). The case for open computer programs. *Nature*, *482*(7386), 485–488.

Isidori, A. (1995). *Nonlinear control systems: An introduction* (3rd ed.). London: Springer.

Knoll, C. (2014a). Python source code for simulation and visualization of the controlled 2-DOF manipulator. Retrieved from http://www.tu-dresden.de/rst/software

Knoll, C. (2014b). Video sequence: Equilibrium transition of the underactuated 2-DOF manipulator. Retrieved from http://www.tu-dresden.de/rst/software

Knoll, C., Leist, B., & Röbenack, K. (2011). Konzeption und prototypische Realisierung eines Versuchsstandes zur Regelung eines unteraktuierten Manipulators. In T. Bertram, B. Corves, & K. Janschek, (Eds.), *Tagungsband Mechatronik 2011* (pp. 241–246), Dresden.

Knoll, C., & Röbenack, K. (2010). Sliding mode control of an underactuated two-link manipulator. *Proc in Applied Mathematics and Mechanics*, *10*(1), 615–616.

Knoll, C., & Röbenack, K. (2011a). *Control of an underactuated manipulator using similarities to the double integrator*. Proceedings of the 18th IFAC World Congress, Milano, Italy (pp. 11501–11507).

Knoll, C., & Röbenack, K. (2011b). Trajectory planning for a non flat mechanical system using time-reversal symmetry. *Proceeding in Applied Mathematics and Mechanics*, *11*(1), 819–820.

Knoll, C., & Röbenack, K. (2013). Stable limit cycles with specified oscillation parameters induced by feedback: Theoretical and experimental results. *Transactions on Systems, Signals and Devices*, *8*(1), 127–147.

Knoll, C., & Röbenack, K. (2014). *On configuration flatness of linear mechanical systems*. Proceeding of the European Control Conference (ECC), 2014, Strasbourg (pp. 1416–1421).

Mahindrakar, A. D., Rao, S., & Banavar, R. N. (2006). Point-to-point control of a 2R planar horizontal underactuated manipulator. *Mechanism and Machine Theory*, *41*, 838–844.

Mareczek, J., Buss, M., & Schmidt, G. (1999). Robust control of a non-holonomic underactuated SCARA robot. In S. G. Tzafestas, & G. Schmidt (Eds.), *Progress in system and robot analysis and control design* (pp. 381–396). Lecture Notes in Control and Information Science, Vol. 243. London: Springer.

Nakamura, Y., Suzuki, T., & Koinuma, M. (1997). Nonlinear behavior and control of a nonholonomic free-joint manipulator. *IEEE Transactions on Robotics and Automation*, *13*(6), 853–862.

Olfati-Saber, R. (2001). *Nonlinear control of underactuated mechanical systems with application to robotics and aerospace vehicles* (Ph.D. thesis). Massachusetts Institute of Technology, Cambridge, MA.

Oriolo, G., & Nakamura, Y. (1991). *Control of mechanical systems with second-order nonholonomic constraints: Underactuated manipulators*. Proceedings of the 30th conference on decision and control, Brighton, England (pp. 2398–2403).

Rathinam, M., & Murray, R. M. (1998). Configuration flatness of Lagrangian systems underactuated by one control. *SIAM J. Control and Optimization*, *36*(1), 164–179.

Riachy, S., Orlov, Y., Floquet, T., Santiesteban, R., & Richard, J. P. (2008). Second-order sliding mode control of underactuated mechanical systems I: Local stabilization with application to an inverted pendulum. *International Journal of Robust and Nonlinear Control*, *18*(4), 529–543.

Sastry, S. (1999). *Nonlinear systems: Analysis, stability, and control*. New York: Springer.

Scherm, N., & Heimann, B. (2000). Dynamics and control of underactuated manipulation systems: A discrete-time approach. *Robotics and Autonomous Systems*, *30*(3), 237–248.

Slotine, J. J. E., & Li, W. (1991). *Applied nonlinear control*. Upper Saddle River, NJ: Prentice-Hall.

Spong, M. W. (1998). Underactuated mechanical systems. In B. Siciliano, & K. P. Valavanis (Eds.), *Control problems in robotics* (pp. 135–150). Lecture Notes in Control and Information Science, Vol. 230. London: Springer.

Spong, M. W. (2000). Some aspects of switching control in robot locomotion. *Automatisierungstechnik*, *48*(4).

Yesiloglu, S. M., & Temeltas, H. (2010). Dynamical modeling of cooperating underactuated manipulators for space manipulation. *Advanced Robotics*, *24*(3), 325–341.

Yoshida, K. (1997). *A general formulation for underactuated manipulators*. Proceedings of the 1997 IEEE/RSJ International Conference on Intelligent Robots and System (IROS '97), Grenoble, France (Vol. 3, pp. 1651–1657).

16

Robust fault diagnosis for an exothermic semi-batch polymerization reactor under open-loop

Abdelkarim M. Ertiame[a*], Dingli Yu[a], Feng Yu[b] and J.B. Gomm[a]

[a]*Control System Research Group, School of Engineering, Liverpool John Moores University, Byrom Street, Liverpool L3 3AF, UK;*
[b]*School of Electronic Information, Changchun Architecture &Civil Engineering College, Changchun, Jilin Province, People's Republic of China*

An independent radial basis function neural network (RBFNN) is developed and employed here for an online diagnosis of actuator and sensor faults. In this research, a robust fault detection and isolation scheme is developed for an open-loop exothermic semi-batch polymerization reactor described by Chylla–Haase. The independent RBFNN is employed here for online diagnosis of faults when the system is subjected to system uncertainties and disturbances. Two different techniques to employ RBFNNs are investigated. Firstly, an independent neural network (NN) is used to model the reactor dynamics and generate residuals. Secondly, an additional RBFNN is developed as a classifier to isolate faults from the generated residuals. Three sensor faults and one actuator fault are simulated on the reactor. Moreover, many practical disturbances and system uncertainties, such as monomer feed rate, fouling factor, impurity factor, ambient temperature and measurement noise, are modelled. The simulation results are presented to illustrate the effectiveness and robustness of the proposed method.

Keywords: robust fault detection; independent RBF model; RBF neural networks; open-loop Chylla–Haase reactor

1. Introduction

In recent years, the task of monitoring complex non-linear processeslts has been intensively studied. Fault detection and isolation (FDI) techniques have attracted much interest due to the increasing demand for good performance and higher standards of safety and reliability of technical plants for improving the supervision and monitoring as part of the overall control of processes (Isermann, 1984). FDI has become a critical issue in the operation of high-performance chemical plants, nuclear plants, airplanes, ships, submarines and space vehicles (Gertler, 1988). In the chemical industry, faults can occur due to sensor failures, equipment failures or changes in process parameters. The occurrence of a fault may cause process performance degradation, or in the worst cases, may cause disastrous accidents. However, FDI can help avoid all these major consequences (Deibert & Isermann, 1992).

Due to several non-linearity and time-varying feature of the reactor dynamics, the observer methods, parity space methods and other first-principle model-based methods cannot be successfully applied for FDI of the Chylla–Haase reactor.

Many research works have been carried out to study neural networks (NNs) for FDI. Yu, Gomm, and Williams (1999) studied sensor fault diagnosis in a chemical process via radial basis function neural networks (RBFNNs); a semi-independent NN was used for sensor fault diagnosis. Moreover, the thins-plate-spline function was used for the neural model and the Gaussian function was used for the neural classifier. Another study was conducted by Gomm and Yu (2000) that introduced the selection of RBF network centres with recursive orthogonal least squares training. Frank and Köppen-Seliger (1997) and Koppen-Seliger and Frank (1995) studied fuzzy logic and NN applications for fault diagnosis. Their paper introduced fuzzy logic for residual evaluation, a dependent NN for residual generation and a NN for residual evaluation by using another dependent NN for generating residuals. All those authors used dependent and semi-dependent mode of NN for FDI. As the residual of these methods is affected by the plant output, the residual is made insensitive to the faults. Although a partial dependent mode is used to enhance the residual to fault sensitivity, the fault detect threshold is still high such that fault with small amplitude cannot be detected.

Patton, Chen, and Siew (1994) proposed an approach for detecting and isolating faults in a non-linear dynamic process using NNs. Firstly, a multi-layer perceptron (MLP) network was trained to predict the future system states, then the residual was generated using the differences between the actual and predicted states. Secondly, another NN was used as a classifier to isolate faults from these

*Corresponding author. Email: A.M.Ertiame@2011.ljmu.ac.uk

state prediction errors. However, this method used the NN model in its so-called dependent mode.

Ferrari, Parisini, and Polycarpou (2008), Xiaodong (2011), Xiaodong, Polycarpou, and Parisini (2002) and Zhang, Polycarpou, and Parisini (2010) studied the design and analysis of a robust FDI scheme for non-linear uncertain dynamic systems, the proposed architecture consists of a bank of non-linear adaptive estimator, one of the estimators is used for the detection and approximation of a fault, whereas the rest are used for an online fault isolation decision scheme which is based on adaptive threshold functions. In their method, they used a state-space non-linear model and then used a simple NN as an estimator for online learning. The model output must be equal to the plant output; however, this method needs to have a plant non-linear dynamic model and sometimes the model needs to be very accurate; this accurate model is difficult to produce. Most of the recent investigations of fault diagnosis for chemical reactors using an independent RBFNN have been studied by Ertiame, Dingli, Feng, and Gomm (2013).

In this research, a new robust FDI scheme is developed for an open-loop Chylla–Haase polymerization reactor using an independent RBFNN. The independent RBFNN is employed here for online diagnosis of faults on the actuator and sensors when the system is subjected to system uncertainties and disturbances. The independent NN mode is developed to generate enhanced residuals for diagnosing faults in the reactor. Then, a second NN is developed as a classifier to isolate these faults. The basis Gaussian function is used for the NN model and for the NN classifier. The K-means clustering algorithm is used to choose the centres of the RBF networks, and a p-nearest-neighbours algorithm is used to choose the widths. Moreover, a recursive least squares (RLS) algorithm is used to update the weights.

2. The Chylla–Haase benchmark reactor

Batch and semi-batch reactors have been widely used in the chemical industry. In this research, a semi-batch polymerization reactor benchmark is considered which is described by Chylla and Haase (1993) and used as a benchmark for process control applications. The schematic diagram of the semi-batch polymerization reactor is shown in Figure 1 (Chylla & Haase, 1993). It consists of a stirred tank reactor with a cooling jacket and a coolant recirculation. The reactor temperature is controlled by manipulating the temperature of the coolant, which is recirculated through the cooling jacket of the reactor. The heat released through the reaction must be removed by circulating cold water through the jacket, where both hot and cold jacket streams are available. When the jacket temperature controller output is between 0% and 50%, the valve is opened and cold water is inserted, and when the jacket controller output is between 50% and 100%, the valve is opened and steam is inserted (Beyer, Grote, & Reinig, 2008; Graichen, Hagenmeyer, & Zeitz, 2005).

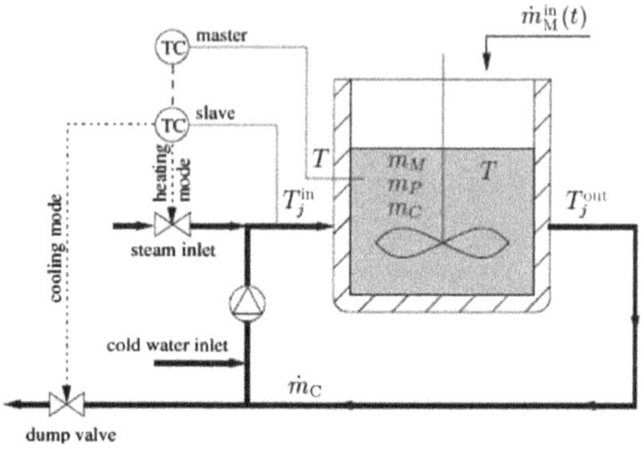

Figure 1. Chylla–Haase reactor schematic.

2.1. *Polymerization reactor model*

The mathematical model of the Chylla–Haase reactor is described by a set of five ordinary differential equations (ODEs) which come from material and heat balances inside the reactor:

$$\frac{dm_M}{dt} = \dot{m}_M^{in}(t) + \frac{Q_{rea}}{\Delta H}, \tag{1}$$

$$\frac{dm_P}{dt} = -\frac{Q_{rea}}{\Delta H}, \tag{2}$$

$$\frac{dT}{dt} = \frac{1}{\sum_i m_i C_{p,i}}[\dot{m}_M^{in}(t)C_{p,M}(T_{amb} - T) - UA(T - T_j) - (UA)_{loss}(T - T_{amb}) + Q_{rea}], \tag{3}$$

$$\frac{dT_{j\,out}}{dt} = \frac{1}{m_C\, C_{p,C}}[\dot{m}_C C_{p,C}(T_{j\,in}(t-\theta_1) - T_{j\,out}) + UA(T - T_j)], \tag{4}$$

$$\frac{dT_{j\,in}}{dt} = \frac{dT_{j\,out}(t-\theta_2)}{dt} + \frac{T_{j\,out}(t-\theta_2) - T_{j\,in}}{\tau_p} + \frac{K_p(c)}{\tau_p}. \tag{5}$$

The reactor model includes the material balances (1) and (2) for the monomer mass $m_M(t)$ and the polymer mass $m_P(t)$, the energy balance (3) with the reactor temperature $T(t)$, and the energy balances (4) and (5) of the cooling jacket and the recirculation loop with the outlet and inlet temperatures $T_{j\,in}(t)$ and $T_{j\,out}(t)$ of the coolant. The available measurements of the process are the temperature of the reactor and the cooling circuitry (Graichen, Hagenmeyer, & Zeitz, 2006):

The heating/cooling function $K_p(c)$ is influenced by an equal-percentage valve with valve position $c(t)$ as shown

Table 1. Variables and parameters of the reactor model.

$\dot{m}_{\mathrm{M}}^{\mathrm{in}}(t)$	Monomer feed rate (kg s^{-1})
$Q_{\mathrm{rea}} = -\Delta H \cdot R_{\mathrm{P}}$	Reaction heat (kW)
R_{P}	Rate of polymerization (kg s^{-1})
$-\Delta H$	Reaction enthalpy (kJ kg^{-1})
U	Overall heat transfer coefficient (kW m^{-2} K^{-1})
A	Jacket heat transfer area (m^2)
$(UA)_{\mathrm{loss}}$	Heat loss coefficient (kW K^{-1})
$C_{p,\mathrm{M}}, C_{p,\mathrm{P}}, C_{p,\mathrm{C}}$	Specific heat at constant pressure (kJ kg^{-1} K^{-1})
θ_1, θ_2	Transport delay (s)
$T_j = (T_{j\,\mathrm{in}} + T_{j\,\mathrm{out}})/2$	Average cooling jacket temperature (K)
$K_{\mathrm{p}}(c)$	Heating/cooling function (K)
τ_{p}	Heating/cooling time constant (s)

in the following equation:

$$K_p(c) = \begin{cases} 0.8 \times 30^{-c/50}(T_{\mathrm{inlet}} - T_{j\,\mathrm{in}}(t)), & c < 50\% \\ 0, & c = 50\%, \\ 0.15 \times 30^{(c/50-2)}(T_{\mathrm{steam}} - T_{j\,\mathrm{in}}(t)), & c > 50\% \end{cases} \tag{6}$$

For $c < 50\%$, cold water with inlet temperature T_{inlet} is injected into the cooling jacket, whereas a valve position $c > 50\%$ leads to a heating of the coolant by injecting steam with temperature T_{steam} into the recirculating water steam. Moreover, the variables and the parameters of the reactor model are listed in Table 1 (Graichen et al., 2006).

2.2. *Uncertainties and disturbances in the process*

In order to model the following practical issues of the control of polymerization reactors, various disturbances and uncertainties are identified:

- The impurity factor $i \in [0.8, 1.2]$ in the polymerization rate R_{P} is random but constant during one batch, which tries to simulate fluctuations in monomer kinetics caused by batch-to-batch variations in reactive impurity.
- The fouling factor $1/h_{\mathrm{f}}$ in the overall heat transfer coefficient U increases with each batch and accounts for the fact that during successive batches a polymer film builds up on the wall resulting in a decrease of U.
- The delay times θ_1 and θ_2 of the cooling jacket and the recirculation loop may vary by $\pm 25\%$ compared to nominal values.
- The ambient temperature T_{amb} is different during summer and winter. This affects the temperature of the monomer feed $\dot{m}_{\mathrm{M}}^{\mathrm{in}}$, as well as the initial conditions $T(0)$, $T_{j\,\mathrm{in}}(0)$ and $T_{j\,\mathrm{out}}(0)$ given by T_{amb} (Graichen et al., 2006).

Table 2 describes the empirical relations for the polymerization rate, the jacket heat transfer area and the overall heat transfer coefficient (Graichen et al., 2006).

3. Residual generation with RBF model

3.1. *Independent model of RBF modelling*

Using RBFNN for modelling, a non-linear dynamic system can be modelled in two modes: a dependent mode and an independent mode as shown in Figures 2 and 3, respectively. The first model referred to is a dependent mode, since the past system output is used as a network input. Thus, the model is dependent on the system output and cannot operate independently from the system. In the independent mode, the past model output is used as a network input. Therefore, the model is not dependent on the system output and can operate independently from the system. The independent model has an advantage in that the model can be used to simulate the system to obtain long-range prediction. In contrast, the dependent model performs as one-step-ahead predication.

Table 2. Empirical relations.

$R_{\mathrm{P}} = ikm_{\mathrm{M}}$	i	Impurity factor (–)
	$k = k_0 \exp(-E/RT) \cdot (k_1\mu)^{k_2}$	First-order kinetic constant (s^{-1})
	$\mu = c_0 \exp(c_1 f) \times 10^{c_2(a_0/T - c_3)}$	Batch viscosity (kg m^{-1} s^{-1})
	$f = m_{\mathrm{P}}/(m_{\mathrm{M}} + m_{\mathrm{P}} + m_{\mathrm{C}})$	Auxiliary variable (–)
	$k_0, k_1, E, R, a_0, c_0, c_1, c_2, c_3$	Constants
	R	Natural gas (kJ kmol^{-1} K^{-1})
$A = \left(\frac{m_{\mathrm{M}}}{\rho_{\mathrm{M}}} + \frac{m_{\mathrm{P}}}{\rho_{\mathrm{P}}} + \frac{m_{\mathrm{W}}}{\rho_{\mathrm{W}}}\right)\frac{P}{B_1} + B_2$	$\rho_{\mathrm{M}}, \rho_{\mathrm{P}}, \rho_{\mathrm{W}}$	Densities (kg m^{-3})
	B_1	Reactor bottom area (m)
	P	Jacket perimeter (m)
	B_2	Jacket bottom area (m^2)
$U = 1/(h^{-1} + h_{\mathrm{f}}^{-1})$	$h = d_0 \exp(d_1 \mu_{\mathrm{wall}})$	heat transfer coefficient (kW m^{-2} K^{-1})
	$\mu_{\mathrm{wall}} = c_0 \exp(c_1 f) \times 10^{c_2(a_0/T_{\mathrm{wall}} - c_3)}$	Wall viscosity (kg m^{-1} s^{-1})
	$T_{\mathrm{wall}} = (T + T_j)/2$	Wall temperature (K)
	h_{f}^{-1}	Fouling factor (m^2 K kW^{-1})
	d_0, d_1	Constants

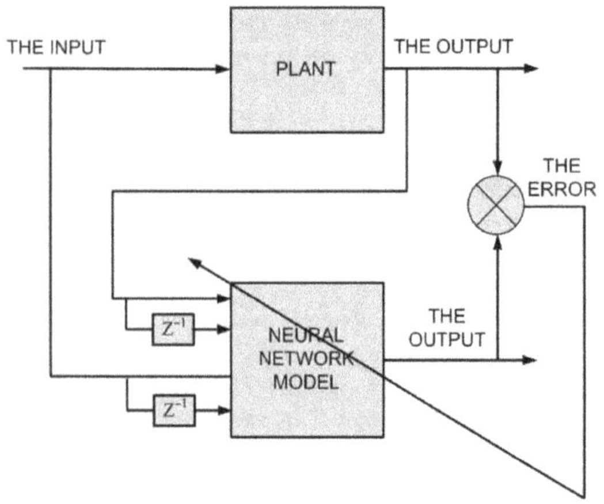

Figure 2. Dependent mode.

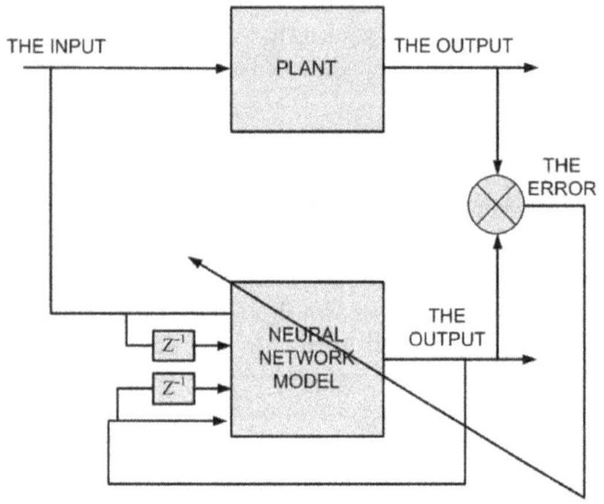

Figure 3. Independent mode.

The non-linear dynamic plant to be modelled is presented by the non-linear autoregressive with exogenous inputs (NARX) model as follows:

$$y(t) = f\,[y(t-1), \ldots, y(t-n_y), u(t-1-d), \ldots, u(t-n_u-d)] + e(t), \tag{7}$$

where $u \in \Re^m$ and $y \in \Re^p$ are the plant input and output, respectively. $e \in \Re^p$ is the random noise, m and $p$are the number of plant inputs and outputs, respectively, n_y and n_u are the maximum lags in the model output and input, respectively, $d$is the time delay in inputs and $f\,(*)$ is a vector-valued non-linear function.

The dependent mode of the network model can be represented by Equation (8), which is referred to dependent mode as the prediction uses the process output and, therefore the model cannot run independent of the process

$$\hat{y}(t) = \hat{f}\,[y(t-1), \ldots, y(t-n_y), u(t-1-d), \ldots, u(t-n_u-d)], \tag{8}$$

where $\hat{f}\,(*)$ is a function approximation of $f\,(*)$. If the past process outputs in the network input are replaced by the network outputs as in Equation (9), then the model is referred to an independent model

$$\hat{y}(t) = \hat{f}\,[\hat{y}(t-1), \ldots, \hat{y}(t-n_y), u(t-1-d), \ldots, u(t-n_u-d)]. \tag{9}$$

The RBF network performs non-linear mapping, and is used because of its advantages over the MLP network of short training time. The RBFNN in this research consists of three layers: an input layer, a hidden layer and an output layer. The hidden layer contains a number of RBF neurons; each of them represents a single RBF, with associated centre and width, and calculates the Euclidean distance between centre c and RBF network input vector x defined by $\| x(t) - c_j(t) \|$, where $c_j(t)$ is jth centre and $x(t)$ is the NN input vector which is given as follows:

$$x(t) = f\,[y(t-1), \ldots, y(t-n_y), u(t-1-d), \ldots, u(t-n_u-d)]. \tag{10}$$

Then, the output of the hidden layer nodes is produced by a so-called non-linear activation function $\varphi_j(t)$. In this work, the Gaussian basis function is chosen as the non-linear activation function

$$\varphi_j(t) = \exp\left(-\frac{\| x(t) - c_j(t) \|^2}{\sigma_j^2} \right), \quad j = 1, \ldots, n_h, \tag{11}$$

where σ_j is a positive scalar called a width and n_h is the number of centres. The network outputs are then computed as a linear weighted sum of the hidden node outputs and bias as follows:

$$\hat{y}_i(t) = \sum_j^{n_h} \varphi_j(t)^T w_{ji}, \quad i = 1, \ldots, q, \tag{12}$$

where w_{ji} is the output layers weight connecting the jth centre output and ith network output, and q is the number of outputs.

3.2. *Input–output determination of RBF model*

The first step towards developing a NN model of the process is to obtain training data. The training data are obtained by designing a set of random amplitude signals (RAS) for the five inputs to the reactor: monomer feed rate, fouling factor, ambient temperature, impurity factor and valve position, as shown in Figure 4. These five inputs are defined as, the system inputs (monomer feed rate and manipulated variable) including the uncertainties and disturbances in the process (fouling factor, ambient temperature and impurity factor). The second step towards developing a NN model of the process is to determine the

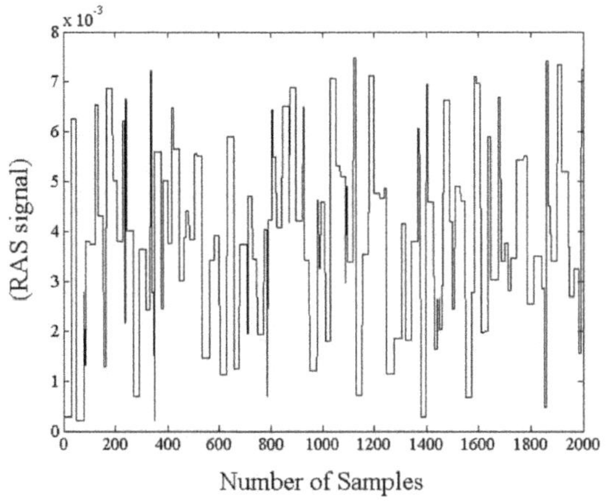

Figure 4. Random amplitude signal.

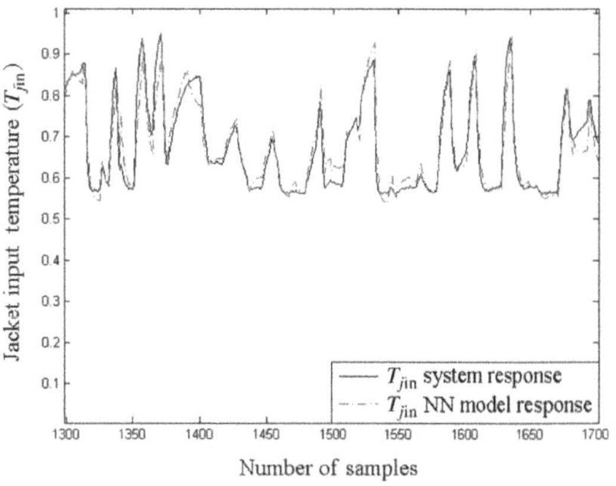

Figure 5. Simulation result of jacket input temperature system output and RBF model.

network input variables and the input vector and the output vector. The network input vector consists of the past values of the five system inputs and the past values of the three system outputs. The determination of the inputs and outputs of the system is based on Equations (1)–(5). A total data set of 2000 samples is collected from the system Simulink model, and 4 s are used as the sampling time. The first 1500 samples are used for training the network model and the remaining 500 samples are used for testing the network model. Before training and testing, the raw data are scaled linearly into the range of [0 1] using the following formulae:

$$u = \begin{bmatrix} m_M \\ 1/h_f \\ T_{amb} \\ i \\ c \end{bmatrix}, \quad y = \begin{bmatrix} T_{j\,in} \\ T_{j\,out} \\ T \end{bmatrix}, \tag{13}$$

$$u_{scaled}(k) = \frac{u(k) - u_{min}}{u_{max} - u_{min}}, \quad y_{scaled}(k) = \frac{y(k) - y_{min}}{y_{max} - y_{min}}. \tag{14}$$

3.3. *RBF model training data acquisition for open-loop and validation*

In this research, an independent RBF network is used to represent the NARX model in Equation (9). Thus, in order to get a good training result with a minimum modelling error, several numbers of maximum lags in the outputs and inputs and several numbers of the maximum time delay in the inputs are tried. The maximum lags in the output were selected as 3, the maximum lags in the input were selected as 2 and the maximum time delay in the inputs is selected as 2, as described in Equation (15). Thus, the RBF model is designed to have 19 inputs and 3 outputs, as shown in Figure 12. The hidden layer nodes are selected as 21. The centres are chosen using a K-means clustering algorithm as 21. Moreover, a p-nearest-neighbours algorithm is used to choose the widths. In the training of the network model, the RLS algorithm is used to update the weight matrix since the weights are linearly related to the output, and the parameters of the RLS algorithm are selected as follows: $\mu = 0.999$, $w(0) = 10^{-6} * U(n_h, 3)$ and $p(0) = 10^6 * I(n_h)$, where μ is the forgetting factor, I is an identity matrix, U is the element unity matrix and n_h is the number of hidden layer nodes:

$$x(t) = [y(t-1) \quad y(t-2) \quad y(t-3) \quad u(t-k-1) \quad u(t-k-2)]^T. \tag{15}$$

Figures 5–7 show the last 200 sample intervals in the training data set and the first 200 sample intervals in the testing data set. It can be clearly seen that the model outputs track the system output with a small modelling error. The mean absolute error (MAE) for the jacket input temperature, jacket output temperature and reactor temperature are 0.004, 0.0054 and 0.0072, respectively.

4. Fault detection

4.1. *Simulating faults*

In this study, after training the independent RBF network model with healthy data, the model will be tested with faulty data. The faulty data are obtained by simulating different faults in the proposed reactor. These faults are classified as three sensor faults and one actuator fault. The sensor faults are jacket input temperature sensor fault, jacket output temperature sensor fault and reactor temperature sensor fault, and the actuator fault is the inlet temperature. These faults are simulated in the following sections.

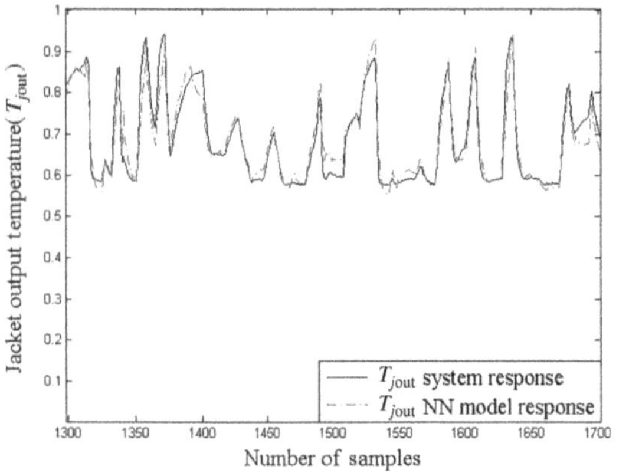

Figure 6. Simulation result of jacket output temperature system output and RBF model output.

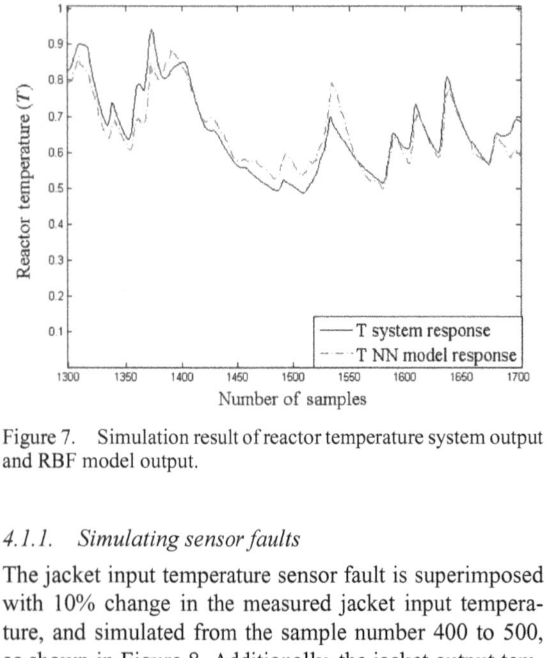

Figure 7. Simulation result of reactor temperature system output and RBF model output.

4.1.1. *Simulating sensor faults*

The jacket input temperature sensor fault is superimposed with 10% change in the measured jacket input temperature, and simulated from the sample number 400 to 500, as shown in Figure 8. Additionally, the jacket output temperature sensor fault is superimposed with 10% change in the measured jacket output temperature, and simulated from the sample number 600 to 700, as shown in Figure 8. Furthermore, the sensor fault of the reactor temperature is superimposed with 10% change in the measured temperature, and simulated from the sample number 800 to 900, as shown in Figure 8.

4.1.2. *Simulating actuator fault*

The heating–cooling function is influenced by an equal-percentage valve with valve position. When the valve position $c < 50\%$, cooling water with inlet temperature (278.71 k) is inserted into the cooling jacket. When the valve position $c > 50\%$, steam with temperature (449.82

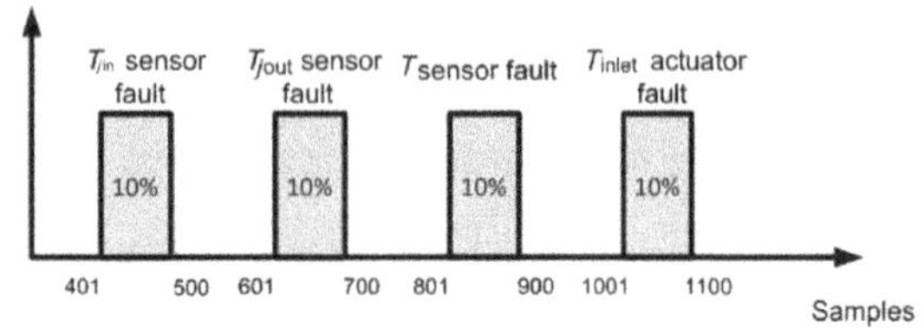

Figure 8. Fault structure with respect to the number of samples.

k) is injected into the recirculating water stream, which will lead to heating up of the coolant. Consequently, it is assumed here that a failure in the pump position of cooling mode has occurred, which leads to an increase in the temperature by 10% change in the measured inlet temperature. This inlet temperature fault is simulated from the sample number 1000 to 1100, as shown in Figure 8.

4.2. *Residual generation*

Figure 9 demonstrates the fault detection approach. An independent model is implemented in parallel with the system to generate the residuals for detecting the sensor and actuator faults in the reactor. After training the network model with healthy random data, as described in the previous section, all four faults were simulated to the reactor model. Then, with another set of 2000 samples, faulty square data are collected. These faulty data are collected by designing a set of square waves for all inputs.

These five inputs are the system inputs (monomer feed rate and manipulated variable) including the uncertainties and disturbances in the process (fouling factor, ambient temperature and impurity factor). The second step towards developing a NN model of the process is to determine the network input variables and the input vector and the output vector. The network input vector consists of the past values of the five system inputs and the past values of the three system outputs, where $m_M(t)$, $1/h_f$, T_{amb}, i and $c(t)$ are the inputs of the system; and jacket input temperature $T_{j\,in}(t)$, jacket output temperature $T_{j\,out}(t)$ and reactor temperature $T(t)$ are the outputs of the system. Moreover, the collected data are scaled linearly. After determining and scaling the input and output vectors of the system, the multivariable NARX is used to represent the non-linear dynamics of the reactor, The maximum lags in the output were selected as 3, the maximum lags in the input were selected as 2 and the maximum time delay in the inputs is selected as 2, as described in Equation (14). Here again the NN is realized by a RBF network with Gaussian basis functions. Moreover, the centres are chosen again using a K-means clustering algorithm and the widths are chosen using p-nearest-neighbours. Different numbers of hidden nodes, such as 21, 31 and 51, are used in order to get good results. The RLS algorithm is used to update the weight matrix. The parameters of the RLS algorithm

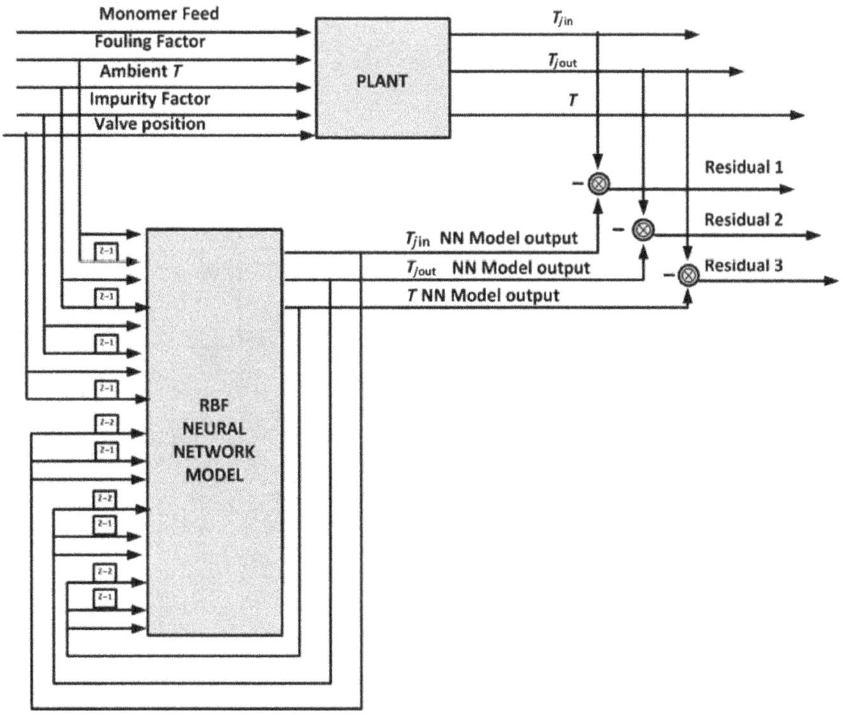

Figure 9. The structure of FD using an independent RBFNN.

are selected as follows: $\mu = 0.999$, $w(0) = 10^{-5} * U(n_h, 5)$ and $p(0) = 10^5 * I(n_h)$, where μ is the forgetting factor, I is an identity matrix, U is the element unity matrix and n_h is the number of hidden layer nodes. The RBF network model is tested with these faulty square data to generate fault-detection residuals. The filtered model prediction errors are shown in Figures 10–12. The first model prediction error of jacket input temperature is shown in Figure 10 and that for jacket output temperature and reactor temperature are shown in Figures 11 and 12, respectively. In this study, the residual ε is generated as the sum-squared filtered modelling error as follows:

$$e(t) = [y(t) - \hat{y}(t)],$$

$$\varepsilon(t) = \sqrt{(e_{T_{j\mathrm{in}}})^2 + (e_{T_{j\mathrm{out}}})^2 + (e_T)^2}.$$

The residuals of testing the neural model are slightly bigger than the residuals of training the neural model. The MAE index is used to evaluate the modelling effects. The MAE for the jacket input temperature, jacket output temperature and reactor temperature are 0.004, 0.0054 and 0.0072, respectively. Figure 10 demonstrates the residuals after using a low-pass filter. It can be observed that the independent network model output is not influenced by any type of fault, because an independent model does not use past faulty measurements as inputs. Thus, it can be clearly noticed that all faults have been clearly detected.

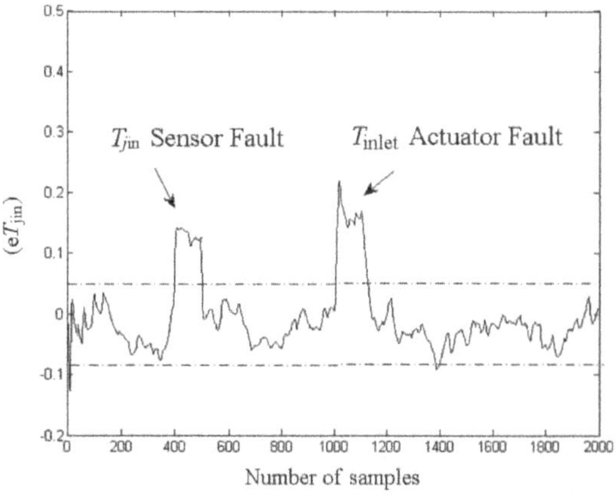

Figure 10. Residual filtered model prediction error of $T_{j\mathrm{in}}$.

Moreover, no false alarms are thereby produced, so this verifies that the proposed scheme has shown an excellent diagnostic performance.

5. Fault isolation

Figure 13 illustrates the fault isolation strategy; an additional NN is applied as a classifier for fault isolation. The application of NNs for fault isolation has been used by many researchers, such as Patton et al. (1994) and Yu et al.

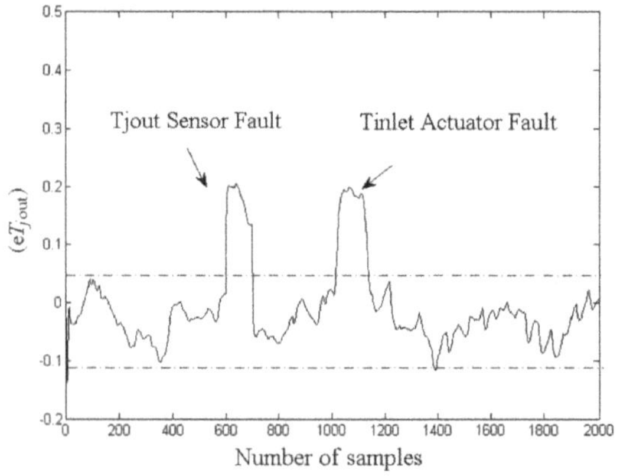

Figure 11. Residual filtered model prediction error of T_{jout}.

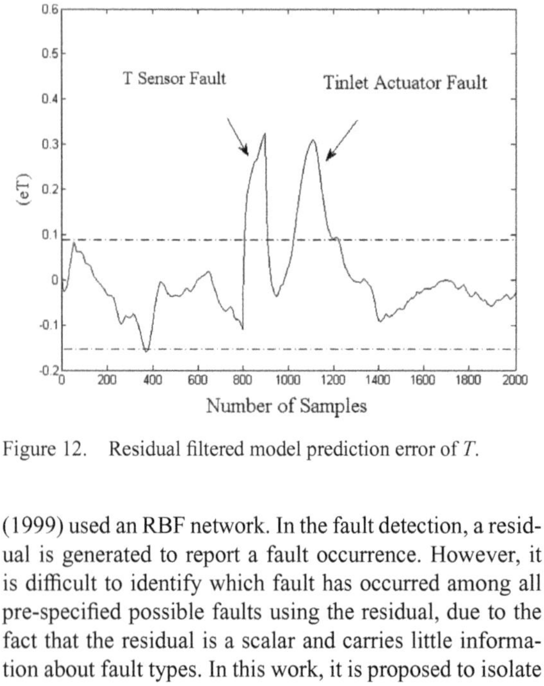

Figure 12. Residual filtered model prediction error of T.

(1999) used an RBF network. In the fault detection, a residual is generated to report a fault occurrence. However, it is difficult to identify which fault has occurred among all pre-specified possible faults using the residual, due to the fact that the residual is a scalar and carries little information about fault types. In this work, it is proposed to isolate faults according to model prediction errors. The model prediction errors are multi-dimensional, three-dimension in this case, and different faults will have different impacts on these vectors in three-dimensional vector space. Classification of these features of different faults on the model prediction error vectors will lead to classification of different faults. Therefore, the faults that have occurred can be isolated. In this work, the neural classifier is developed by an RBF network with Gaussian basis functions. The residuals shown in Figures 10–12 which are the difference between the real system output and the tested neural output were used as inputs for the RBF network classifier. Moreover, the neural classifier was developed with five outputs, with four outputs associated with the four faults and one output for no-fault case. The centres are chosen

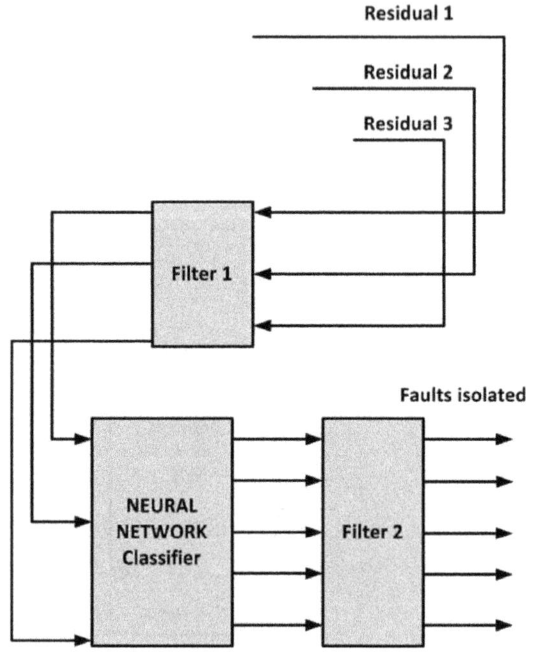

Figure 13. Block diagram for fault isolation.

again using a K-means clustering algorithm and the widths are chosen using p-nearest-neighbours. Different numbers of hidden nodes, such as 51, 151 and 251, are used in order to get good results. Finally, 51 hidden layer nodes are selected and the centres are chosen as 51. The parameters of the RLS algorithm are selected as follows: $\mu = 0.9999$, $w(0) = 10^{-6} * U(n_h, 5)$ and $p(0) = 10^6 * I(n_h)$. The samples arranged for fault occurrence are illustrated in Table 3. Moreover, the target is set such that all four outputs are set as zero for the healthy condition data, and one output is set as 1 for a specific fault, with the others remaining at zero. Thus, once the first output is 1 and the other outputs are zero, this means that the jacket input temperature sensor fault with 10% change has occurred. In the same way, the jacket output temperature sensor fault with 10% is believed to have occurred when the second output is 1, while the others remain at zero. Similarly, the reactor temperature sensor fault and the inlet temperature actuator fault with 10% changes will have occurred when the third and the fourth outputs are 1. After training, the RBF network classifier is tested with another set of faulty data with the same arrangement of training data. The samples arranged for fault occurrence can be different from those of the training data. Table 1 shows the classification of faults with respect to the number of samples. The four outputs of the neural classifier after use of a filter are displayed in Figures 14–17. It can be clearly noticed that all faults have been clearly detected and isolated.

Robust is that the fault detection always see residual sensitive to the fault but insensitive to the disturbances.

When the disturbances come in it will not affect the report of the fault, and will not increase false alarm reading. False alarm reading is that when there is no fault but fault is reported and when there is a fault but is not reported. False alarm reading should be zero percentage when all faults are reported, if there is no fault but report affected by disturbances this should be zero, but if not zero then should

Table 3. Classification of faults with respect to the number of samples.

Faults	Number of samples
No fault	0–400
$T_{j\,in}$ sensor fault	401–400
No fault	501–600
$T_{j\,out}$ sensor fault	601–700
No fault	701–800
Reactor temperature sensor fault	801–900
No fault	901–1000
Inlet temperature actuator fault	1001–1100
No fault	1101–2000

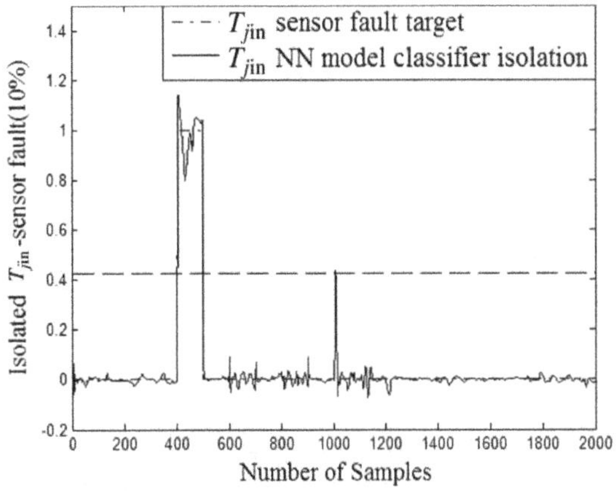

Figure 14. Classifier output 1.

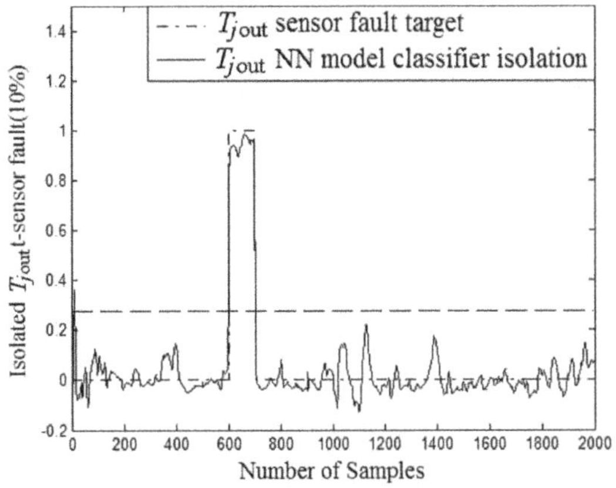

Figure 15. Classifier output 2.

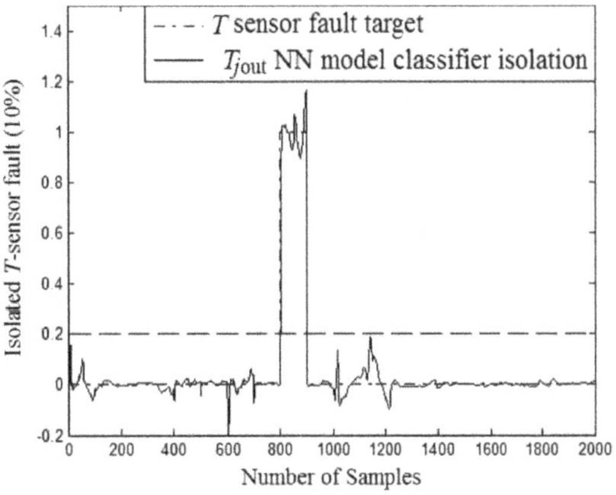

Figure 16. Classifier output 3.

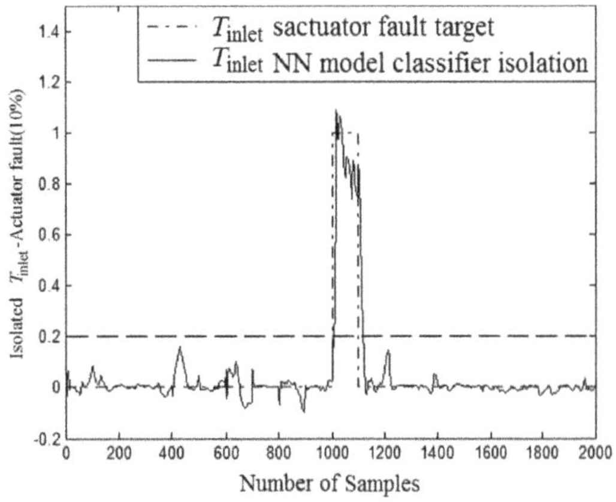

Figure 17. Classifier output 4.

be reduced as small as possible. In this research work, when collecting training data all disturbances are simulated, because of this the model is trained considering the disturbances. When disturbances occurred will not affect the residual that because the disturbances in this system are not big enough to make the residual high, in this process the disturbances just change the non-linear function of the system and that are big enough from the control point of view. It is observed from simulation results that all faults have been clearly detected and isolated, and no false alarm was thereby produced, so this verifies that the proposed scheme has shown an excellent performance. Note that the outputs are not zero when no faults occur, as a result of the effects of the disturbances.

6. Conclusion

A new robust fault diagnosis scheme has been developed for an open-loop Chylla–Haase reactor using an independent RBFNN. Three sensor faults and one actuator fault have been simulated on the Chylla–Haase reactor. All the simulated faults are superimposed with 10% changes in the measured temperatures, and simulated for different numbers of samples. Moreover, the uncertainties and disturbances in the process, such as fouling factor, impurity factor and measurement noise, have been simulated. Two different techniques to employ RBF NNs for fault diagnosis have been investigated. The first technique is implementing an independent RBNN for residual generation. Moreover, the generated residuals were used for detecting actuator and sensor faults. The second technique is applying an additional RBFNN as a classifier to perform the classification task for residual evaluation and therefore to diagnose and isolate the actuator and sensor faults from the generated residuals. The simulation results show that all faults were clearly detected and isolated. Moreover, no false alarms are thereby produced, so this verifies that the proposed scheme has shown an excellent diagnosis performance. The main contribution of this work is to show how to apply an independent RBFNN to open-loop Chylla–Haase benchmark polymerization reactor fault diagnosis. Therefore, this proposed method can contribute to the safety of chemical reactors. In future work, we propose to develop a new robust FDI method for closed-loop reactor using an independent RBFNN. Due to the high non-linearity and the noise in the process, we will propose to develop a new FDI method for closed-loop reactor using the extended Kalman filter.

Disclosure statement

No potential conflict of interest was reported by the author(s).

References

Beyer, M.-A., Grote, W., & Reinig, G. (2008). Adaptive exact linearization control of batch polymerization reactors using a sigma-point Kalman filter. *Journal of Process Control*, *18*(7–8), 663–675.

Chylla, R. W., & Haase, D. R. (1993). Temperature control of semibatch polymerization reactors. *Computers & Chemical Engineering*, *17*(3), 257–264.

Deibert, R., & Isermann, R. (1992). Examples for fault detection in closed loops. *Annual Review in Automatic Programming*, *17*, 235–240.

Ertiame, A. M., Dingli, Y., Feng, Y., & Gomm, J. B. (2013). *Fault detection and isolation for open-loop Chylla–Haase polymerization reactor*. 19th international conference on automation and computing (ICAC), Brunel University, London, UK (pp. 1–6).

Ferrari, R. M. G., Parisini, T., & Polycarpou, M. M. (2008). *A robust fault detection and isolation scheme for a class of uncertain input-output discrete-time nonlinear systems*. American control conference, Seattle, Washington, USA (pp. 2804–2809).

Frank, P. M., & Köppen-Seliger, B. (1997). Fuzzy logic and neural network applications to fault diagnosis. *International Journal of Approximate Reasoning*, *16*(1), 67–88.

Gertler, J. J. (1988). Survey of model-based failure detection and isolation in complex plants. *IEEE Control Systems Magazine*, *8*(6), 3–11.

Gomm, J. B., & Yu, D. L. (2000). Selecting radial basis function network centers with recursive orthogonal least squares training. *IEEE Transactions on Neural Networks*, *11*(2), 306–314.

Graichen, K., Hagenmeyer, V., & Zeitz, M. (2005). *Adaptive feedforward control with parameter estimation for the Chylla–Haase polymerization reactor*. 44th IEEE conference on decision and control, and the European Control Conference, Seville, Spain (pp. 3049–3054).

Graichen, K., Hagenmeyer, V., & Zeitz, M. (2006). Feedforward control with online parameter estimation applied to the Chylla–Haase reactor benchmark. *Journal of Process Control*, *16*(7), 733–745.

Isermann, R. (1984). Process fault detection based on modeling and estimation methods—A survey. *Automatica*, *20*(4), 387–404.

Koppen-Seliger, B., & Frank, P. M. (1995). *Fault detection and isolation in technical processes with neural networks*. 34th IEEE conference on decision and control, New Orleans, LA (Vol. 3, pp. 2414–2419).

Patton, R. J., Chen, J., & Siew, T. M. (1994). *Fault diagnosis in nonlinear dynamic systems via neural networks*. International conference on control, Coventry, UK (Vol. 2, pp. 1346–1351).

Xiaodong, Z. (2011). Sensor bias fault detection and isolation in a class of nonlinear uncertain systems using adaptive estimation. *IEEE Transactions on Automatic Control*, *56*(5), 1220–1226.

Xiaodong, Z., Polycarpou, M. M., & Parisini, T. (2002). A robust detection and isolation scheme for abrupt and incipient faults in nonlinear systems. *IEEE Transactions on Automatic Control*, *47*(4), 576–593.

Yu, D. L., Gomm, J. B., & Williams, D. (1999). Sensor fault diagnosis in a chemical process via RBF neural networks. *Control Engineering Practice*, *7*(1), 49–55.

Zhang, X., Polycarpou, M. M., & Parisini, T. (2010). Fault diagnosis of a class of nonlinear uncertain systems with Lipschitz nonlinearities using adaptive estimation. *Automatica*, *46*(2), 290–299.

17

On the numerical solution of the nonlinear Korteweg–de Vries equation

Maryam Sarboland[a] and Azim Aminataei[b]*

[a]*Department of Mathematics, Saveh Branch, Islamic Azad University, PO Box 39187-366, Saveh, Iran;* [b]*Faculty of Mathematics, Department of Applied Mathematics, K. N. Toosi University of Technology, PO Box 16315-1618, Tehran, Iran*

In this paper, we present a new method for solving a nonlinear third-order Korteweg–de Vries equation. This method is based on the multiquadric (MQ) quasi-interpolation operator $\mathcal{L}_{\mathcal{W}_2}$ and an integrated radial basis function networks scheme. In the present scheme, the second-order central divided difference of the spatial derivative is used to approximate the third-order spatial derivative, and the Taylors series expansion to discretize the temporal derivative. Then, the spatial derivative is approximated by the MQ quasi-interpolation operator $\mathcal{L}_{\mathcal{W}_2}$. This method is applied on some test experiments and the numerical results have been compared with the exact solutions and the solutions of other numerical methods. The L_∞, L_2 and root-mean-square errors of the solutions show the efficiency and the accuracy of the method. Furthermore, the stability analysis of the method is surveyed.

Keywords: nonlinear KdV equation; multiquadric quasi-interpolation scheme; integrated radial basis function networks scheme; Taylors series expansion

1. Introduction

In this paper, we concentrate on the numerical solution of one of the well-known equation named as Korteweg–de Vries (KdV) equation:

$$u_t + \varepsilon u u_x + \mu u_{xxx} = 0, \quad x \in \Omega = [a, b],\ t \geqslant t_0, \qquad (1)$$

where ε and μ are positive parameters. The KdV equation expresses a balance between dispersion form from its third derivative term u_{xxx} and the shock forming tendency of its nonlinear term uu_x.

The most important property of Equation (1) is that solutions may exhibit solitons. Solitons are localized waves that propagate without change of its shape and velocity and are stable in mutual interaction just like the phenomenon of totally elastic collision in kinetics. Over the years, the KdV equation has found wide applications in many fields such as waves in enharmonic crystals, bubble liquid mixtures, ion acoustic wave and magneto-hydrodynamic waves in a warm plasma as well as shallow water waves (Dodd, Eilbeck, Gibbon, & Morris, 1982; Gardner & Marikawa, 1965; Korteweg–de Vries & de Vries, 1895; Washimi & Taniuti, 1966; Wijngaarden, 1968).

The KdV equation is a completely integrable Hamiltonian system which can be solved explicitly. Thus, some analytical solutions of it are found, and for appropriate initial conditions, their existence and uniqueness have been shown by Gardner, Grrne, and Kruskal (1967). The numerical solutions of the KdV equation can be found by using the techniques known as finite difference schemes, finite element schemes, and Fourier spectral methods (Geyikli & Kaya, 2005; Helal & Mehanna, 2006; Li & Wang, 2007; Yan, 2006). The difficulty of mesh generation, especially in two or more dimensions, makes these methods hard to implement. In the past two decades, the radial basis function (RBF) meshless method has been considered for the numerical solution of the various types of partial differential equations (PDEs). These methods do not require the structured grid, that is, that are truly meshless methods. Kansa (1990a, 1990b) was the first researcher to approximate PDEs using RBFs. He has directly collocated the RBFs, particularly the multiquadric (MQ), for the approximated solutions of the PDEs. The MQ–RBF was first developed by Hardy (1971) as a multi-dimensional scattered interpolation method. Franke (1982) published a detailed comparison of 29 different scattered data schemes for analytic problems. Of all the techniques tested, he concluded that the MQ approximation scheme performed the best in accuracy, visual appeal, and ease of implementation.

In most of the known methods of solving PDEs using the MQ approximation scheme, one must resolve a linear system of equations at each time step. Hon and Wu (2000) and Wu (2004, 2005) and others have provided some successful examples using the MQ quasi-interpolation scheme for solving differential equations.

*Corresponding author. Email: ataei@kntu.ac.ir

Beatson and Powell (1992) proposed three univariate MQ quasi-interpolations, namely, $\mathcal{L}_{\mathcal{A}}$, $\mathcal{L}_{\mathcal{B}}$ and $\mathcal{L}_{\mathcal{C}}$. Wu and Schaback (1994) presented the univariate MQ quasi-interpolation $\mathcal{L}_{\mathcal{D}}$ and proved that the scheme is shape preserving and convergent. Chen and Wu (2006, 2007) used MQ quasi-interpolation to solve Burgers' equation and hyperbolic conservation laws. Also, Xiao, Wang, and Zhu (2011) presented the numerical method based on Chen and Wu's method for solving the third-order KdV equation. Recently, Jiang, Wang, Zhu, and Xu (2011) have introduced a new multi-level univariate MQ quasi-interpolation approach with high approximation order compared with the initial MQ quasi-interpolation scheme. This approach is based on inverse multiquadric (IMQ) RBF interpolation, and Wu and Schaback's MQ quasi-interpolation operator $\mathcal{L}_{\mathcal{D}}$ that have the advantages of high approximation order.

In the Kansa's RBF method and the other conventional RBF method, the RBF approximation is directly applied to the solution function. As such, any derivative of the solution function can be obtained by differentiating the RBF expression. But recently, Mai-Duy, Mai-Cao, and Tran-Cong (2007) and Mai-Cao and Tran-Cong (2005) have introduced an integrated radial basis function networks (IRBFNs) scheme for the approximation of the solution function and its derivatives. In the IRBFNs approach, the RBF approximation is applied to a targeted derivative (first-order or second-order) of the solution function. Consequently, the solution function is obtained by integrating the derivative with the RBF expression. Numerical experiments and theoretical analysis indicate that the IRBF scheme is more accurate than the direct radial basis function (DRBF) for solving PDEs (Aminataei & Mazarei, 2008).

The purpose of this paper is to present a new numerical scheme to solve the nonlinear third-order KdV equation, based on the IRBF scheme and MQ quasi-interpolation scheme. In our scheme, we use the MQ quasi-interpolation scheme and the integration of it to approximate the first-order spatial derivative and the solution function, respectively, similar to the work that we did in Sarboland and Aminataei (2014). Besides, we use the second-order central divided difference of the first-order spatial derivative to approximate the third-order spatial derivative. Also, we employ Taylors series expansion to approximate the temporal derivative as Dağ, Conlvar, and Şahin (2011) did for Burgers' equation.

The organization of this paper is as follows. Section 2 gives brief information about RBFs and the MQ quasi-interpolation scheme. In Section 3, an IRBFNs scheme is presented. In Section 4, we apply the method on the nonlinear third-order KdV equation. The stability analysis of the methods is discussed in Section 5. The results of three numerical experiments are presented in Section 6 and are compared with the analytical solutions and the results in Xiao et al. (2011) and Siraj, Khattak, and Tirmizi (2008). Finally, a brief discussion and conclusion are presented in Section 7.

2. The RBFs and the MQ quasi-interpolation scheme

In this section, we present some elementary knowledge about RBF interpolation and introduce three univariate MQ quasi-interpolation schemes, namely, $\mathcal{L}_{\mathcal{D}}, \mathcal{L}_{\mathcal{W}}$ and $\mathcal{L}_{\mathcal{W}_2}$. For more details about MQ quasi-interpolation operators, one can see Beatson and Powell (1992), Wu and Schaback (1994), Chen and Wu (2006, 2007) and Powell (1992).

For a given region $\Omega = [a, b]$ and a finite set of distinct points

$$a = x_0 < x_1 < \cdots < x_N = b, \quad h = \max_{1 \leqslant i \leqslant N} (x_i - x_{i-1}),$$

quasi-interpolation of a univariate function $f : [a, b] \to \mathbb{R}$ has the form

$$\mathcal{L}(f) = \sum_{i=0}^{N} f(x_i)\varphi_i(x),$$

where each function $\varphi_i(x)$ is a linear combination of the MQs

$$\psi_i(x) = \sqrt{c^2 + (x - x_i)^2},$$

and $c \in \mathbb{R}^+$ is a shape parameter. Wu and Schaback (1994) presented the univariate MQ quasi-interpolation operator $\mathcal{L}_{\mathcal{D}}$ that is defined as

$$\mathcal{L}_{\mathcal{D}}f(x) = \sum_{i=0}^{N} f(x_i)\tilde{\psi}_i(x), \tag{2}$$

where

$$\begin{aligned}
\tilde{\psi}_0(x) &= \frac{1}{2} + \frac{\psi_1(x) - (x - x_0)}{2(x_1 - x_0)}, \\
\tilde{\psi}_1(x) &= \frac{\psi_2(x) - \psi_1(x)}{2(x_2 - x_1)} - \frac{\psi_1(x) - (x - x_0)}{2(x_1 - x_0)}, \\
\tilde{\psi}_i(x) &= \frac{\psi_{i+1}(x) - \psi_i(x)}{2(x_{i+1} - x_i)} - \frac{\psi_i(x) - \psi_{i-1}(x)}{2(x_i - x_{i-1})}, \\
& \quad 2 \leqslant i \leqslant N - 2, \\
\tilde{\psi}_{N-1}(x) &= \frac{(x_N - x) - \psi_{N-1}(x)}{2(x_N - x_{N-1})} - \frac{\psi_{N-1}(x) - \psi_{N-2}(x)}{2(x_{N-1} - x_{N-2})},
\end{aligned} \tag{3}$$

and

$$\tilde{\psi}_N(x) = \frac{1}{2} + \frac{\psi_{N-1}(x) - (x_N - x)}{2(x_N - x_{N-1})}.$$

In RBFs interpolation, we can get high approximation order by increasing the number of interpolation centers but we have to solve unstable linear system of equations. By using the MQ quasi-interpolation scheme, we can avoid this problem, but the approximation order is not good. Hence, Jiang et al. (2011) have defined two

MQ quasi-interpolation operators denoted as $\mathcal{L}_{\mathcal{W}}$ and $\mathcal{L}_{\mathcal{W}_2}$, which pose the advantages of RBFs interpolation and MQ quasi-interpolation schemes. The process of MQ quasi-interpolation of $\mathcal{L}_{\mathcal{W}}$ and $\mathcal{L}_{\mathcal{W}_2}$ is as follows and is described in Jiang et al. (2011).

Suppose that $\{x_{k_j}\}_{j=1}^{\bar{N}}$ is a smaller set from the given points $\{x_i\}_{i=0}^{N}$ where $\bar{N}$ is a positive integer satisfying $\bar{N} < N$ and $0 = k_0 < k_1 < \cdots < k_{\bar{N}+1} = N$. Using the IMQ–RBF, the second derivative of $f(x)$ can be approximated by the RBF interpolant $S_{f''}$ as

$$S_{f''}(x) = \sum_{j=1}^{\bar{N}} \alpha_j \bar{\varphi}(|x - x_{k_j}|), \tag{4}$$

where

$$\bar{\varphi}(r) = \frac{s^2}{(s^2 + r^2)^{3/2}},$$

and $s \in \mathbb{R}^+$ is a shape parameter. The coefficients $\{\alpha_j\}_{j=1}^{\bar{N}}$ are uniquely determined by the interpolation condition

$$S_{f''}(x_{k_i}) = \sum_{j=1}^{\bar{N}} \alpha_j \bar{\varphi}(|x_{k_i} - x_{k_j}|) = f''(x_{k_i}), \quad 1 \leqslant i \leqslant \bar{N}. \tag{5}$$

Since, Equation (5) is solvable (Madych & Nelson, 1990), so

$$\alpha = A_X^{-1} \cdot f_X'', \tag{6}$$

where

$$X = \{x_{k_1}, \ldots, x_{k_{\bar{N}}}\}, \quad \alpha = [\alpha_1, \ldots, \alpha_{\bar{N}}]^{\mathrm{T}},$$
$$A_X = [\bar{\varphi}(|x_{k_i} - x_{k_j}|)], \quad f_X'' = [f''(x_{k_1}), \ldots, f''(x_{k_{\bar{N}}})]^{\mathrm{T}}.$$

By using f and the coefficient α defined in Equation (6), a function $e(x)$ is constructed in the following form:

$$e(x) = f(x) - \sum_{i=1}^{\bar{N}} \alpha_i \sqrt{s^2 + (x - x_{k_i})^2}. \tag{7}$$

Then the MQ quasi-interpolation operator $\mathcal{L}_{\mathcal{W}}$ by using $\mathcal{L}_{\mathcal{D}}$ defined by Equations (2) and (3) on the data $(x_i, e(x_i))_{0 \leqslant i \leqslant N}$ with the shape parameter c is defined as follows:

$$\mathcal{L}_{\mathcal{W}} f(x) = \sum_{i=1}^{\bar{N}} \alpha_i \sqrt{s^2 + (x - x_{k_i})^2} + \mathcal{L}_{\mathcal{D}} e(x). \tag{8}$$

The shape parameters c and s should not be the same constant as in Equation (8).

In Equation (5), $f''_{x_{k_j}}$ can be replaced by

$$f''_{x_{k_j}} = \frac{2f(x_{k_{j+1}})}{(x_{k_{j+1}} - x_{k_j})(x_{k_{j+1}} - x_{k_{j-1}})} - \frac{2f(x_{k_j})}{(x_{k_j} - x_{k_{j-1}})(x_{k_{j+1}} - x_{k_j})} + \frac{2f(x_{k_{j-1}})}{(x_{k_j} - x_{k_{j-1}})(x_{k_{j+1}} - x_{k_{j-1}})},$$

when the data $(x_{k_i}, f(x_{k_i}))_{0 \leqslant i \leqslant \bar{N}}$ are given. So, if f_{X}'' in Equation (6) can be replaced by

$$F_X'' = (f''_{x_{k_1}}, \ldots, f''_{x_{k_{\bar{N}}}})^{\mathrm{T}}, \tag{9}$$

then the quasi-interpolation operator defined by Equations (7) and (8) is denoted by $\mathcal{L}_{\mathcal{W}_2}$. For more details about the properties and accuracy of $\mathcal{L}_{\mathcal{W}}$ and $\mathcal{L}_{\mathcal{W}2}$, one can see Jiang et al. (2011). In this paper, we use the MQ quasi-interpolation operator $\mathcal{L}_{\mathcal{W}_2}$ and $N = 4\bar{N}$.

The operator $\mathcal{L}_{\mathcal{W}_2}$ can be written in the compact form

$$\mathcal{L}_{\mathcal{W}_2} f(x) = \sum_{i=0}^{N} f(x_i) \widehat{\psi}_i(x),$$

where the basis functions $\widehat{\psi}_i(x)$ are obtained by substituting Equations (6), (7) and (9) into Equation (8), see Sarboland and Aminataei (2014).

3. The IRBFNs scheme

In the past two decades, there has been some developments in applying the RBFs for the numerical solutions of various types of PDEs. In the RBF-based methods or DRBF methods, a RBF approximation is directly applied to the solution function. As such, any derivative of the solution function can be obtained by differentiating the RBF expression. Although the method has the ability to represent any continuous function to a prescribed degree of accuracy, the process of differentiation magnifies any errors that might arise from approximating the original function and thus result in inaccurate derivatives. But the IRBF method starts with the decomposition of the highest derivative (one-order in this paper) of the solution function into a linear combination of RBFs. Then, the solution function is obtained by symbolic integration. In contrast to the process of differentiation where any errors associated with the function approximation might be amplified, the integration has the effect of averaging out such errors. The IRBF method, therefore, results in a better approximation accuracy than the usual approach as in the DRBF method with the same numerical configuration. The IRBF method has been described completely in Mai-Duy et al. (2007), Mai-Cao and Tran-Cong (2005) and Aminataei and Mazarei (2008).

Due to the aforementioned advantages of the IRBF method over the DRBF method, it is decided to apply the MQ quasi-interpolation method in the indirect form. Hence, the MQ quasi-interpolation operator $\mathcal{L}_{\mathcal{W}2}$ is used to approximate the highest derivative of the solution function that is described in Section 2. This method is as follows.

Let $u(x)$ be an unknown function defined on a bounded domain $[a,b] \subset \mathbb{R}$. Then

$$\frac{\partial u}{\partial x} = \sum_{i=0}^{N} \frac{\partial u}{\partial x}(x_i)\widehat{\psi}_i(x), \tag{10}$$

$$u(x) = \sum_{i=0}^{N} \frac{\partial u}{\partial x}(x_i) \int \widehat{\psi}_i(x)\,\mathrm{d}x + C_1. \tag{11}$$

Equations (10) and (11) can be rewritten in the compact forms as follows:

$$\frac{\partial u}{\partial x} = \sum_{i=0}^{N+1} w_i \varphi_i(x), \tag{12}$$

$$u(x) = \sum_{i=0}^{N+1} w_i \tilde{\varphi}_i(x), \tag{13}$$

where

$$\varphi_i(x) = \widehat{\psi}_i(x), \quad 0 \leqslant i \leqslant N, \quad \varphi_{N+1}(x) = 0,$$

$$\tilde{\varphi}_i(x) = \int \widehat{\psi}_i(x)\,\mathrm{d}x, \quad 0 \leqslant i \leqslant N, \quad \tilde{\varphi}_{N+1}(x) = 1,$$

and

$$w_i = \frac{\partial u}{\partial x}(x_i), \quad 0 \leqslant i \leqslant N, \quad w_{N+1} = C_1. \tag{14}$$

4. The numerical method on the nonlinear third-order KdV equation

Consider the nonlinear third-order KdV equation:

$$\frac{\partial u(x,t)}{\partial t} + \varepsilon u(x,t)\frac{\partial u(x,t)}{\partial x} + \mu \frac{\partial^3 u(x,t)}{\partial x^3} = 0, \tag{15}$$

with the initial condition

$$u(x,t) = u_0(x), \quad t = 0, \tag{16}$$

and the boundary conditions

$$u(x,t) = f(t), \quad x \in \partial\Omega,\ t > 0, \tag{17}$$

$$u_x(b,t) = g(t), \quad t > 0, \tag{18}$$

where $\Omega = [a,b] \subset \mathbb{R}$, ε and μ are positive parameters, $u_0(x)$,$f(t)$ and $g(t)$ are known functions.

At first, Equation (15) is discretized in time with time step Δt by using Taylors series expansion. In this approach, the term $u_t^n = u_t(x,t_n)$ can be arranged with the help of Taylors series expansion as follows:

$$u_t^n = \frac{u^{n+1} - u^n}{\Delta t} - \frac{\Delta t}{2} u_{tt}^n + O(\Delta t^2). \tag{19}$$

Differentiating Equation (15) with respect to time, u_{tt}^n may be written as follows:

$$\begin{aligned} u_{tt}^n &= (-\varepsilon u^n u_x^n - \mu u_{xxx}^n)_t \\ &= -\varepsilon u_x^n u_t^n - \varepsilon u^n (u_t^n)_x - \mu (u_t^n)_{xxx}. \end{aligned} \tag{20}$$

Using the forward difference formula for the time derivative u_t^n in Equation (20), u_{tt}^n can be rewritten as

$$\begin{aligned} \Delta t u_{tt}^n = &-\varepsilon u_x^n (u^{n+1} - u^n) - \varepsilon u^n (u_x^{n+1} - u_x^n) \\ &- \mu (u_{xxx}^{n+1} - u_{xxx}^n). \end{aligned} \tag{21}$$

Substituting Equation (21) into Equation (19) and using the resulted expression in Equation (15) lead to the following time discretized form of the KdV equation

$$\begin{aligned} &u^{n+1} + \varepsilon \frac{\Delta t}{2}(u^n u_x^{n+1} + u_x^n u^{n+1}) + \mu \frac{\Delta t}{2} u_{xxx}^{n+1} \\ &= u^n - \mu \frac{\Delta t}{2} u_{xxx}^n. \end{aligned} \tag{22}$$

Now, we apply the collocation method by using Equations (12) and (13). Also, we use the second-order central divided difference of the first-order spatial derivative to approximate the third-order spatial derivative

$$\begin{aligned} (u_{xxx})_j^n = &\frac{2(u_x)_{j+1}^n}{(x_{j+1} - x_j)(x_{j+1} - x_{j-1})} \\ &- \frac{2(u_x)_j^n}{(x_j - x_{j-1})(x_{j+1} - x_j)} \\ &+ \frac{2(u_x)_{j-1}^n}{(x_j - x_{j-1})(x_{j+1} - x_{j-1})}. \end{aligned} \tag{23}$$

Therefore, we obtain the following equations:

$$\begin{aligned} &\sum_{j=0}^{N+1} w_j^{n+1} \tilde{\varphi}_j(x_i) + \varepsilon \frac{\Delta t}{2} \left(\sum_{j=0}^{N+1} w_j^n \tilde{\varphi}_j(x_i) \sum_{j=0}^{N+1} w_j^{n+1} \varphi_j(x_i) \right. \\ &\left. + \sum_{j=0}^{N+1} w_i^n \varphi_j(x_i) \sum_{j=0}^{N+1} w_j^{n+1} \tilde{\varphi}_j(x_i) \right) \\ &+ \frac{\mu \Delta t}{2} \left(\alpha_i \sum_{j=0}^{N+1} w_j^{n+1} \varphi_j(x_{i+1}) - \beta_i \sum_{j=0}^{N+1} w_j^{n+1} \varphi_j(x_i) \right. \\ &\left. + \gamma_i \sum_{j=0}^{N+1} w_j^{n+1} \varphi_j(x_{i-1}) \right) \end{aligned}$$

$$= \sum_{j=0}^{N+1} w_j^n \tilde{\varphi}_j(x_i) - \frac{\mu \Delta t}{2}\left(\alpha_i \sum_{j=0}^{N+1} w_j^n \varphi_j(x_{i+1}) - \beta_i \sum_{j=0}^{N+1} w_j^n \varphi_j(x_i) + \gamma_i \sum_{j=0}^{N+1} w_j^n \varphi_j(x_{i-1}) \right), \quad (24)$$

for $1 \leqslant j \leqslant N-1$ and

$$\sum_{j=0}^{N+1} w_j^{n+1} \tilde{\varphi}_j(x_i) = f^{n+1}, \quad i = 0, N, \quad (25)$$

$$\sum_{j=0}^{N+1} w_j^{n+1} \varphi_j(x_N) = g^{n+1}, \quad (26)$$

where

$$\alpha_i = \frac{1}{(x_{i+1} - x_i)(x_{i+1} - x_{i-1})},$$

$$\beta_i = \frac{1}{(x_i - x_{i-1})(x_{i+1} - x_i)},$$

and

$$\gamma_i = \frac{1}{(x_i - x_{i-1})(x_{i+1} - x_{i-1})}.$$

Hence, Equations (24)–(26) generate a system of $N+1$ linear equations in $N+1$ unknown parameters w_i^{n+1}. In the above system, the centers and the collocation points have been chosen as the same. The approximate solution can be obtained from Equation (13) at any point in the problem domain after finding the unknown coefficients w_i^n in each time steps.

Equations (24)–(26) can be written in the matrix form as

$$\left[\mathbf{A}_d + \mathbf{A}_b + \frac{\varepsilon \Delta t}{2}(\mathbf{u}^n * \mathbf{D}_2 + \mathbf{u}_x^n * \mathbf{A}_d) + \frac{\mu \Delta t}{2}(\alpha * \mathbf{D}_1 - \beta * \mathbf{D}_2 + \gamma * \mathbf{D}_3) \right] \mathbf{w}^{n+1} = \left[\mathbf{A}_d - \frac{\mu \Delta t}{2}(\alpha * \mathbf{D}_1 - \beta * \mathbf{D}_2 + \gamma * \mathbf{D}_3) \right] \mathbf{w}^n + \mathbf{G}^{n+1}, \quad (27)$$

where symbol * stands for component-by-component multiplication,

$$\mathbf{A}_{d(i+1)(j+1)} = \tilde{\varphi}_j(x_i), \quad \mathbf{A}_{b(i+1)(j+1)} = 0,$$
$$\mathbf{D}_{1(i+1)(j+1)} = \varphi_j(x_{i+1}), \quad \mathbf{D}_{2(i+1)(j+1)} = \varphi_j(x_i),$$
$$\mathbf{D}_{3(i+1)(j+1)} = \varphi_j(x_{i-1}),$$

for $i = 1, \ldots, N$; $j = 0, 1, \ldots, N+1$ and

$$\mathbf{A}_{d(i+1)(j+1)} = 0, \quad \mathbf{D}_{1(i+1)(j+1)} = 0, \quad \mathbf{D}_{2(i+1)(j+1)} = 0,$$
$$\mathbf{D}_{3(i+1)(j+1)} = 0,$$

for $i = 0, N, N+1$; $j = 0, 1, \ldots, N+1$ and

$$\mathbf{A}_{b(1)(j+1)} = \tilde{\varphi}_j(x_0), \quad \mathbf{A}_{b(N+1)(j+1)} = \tilde{\varphi}_j(x_N),$$
$$\mathbf{A}_{b(N+2)(j+1)} = \varphi_j(x_N),$$

and

$$\mathbf{G}^{n+1} = [f^{n+1}(x_0), 0, \ldots, 0, f^{n+1}(x_N), g^{n+1}]^{\mathrm{T}}.$$

Subsequently, Equation (27) can be written as

$$\mathbf{w}^{n+1} = \mathbf{M}^{-1}\mathbf{N}\mathbf{w}^n + \mathbf{M}^{-1}\mathbf{G}^{n+1}, \quad (28)$$

where

$$\mathbf{M} = \mathbf{A} + \frac{\varepsilon \Delta t}{2}(\mathbf{u}^n * \mathbf{D} + \mathbf{u}_x^n * \mathbf{A}_d) + \frac{\mu \Delta t}{2}(\alpha * \mathbf{D}_1 - \beta * \mathbf{D}_2 + \gamma * \mathbf{D}_3), \quad (29)$$
$$\mathbf{N} = \mathbf{A}_d - \frac{\mu \Delta t}{2}(\alpha * \mathbf{D}_1 - \beta * \mathbf{D}_2 + \gamma * \mathbf{D}_3).$$

and $\mathbf{A} = \mathbf{A}_d + \mathbf{A}_b$. From Equations (13) and (26), it can be written as

$$\bar{\mathbf{u}}^n = \mathbf{A}\mathbf{w}^n, \quad (30)$$

where $\bar{\mathbf{u}}^n = [u^n(x_0), \ldots, u^n(x_N), g^n]^{\mathrm{T}}$. Hence, the combination of Equations (29) and (30) is given as

$$\bar{\mathbf{u}}^{n+1} = \mathbf{A}\mathbf{M}^{-1}\mathbf{N}\mathbf{A}^{-1}\bar{\mathbf{u}}^n + \mathbf{A}\mathbf{M}^{-1}\mathbf{G}^{n+1}. \quad (31)$$

It is evident that to have a unique solution for the unknown vector u^{n+1} in each time step, the matrices **M** and **A** must be non-singular and invertible. The non-singularity of these matrices depends on the situation of RBFs and cannot be proven in general. For many of the RBFs, the interpolation matrices are non-singular or specially positive-definite matrices. Notice that in many of the practical problems the singularities are rare. In the mentioned meshfree method, the accuracy and stability of the method depend on the shape parameters c and s and also on the distance between any two collocation points. The effects of this parameter on stability of the method will be investigated in the next section.

5. The stability analysis

In this section, the stability analysis from integrated quasi-interpolation scheme is presented by using spectral radius of the amplification matrix similar to work that ul-Islam, Haq, and Uddin (2009) did. Let $\mathbf{u}$ be the exact, $\mathbf{u}^*$ the numerical solution and $\mathrm{e}^{n+1} = \mathbf{u}^{n+1} - \mathbf{u}^{*n+1}$ the error vector of Equation (1). Also, let $\bar{e} = [e_0, \ldots, e_N, 0]$, then it can be written as

$$\bar{e}^{n+1} = \bar{\mathbf{u}}^{n+1} - \bar{\mathbf{u}}^{*n+1} = \mathbf{A}\mathbf{M}^{-1}\mathbf{N}\mathbf{A}^{-1}\,\bar{e}^n = \mathbf{E}\,\bar{e}^n, \quad (32)$$

where $\mathbf{E} = \mathbf{A}\mathbf{M}^{-1}\mathbf{N}\mathbf{A}^{-1}$. For the stability of the numerical scheme, we must have $\mathrm{e}^n \to 0$ as $n \to \infty$, that is,

$\rho(\mathbf{E}) \leqslant 1$ which is the necessary and sufficient condition for the numerical scheme to be stable, where $\rho(\mathbf{E})$ denotes the spectral radius of the amplification matrix $\mathbf{E}$. Equation (32) can be written as

$$\mathbf{MA}^{-1}\bar{e}^{n+1} = \mathbf{NA}^{-1}\bar{e}^{n}, \tag{33}$$

Equation (33) can be written in the following form by using the values of $\mathbf{M}$ and $\mathbf{N}$ defined in Equation (29):

$$\left[\mathbf{I}+\left(\frac{\Delta t}{2}\right)\mathbf{R}_1\right]\bar{e}^{n+1} = \left[\mathbf{K}+\left(\frac{\Delta t}{2}\right)\mathbf{R}_2\right]\bar{e}^{n},$$

where

$$\mathbf{R}_1 = [\varepsilon(\mathbf{u}^n * \mathbf{D} + \mathbf{u}_x^n * \mathbf{A}_d) + \mu(\alpha * \mathbf{D}_1 - \beta * \mathbf{D}_2 + \gamma * \mathbf{D}_3)]\mathbf{A}^{-1}, \quad \mathbf{K} = \mathbf{A}^{-1}\mathbf{A}_d,$$

and

$$\mathbf{R}_2 = -\mu(\alpha * \mathbf{D}_1 - \beta * \mathbf{D}_2 + \gamma * \mathbf{D}_3)\mathbf{A}^{-1}.$$

The condition of stability will be satisfied if the maximum eigenvalue of the matrix $\mathbf{E} = [\mathbf{I} + (\Delta t/2)\mathbf{R}_1]^{-1}[\mathbf{K} + (\Delta t/2)\mathbf{R}_2]$ is less than unity, that is,

$$\left|\frac{\eta_K + (\Delta t/2)\eta_{R_2}}{1 + (\Delta t/2)\eta_{R_1}}\right| \leqslant 1. \tag{34}$$

where η_{R_1}, η_{R_2} and η_K denote the eigenvalues of the matrices $\mathbf{R}_1$, $\mathbf{R}_2$ and $\mathbf{K}$, respectively.

It is clear from Equation (34) that the stability of the method depends on the time step Δt and eigenvalues of the matrices η_{R_1}, η_{R_2} and η_K. The condition numbers and magnitude of the eigenvalues of the matrices $\mathbf{R}_1$, $\mathbf{R}_2$ and $\mathbf{K}$ depend on the shape parameter and the number of collocation points. Hence, the condition number and the spectral radius of the matrix $\mathbf{E}$ are dependent on the shape parameter and the number of collocation points. It is shown that when the shape parameter c is very large, the RBFs system error is of exponential order. But there is a certain limit for the value of c after which the solution breaks down. For the limiting value of c , the condition number of the RBFs system becomes so large that the system leads to ill-conditioning. In case of an ill-conditioned system, the spectral radius of the matrix $\mathbf{E}$ becomes bigger than 1 so the numerical solution thus produced is not stable. Since it is not possible to find an explicit relationship among the spectral radius of the matrix $\mathbf{E}$ and the shape parameter, this dependency is approximated numerically by keeping the number of collocation points fixed.

6. The numerical experiments

Three experiments are studied to investigate the robustness and the accuracy of the proposed method. We compare the numerical results of the KdV equation by using this scheme with the analytical solutions and solutions in Xiao et al. (2011) and Siraj et al. (2008). These methods include the MQ quasi-interpolation method (MQQI) (Xiao et al., 2011) and RBFs methods (MQ, IMQ) (Siraj et al., 2008). We denote our scheme by IMQQI. The L_2, L_∞ and root-mean-square (RMS) error norms defined by

$$L_2 = \|u - u^*\|_2 = \sqrt{\sum_{j=0}^{N}(u(x_j) - u^*(x_j))^2},$$

$$L_\infty = \|u - u^*\|_\infty = \max_{0\leqslant j\leqslant N} |u(x_j) - u^*(x_j)|,$$

$$\text{RMS} = \sqrt{\frac{(\sum_{j=0}^{N}(u(x_j) - u^*(x_j))^2)}{(N+1)}},$$

are used to measure the accuracy of our scheme. Also, the stability analysis of the methods is considered for the first experiment. In all experiments, the shape parameter s is considered fourfold of the shape parameter c. Moreover, the centers and the collocation points have been chosen as the same and equidistant. The rate of convergence in space is calculated by using the following formulae:

$$\frac{\log(\|u - u^*_{h_i}\|/\|u - u^*_{h_{i+1}}\|)}{\log(h_i/h_{i+1})},$$

$$\frac{\log(\|u - u^*_{\Delta t_i}\|/\|u - u^*_{\Delta t_{i+1}}\|)}{\log(\Delta t_i/\Delta t_{i+1})},$$

whereas $u^*_{h_i}$ and $u^*_{\Delta t_i}$ are the numerical solutions with the spatial step size h_i and time step size Δt_i.

We perform the computations associated with our experiments in Maple 16 on a PC with a CPU of 2.4 GHZ.

Experiment 1 Propagation of the single solitary wave (Dehghan & Shokri, 2007). In this experiment, we consider a single solitary wave propagation of nonlinear third-order KdV Equation (15) with $\varepsilon = 6$, $\mu = 1$ and $[a, b] = [0, 40]$. The initial condition at $t = 0$ is given by

$$u_0(x) = \frac{r}{2}\,\text{sech}^2\left(\frac{\sqrt{r}}{2}x - 7\right), \quad r = 0.5.$$

The exact solution is given by

$$u(x,t) = \frac{r}{2}\,\text{sech}^2\left(\frac{\sqrt{r}}{2}(x - rt) - 7\right), \quad r = 0.5.$$

The boundary functions $f(t)$ and $g(t)$ are extracted from the exact solution. The L_∞, L_2 and RMS of errors are listed in Table 1 at different times by taking time step $\Delta t = 0.01$ and 0.1 with $N = 200$ and $c = 4.075 \times 10^{-3}$.

It can be observed from Table 2 that the L_∞, L_2 and RMS error norms increase slightly by increasing time step Δt. Also, the L_∞ error is compared with the results in Xiao et al. (2011) and Siraj et al. (2008) in Table 2 and the numerical solution is compared with the exact solution and the result in Xiao et al. (2011) in Table 3 with

Table 1. L_∞, L_2 and RMS error norms at different times with $\Delta t = 0.01$, $\Delta t = 0.1, N = 200$ and $c = 4.075 \times 10^{-3}$ of Experiment 1.

	$\Delta t = 0.01$			$\Delta t = 0.1$		
t	L_∞	L_2	RMS	L_∞	L_2	RMS
1	1.6728×10^{-4}	5.9993×10^{-4}	4.2316×10^{-5}	1.7111×10^{-4}	6.2137×10^{-4}	4.3828×10^{-5}
2	2.3758×10^{-4}	9.2242×10^{-4}	6.5063×10^{-5}	2.3897×10^{-4}	9.4334×10^{-4}	6.6538×10^{-5}
3	2.3758×10^{-4}	1.1337×10^{-4}	7.9967×10^{-5}	2.7857×10^{-4}	1.1457×10^{-3}	8.0811×10^{-5}
4	3.1384×10^{-4}	1.2931×10^{-3}	9.1213×10^{-5}	3.0797×10^{-4}	1.2920×10^{-3}	9.1133×10^{-5}
5	3.4136×10^{-4}	1.4216×10^{-3}	1.0027×10^{-4}	3.3183×10^{-4}	1.4046×10^{-3}	9.9075×10^{-5}

Table 2. The comparison of L_∞ error between the numerical solution using our method and the solutions in Xiao et al. (2011) and Siraj et al. (2008) with $N = 200$ and $c = 4.075 \times 10^{-3}$ of Experiment 1.

	$\Delta t = 0.1$	$\Delta t = 0.001$		
t	IMQQI	MQQI	MQ	IMQ
1	1.71116×10^{-4}	1.5259×10^{-3}	1.7923×10^{-5}	6.9584×10^{-5}
2	2.38973×10^{-4}	2.8672×10^{-3}	3.0151×10^{-5}	1.9553×10^{-4}
3	2.78578×10^{-4}	4.1428×10^{-3}	3.9839×10^{-5}	3.8286×10^{-3}
4	3.07970×10^{-4}	5.3859×10^{-3}	4.7835×10^{-5}	5.9098×10^{-3}
5	3.31838×10^{-4}	6.8141×10^{-3}	5.4599×10^{-5}	8.3667×10^{-3}

Table 3. The comparison between the numerical solutions using our method, the MQQI method and the exact solution with $\Delta t = 0.1, N = 200$ and $c = 4.075 \times 10^{-3}$ of Experiment 1.

	$t = 1$			$t = 3$			$t = 5$		
x	IMQQI	MQQI	Exact	IMQQI	MQQI	Exact	IMQQI	MQQI	Exact
17	0.080645	0.081817	0.080625	0.043585	0.045284	0.043573	0.022560	0.024489	0.022515
18	0.137257	0.137899	0.137393	0.080469	0.082636	0.080625	0.043493	0.045640	0.043573
19	0.203742	0.204288	0.203886	0.137116	0.139860	0.137393	0.080386	0.084426	0.080625
20	0.247270	0.247960	0.247227	0.203713	0.206787	0.203886	0.137061	0.142601	0.137393
21	0.235375	0.235194	0.235251	0.247339	0.249011	0.247227	0.203699	0.208712	0.203886
22	0.177661	0.176306	0.177627	0.235452	0.233662	0.235251	0.247364	0.249883	0.247227
23	0.112325	0.110928	0.112353	0.177692	0.173616	0.177627	0.235492	0.231660	0.235251
24	0.063395	0.062591	0.063421	0.112322	0.108716	0.112353	0.177721	0.171589	0.177627
25	0.033535	0.033223	0.033545	0.063384	0.061304	0.063420	0.112335	0.106221	0.112353

$N = 200$ and $c = 4.075 \times 10^{-3}$ at $t = 1, 3$ and 5. In Xiao et al. (2011) and Siraj et al. (2008), time step $\Delta t = 0.001$ was used, whereas in IMQQI $\Delta t = 0.1$ is used. Table 2 shows that the accuracy of the IMQQI method is better than MQQI and IMQ methods when the time goes ahead. Moreover, Table 4 presents the errors and computation orders obtained using our scheme with $\Delta t = 0.1$ and different values of N for $t = 2$. It can be concluded from Table 4 that the convergence rate decreases with the smaller spatial step size, but the error norms decrease slightly by increasing N. Table 5 shows the time rate of convergence obtained for Experiment 1 with $N = 100$, $c = 8.15 \times 10^{-3}$ and different values of Δt. It can be noted from Table 5 that the rate of convergence decreases with a smaller time step size. We also plot the graphs of absolute error and the estimated and analytical functions at $t = 5$ in Figure 1.

Table 4. The spatial rate of convergence at $t = 2$ with $\Delta t = 0.1$ of Experiment 1.

N	L_∞	Order	L_2	Order
40	1.72007×10^{-3}	–	1.10138×10^{-2}	–
80	1.48476×10^{-3}	2.15182	3.70414×10^{-3}	1.57211
120	6.59904×10^{-4}	1.99997	2.00839×10^{-3}	1.50967
160	3.73295×10^{-4}	1.98039	1.30908×10^{-3}	1.48778
200	2.38973×10^{-4}	1.99881	9.43347×10^{-4}	1.46831

Table 5. The time rate of convergence at $t = 2$ with $N = 100$ of Experiment 1.

Δt	L_∞	Order	L_2	Order
1	1.77413×10^{-3}	–	1.10138×10^{-2}	–
0.5	1.00571×10^{-3}	0.81890	3.70414×10^{-3}	0.80321
0.25	9.70341×10^{-4}	0.05164	2.00839×10^{-3}	0.19195
0.125	9.63649×10^{-4}	0.00998	1.30908×10^{-3}	0.04014
0.0625	9.61999×10^{-4}	0.00247	4.83991×10^{-4}	0.00944

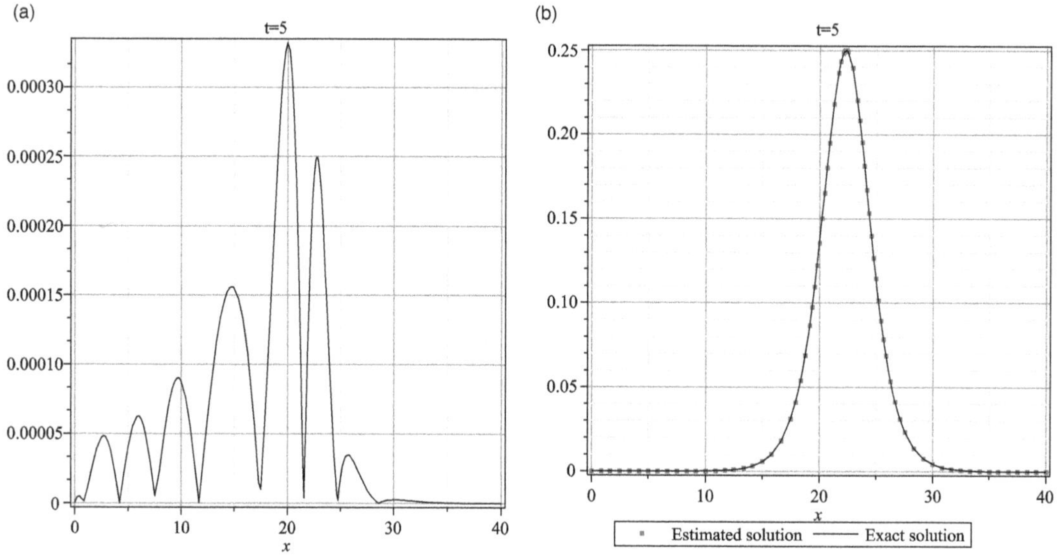

Figure 1. Absolute error (a) and the analytical and estimated functions (b) at $t = 5$, with $\Delta t = 0.1$ and $N = 200$, of Experiment 1.

Table 6. The spectral radius and L_∞ and L_2 error norms versus shape parameter c when $\Delta t = 0.01$ and $N = 100$ at $t = 5$ of Experiment 1.

c	$\rho(\mathbf{E})$	L_∞	L_2	RMS
0.00001	0.99994	2.15966×10^{-3}	6.49144×10^{-3}	6.45923×10^{-4}
0.001	0.99994	2.14819×10^{-3}	4.50147×10^{-3}	6.41813×10^{-4}
0.01	0.99994	2.04674×10^{-3}	6.08820×10^{-3}	6.05799×10^{-4}
0.05	0.99994	1.70590×10^{-3}	4.95110×10^{-3}	4.92652×10^{-4}
0.10	0.99994	1.50597×10^{-3}	4.37502×10^{-3}	4.35331×10^{-4}
0.50	0.99994	1.37120×10^{-3}	4.04434×10^{-3}	4.02427×10^{-4}
1.00	0.99994	1.37095×10^{-3}	4.04385×10^{-3}	4.02378×10^{-4}
1.20	0.99994	1.37095×10^{-3}	4.04384×10^{-3}	4.02377×10^{-4}
1.30	1.01481	1.37095×10^{-3}	4.04383×10^{-3}	4.02376×10^{-4}
1.40	1.06072	1.37096×10^{-3}	4.04381×10^{-3}	4.02374×10^{-4}
1.50	5.33193	2.16534×10^{0}	5.65118×10^{0}	5.62314×10^{-1}

The relation between the spectral radius of the matrices $\mathbf{E}$ and the different values of the shape parameter c is shown in Table 6 by keeping the number collocation points fixed. It can be seen from Table 6 that the accuracy of the IMQQI method is reasonably good by keeping the shape parameter c in the interval $(0, 1.4)$. But, it is clear that if the values of shape parameter c are greater than the critical value $c = 1.2$, then the spectral radius of the matrix $\mathbf{E}$ becomes bigger than 1 and hence the IMQQI method becomes unstable. Therefore, the interval stability of the IMQQI scheme is $(0, 1.2)$.

Experiment 2 Propagation of two solitary waves (Dehghan & Shokri, 2007). In this experiment, we consider two solitary waves propagation of nonlinear third-order

KdV Equation (15) with $\varepsilon = 6$ and $\mu = 1$. The initial condition at $t = 0$ is given by

$$u_0(x) = 12\frac{\{3 + 4\cosh(2x) + \cosh(4x)\}}{\{3\cosh(x) + \cosh(3x)\}^2}.$$

The exact solution is given by

$$u(x,t) = 12\frac{\{3 + 4\cosh(2x - 8t) + \cosh(4x - 64t)\}}{\{3\cosh(x - 28t) + \cosh(3x - 36t)\}^2},$$

and the boundary functions $f(t)$ and $g(t)$ can be obtained from the exact solution. In this experiment, we consider $\Delta t = 0.0001, 0.001$, $N = 200$ and $[a,b] = [-5,15]$. We compare the numerical solutions using our scheme and the MQQI scheme (Xiao et al., 2011) with the exact solution in Table 7 and compare the L_∞-error with the results in Xiao et al. (2011) and Siraj et al. (2008) in Table 8 for $t = 0.01, 0.05$ and 0.10. In Xiao et al. (2011) and Siraj et al. (2008), $\Delta t = 0.00001$ was used but we use $\Delta t = 0.001$. It is observable that the proposed method is more accurate in comparison with MQQI and IMQ methods.

The norms L_∞, L_2 and RMS of errors are obtained in Table 9 for $t = 0.01, 0.05, 0.10, 0.15$ and 0.20. The graphs of analytical and estimated functions at $t = 0.1$ and 0.2 and theirs absolute error are given in Figure 2.

Table 7. The comparison between the numerical solutions using our method, the MQQI method and the exact solution with $\Delta t = 0.001$ and $N = 200$ of Experiment 2.

	$t = 0.01$			$t = 0.05$			$t = 0.1$		
x	IMQQI	MQQI	Exact	IMQQI	MQQI	Exact	IMQQI	MQQI	Exact
0	5.634246	5.640733	5.628245	2.593411	2.597785	2.574829	2.030865	1.990040	2.000572
1	3.192235	3.186663	3.192964	6.893042	6.922214	6.881609	1.700870	1.764999	1.717101
2	0.478502	0.478633	0.478495	1.204642	1.191924	1.207024	7.230098	7.139714	7.171392
3	0.064535	0.064570	0.064520	0.101147	0.101458	0.100955	0.465253	0.452621	0.464299
4	0.008725	0.008730	0.008723	0.012259	0.012227	0.012239	0.024487	0.024443	0.024308
5	0.001181	0.001181	0.001180	0.001632	0.001559	0.001630	0.002553	0.002261	0.002542

Table 8. The comparison of L_∞-error between the numerical solution using our method and the solutions in Xiao et al. (2011) and Siraj et al. (2008) of Experiment 2.

	$\Delta t = 0.001$	$\Delta t = 0.00001$		
t	IMQQI	MQQI	MQ–RBF	IMQ–RBF
0.01	3.99855×10^{-3}	7.7405×10^{-3}	9.2114×10^{-4}	2.2071×10^{-2}
0.05	3.98427×10^{-2}	6.3762×10^{-2}	2.9608×10^{-3}	7.2316×10^{-2}
0.10	8.79034×10^{-2}	1.6196×10^{-1}	1.2806×10^{-2}	1.0121×10^{-1}

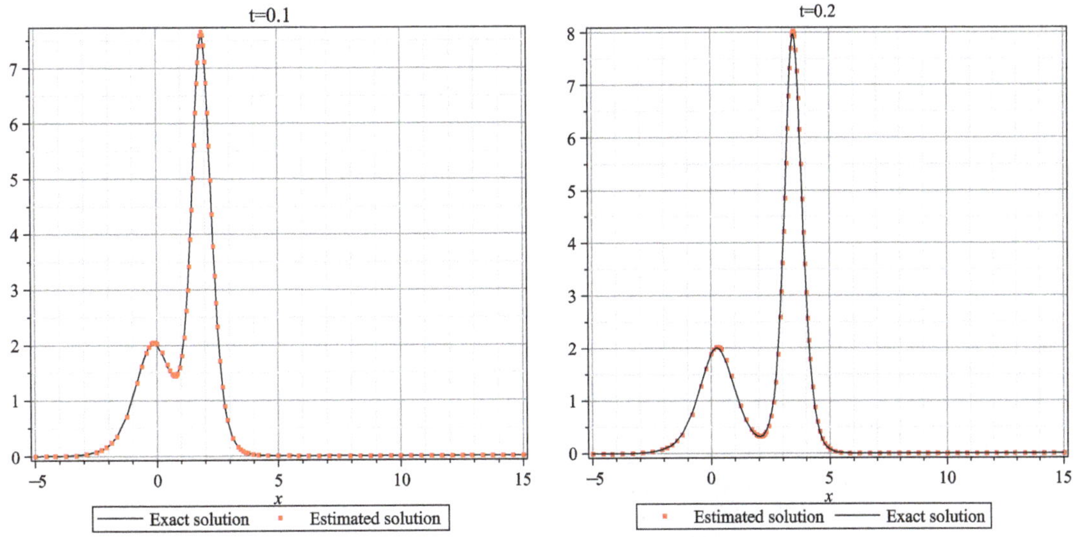

Figure 2. Analytical and estimated functions at $t = 0.1, 0.2$ with $\Delta t = 0.001$ and $N = 200$ of Experiment 2.

Table 9. L_∞, L_2 and RMS error norms at different times with $\Delta t = 0.0001, 0.001$ and $N = 200$ of Experiment 2.

	$\Delta t = 0.0001$			$\Delta t = 0.001$		
t	L_∞	L_2	RMS	L_∞	L_2	RMS
0.01	4.0579×10^{-3}	9.9105×10^{-3}	6.9903×10^{-4}	3.9985×10^{-3}	9.7257×10^{-3}	6.8600×10^{-4}
0.05	4.1003×10^{-2}	1.0295×10^{-1}	7.2619×10^{-3}	3.9842×10^{-2}	1.0072×10^{-2}	7.1048×10^{-3}
0.10	9.1691×10^{-2}	2.3373×10^{-1}	1.6486×10^{-2}	8.7903×10^{-2}	2.2546×10^{-1}	1.5903×10^{-2}
0.15	1.3257×10^{-1}	3.4201×10^{-1}	2.4124×10^{-2}	1.2588×10^{-1}	3.2553×10^{-1}	2.2961×10^{-2}
0.20	1.6644×10^{-1}	4.3607×10^{-1}	3.0758×10^{-2}	1.5682×10^{-1}	4.1073×10^{-1}	2.8970×10^{-2}

Table 10. The percentage error using different schemes at time $t = 0.005$ with $N = 160$ and $\Delta t = 0.001$ of Experiment 3.

x	IMQQI	MQQI	MQ	FEM	ANS	HBIM
0.1	0.0000	0.1082	0.0000	0.0156	0.0078	3.8033
0.2	0.0001	0.1019	0.0003	0.0107	0.0272	3.7984
0.3	0.0005	0.0597	0.0003	0.0105	0.0634	3.7243
0.4	0.0144	0.1095	0.0103	0.0104	0.1105	2.9326
0.5	0.0374	0.0691	0.0060	0.0076	0.0603	0.7865
0.6	0.0050	0.0821	0.0101	0.0104	0.0276	3.2960
0.7	0.0003	0.0534	0.0015	0.0101	0.0223	6.6331
0.8	0.0000	0.0748	0.0007	0.0096	0.0470	3.6626
0.9	0.0000	0.0767	0.0080	0.0088	0.0442	3.6656
1.0	0.0000	0.0769	0.0000	0.0000	0.1067	3.7353

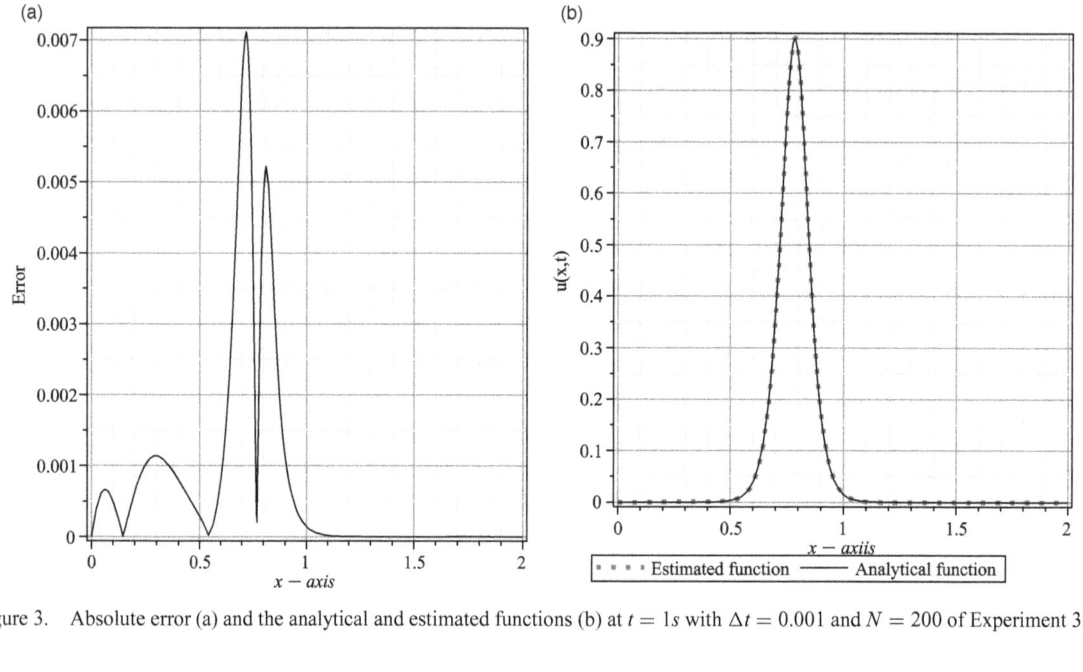

Figure 3. Absolute error (a) and the analytical and estimated functions (b) at $t = 1s$ with $\Delta t = 0.001$ and $N = 200$ of Experiment 3.

Similar to Experiment 1, we observe that by an increase in Δt from 0.0001 to 0.001, the L_∞, L_2 and RMS error norms do not increase as t moves ahead from 0.01 to 0.20, in the monitored time at Table 9.

Experiment 3 A special model problem of the KdV equation was investigated in Xiao et al. (2011), Siraj et al. (2008), Aksan and Özdes (2006), Özer and Kutluay (2005) and Kutluay, Bahadir, and Özer (2000). Consider KdV Equation (15) with the initial condition

$$u_0(x) = 3C\,\mathrm{sech}(Ax + D), \quad 0 \leqslant x \leqslant 2,$$

and boundary conditions

$$u(0,x) = u(2,t) = u_x(2,t) = 0, \quad t > 0.$$

The exact solution of this problem is taken from Alexander and Morris (1979) and is given by

$$u(x,t) = 3C\,\text{sech}^2(Ax - Bt + D), \quad 0 \leqslant x \leqslant 2,$$

where C, D are real constants, $A = \frac{1}{2}\sqrt{\varepsilon C/\mu}$ and $B = \varepsilon AC$.

The new method based on the MQ quasi-interpolation scheme is applied to Experiment 3 and the results of percent error are compared with those given in Xiao et al. (2011), Siraj et al. (2008), Aksan & Özdes (2006), Özer and Kutluay (2005) and Kutluay et al. (2000) in Table 10. These methods include the MQQI method (Xiao et al., 2011), RBF(MQ) method (Siraj et al., 2008), finite element method (FEM) (Aksan and Özdes, 2006), analytical–numerical solution (ANS) (Özer & Kutluay, 2005) and HBIM (Kutluay et al., 2000). In this experiment, we consider $\varepsilon = 1, \mu = 4.84 \times 10^{-4}$, $C = 0.3, D = -6, N = 160$ and $\Delta t = 0.001$. Moreover, we plot the graphs of absolute error and estimated and analytical functions at $t = 1$s in Figure 3.

7. Conclusion

In this paper, we have presented a new numerical scheme based on the high accuracy MQ quasi-interpolation scheme and IRBF approximation scheme for solving the nonlinear third-order KdV equation.

The numerical results which are given in the previous section demonstrate the good accuracy of the present scheme. Also, the tables show that this scheme performs better than the MQQI method and IMQ–RBF method in more cases and Table 8 shows that it even performs better than the MQ–RBF method. Moreover, we have used a bigger time step Δt, in comparison with Xiao et al. (2011) and Siraj et al. (2008).

The difficulty involved in using this approach is solving a system of algebraic equations at each time step which the dimension of the coefficient matrix is equal to the number of interpolant points and finding the appropriate value for the shape parameters c and s.

Therewith, we would like to emphasize that the scheme introduced in this paper can be well studied for any other nonlinear PDEs.

Disclosure statement

No potential conflict of interest was reported by the author(s).

References

Aksan, E. N., & Özdes, A. (2006). Numerical solution of Korteweg–de Vries equation by Galerkin B-spline finite element method. *Applied Mathematics and Computation*, *175*, 1256–1265.

Alexander, M. E., & Morris, J. L. I. (1979). Galerkin methods applied to some model equations for non-linear dispersive waves. *Journal of Computational Physics*, *30*(3), 428–451.

Aminataei, A., & Mazarei, M. M. (2008). Numerical solution of Poisson's equation using radial basis function networks on the polar coordinate. *Computer and Mathemathics with Applications*, *56*(11), 2887–2895.

Beatson, R. K., & Powell, M. J. D. (1992). Univariate multiquadric approximation: quasi-interpolation to scattered data. *Constructive Approximation*, *8*(3), 275–288.

Chen, R. H., & Wu, Z. M. (2006). Solving hyperbolic conservation laws using multiquadric quasi-interpolation. *Numerical Methods for Partial Differential Equations*, *22*(4), 776–796.

Chen, R. H., & Wu, Z. M. (2007). Solving partial differential equation by using multiquadric quasi-interpolation. *Applied Mathematics and Computation*, *186*(2), 1502–1510.

Dağ İ., Conlvar, A., & Şahin, A. (2011). Taylor-Galerkin and Taylor-collocation methods for the numerical solutions of Burgers' equation using B-splines. *Communications in Nonlinear Science and Numerical Simulation*, *16*, 2696–2708.

Dehghan, M., & Shokri, A. (2007). A numerical method for KdV equation using collocation and radial basis functions. *Nonlinear Dynamics*, *50*, 111–120.

Dodd, R. K., Eilbeck, J. C., Gibbon, J. D., & Morris, H.C. (1982). *Solitons and nonlinear wave equations*. New York: Acedemic Press.

Franke, R. (1982). Scattered data interpolation: Test of some methods. *Mathematics and Computers*, *38*, 181–200.

Gardner, C. S., Grrne, J. M., & Kruskal, M.D. (1967). Method for solving Korteweg–de Vries equation. *Physical Review Letters*, *19*, 1095–1097.

Gardner, C. S., & Marikawa, G. K. (1965). The effect of temperature of the width of a small amplitude solitary wave in a collision free plasma. *Communications on Pure and Applied Mathematics*, *18*, 35–49.

Geyikli, T., & Kaya, D. (2005). An application for a modified KdV equation by the decomposition method and finite element method. *Applied Mathematics and Computation*, *169*, 971–81.

Hardy, R. L. (1971). Multiquadric equations of topography and other irregular surfaces. *Journal of Geophysical Research*, *176*, 1905–1915.

Helal, M. A., & Mehanna, M. A. (2006). A comparison between two different methods for solving KdV-Burgers' equation. *Chaos, Solitons & Fractals*, *28*, 320–326.

Hon, Y. C., & Wu, Z. M. (2000). A quasi-interpolation method for solving stiff ordinary differential equations. *International Journal for Numerical Methods in Engineering*, *48*(8), 1187–1197.

Jiang, Z. W., Wang, R. H., Zhu, C. G., & Xu, M. (2011). High accuracy multiquadric quasi-interpolation. *Applied Mathematical Modelling*, *35*, 2185–2195.

Kansa, E. J. (1990a). Multiquadric – a scattered data approximation scheme with applications to computational fluid dynamics I. *Computers & Mathematics with Applications*, *19*, 127–145.

Kansa, E. J. (1990b). Multiquadric – a scattered data approximation scheme with applications to computational fluid dynamics II. *Computers & Mathematics with Applications*, *19*, 147–161.

Korteweg–de Vries, D. J., & de Vries, G. (1895). On the change in form of long waves advancing in rectangular canal and on a new type of long stationary waves. *Philosophical Magazine*, *39*, 422–443.

Kutluay, S., Bahadir, A. R., & Özer, A. (2000). A small time solutions for the Korteweg–de Vries equation. *Applied Mathematics and Computation*, *107*, 203–210.

Li, X., & Wang, M. (2007). A sub-ODE method for finding exact solutions of a generalized KdV–mKdV equation with high order nonlinear terms. *Physics Letters A*, *361*, 115–118.

Madych, W. R., & Nelson, S. A. (1990). Multivariate interpolation and conditionally positive definite functions. *Mathematics of Computation*, *54*, 211–230.

Mai-Cao, L., & Tran-Cong, T. (2005). A meshless IRBFN-based method for transient problems. *Computer Modeling in Engineering and Science*, *7*(2), 149–171.

Mai-Duy, N., Mai-Cao, L., & Tran-Cong, T. (2007). Computation of transient viscous flows using indirect radial basis function networks. *Computer Modeling in Engineering and Science*, *18*(1), 59–77.

Özer, A., & Kutluay, S. (2005). An analytical-numerical method for solving the Korteweg–de Vries equation. *Applied Mathematics and Computation*, *164*, 789–797.

Powell, M. J. D. (1992). The theory of radial basis function approximation in 1990. In W. Light (Ed.), *Advances in numerical analysis* (Vol. II, pp. 105–210.). Oxford: Oxford Science Publications.

Sarboland, M., & Aminataei, A. (2014). On the numerical solution of one-dimensional nonlinear nonhomogeneous Burgers' equation. *Journal of Applied Mathematics*, 15 pages. doi:10.1155/2014/598432.

Siraj, U. I., Khattak, A. J., & Tirmizi, I. A. (2008). A meshfree method for numerical solution of KdV equation. *Engineering Analysis with Boundary Elements*, *32*, 849–855.

ul-Islam, S., Haq, S., & Uddin, M. (2009). A meshfree interpolation method for the numerical solution of the coupled nonlinear partial differential equations. *Engineering Analysis with Boundary Elements*, *33*, 399–409.

Washimi, H., & Taniuti, T. (1966). Propogation of ion acoustic solitary waves of small amplitude. *Physical Review Letters*, *17*, 996–998.

Wijngaarden, L. V. (1968). On the equation of motion for mixtures of liquid and gas bubbles. *Journal of Fluid Mechanics*, *33*, 465–474.

Wu, Z. M. (2004). Dynamically knots setting in meshless method for solving time dependent propagations equation. *Computer Methods in Applied Mechanics and Engineering*, *193*(12–14), 1221–1229.

Wu, Z. M. (2005). Dynamically knot and shape parameter setting for simulating shock wave by using multiquadric quasi-interpolation. *Engineering Analysis with Boundary Elements*, *29*, 354–358.

Wu, Z. M., & Schaback, R. (1994). Shape preserving properties and convergence of univariate multiquadric quasi- interpolation. *Acta Mathematicae Applicatae Sinica (English Series)*, *10*(4), 441–446.

Xiao, M. L., Wang, R. H., & Zhu, C. H. (2011). Applying multiquadric quasi-interpolation to solve KdV equation. *Mathematical Research Exposition*, *31*, 191–201.

Yan, Z. (2006). New compacton-like and solitary patterns-like solutions to nonlinear wave equations with linear dispersion terms. *Nonlinear Analysis*, *64*, 901–909.

18

Dynamically dual vibration absorbers: a bond graph approach to vibration control

Peter Gawthrop[a]*, S.A. Neild[b] and D.J. Wagg[c]

[a]Systems Biology Laboratory, Melbourne School of Engineering, University of Melbourne, Victoria 3010, Australia; [b]Department of Mechanical Engineering, Queens Building, University of Bristol, Bristol BS8 1TR., UK; [c]Department of Mechanical Engineering, Sir Frederick Mappin Building, University of Sheffield, Mappin Street Sheffield S1 3JD, UK

This paper investigates the use of an actuator and sensor pair coupled via a control system to damp out oscillations in resonant mechanical systems. Specifically the designs emulate passive control strategies, resulting in controller dynamics that resemble a physical system. Here, the use of the novel dynamically dual approach is proposed to design the vibration absorbers to be implemented as the controller dynamics; this gives rise to the dynamically dual vibration absorber (DDVA). It is shown that the method is a natural generalisation of the classical single-degree of freedom mass–spring–damper vibration absorber and also of the popular acceleration feedback controller. This generalisation is applicable to the vibration control of arbitrarily complex resonant dynamical systems. It is further shown that the DDVA approach is analogous to the hybrid numerical-experimental testing technique known as substructuring. This analogy enables methods and results, such as robustness to sensor/actuator dynamics, to be applied to dynamically dual vibration absorbers. Illustrative experiments using both a hinged rigid beam and a flexible cantilever beam are presented.

Keywords: vibration absorber; bond graph; acceleration feedback control; dynamic dual

1. Introduction

The use of a secondary resonant mechanical systems to damp out oscillations in a resonant mechanical system by absorbing and dissipating energy has a long history and early work is summarised in the classical textbook by Den Hartog (1985). An alternative method for damping unwanted oscillations is to use some form of active vibration control. To achieve this some type of actuator and sensor system needs to be used. For example, vibrations can be damped from a mechanical system using a piezo-electric transducer and an associated electrical circuit (Hagood & von Flotow, 1991). This can have considerable advantages, although, as discussed by Moheimani & Behrens (2004) multi-modal resonant structures require sophisticated circuit synthesis.

The adjective 'passive' applied to 'system' has two different but related meanings: a physical system not containing a power source and a mathematical expression imposing the corresponding property on the input and output variables of a set of equations (Hogan, 1985; Sharon, Hogan, & Hardt, 1991; Slotine & Li, 1991). In general, this means that passive mechanical (or electrical) vibration absorbers can be replaced by a computer and associated sensor-actuator pairs which emulate the physical passivity in the equivalent mathematical sense. The algorithm implemented in the computer does not have to represent a physical system and can be designed using conventional control-theoretic methods (Balas, 1978; Fleming & Moheimani, 2005; Hong & Bernstein, 1998; Hogsberg & Krenk, 2006; Moheimani & Fleming, 2006), optimisation (Krenk & Hogsberg, 2009) or via system inversion (Ali & Padhi, 2009).

However, it can be argued that there are advantages in implementing *physical* systems within the digital computer (Gawthrop, 1995; Gawthrop, Bhikkaji, & Moheimani, 2010; Hogan, 1985; Lozano, Brogliato, Egelund, & Maschke, 2000; Ortega, Loria, Nicklasson, & Sira-Ramirez, 1998; Ortega, van der Schaft, Mareels, & Maschke, 2001; Sharon et al., 1991; Slotine & Li, 1991); this is the approach explored in this paper. In particular, the well-known relationship between dissipativity, passivity and physical systems (Lozano et al., 2000; Ortega et al., 1998, 2001; Willems, 1972) is exploited. Such energy based concepts rely on the properties of physical connections. In particular, the concept of *collocation* is a key system property in the context of active vibration control (Gawthrop et al., 2010; Preumont, 2002).

Replacing a mechanical vibration absorber by a digital computer is analogous to the well-known hybrid numerical-experimental testing technique where the structure under consideration is split into an experimental test piece (or *physical substructure*) and a numerical model

*Corresponding author. Email: peter.gawthrop@unimelb.edu.au

describing the remainder of the structure (or *numerical substructure*). Although the two coupled passive subsystems resulting from this process are stable (Anderson & Vongpanitlerd, 2006; Desoer & Vidyasagar, 1975; Lozano et al., 2000; Ortega, Praly, & Landau, 1985; Ortega et al., 2001), this stability can be destroyed by the digital implementation of the numerical substructure and the corresponding actuator and sensor dynamics that couple the substructures (Gawthrop, Wallace, Neild, & Wagg, 2007). Fortunately, this problem of coupling the numerical and experimental substructures using real-time digital implementation has been solved and a suite of techniques for robust *numerical-experimental substructuring* is now available (Blakeborough, Williams, Darby, & Williams, 2001; Gawthrop, Wallace, & Wagg, 2005; Wagg, Neild, & Gawthrop, 2008). Furthermore, the substructuring approach is particularly suitable for the type of resonant systems that are the focus of this paper (Gawthrop et al., 2007).

As discussed by Gawthrop et al. (2005), the bond-graph approach (Borutzky, 2011; Gawthrop & Bevan, 2007; Gawthrop & Smith, 1996; Karnopp, Margolis, & Rosenberg, 2012; Mukherjee, Karmaker, & Samantaray, 2006) gives a natural and convenient formulation of substructuring and control (Gawthrop, 2004; Gawthrop et al., 2005; Vink, Ballance, & Gawthrop, 2006) and the concept of actuator/sensor collocation has a clear bond graph interpretation. For these reasons, the bond-graph approach is adopted in this paper.

As discussed by Den Hartog (1985), choosing the structure of a vibration absorber for a single degree of freedom system is straightforward. However, choosing the structure for multi-degree of freedom systems such as those arising from modal decomposition is considerably more involved (Moheimani & Behrens, 2004). This complexity motivates the novel approach of this paper based on the concept of of a *dynamically dual*[1] system.

As discussed in more detail in Section 3, a dynamically dual mechanical system is obtained by interchanging the rôles of velocity and force. This concept of duality has been used for analysis of dynamical systems (Cellier, 1991; Karnopp, 1966; Samanta & Mukherjee, 1985, 1990; Shearer, Murphy, & Richardson, 1971), and this paper uses the concept to design *dynamically dual vibration absorbers* (DDVA).

Although the DDVA method originated an extension of the physically based design of Den Hartog (1985), it will be shown that the method also includes the well-established *acceleration feedback* approach (Preumont, 2002).

In summary, placing both the traditional Den Hartog mechanical vibration absorber and acceleration feedback into the wider context of the DDVA of this paper has two advantages: the method immediately extends to multi degree of freedom systems and the implementation and theoretical results (including robustness to sensor/actuator dynamics) from substructuring can be directly applied.

Section 2 reviews the substructuring approach to provide a framework for the paper. Section 3 gives the foundations of the DDVA approach; Section 3.2 focuses on the Den Hartog (1985) absorber version and Section 3.3 focuses on the acceleration feedback controller (Preumont, 2002). Section 4 discusses a number of multi-mode examples. Section 5 gives illustrative experimental results; Sections 5.1 and 5.2 experimentally verify the approach when applied to a rigid beam with an flexible joint and a flexible cantilever beam, respectively. Section 6 concludes the paper.

2. The substructuring formulation

Substructuring is a novel dynamic testing technique that allows the experimental testing of a component within the context of a larger system. This is achieved through the coupling of the physical component with a controller that numerically simulates the dynamics of the remainder of the system. Note that as the controller dynamics are designed to simulate part of a real system, the dynamics are physically realisable.

Figure 1 summarises the basic substructuring formulation Gawthrop et al. (2005) and Gawthrop, Wagg, & Neild (2009). For simplicity, Figure 1 will be assumed to represent a system with scalar quantities, although this can readily be extended to vectors. The three key parts are shown in Figure 1:

(1) **Phy** representing the physical component, with transfer function $p(s)$, to be controlled,
(2) **Num** representing the controller, with physically realisable dynamics, which is implemented numerically and has a transfer function $n(s)$,
(3) **Se**:$\mathbf{F_0}$ representing a disturbing external force, F_0,

where s is the Laplace domain independent variable. Firstly the velocity feedback case is shown, Figure 1(a) as a bond graph[2] and in Figure 1(b) as a block diagram. Here, the physical component **Phy** has a force input, $F_0 - F$, and a measured velocity output, v. In addition, the parameters in the physical system are represented by a vector, θ_p, Similarly, θ_n, represents the vector of parameters in the numerical system.

An advantage of the bond graph representation is that it emphasises the fact that the physical system **Phy** and the controller **Num** are connected by *power bonds* and thus the control system is collocated – meaning that the actuator and sensor are located at the same point. In Figure 1(a) the parts are joined by a common flow (velocity) junction denoted as **1**. The bond graph also indicates causality and **Phy** and **Num** are represented by the positive real transfer functions $p(s, \theta_p)$ and $n(s, \theta_n)$, respectively. The transfer

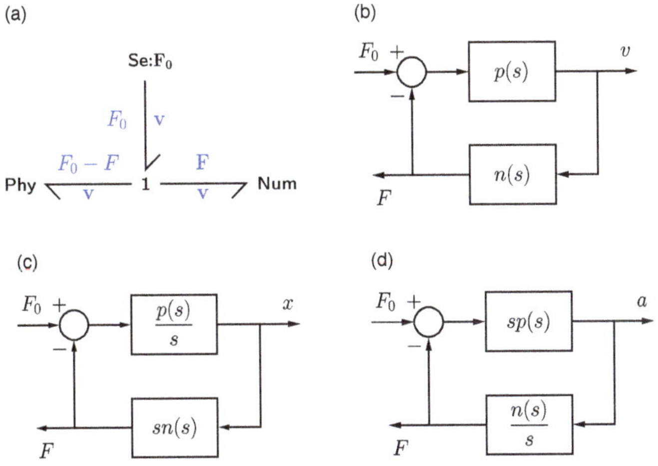

Figure 1. The substructuring formulation. The mathematically equivalent formulations (b)–(d) allow for a choice of sensors. (a) Bond graph, (b) block-diagram: velocity formulation, (c) block-diagram: displacement formulation and (d) block-diagram: acceleration formulation.

functions are related by the following relationships:

$$v = p(s, \theta_p)(F_0 - F), \tag{1}$$

$$F = n(s, \theta_n)v. \tag{2}$$

Although it is natural to work in terms of velocity v rather than displacement x, Figure 1(b) can be easily rewritten in terms of displacement as Figure 1(c) where $dx/dt = v$ or in terms of acceleration as Figure 1(d) where $a = dv/dt$. The choice of formulation (displacement, velocity or acceleration) does not change the theoretical closed-loop stability properties defined by the loop-gain $L(s) = n(s, \theta_n)p(s, \theta_p)$, but allows flexibility in the choice of sensor and actuator. As well as providing a conceptual basis for this paper, the substructuring approach links to classical control system concepts useful for stability and robustness analysis. Details can be found elsewhere (Gawthrop et al., 2007, 2009; Wagg et al., 2008).

The substructuring formulation of Figure 1(a) assumes an inertia-like physical component driven by a force; as discussed by Gawthrop et al. (2005), compliance-like physical components can be treated by the formulation of Figure 2(a) where the external force F_0 is replaced by an external velocity v_0 and the three components are now connected by a common force, or **0**, junction. To distinguish this velocity-driven formulation from the force-driven formulation in Figure 1(a) an over-bar is used:

$$F = \bar{p}(s, \theta_p)(v_0 - v), \tag{3}$$

$$v = \bar{n}(s, \theta_n)F. \tag{4}$$

Using the definitions of Equations (3) and (4), the block diagram equivalent of Figure 2(a) is Figure 2(b). Once again, displacement and acceleration formulations are given by Figure 2(c) and 2(d), respectively.

3. Dynamically dual design

As already discussed, in this paper a vibration absorber attached to a system is considered. This vibration absorber, while based on a physical component thus ensuring that the system is passive, is implemented as a controller. This setup can be considered within the substructuring framework, with the vibration absorber forming **Num** and the system which the vibration absorber is attached being **Phy**. One possible absorber is the Den Hartog resonant vibration absorber, which is usually represented by a conventional mass–spring–damper schematic. However, as pointed out by Den Hartog (1985), and discussed in greater depth by Shearer et al. (1971), can equally well be described by an equivalent electrical circuit analogue.

Here, the use of DDVAs is proposed as a method for generating suitable **Num** dynamics. As discussed in Section 3.2, the resonant vibration absorber of Den Hartog (1985) is an example of a DDVA and provides the motivation for this approach. In formulating the DDVA approach the following features of the Den Hartog resonant vibration absorber are abstracted and generalised:

(1) it is a one-port[3] passive[4] physical system,
(2) it is causally compatible with the system, in that, the output velocity of the system provides the input to the absorber and the force output of the absorber provides the input to the system,
(3) there is a variable coupling parameter,
(4) the absorber has the *same* resonant frequency as the system, and
(5) the damping ratio of the absorber is *greater* than that of system.

The DDVA design approach is to set **Num** to be a dynamic-dual of the key mode or modes of the system that

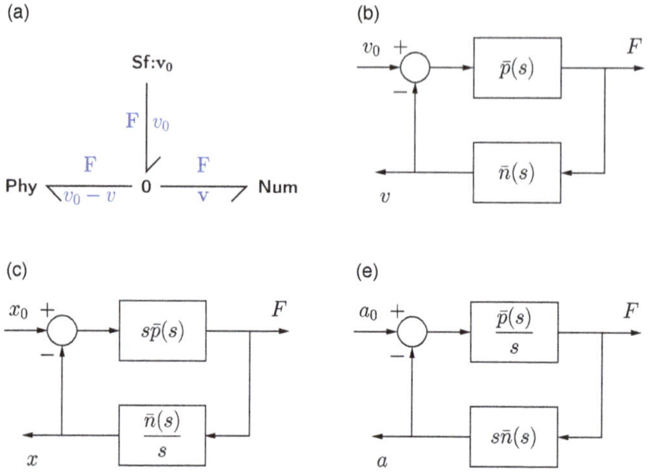

Figure 2. The velocity-driven substructuring formulation. (a) Bond graph, (b) block-diagram: velocity formulation, (c) block-diagram: displacement formulation and (d) block-diagram: acceleration formulation.

the absorber is attached to (which is contained in **Phy**). The method of obtaining a dynamic dual is now discussed. This is followed by discussions of two common absorber strategies which are DDVA; the Den Hartog absorber and the acceleration feedback method proposed by Preumont (2002).

3.1. A dynamic-dual

A dynamic-dual of a system is obtained by interchanging the rôles of velocity and force, is defined in Shearer et al. (1971). An extended version of this concept, the *scaled dual* is used here and, in the context of mechanical systems is defined as follows:

(1) Each force F_i, and each velocity v_i, in the original system has a scaled dual v_i^D, and F_i^D in the dual system given by:

$$v_i^D = \frac{1}{g} F_i, \tag{5}$$

$$F_i^D = g v_i, \tag{6}$$

where g is the scaling factor and $g = 1$ corresponds to the unscaled dual.

(2) Each mass component with mass m_i is replaced in the dual system with a spring component of stiffness K_i, each spring component with stiffness k_i is replaced in the dual system with a mass component of mass M_i, and each damper component with damping coefficient r_i is replaced in the dual system with a damper component with damping coefficient R_i where:

$$K_i = \frac{g^2}{m_i}, \tag{7}$$

$$M_i = \frac{g^2}{k_i}, \tag{8}$$

$$R_i = \frac{g^2}{r_i}. \tag{9}$$

(3) Common force connections become common velocity connections and common velocity connections become common force connections in the dual system.

(4) If the system transfer function $h(s)$ has force F as input and velocity v as output, then the dual transfer function $H(s)$ has velocity v^D as input and force F^D as output and is given by:

$$F^D = g v, \tag{10}$$

$$v^D = \frac{1}{g} F, \tag{11}$$

$$H(s) = \frac{F^D}{v^D} = \frac{g v}{(1/g) F} = g^2 h(s). \tag{12}$$

Equations (10) and (11) are *power conserving* in the sense that

$$F^D v^D = F v. \tag{13}$$

As these Equations (10) and (11) also interchange the roles of force and velocity, they correspond to the bond graph *gyrator* (**GY**) component of Figure 3.

Equations (7) and (8) ensure that the scaled dual retains the same natural frequencies as the system; in the sequel, the value R_i given using Equation (9) is not used, instead it

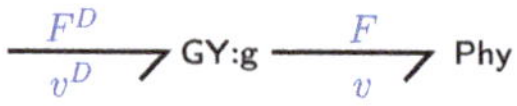

Figure 3. Gyrator interpretation of dynamic-dual.

is replaced by the user-selected value R'_i. This allows the damping of the modified scaled dual system, which forms the controller implemented in **Num**, to be adjusted.

Although not essential to the approach of this paper, the bond graph formulation provides a clear exposition of the notion of a scaled dual. In particular, the scaled dual system can be described in two different but equivalent ways as:

(1) the bond graph dual where the component moduli are given by Equations (7)–(9) or
(2) following Equation (13), the system obtained by appending a gyrator of modulus g to the system **Phy** port as in Figure 3. This point is also discussed by Gawthrop et al. (2010).

3.2. *Den Hartog absorber*

In his classical text book (Den Hartog, 1985, Section 3.3), Den Hartog considers the design of a damped vibration absorber for an undamped mass–spring system which is subject to a force disturbance. The specifications

(1) 'the main mass is 20 times greater than the damper mass',
(2) 'the frequency of the damper is equal to the frequency of the main system',
(3) The damping ratio of the damper is $\zeta = 0.1$.

were considered.

In the terminology used in this paper, the physical system requiring vibration suppression (**Phy**) is the undamped mass–spring oscillator. In Den Hartog (1985) the vibration absorber was considered to be a physical mechanical device but here it is considered to be a controller (with sensor and actuator) with the same dynamics as the absorber and forms **Num**. The disturbance force acts on the undamped mass–spring oscillator **Phy**, as does a force due to the presence of the absorber **Num**, therefore the system can be represented by the block diagram given in Figure 1(b).

Figure 4(b) and 4(a) gives the schematic diagram of the damped vibration absorber **Num** and the undamped oscillator **Phy**, respectively. A damper with $r = \infty$ is included in the subsystem **Phy** of Figure 4(a) to allow for the corresponding component in the subsystem **Num**. Using standard manipulations, the transfer function of the physical system, **Phy**, of Figure 4(a) is:

$$p(s, \theta_p) = \frac{s(ms + r)}{mrs^2 + kms + kr}. \tag{14}$$

Letting $r \to \infty$ gives:

$$p(s, \theta_p) = \frac{s}{ms^2 + k}. \tag{15}$$

Similarly, from Figure 4(b), the transfer function for **Num**, which represents the Den Hartog absorber, is

$$n(s, \theta_n) = \frac{Ms(Rs + K)}{Ms^2 + Rs + K}. \tag{16}$$

It can be shown that this absorber is a scaled dynamic-dual of the system, **Phy**, it is applied to. Considering the system **Phy**, given in Equation (14), and applying the dual transforms, given in Equations (7)–(9), the parameters m, k and r can be rewritten in terms of M,K and D to give:

$$\begin{aligned} p(s, \theta_p) &\\ &= \frac{s((g^2/k)s + (g^2/R))}{(g^2/k)(g^2/R)s^2 + (g^2/M)(g^2/K)s + (g^2/M)(g^2/R)} \\ &= \left(\frac{1}{g^2}\right)\frac{Ms(Rs + K)}{Ms^2 + Rs + Kr}. \end{aligned} \tag{17}$$

Applying the scaling given in Equation (12), the scaled dual of **Phy** is

$$P(s, \Theta_p) = g^2 p(s, \theta_p) = \frac{Ms(Rs + K)}{Ms^2 + Rs + Kr}. \tag{18}$$

Thus, by comparing this to Equation (16), it can be seen that the Den Hartog absorber in **Num** corresponds to the scaled dual of **Phy**:

$$n(s, \theta_n) = P(s, \Theta_p). \tag{19}$$

The first part of the Den Hartog specifications is achieved by setting:

$$M = \alpha m \quad \text{where } \alpha = \frac{1}{20}. \tag{20}$$

The second part of the specification is achieved by setting

$$\frac{K}{M} = \frac{k}{m}. \tag{21}$$

Equations (20) and (21) imply that

$$K = \alpha k. \tag{22}$$

Moreover, using Equations (7) and (22), the scaling gain g is given by

$$g^2 = Km = \alpha mk. \tag{23}$$

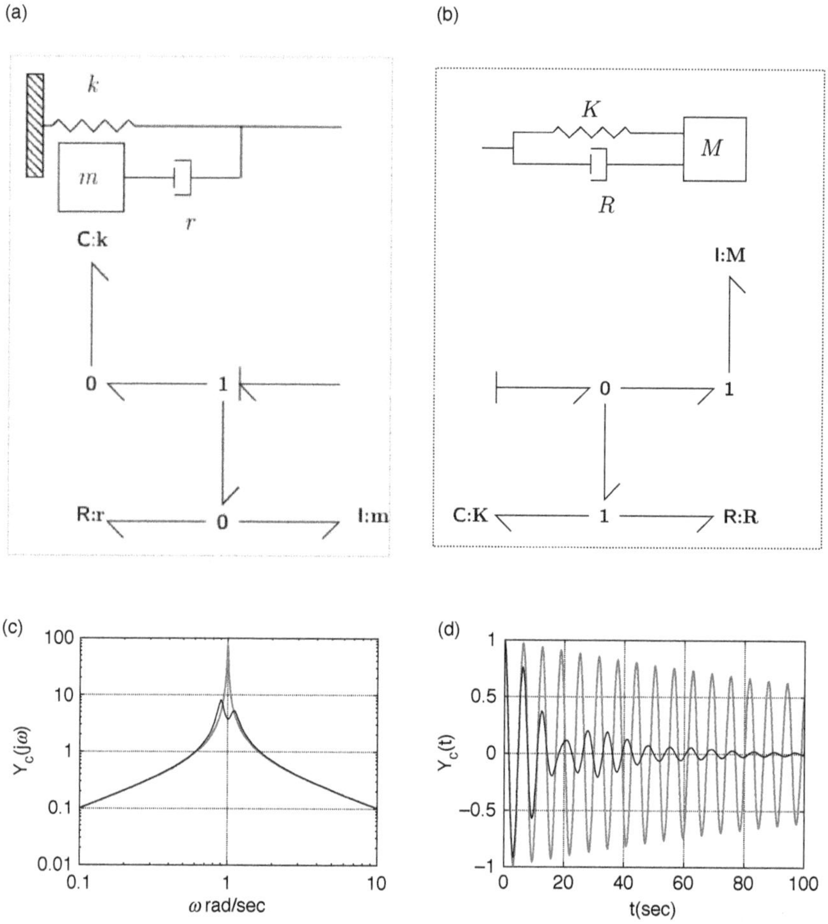

Figure 4. Den Hartog absorber. The physical system (a) and its vibration absorber (and dynamic-dual) (b) are first-order mass–spring–damper systems. (c) The closed-loop frequency response (black line) has a lower resonant peak than the open-loop response (grey line). (d) The closed-loop impulse response (black line) exhibits more damping than the open-loop impulse response (grey line). (a) **Phy**: the physical system, (b) **Num**: the dual physical system, (c) frequency response $H(j\omega)$ and (d) impulse response $h(t)$.

Finally, the third part of the specification is achieved by replacing the damping coefficient R of $n(s, \theta_n)$ by

$$R' = 2\zeta\sqrt{MK}. \tag{24}$$

To illustrate the properties of this particular vibration absorber, the unit system with $m = k = 1$ was used. Using the specification described above, this gives the numerical system parameters $M = K = 0.05$ and $R' = 0.01$, and so the DDVA is given by:

$$n(s, \theta_n') = \frac{Ms(R's + K)}{Ms^2 + R's + K} = \frac{0.01s^2 + 0.05s}{s^2 + 0.2s + 1}. \tag{25}$$

The corresponding closed-loop frequency response appears in Figure 4(c); this shows the 'split peak' phenomenon described by Den Hartog (1985). The corresponding closed-loop impulse response appears in Figure 4(d); this decays exponentially over the time scales determined by the specified damping ratio.

3.3. *Acceleration feedback*

The acceleration feedback method has been proposed by Preumont (2002). This section rederives the algorithm from the DDVA point of view. In particular, the undamped physical system of Figure 4(a) (with $1/r = 0$) can equally well be represented in Figure 5(a) with $r = 0$. This system has a different modified dual and thus gives a different form of control; this turns out to be a form of acceleration feedback.

As with the last example the vibration absorber is acting on an undamped mass–spring oscillator. The undamped oscillator forms **Phy** as shown in Figure 5(a). Note that a damper with $r = 0$ is included to allow a dynamic-dual to be formulated. Using standard manipulations, the transfer function of the physical system **Phy** of Figure 5(a) is

$$p(s, \theta_p) = \frac{s}{ms^2 + rs + k}. \tag{26}$$

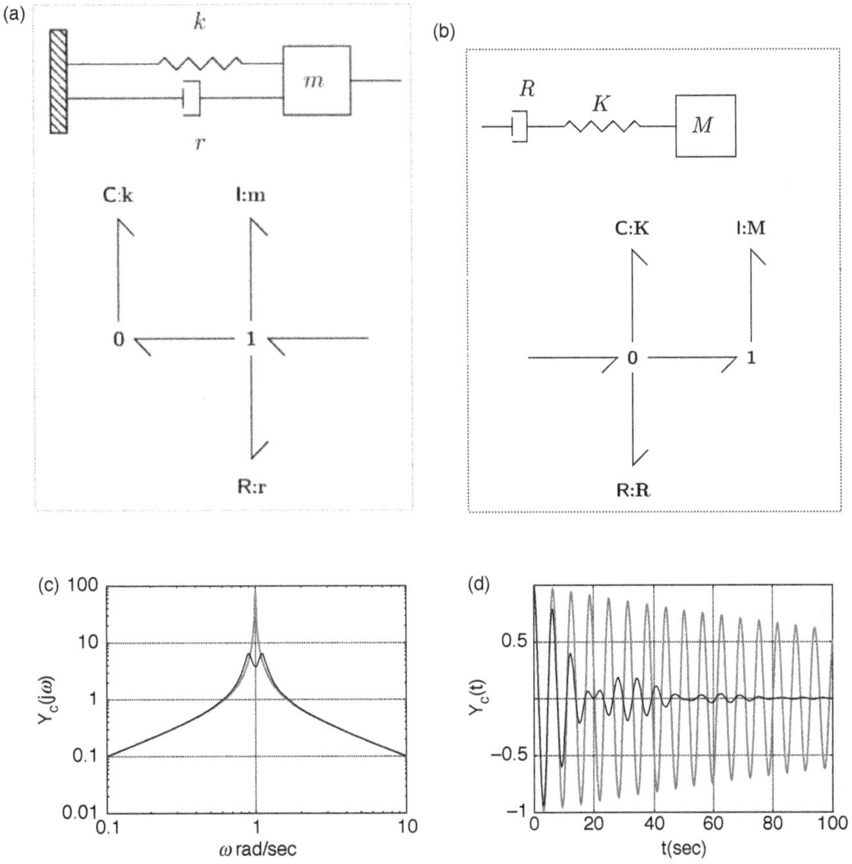

Figure 5. Acceleration feedback. Like the Den Hartog absorber of Figure 4, the physical system and the dual are mass–spring–damper systems, but the configuration is different. The closed and open-loop responses are similar to those of the Den Hartog absorber of Figure 4, but the split peaks are more symmetrical. (a) **Phy**: the physical system, (b) **Num**: the dual physical system, (c) frequency response $H(j\omega)$ and (d) impulse response $h(t)$.

Setting $r = 0$ gives

$$p(s, \theta_p) = \frac{s}{ms^2 + k}. \tag{27}$$

The controller transfer function, forming **Num**, for the acceleration feedback method (Preumont, 2002) is given by

$$n(s, \theta_n) = g^2 p(s, \theta_p) = \frac{g^2 s}{ms^2 + rs + k}. \tag{28}$$

Using Equation (12), it can be seen that the numerical system **Num** is the scaled dynamic-dual of **Phy**. Figure 5(b) gives a physical representation of the acceleration feedback controller, where the component values R, K and M can be calculated using Equations (7)–(9) but are not required.

To give a direct comparison with Section 3.2, the same system and design considerations are used to give the DDVA:

$$n(s, \theta_n') = \frac{0.05s}{s^2 + 0.2s + 1}. \tag{29}$$

This is similar to the DDVA of Equation (25) except that the numerator s^2 term is not present.

The corresponding closed-loop frequency response appears in Figure 5(c); this is similar to Figure 4(c) except that the peaks have a more similar amplitude. The corresponding closed-loop step response appears in Figure 5(d); again, this decays exponentially over the time scales determined by the specified damping ratio.

4. Systems with multiple modes of vibration

The examples discussed in the previous section demonstrate that the DDVA approach gives the same type of vibration absorber for the well known cases associated with mass–spring–damper systems. The real advantage of the DDVA approach is when using it to reduce vibrations in systems with multiple modes of vibration. The steps involved are the same as above: (i) define a physical model of the system **Phy**, (ii) set **Num** as the modified scaled dual of **Phy**, and (iii) connect the systems via a single (one-port) connection. As discussed in Section 3.1, step (ii) can either be accomplished directly or indirectly using the **GY** approach of Figure 3. In this paper, attention is focused

Table 1. Modal system: resonant frequencies.

n	ω_n (rad/s)
1	1.0
2	2.0

Table 2. Cantilever beam model modal frequencies.

n	ω_n (rad/s)
1	2.919
2	18.28
3	50.37
4	95.59
5	150.4
6	210.0
7	269.1
8	322.0
9	363.9
10	390.8

Note: Not all frequencies appear in Figures 7 and 8 due to coincident zeros.

on linear models thus giving rise to transfer-function representations.

Two examples are considered here. The first is a two degree-of-freedom lumped mass system which is shown schematically in Figure 6(a) and 6(b). Figure 6(a) and 6(b) are similar to Figure 5(a) and 5(b) except that there are *two* coupled mass–spring damper systems involved. Thus **Phy** can be regarded as the modal decomposition of a 2DOF system and **Num** the corresponding vibration absorber. For the purposes of illustration, each subsystem of **Num** has the same parameters as in the example of Figure 5 of Section 3.3 (Table 1).

Figure 6(c) shows the open (without the vibration absorber) and closed-loop (with the vibration absorber) frequency response magnitudes. The magnitude of the closed-loop response (black line) is clearly smaller than the corresponding open-loop response (grey line) at the two resonant frequencies. Figure 6(d) shows the equivalent impulse response; as predicted by the frequency responses, the closed-loop impulse response decays more rapidly than the equivalent open-loop response.

As a second example, a uniform Euler-Bernoulli[5] cantilever beam (with one end fixed and the other free) modelled using a 10 element, finite-element bond graph model (Karnopp, Margolis, & Rosenberg, 2000; Margolis, 1985) is considered. Such beam models are undamped; but the DDVA approach needs to include damping in **Num**. For the purposes of this example, Rayleigh damping is assumed; in particular, each compliant element in the lumped model has an associated damping term represented by a damper connected across the ends of the compliant elements.

Following Balas (1978, Section V) who considers a 'unit beam', the cantilever beam is normalised to have unit mass per unit length and unit compliance per unit length. **Phy** is assumed to have a small (but non-zero) damping of 10^{-6} per unit length. The 10 modal frequencies appear in Table 2. For the purposes of illustration, the vibration absorber was applied to the beam using a collocated point Force/Velocity actuator/sensor halfway along the beam. As discussed in the sequel, this point corresponds to a nodal point of the third-resonance and thus this mode cannot be controlled with this choice. The choice of actuator/sensor location is an interesting topic not considered in this paper.

Two versions of DDVA were used. The first DDVA was obtained by considering the complete dynamic-dual of **Phy**. The scaled dual **Num** was obtained using the **GY** approach of Figure 3. A feature of this approach is that there are only two control parameters. These were chosen as the gyrator gain $g^2 = 0.05$ and the damping of the cantilever beam model in **Num** as 2 per unit length.

Figure 7(a) shows the open (without the vibration absorber) and closed-loop (with the vibration absorber) frequency response magnitudes. This figure has been expanded to show the frequency responses close to the first–fourth resonances in Figure 7(c)–7(f), respectively. Near the first two resonances (Figure 7(c) and 7(d)), the magnitude of the closed-loop response (black line) is clearly smaller than the corresponding open-loop response (grey line) at the two resonant frequencies. The third resonance corresponds to a node at the sensor/actuator and the fourth is well damped anyway. Thus this controller design naturally applies control authority at the important resonances. Figure 7(b) shows the equivalent impulse response; as predicted by the frequency responses, the closed-loop impulse response decays more rapidly than the equivalent open-loop response. As noted above, the third resonance is not controlled using this approach. However, it could be controlled either by moving the sensor/actuator away from the node or by having a second sensor/actuator away from the node.

The second approach is to use the scaled dynamic-dual of a two mode modal model (as in Figure 6(b)), capturing the dynamics of the first two modes of the cantilever. This is then connected to the same cantilever beam. The parameters of **Num** are the same as in the example of Figure 6 and those of **Phy** the same as those of the example of Figure 7.

Figure 8(a) shows the open (without the vibration absorber) and closed-loop (with the vibration absorber) frequency response magnitudes. This figure has been expanded to show the frequency responses close to the first–fourth resonances in Figure 8(c)–8(f), respectively. Near the first two resonances (Figure 8(c) and 8(d)), the magnitude of the closed-loop response (black line) is clearly smaller than the corresponding open-loop response (grey line) at the two resonant frequencies; these Figures are not the same as Figure 7(c) and 7(d) because

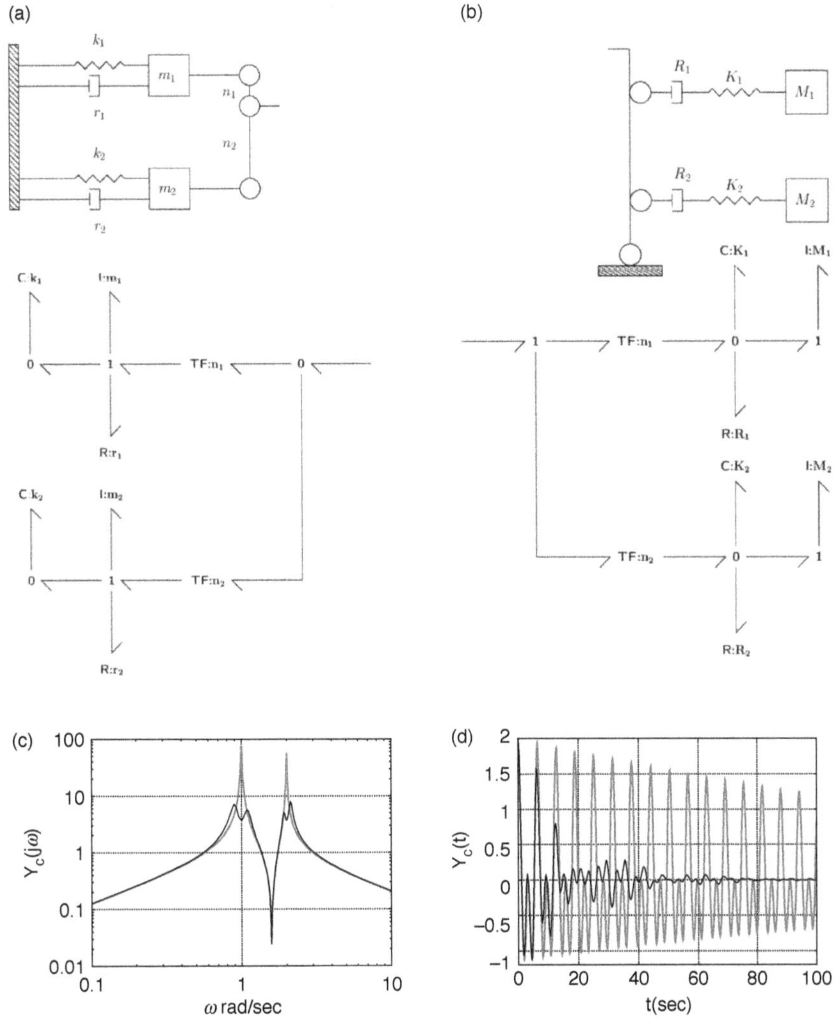

Figure 6. Modal system. Both the physical system and its dual are coupled mass–spring–damper systems and are thus fourth-order. (c) The closed-loop frequency response (black line) has both resonant peaks lower than the open-loop response (grey line). (d) Again, the closed-loop impulse response (black line) exhibits more damping than the open-loop impulse response (grey line). (a) **Phy**: the physical system, (b) **Num**: the dual physical system, (c) frequency response $H(j\omega)$ and (d) impulse response $h(t)$.

the controller is different; but the effect is similar. The third and fourth resonances are explicitly not controlled with this method; but, in this case, the effect is the same as that of the controller of the example of Figure 7. In particular Figure 8(b) shows that the closed-loop impulse response decays more rapidly than the equivalent open-loop response in a similar fashion to that of Figure 7(b). In this particular example, the performance of the two controllers is quite similar.

5. Experimental results

As indicated in Figure 9, the experiments were based on the Quanser (Apkarian, 1995) SRV02 rotational servo-motor and associated UPM-15-03-240 power and instrumentation module. The SRV02 was firmly clamped to a rigid bench and interfaced to a Intel CoreTM 2 Duo Processor (2.66 GHz) based computer via a National Instruments PCI-8024E analogue-digital conversion card and cable and the corresponding Quanser interface board.

In the experiment described here, the computer used the real-time Linux operating system RTAI together with the control-orientated software RTAI-Lab (Bucher & Balemi, 2006) running at a sampling frequency of 500 Hz. Using this software, the SRV02 rotational servo motor, rotational position sensor and associated power supply were controlled to give high-gain position control using a proportional and derivative (PD) controller. The servo angle was measured using a potentiometer and scaled within the computer to measure angular position in radians.

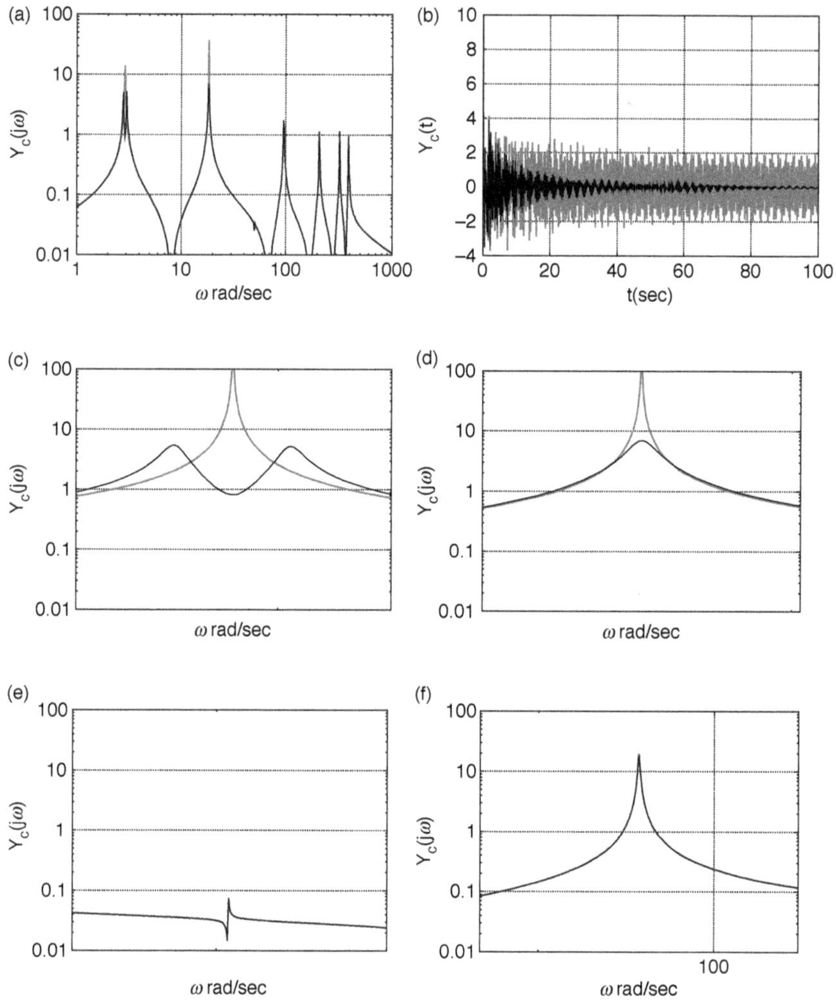

Figure 7. Cantilever beam with dual feedback. (a) Frequency responses $H(j\,\omega)$, (b) impulse responses $h(t)$, (c) $H(j\,\omega)$ – first resonance, (d) $H(j\,\omega)$ – second resonance, (e) $H(j\,\omega)$ – third resonance and (f) $H(j\,\omega)$ – fourth resonance.

5.1. Flexible joint

The Quanser cantilever beam experiment (Apkarian, 1995) has two parts that may be considered using the substructuring configuration shown in Figure 2(c). The physical component, **Phy**, consists of a rigid arm which is mounted to a platform via a pivot. This pivot exhibits a stiffness due to two linear springs mounted between the platform and the arm. A position disturbance is provided to the system via the rotational servo motor on which the platform is mounted (the pivot is directly above the motor). The vibration absorber, **Num**, has a torque input F. Because of the springs in **Phy**, this torque is proportional to the joint deflection angle θ (the arm rotation relative to the platform rotation), and so is generated from this measurement. The output of **Num** is a rotational displacement x which, along with the disturbance x_0, is imposed on **Phy** using the servo motor by setting the servo motor PD controller demand to $x_0 - x$ (Figure 10).

The open-loop properties of the system were investigated by applying a square-wave reference signal with a period of 10 s to the servo and measuring both the servo angle x_a and the joint angle θ for 5 periods. Because all of the signals are periodic, the methods of Pintelon & Schoukens (2001) were used to generate the frequency response of the system at the discrete frequencies corresponding to the periodic input. Figure 11 gives two measured frequency responses:

(1) $+$ indicates the response from servo angle x_a to joint angle θ.
(2) $\circ$ indicates the response from servo reference to θ.

These responses match at low frequencies, but the gain of the second transfer function falls at the higher frequencies due to the limited servo bandwidth of about 10 Hz.

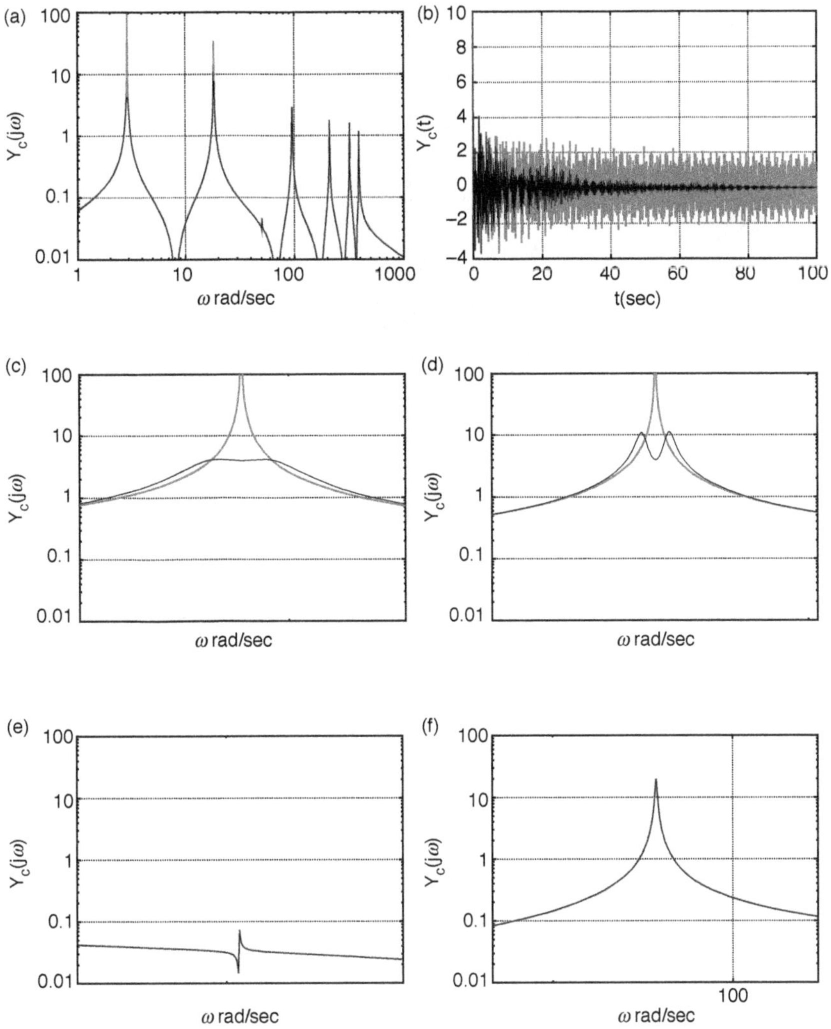

Figure 8. Cantilever with dual modal feedback. (a) Frequency responses $H(j\omega)$, (b) impulse responses $h(t)$, (c) $H(j\omega)$ – first resonance, (d) $H(j\omega)$ – second resonance, (e) $H(j\omega)$ – third resonance and (f) $H(j\omega)$ – fourth resonance.

With reference to Figures 2(c) and 10(a) the physical system **Phy** relating input displacement to output force is of the form

$$s\bar{p}(s, \theta_p) = \frac{g_0 s^2}{s^2 + 2\xi_0\omega_0 s + \omega_0^2}. \tag{30}$$

The parameters θ_p were fitted to the first measured frequency response with $\omega_0 = 15.1\,\mathrm{rad}\,s^{-1}$ and $\xi_o = 0.02$.

Following the methodology of Section 3.3 in the dual version of Figure 2(c), the feedback transfer function was chosen to be of the form:

$$\frac{\bar{n}(s, \theta_n)}{s} = \frac{g_c}{s^2 + 2\xi_c\omega_0 s + \omega_0^2}, \tag{31}$$

where $g_c = g^2 g_0$ is a variable positive gain factor and the damping ratio $\xi_c = 0.3$.

The periodic input experimental method described above was used. Figure 12 shows the experimental frequency results for three values of g_c: $g_c = 0$, $g_c = 20$, and $g_c = 40$. $g_c = 0$ corresponds to Figure 11. The height of the resonant peak is reduced in the two non-zero cases and the peak splitting of Figure 5(c) is evident in Figure 12 for the highest gain of $g_c = 40$. Figure 13 shows the periodic data corresponding to the joint angle θ for the three gain values. The five consecutive periods have been superimposed to form the figures; the variability between periods is essentially high-frequency noise. As indicated by the frequency responses, the time responses show damping increasing with gain.

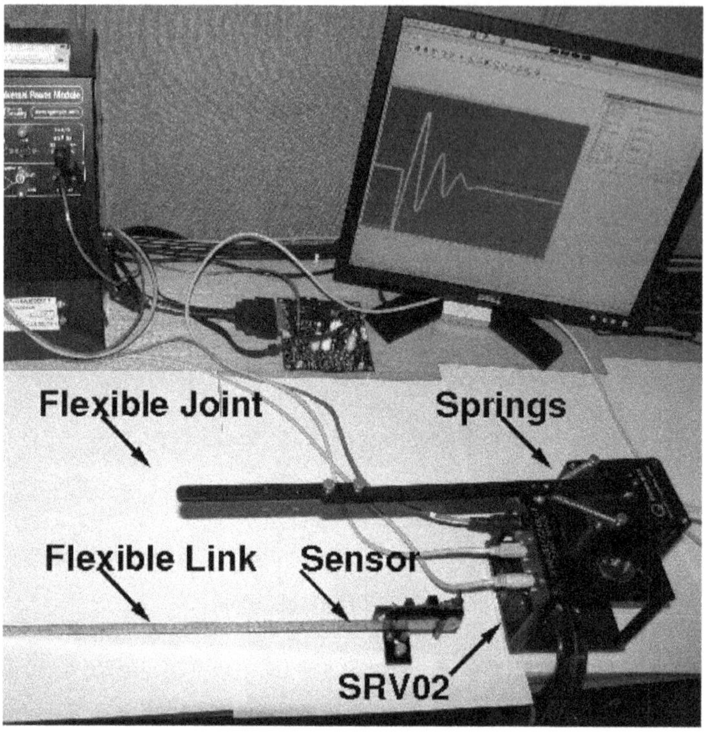

Figure 9. Experimental systems. Experimental systems. The SRV02 servomotor module is in the bottom right-hand corner and the associated power module in the top left-hand corner. The computer display is at the top right and the computer interface board near the centre. The flexible joint module is shown mounted on the SRV02 and rotates about a vertical axis driven though the two springs. The cantilever beam module is shown unmounted and replaces the flexible joint module in the second set of experiments.

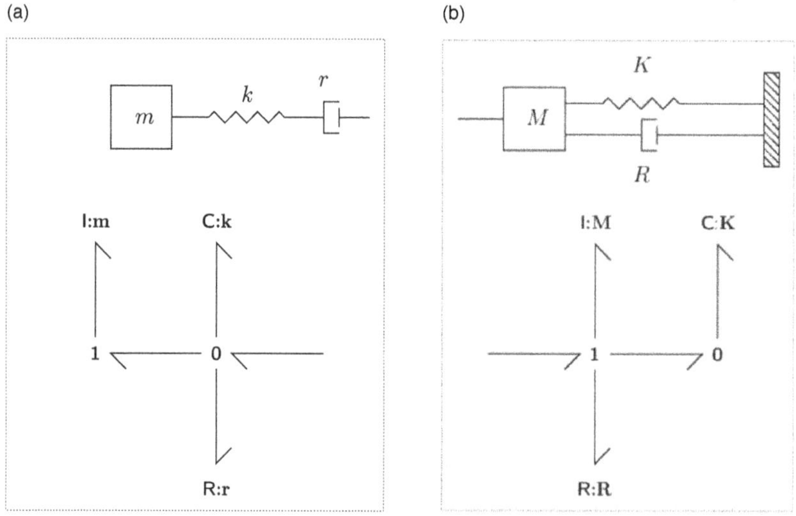

Figure 10. Rotational joint experiment. (a) With the components interpreted in a rotational sense and $r \to \infty$, **Phy** represents the rotating arm with the attached springs. (b) **Num** is the modified scaled dual of **Phy**. (a) **Phy**: the physical system and (b) **Num**: the dual physical system.

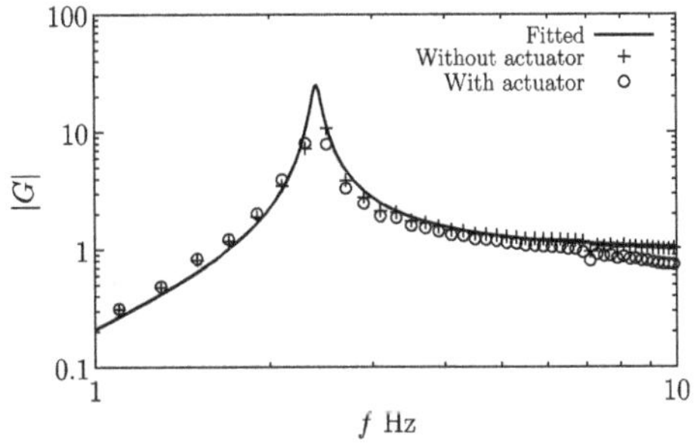

Figure 11. Flexible joint: open-loop frequency responses. Flexible joint: open-loop frequency responses. + indicates the response from servo angle x to joint angle θ; ∘ indicates the response from servo reference to θ. The solid line is the fitted frequency response.

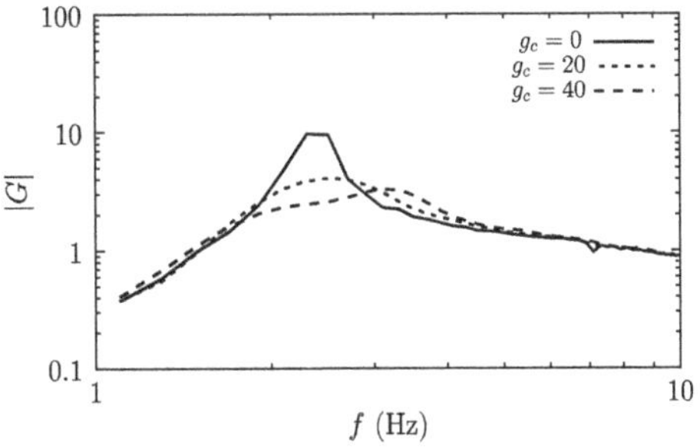

Figure 12. Flexible joint: closed-loop frequency responses.

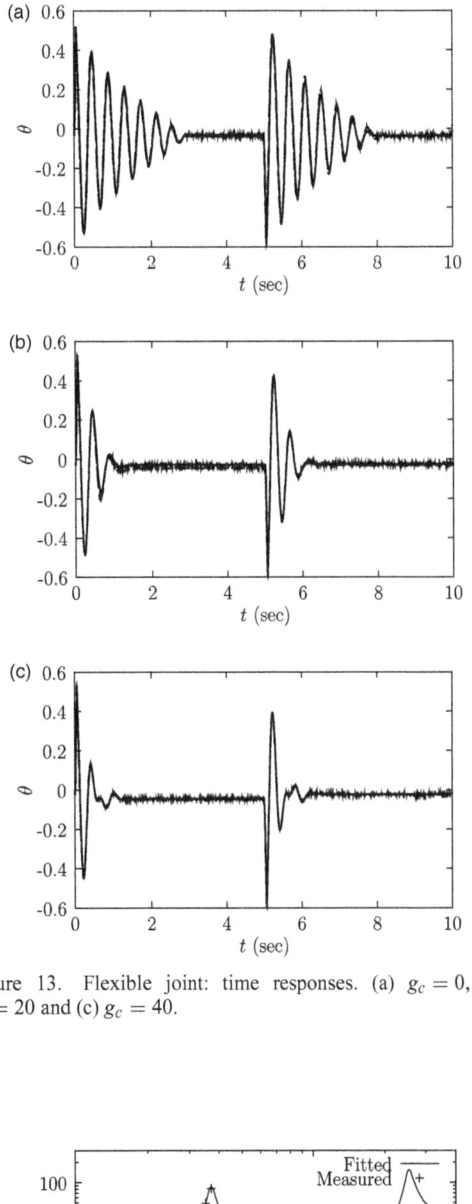

Figure 13. Flexible joint: time responses. (a) $g_c = 0$, (b) $g_c = 20$ and (c) $g_c = 40$.

5.2. Cantilever beam

The flexible joint module was replaced by the cantilever beam module in Figure 9. A strain gauge measures the curvature at the root of the cantilever beam. In the same way as the joint potentiometer of Section 5.1 provided a voltage proportional to torque F, the strain gauge sensor provides a voltage proportional to torque F. The open-loop response was measured using the same methods. Two resonances and one anti-resonance appear in the measured frequency response and similarly to Section 5.1, this was fitted by a transfer function of the form:

$$s\bar{p}(s, \theta_p) = g_0 \left[\frac{\kappa_1 s^2}{s^2 + 2\xi_1 \omega_1 s + \omega_1^2} + \frac{\kappa_2 s^2}{s^2 + 2\xi_2 \omega_2 s + \omega_2^2} \right] \tag{32}$$

with $\omega_1 = 23.25$ rad s−1,$\omega_2 = 159.00$ rad s−1, $\xi_1 = \xi_2 = 0.04$, $\kappa_1 = 0.36$ and $\kappa_2 = 1 - \kappa_1 = 0.64$. Because of the 10 Hz servo bandwidth, the discrepancy between measured and fitted transfer function is large above 10 Hz.

Following the methodology of Section 4 a two-mode transfer function corresponding to Equation (33) was

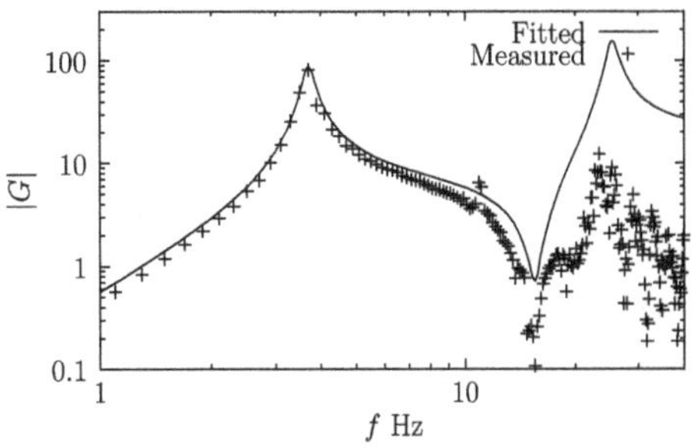

Figure 14. Cantilever beam: open-loop frequency responses. Cantilever beam: open-loop frequency responses. + indicates the response from servo angle x to joint angle θ. The firm line is the fitted frequency response.

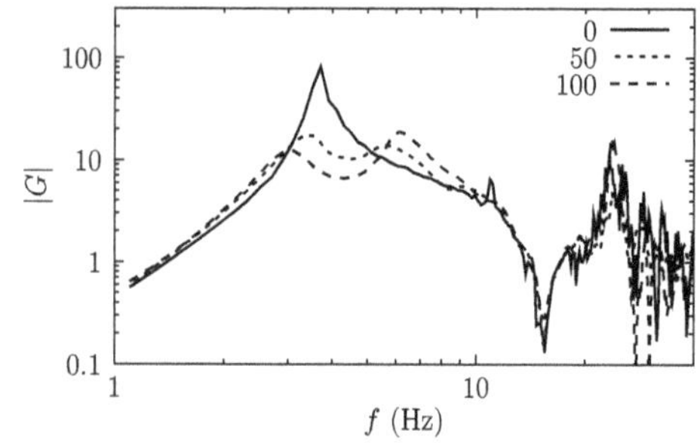

Figure 15. Cantilever beam: closed-loop frequency responses for the cases where $g_c = 0$, 50 and 100.

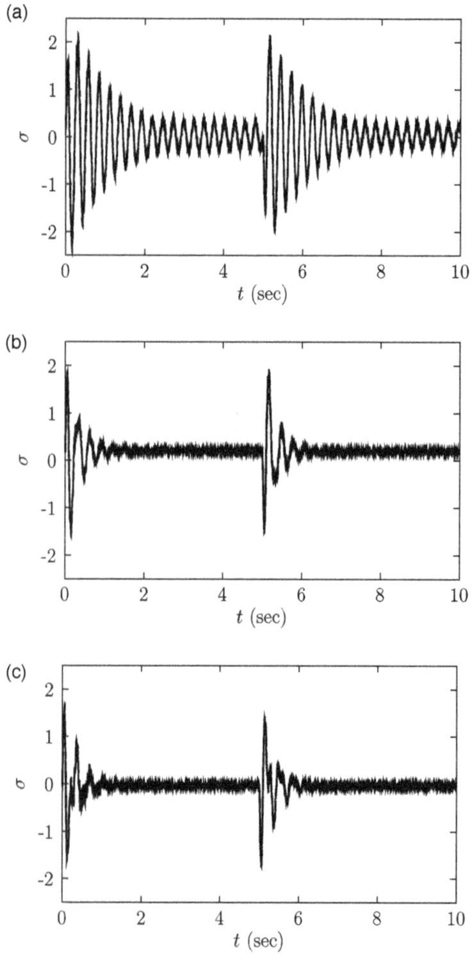

Figure 16. Cantilever beam: time responses. (a) $g_c = 0$, (b) $g_c = 50$ and (c) $g_c = 100$.

chosen as:

$$\frac{\bar{n}(s,\theta_n)}{s} = g_c \left[\frac{\kappa_1 s^2}{s^2 + 2\xi_c \omega_1 s + \omega_1^2} + \frac{\kappa_2 s^2}{s^2 + 2\xi_c \omega_2 s + \omega_2^2} \right] \quad (33)$$

with $\xi_c = 0.3$.

The periodic input experimental method described above was used. Figure 15 shows the experimental frequency results for three values of g_c: $g_c = 0$, $g_c = 50$, and $g_c = 100$. $g_c = 0$ corresponds to Figure 14. The height of the first resonant peak is reduced in the two non-zero cases and the peak splitting of Figure 8(a) is evident in Figure 15 for both cases. The second resonance is largely unaffected; we attribute this to the limited actuator bandwidth. Figure 16 shows the periodic data corresponding to the measured strain voltage σ for the three gain values. As with Figure 13, five consecutive periods have been superimposed to form the figures showing that the variability between periods is essentially high-frequency noise. Again, as indicated by the frequency responses, the time responses show damping increasing with gain.

6. Conclusion

The DDVA approach has been shown to provide a novel method to design vibration absorbers in the physical domain. In particular, the method is a natural generalisation of not only the classical single-degree of freedom vibration absorber of Den Hartog (1985, Section 3.3) but also of acceleration feedback (Preumont, 2002). Placing these two well-known design methods into the wider context of the DDVA of this paper has the following advantages: the methods immediately extend to multi degree of freedom systems and the implementation and theoretical results (including robustness to sensor/actuator dynamics) from substructuring can be directly applied.

The DDVA approach has been illustrated using numerical simulations of single mode and multi-mode systems and verified using two experimental systems: a rigid beam with an flexible joint and a flexible cantilever beam. Future work will apply the results to more complex dynamical systems including those with multiple sensor-actuator pairs.

The location of the sensor-actuator pairs has not been considered in this paper even though it certainly affects controllability and observability issues (Balas, 1978). Future work in this area will extend bond graph approaches (for example those of Marquis-Favre & Jardin (2011) and Gawthrop & Rizwi (2011)) to sensor/actuator placement in this context.

In principle, the method is equally applicable to the control nonlinear vibrations where dynamical dual of the nonlinear physical system provides the basis for a nonlinear controller. This is also an area for future work.

Acknowledgements

Peter Gawthrop was a Visiting Research Fellow at The University of Bristol when this work was accomplished; he is now supported by a Professorial Fellowship at the University of Melbourne.

Disclosure statement

No potential conflict of interest was reported by the authors.

Funding

Simon Neild is supported by EPSRC fellowship EP/K005375/1. David Wagg is supported by EPSRC [grant EP/K003836/2].

Notes

1. As the word 'dual' has many meanings, the term dynamically dual is used for the specific meaning of this paper.
2. The bond directions have been changed for this paper to correspond to the usual sign convention for feedback control block diagrams.
3. 'One-port' refers to the single *energy port* associated with force and velocity
4. In the sense that it consumes but does not produce energy.
5. Other models such as the Timoshenko model, as well as non-uniform beams, could similarly be handled using this approach

References

Ali, S. F., & Padhi, R. (2009). Active vibration suppression of non-linear beams using optimal dynamic inversion. *Journal of Systems and Control Engineering*, *223*(5), 657–672.

Anderson, B. D. O., & Vongpanitlerd, S. (2006). *Network analysis and synthesis a modern systems theory approach*. Dover. First published 1973 by Prentice-Hall.

Apkarian, J. (1995). *A comprehensive and modular laboratory for control systems design and implementation*. Markham, Ontario: Quanser Consulting.

Balas, M. J. (1978). Feedback control of flexible systems. *IEEE Transactions on Automatic Control*, *23*(4), 673–679.

Blakeborough, A., Williams, M. S., Darby, A. P., & Williams, D. M. (2001). The development of real-time substructure testing. *Philosophical Transactions of the Royal Society Part A*, *359*, 1869–1891.

Borutzky, W. (2011). *Bond graph modelling of engineering systems: Theory, applications and software support*. New York, NY: Springer.

Bucher, R., & Balemi, S. (2006). Rapid controller prototyping with matlab/simulink and linux. *Control Engineering Practice*, *14*(2), 185–192.

Cellier, F. E. (1991). *Continuous system modelling*. Berlin: Springer-Verlag.

Den Hartog, J. P. (1985). *Mechanical vibrations*. Dover. Reprint of 4th ed. Published by McGraw-Hill 1956.

Desoer, C. A., & Vidyasagar, M. (1975). *Feedback systems: Input-output properties*. London: Academic Press.

Fleming, A. J., & Moheimani, S. O. R. (2005). Control orientated synthesis of high-performance piezoelectric shunt impedances for structural vibration control. *IEEE Transactions on Control Systems Technology*, *13*(1), 98–112.

Gawthrop, P. J. (1995). Physical model-based control: A bond graph approach. *Journal of the Franklin Institute*, *332B*(3), 285–305.

Gawthrop, P. J. (2004). Bond graph based control using virtual actuators. *Proceedings of the Institution of Mechanical Engineers Pt. I: Journal of Systems and Control Engineering*, *218*(4), 251–268.

Gawthrop, P. J., & Bevan, G. P. (2007). Bond-graph modeling: A tutorial introduction for control engineers. *IEEE Control Systems Magazine*, *27*(2), 24–45.

Gawthrop, P. J., & Rizwi, F. (2011). Coaxially coupled inverted pendula: Bond graph-based modelling, design and control. In W. Borutzky (Ed.), *Bond graph modelling of engineering systems* (pp. 179–194). New York, NY: Springer.

Gawthrop, P. J., & Smith, L. P. S. (1996). *Metamodelling: Bond graphs and dynamic systems*. Hemel Hempstead: Prentice Hall.

Gawthrop, P. J., Bhikkaji, B., & Moheimani, S. O. R. (2010). Physical-model-based control of a piezoelectric tube for nano-scale positioning applications. *Mechatronics*, *20*(1), 74–84. Available online 13 October 2009.

Gawthrop, P. J., Wagg, D. J., & Neild, S. A. (2009). Bond graph based control and substructuring. *Simulation Modelling Practice and Theory*, *17*(1), 211–227. Available online 19 November 2007.

Gawthrop, P. J., Wallace, M. I., Neild, S. A., & Wagg, D. J. (2007). Robust real-time substructuring techniques for under-damped systems. *Structural Control and Health Monitoring*, *14*(4), 591–608. Published on-line: 19 May 2006.

Gawthrop, P. J., Wallace, M. I., & Wagg, D. J. (2005). Bond-graph based substructuring of dynamical systems. *Earthquake Engineering & Structural Dynamics*, *34*(6), 687–703.

Hagood, N. W., & von Flotow, A. (1991). Damping of structural vibrations with piezoelectric materials and passive electrical networks. *Journal of Sound and Vibration*, *146*(2), 243–268.

Hogan, N. (1985). Impedance control: An approach to manipulation. part I – theory. *ASME Journal of Dynamic Systems, Measurement and Control*, *107*, 1–7.

Høgsberg, J. R., & Krenk, S. (2006). Linear control strategies for damping of flexible structures. *Journal of Sound and Vibration*, *293*(1–2), 59–77.

Hong, J. H., & Bernstein, D. S. (1998). Bode integral constraints, colocation, and spillover in active noise and vibration control. *IEEE Transactions on Control Systems Technology*, *6*(1), 111–120.

Karnopp, D. (1966). Coupled vibratory-system analysis, using the dual formulation. *The Journal of the Acoustical Society of America*, *40*(2), 380–384.

Karnopp, D., Margolis, D. L., & Rosenberg, R. C. (2000). *System dynamics: Modeling and simulation of mechatronic systems* (3rd ed.). New York, NY: Horizon Publishers and Distributors Inc.

Karnopp, D. C., Margolis, D. L., & Rosenberg, R. C. (2012). *System dynamics: Modeling, simulation, and control of mechatronic systems* (5th ed.). John Wiley & Sons.

Krenk, S., & Hogsberg, J. (2009). Optimal resonant control of flexible structures. *Journal of Sound and Vibration. 323* (3–5), 530–554.

Lozano, R., Brogliato, B., Egelund, O., & Maschke, B. (2000). *Dissipative systems: Analysis and control*. New York, NY: Springer.

Margolis, D. L. (1985). A survey of bond graph modelling for interacting lumped and distributed systems. *Journal of the Franklin Institute*, *319*, 125–135.

Marquis-Favre, W., & Jardin, A. (2011). Bond graphs and inverse modeling for mechatronic system design. In W. Borutzky (Ed.), *Bond graph modelling of engineering systems* (pp. 195–226). New York, NY: Springer.

Moheimani, S. O. R, & Behrens, S. (2004). Multimode piezoelectric shunt damping with a highly resonant impedance. *IEEE Transactions on Control Systems Technology*, *12*(3), 484–491.

Moheimani, S. O. R., & Fleming, A. J. (2006). *Piezoelectric transducers for vibration control and damping*. Advances in Industrial Control. New York, NY: Springer.
Mukherjee, A., Karmaker, R., & Samantaray, A. K. (2006). *Bond graph in modeling, simulation and fault indentification*. New Delhi: I.K. International.
Ortega, R., Loria, A., Nicklasson, P. J., & Sira-Ramirez, H. (1998). *Passivity-based control of Euler-lagrange systems*. London: Springer.
Ortega, R., Praly, L., & Landau, I. D. (1985). Robustness of discrete-time direct adaptive controllers. *IEEE Transactions on Automatic Control, AC-30*(12), 1179–1187.
Ortega, R., van der Schaft, A. J., Mareels, I., & Maschke, B. (2001). Putting energy back in control. *IEEE Control Systems Magazine, 21*(2), 18–33.
Pintelon, R., & Schoukens, J. (2001). *System identification. A frequency domain approach*. New York, NY: IEEE Press.
Preumont, A. (2002). Vibration control of active structures: An introduction, volume 96 of *Solid mechanics and its applications*. Dordrecht: Kluwer.
Samanta, B., & Mukherjee, A. (1985). A bond graph based analysis of coupled vibratory systems taking advantage of the dual formulation. *Journal of the Franklin Institute, 320*, 111–131.
Samanta, B., & Mukherjee, A. (1990). Analysis of acoustoelastic systems using modal bond graphs. *Journal of Dynamic Systems, Measurement, and Control, 112*(1), 108–115.
Sharon, A., Hogan, N., & Hardt, D. E. (1991). Controller design in the physical domain. *Journal of the Franklin Institute, 328*(5–6), 697–721.
Shearer, J. L., Murphy, A. T., & Richardson, H. H. (1971). *Introduction to system dynamics*. Reading, MA: Addison-Wesley.
Slotine, J. E., & Li, W. (1991). *Applied nonlinear control*. Englewood Cliffs: Prentice-Hall.
Vink, D., Ballance, D., & Gawthrop, P. (2006). Bond graphs in model matching control. *Mathematical and Computer Modelling of Dynamical Systems, 12*(2–3), 249–261.
Wagg, D., Neild, S., & Gawthrop, P. (2008). Real-time testing with dynamic substructuring. In O. S. Bursi and D. Wagg (Eds.), *Modern testing techniques for structural systems*, volume 502 of *CISM courses and lectures*, Chapter 7 (pp. 293–342). Wien, NY: Springer.
Willems, J. C. (1972). Dissipative dynamical systems, part I: General theory, part II: Linear system with quadratic supply rates. *Arch. Rational Mechanics and Analysis, 45*(5), 321–351.

Appendix. Derivation of Equation (14)

Equation (14) can be derived directly from the bond graph of Figure 4(a). Letting F and v be the force and velocity at the component interface, letting F_{mr} be the force acting on the mass and damper and F_c the spring force, it follows that the components represented by **I**:**m**, **C**:**k** and **R**:**r** have equations:

$$m\frac{\mathrm{d}v_m}{\mathrm{d}t} = F_{mr}, \tag{A1}$$

$$\frac{\mathrm{d}F_c}{\mathrm{d}t} = kv, \tag{A2}$$

$$v_r = \frac{1}{r}F_{mr}. \tag{A3}$$

Taking Laplace transforms (with zero initial conditions) it follows that:

$$\begin{aligned} v &= v_r + v_m \\ &= \left[\frac{1}{r} + \frac{1}{ms}\right]F_{mr} \\ &= \left[\frac{1}{r} + \frac{1}{ms}\right](F - F_c) \\ &= \left[\frac{1}{r} + \frac{1}{ms}\right]\left[F - \frac{k}{s}v\right] \\ &= \frac{ms+r}{mrs}\left[F - \frac{k}{s}v\right]. \end{aligned} \tag{A4}$$

Collecting terms in Equation (A4) gives:

$$\frac{k(ms+r) + mrs^2}{s(ms+r)}v = F. \tag{A5}$$

Hence, rearranging Equation (A5):

$$\frac{v}{F} = \frac{s(ms+r)}{mrs^2 + kms + kr}. \tag{A6}$$

The right-hand side of Equation (A6) corresponds to the transfer function of Equation (14).

19

Optimal control for a shield machine subject to multi-point earth pressure balance

Kairu Li[a,b] and Cheng Shao[a]*

[a]*School of Control Science and Engineering, Dalian University of Technology, Dalian 116024, People's Republic of China;* [b]*School of Engineering, The University of Warwick, Coventry CV4 7AL, UK*

This paper discusses multi-point earth pressure balance (EPB) optimal control for EPB shield machine with five pressure sensors on its chamber clapboard. The predictive models taking into account advance speed, rotating speed of the cutter head, screw conveyor speed, and the earth pressure measurements are established for five shield chamber pressure points with the adaptive neuro-fuzzy inference system by minimizing the deviation between the corresponding point's predicted pressures and the settings. Then, the ant colony system algorithm is employed to get the optimal screw conveyor speed to control the EPB during the tunnelling process. Simulation results show that the optimal control method gives better performance with small tracking error and fast tracking speed.

Keywords: shield machine; earth pressure balance; adaptive neuro-fuzzy inference system (ANFIS); ACS algorithm

1. Introduction

The earth pressure balance (EPB) shield is a large engineering machine widely applied in underground construction, such as metro tunnel, municipal construction, resources exploitation, hydraulic engineering, and so on. However, the imbalance of earth pressure in front of the tunnel face can lead to unpredictable and serious geological and hydrogeological accidents or even disasters, such as ground subsidence and heave. Thus, the stability of the tunnel face is a critical factor to ensure the safety of an excavating process.

Hongxin and Deming (2007) established a mathematical model for the excavating process based on physical analysis. However, a mathematical–physical model inevitably has some deviation from the practical condition. Then the analysis based on online measurement data and intelligent strategy is introduced in the study of EPB shield control. Cheng and Dongsheng (2012) designed a numerical method based on pressure field surface normal vector to judge the tunnel face stability, and optimal control with guarantee of tunnel face stability was suggested (2014). Yeh (1997) modelled earth pressure using back propagation neural network, which only considers the speed of the shield jack and the screw conveyor. Hu, Guofang, and Huayong (2008) applied adaptive fuzzy artificial neural network to EPB control based on a single point on the chamber pressure field. However, the studies mentioned above ignored other factors' influences on the tunnel face, such as cutter speed and total thrust. Due to the complexity of the pressure distribution surface, only considering the pressure at one point cannot represent the working condition of the whole tunnel face and is hard to ensure the construction safety and stability.

The pressure balance between the tunnel face and the shield chamber is mainly controlled by adjusting the shield's advance speed, the cutter head rotating speed and the speed of the screw conveyor which discharges the excavated soil out of the chamber. Given that the advance speed and cutter head speed need to be designed according to the local geological and hydrogeological conditions, the rotating speed of the screw conveyor is chosen as the control variable.

This paper proposes an optimal EPB control method based on the ant colony system (ACS) algorithm for the screw conveyor rotating speed to keep the chamber pressure following its set values and ensure the stability of the tunnel face. Previous studies on EPB control are generally based on the central pressure point which cannot give the characteristic of pressure distribution on the whole pressure field. In this case, this paper, considering multiple pressure monitoring points, establishes an ANFIS predictive model for each of the five pressure points which facilitates a more comprehensive understanding of the pressure field before applying proper control strategy. Compared with the data of four sensors analysed by Kairu, Yan, and Cheng (2014), a new set of data measured in a practical construction project is introduced. This certain type of EPB shield has five pressure sensors on its chamber clapboard. Additionally, all of the five relative variables, including advance speed, total thrust, cutter speed, screw conveyor

*Corresponding author. Email: cshao@dlut.edu.cn

speed, and the real-time earth pressure, are considered in the ANFIS model. Then, with online measurement data from pressure sensors, the ACS algorithm is used to find the optimal screw conveyor speed, which aims to minimize the difference between the earth pressure in the chamber and the corresponding pressure set value. Finally, simulation results are presented to demonstrate the effectiveness of the proposed method.

2. Optimal control of multi-point EPB in pressure chamber

2.1. Control strategy for multi-point EPB

Generally, the EPB in the pressure chamber is only controlled by monitoring the sensor on the centre horizontal line in the chamber clapboard. In this paper, the control of EPB is based on five sensors on different points on the chamber clapboard. Figure 1 shows the distribution of the five pressure sensors on the chamber clapboard of a certain type of EPB shield which is used in a real tunnel construction project.

When a shield machine excavates a homogeneous and stable soil layer, the EPB in the chamber is controlled by adjusting the advance speed or screw conveyor speed based on the discharge control mode. The cutter speed, total thrust, and advance speed decide the amount of earth charged, and the screw conveyor speed influences the discharge. They directly influence the balance of earth pressure in the chamber.

Based on the consideration above, this paper proposes an optimal control strategy for EPB: firstly, the ANFIS pressure predictive model is established, taking advance speed v_a, total thrust F, cutter speed v_c, screw conveyor speed v_s, and the real-time earth pressure $p_i(k)$ as the inputs and next time earth pressure $p_i(k+1)$ as the output variable; secondly, the optimal output variable (the screw conveyor speed) is obtained by using the ACS algorithm which uses the minimization of the pressure deviation between the real-time pressures and corresponding set values. The control strategy is as shown in Figure 2, where $p_i(k)$ $(i=1, \ldots ,5)$ is the real-time earth pressure, $p_i(k+1) = f(v_s, F, v_a, v_c, p_i(k))$ is the next time earth pressure; and $p_{d1}, p_{d2}, p_{d3} \cdot p_{d4}, p_{d5}$ are the set earth pressures of the points of sensors.

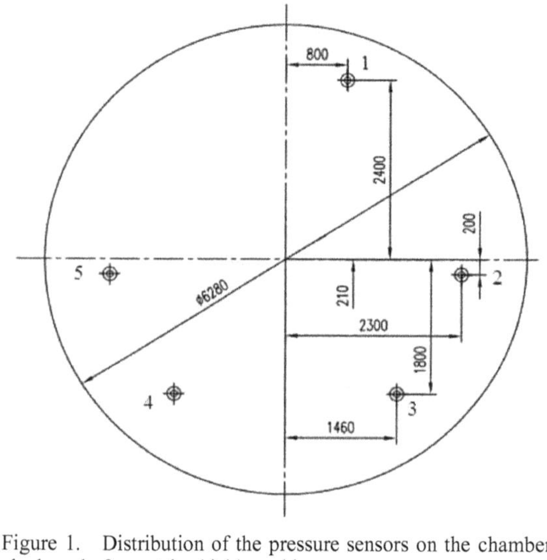

Figure 1. Distribution of the pressure sensors on the chamber clapboard of a certain shield machine.

2.2. Establishment of ANFIS model for chamber earth pressure

Due to the complexity of the tunnelling process, the earth pressure is hard to be accurately predicted based on physical mechanism analysis. Fuzzy inference system (FIS) is a non-linear, strong coupling system. It is an effective method to predict systems that are hard to be modelled. The selection of a fuzzy membership function and the generation of fuzzy rules in the FIS generally depend on expert experience. However, insufficient expert experience and wrong judgements may strongly influence the prediction. Thus, a Takagi-Sugeno-Kang fuzzy inference model (Michio & Kang, 1988; Tomohiro & Michio, 1985) is applied to build an adaptive neuro-fuzzy inference system (ANFIS) for identifying parameters. In the ANFIS system, the learning mechanism of a neural network is used to obtain the fuzzy membership function and fuzzy rules instead of depending on expert experience, which could overcome the deficiency of a fuzzy system and make the model more objective.

The ANFIS model is based on a multilayer feed-forward neural network. Each layer has a certain task, and then transfers information to the next layer.

The first layer receives the data of five input parameters, including total thrust, advance speed, cutter speed, screw conveyor speed, and real-time earth pressure.

The second layer completes the blurring process and transfers the numerical value to fuzzy value, a membership belonging to a certain fuzzy subset.

The third layer and the fourth layer complete the fuzzy inference process together. The third layer completes the rule antecedent; the fourth layer completes the rule consequently and output fuzzy value after fuzzy inference.

The fifth layer completes defuzzification and then output results.

Based on this theory, five 'five-input and single-output' ANFIS systems are obtained. The input values, screw conveyor speed, total thrust, thrust speed, cutter speed, and real-time earth pressure $p_i(k)$ $(i = 1, 2, \ldots, 5)$, are obtained by data training. The outputs, next time earth pressures, are $p_i(k+1)$. $p_{d1}, p_{d2}, p_{d3}, p_{d4}, p_{d5}$ are the set earth pressures of the five sensors on the chamber.

ANFIS is a Sugeno fuzzy non-linear model. It can be used to indicate the dynamic response of complicated

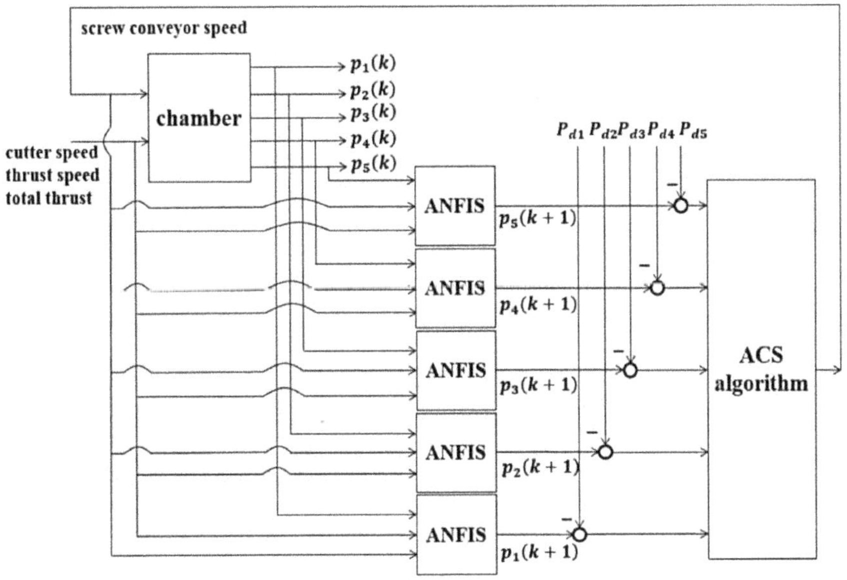

Figure 2. Control strategy of the EPB shield.

systems. The most common fuzzy inference rule is: if v_s is A_1, F is A_2, v_a is A_3, v_c is A_4, $p_i(k)$ is A_5, then $p_i(k+1) = f(v_s, F, v_a, v_c, p_i(k))$, where $A_1, \ldots, A_5$ are fuzzy sets. Finally, $p_i(k+1) = f(v_s, F, v_a, v_c, p_i(k))$, the model of each pressure point is obtained.

2.3. *Optimal control mechanism and solution*

2.3.1. *Establishing the optimization model*

To realize dynamic balance control of the earth pressure, an EPB optimal control strategy is proposed in this paper. The control object is to make the pressure in the chamber track the desired value in the future, which we can achieve through the optimization function

$$\min \sum_{i=1}^{5} |p_i(k+1) - p_{di}|,$$
$$\text{s.t. } V_{S_{\min}} \leq V_S(k) \leq V_{S_{\max}}, \qquad (1)$$

where $V_{S_{\min}}$, $V_{S_{\max}}$ are the minimum and maximum values of screw conveyor speed, respectively, and p_{di} is the set earth pressure of the point of a sensor.

2.3.2. *The solution of the optimization model*

Because the optimization model (1) is smooth, non-convex, and constrained, it is difficult to solve by using the conventional numerical optimization method. However, the ant algorithm can be easily implemented with fewer parameters to adjust, and it is very suitable for solving the complicated non-linear problems and can get the global optimal solution (Haibin, Daobo, & Xiufen, 2005). Therefore, an ACS algorithm is employed to solve the optimization model (1). The specific process for searching the optimal solution is as follows:

2.3.2.1. Formation of the ant moving path. The optimal screw conveyor speed is represented by $V_S(k)$. Because the rotating speed of the screw conveyor is usually less than 22.4 rpm, each of them is represented by three decimal significant figures and, according to actual values, two figures are placed in front of the decimal point and the third one after the decimal point. In order to use the ACS algorithm conveniently, we express the values on the plane. As shown in Figure 3, $x_i (i = 1, 2, 3)$ represents the significant digit of the values of the control variables from the first one to the third one, respectively. y_{ij} $(i = 0, 1, \ldots, 9)$ represents the possible value of each significant digit. Especially, taking the dimension parameters of the shield machine into consideration, the screw conveyor rotational speed is usually less than 22.4 rpm, which means that the value of the first digit of each control value may be 0, 1, or 2; therefore, there are only 3 nodes on line L_i as shown in Figure 3. We use (x_i, y_{ij}) to denote the node j on line L_i, whose value is y_{ij}. Let an ant depart from the starting point 0. In each step forward, it chooses a node from the next line L_i and then moves to this node along the straight line. The ant travels through in turn L_1, L_2, L_3; when it arrives at any node on line L_3, it completes one tour which is also called a cycle. Its moving path is expressed as

$$\text{Path } k = \{0, (x_1, y_{1j}), (x_2, y_{2j}), (x_3, y_{3j})\}.$$

Obviously, the values of the control variable on this path can be computed according to the following formula:

$$V_S(k) = y_{1j} \times 10^1 + y_{2j} \times 10^0 + y_{3j} \times 10^{-1}. \qquad (2)$$

2.3.2.2. Selecting nodes on a moving path. Ants decide their direction according to the pheromone intensity on each node. $p_{ij}^{k}(t)$ denotes the state transition probability from position i to f for ant k at time t:

$$p^{k}(x_i, y_{ij}, t) = \begin{cases} \dfrac{[\tau(x_i, y_{ij}, t)]^{\alpha}}{\sum_{w \in \text{allowed}_k} [\tau(x_i, y_{iw}, t)]^{\alpha}} & \text{if } j \in \text{allowed}_k, \\ 0 & \text{else}, \end{cases} \tag{3}$$

where allowed_k is the set of nodes that are allowed to choose in the next step for ant k, and α is a parameter that weighs the relative importance of the pheromone values.

2.3.2.3. Updating the pheromone intensity. After n steps, each ant has travelled all lines and completed a cycle. The pheromone intensity of each node on the best tour is updated according to the following formula:

$$\tau(x_i, y_{ij}, t+n) = (1-\rho)\tau(x_i, y_{ij}, t) + \Delta\tau(x_i, y_{ij}, t+n), \tag{4}$$

where $\rho(0 < \rho < 1)$ is the evaporation factor, $\tau(x_i, y_{ij}, 0) = \tau_0$, and $\Delta\tau$ is the increment of pheromone on the node (x_i, y_{ij}), which is calculated in term of the following rule:

$$\Delta\tau(x_i, y_{ij}, t+n) = \begin{cases} \dfrac{Q}{F_k} & \text{if } (x_i, y_{ij}) \text{ belongs to the best tour}, \\ 0 & \text{else}, \end{cases}$$

where Q is a constant and F_k is the objective function value corresponding to the optimal path done by the best ant k. To avoid the algorithm converging to a non-global optimization solution prematurely, the pheromone intensity on each path is limited in $[\tau_{\min}, \tau_{\max}]$.

The procedures of the ACS algorithm used in this paper are summarized as follows:

Step 1: Establish a free solution space for the ant searching the path as shown in Figure 3, and define a one-dimensional array

$$\text{Path } k = \{0, (x_1, y_{1j}), (x_2, y_{2j}), (x_3, y_{3j})\}$$

with four elements to store the ordinate values of the nodes that the ant k passes through in a tour.

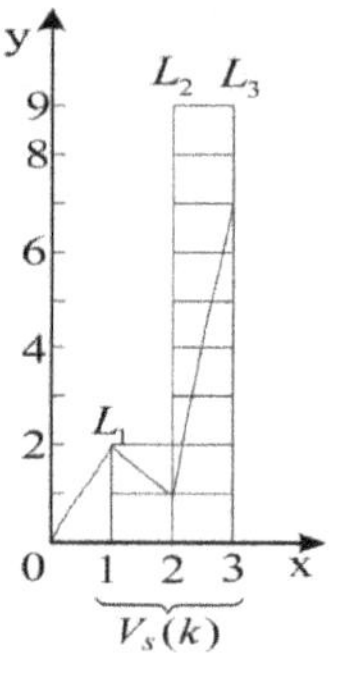

Figure 3. The searching path of the ants.

Step 2: Define the number of ants m and the maximum number of iterations $NC_{\max}$; set time counter $t = 0$ and the number of cycle $N = 0$; specify the values of parameters $\alpha, \rho, \tau_0, \tau_{\min}, \tau_{\max}$, and Q; and then place all the m ants at the start point 0.

Step 3: Set $i = 0, k = 1$.

Step 4: Ant k selects a node on line L_i to move according to Equations (3–5) and the path rule shown in Figure 3; then, store the ordinate value y_{ij} of this node into the ith element of Path k.

Step 5: Set $k = k + 1$, if $k \leq m$, then go to step 4; otherwise, continue.

Step 6: Set $i = i + 1$, if $i \leq 3$, then go to step 4; otherwise, continue.

Step 7: According to the array path k of each ant k, we can obtain the ordinate values y_{ij} of the nodes that it passes through, and compute the values of $V_S(k)$ by using Equation (2).

Step 8: For all the m ants, compute the objective function values according to Equation (1), find the minimum F_k which corresponds to the optimal path done by the best ant k, and then update the pheromone intensity according to Equation (4).

Step 9: Define path_b as the optimal path till the previous iteration $N - 1$, compare the path_b with the optimal path k obtained at this iteration to N and find the optimal one, and then save the nodes of the optimal path into path_b.

Step 10: Set each element of path k to 0, and $t = t + d, N = N + 1$; if $N > NC_{\max}$ or m ants make the same tour, the iteration is over, then output the optimal path path_b and its corresponding control value $V_S(k)$; otherwise, return to step 3.

3. Simulation

3.1. Simulation of the ANFIS model

The simulation is based on a set of practical excavation data measured from an EPB shield of a construction project. A total of 100 groups of monitoring data are taken as the training data, and the other 100 groups of data are taken as the testing data. Then sub-clustering is used

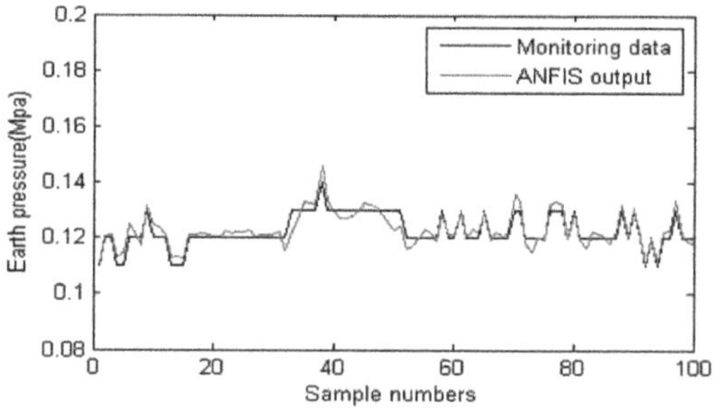

Figure 4. P_1's pressure contrast of monitoring data and ANFIS output.

to initialize FIS, using Gauss subordination function. The model is trained for 10 times based on a hybrid algorithm. As in Figures 4–8, the black curves represent the real-time monitoring data, while the blue curves show the model outputs of the five points' earth pressures, and the mean square errors of model outputs are detailed in Table 1. It is observed that the ANFIS model has small errors, which indicate a high precision for the proposed model.

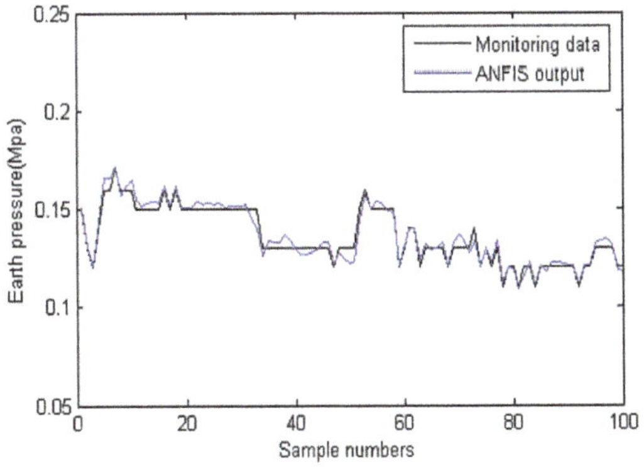

Figure 5. P_2's pressure contrast of monitoring data and ANFIS output.

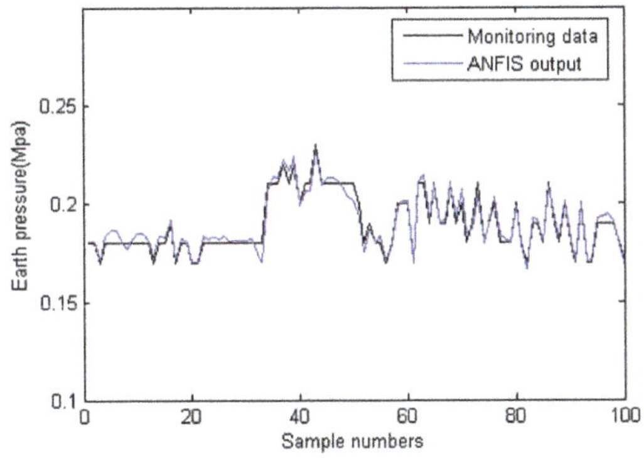

Figure 6. P_3's pressure contrast of monitoring data and ANFIS output.

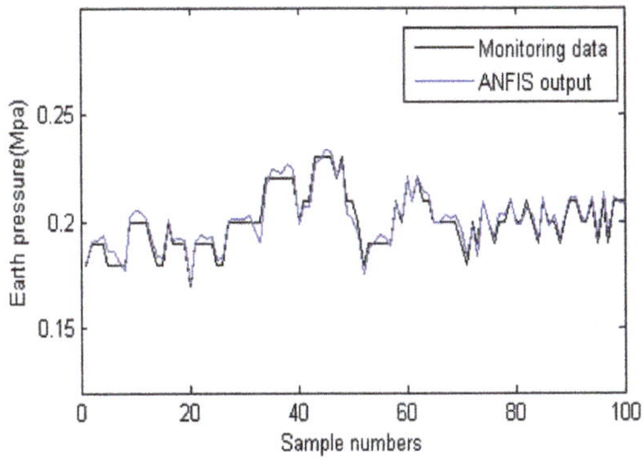

Figure 7. P_4's pressure contrast of monitoring data and ANFIS output.

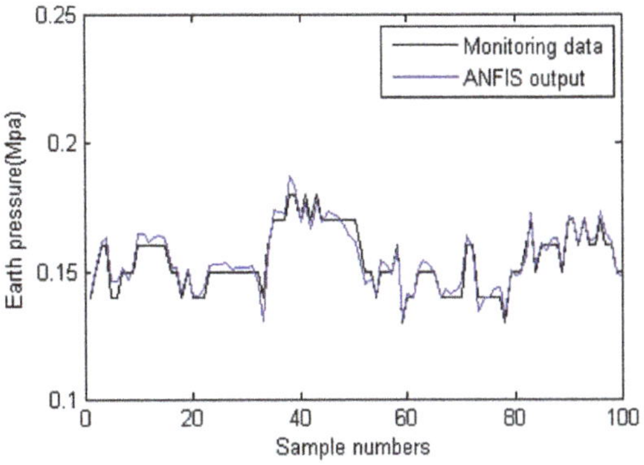

Figure 8. P_5's pressure contrast of monitoring data and ANFIS output.

Table 1. The mean square error of ANFIS model's output.

Pressure point	Mean square error
P_1	0.004636
P_2	0.004867
P_3	0.005158
P_4	0.005039
P_5	0.004723

3.2. *Effect simulation of optimal control algorithm*

A certain type of EPB shield is used to train and test the control model in order to verify the effectiveness of the control method. The diameter of the shield cutter is 9.75 m and the tunnel length is 3.065 km. The first 50 groups of data are obtained when tunnelling in sand pebble. Correspondingly, the set pressures of each point are as follows: $p_{d1} = 0.07\,\text{MPa}, p_{d2} = p_{d5} = 0.10\,\text{MPa}$,

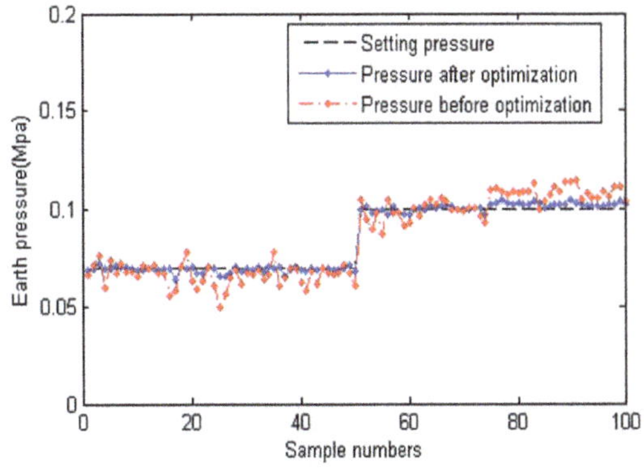

Figure 9. The contrast of P_1's earth pressure value before and after optimization.

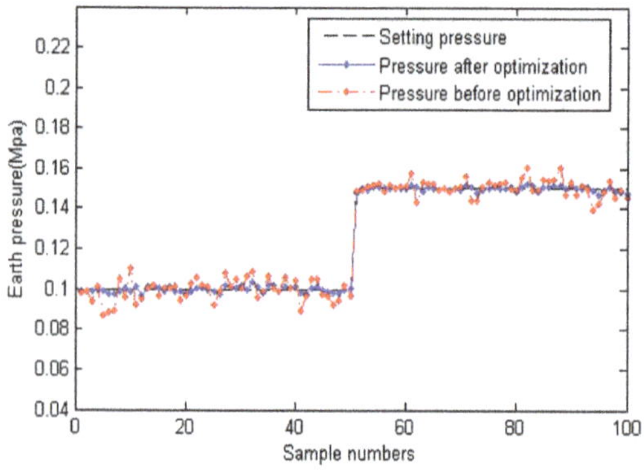

Figure 10. The contrast of P_2's earth pressure value before and after optimization.

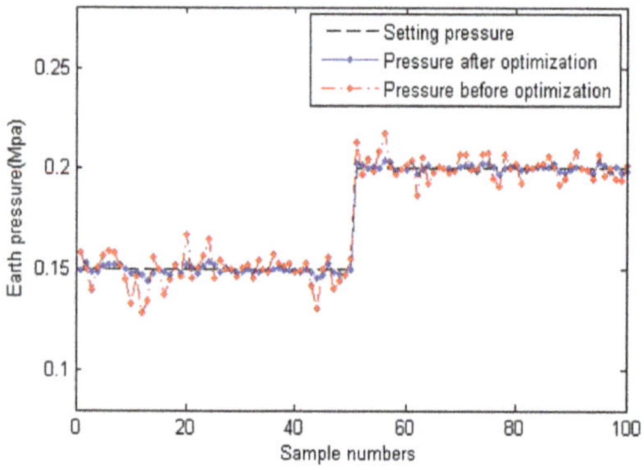

Figure 11. The contrast of P_3's earth pressure value before and after optimization.

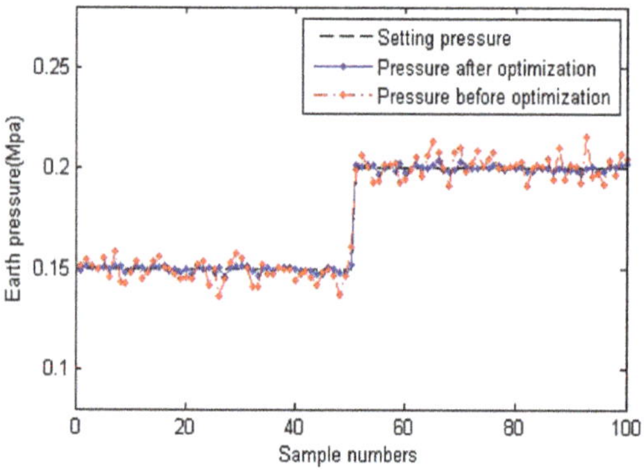

Figure 12. The contrast of P_4's earth pressure value before and after optimization.

$p_{d3} = p_{d4} = 0.15$ MPa. The second 50 groups of data are obtained when tunnelling in silty clay; in this condition, the set pressures of each point are as follows: $p_{d1} = 0.10$ MPa, $p_{d2} = p_{d5} = 0.15$ MPa, $p_{d3} = p_{d4} = 0.20$ MPa. The control simulation results as in Figures 9–13 show that the method proposed in this paper has better control effect than

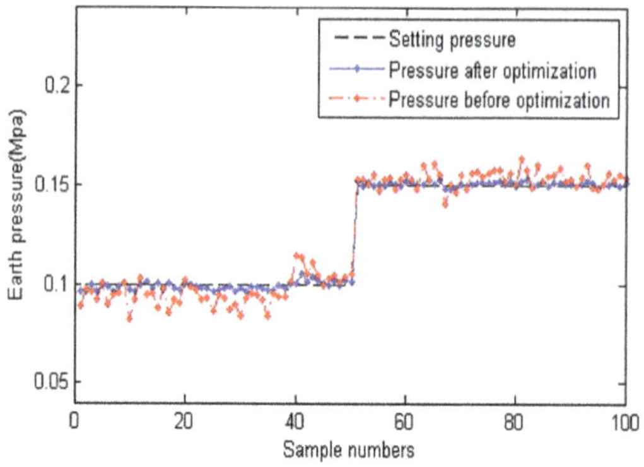

Figure 13. The contrast of P_5's earth pressure value before and after optimization.

that without optimization. It is confirmed that the presented method has good tracking performance even when the geological condition is changed. Through simulation analysis, the effectiveness of the presented method is verified.

4. Conclusions and future work

The five-point pressure data of an EPB shield chamber is used to learn and train the parameters of fitness function and fuzzy rules of the fuzzy system by means of a hybrid algorithm in ANFIS, and then the five points' earth pressure prediction models are established. Later, an optimal control strategy of multi-point EPB for EPB shield is presented. The optimization function is established and solved by ACS algorithms. The simulation results show that the prediction model established by ANFIS has small errors and can achieve high precision, and the optimal control method has better control performance even if the geological conditions are changed, which indicates that the method has better tracking control performance than that without optimization.

Currently, only the discharge system is controlled in this paper to obtain the balance of an earth pressure field. However, the pressure balance is also influenced by the thrust system and cutter head system. Although the advance speed and the cutter head speed are usually decided based on the local geological and hydrogeological conditions, a coordinating optimization control of all the three systems with an integration of a geological identification system could be researched in the future to obtain a more accurate control performance of the tunnelling process.

Disclosure statement

No potential conflict of interest was reported by the authors.

Funding

This work was supported by National Natural Science Foundation of China [No. 61074020].

References

Cheng, S., & Dongsheng, L. (2012). Tunnel face stability analysis based on pressure field surface's normal vector for earth pressure balanced shield. *International Journal of Modelling, Identification and Control*, *17*(2), 143–150.

Cheng, S., & Dongsheng, L. (2014). Optimal control of an earth pressure balance shield with tunnel face stability. *Automation in Construction*, *46*(10), 22–29.

Haibin, D., Daobo, W., & Xiufen, Y. (2005). Research on the optimum configuration strategy for the adjustable parameters in ant colony algorithm. *Journal of Communication and Computer*, *2*(9), 32–35.

Hongxin, W., & Deming, F. (2007). Theoretical and test studies on balance control of EPB shields. *China Civil Engineering Journal (in Chinese)*, *40*(5), 61–68.

Hu, S., Guofang, G., & Huayong, Y. (2008). Control model of EPB shield machine. *Journal of Coal Science & Engineering (in Chinese)*, *33*(3), 343–346.

Kairu, L., Yan, Z., & Cheng, S. (2014). *Earth pressure multipoint forecasts and optimal control for EPB shield*. The 20th IEEE International Conference on Automation and Computing (ICAC), 272–276.

Michio, S., & Kang, G. T. (1988). Structure identification of fuzzy model. *Fuzzy sets and systems*, *28*(1), 15–33.

Tomohiro, T., & Michio, S. (1985). Fuzzy identification of systems and its applications to modeling and control. *IEEE Transactions on Systems, Man and Cybernetics*, *15*(1), 116–132.

Yeh, I.-C. (1997). Application of neural networks to automatic soil pressure balance control for shield tunneling. *Automation in Construction*, *5*(5), 421–426.

Permissions

All chapters in this book were first published in SSCR, by Taylor & Francis Online; hereby published with permission under the Creative Commons Attribution License or equivalent. Every chapter published in this book has been scrutinized by our experts. Their significance has been extensively debated. The topics covered herein carry significant findings which will fuel the growth of the discipline. They may even be implemented as practical applications or may be referred to as a beginning point for another development.

The contributors of this book come from diverse backgrounds, making this book a truly international effort. This book will bring forth new frontiers with its revolutionizing research information and detailed analysis of the nascent developments around the world.

We would like to thank all the contributing authors for lending their expertise to make the book truly unique. They have played a crucial role in the development of this book. Without their invaluable contributions this book wouldn't have been possible. They have made vital efforts to compile up to date information on the varied aspects of this subject to make this book a valuable addition to the collection of many professionals and students.

This book was conceptualized with the vision of imparting up-to-date information and advanced data in this field. To ensure the same, a matchless editorial board was set up. Every individual on the board went through rigorous rounds of assessment to prove their worth. After which they invested a large part of their time researching and compiling the most relevant data for our readers.

The editorial board has been involved in producing this book since its inception. They have spent rigorous hours researching and exploring the diverse topics which have resulted in the successful publishing of this book. They have passed on their knowledge of decades through this book. To expedite this challenging task, the publisher supported the team at every step. A small team of assistant editors was also appointed to further simplify the editing procedure and attain best results for the readers.

Apart from the editorial board, the designing team has also invested a significant amount of their time in understanding the subject and creating the most relevant covers. They scrutinized every image to scout for the most suitable representation of the subject and create an appropriate cover for the book.

The publishing team has been an ardent support to the editorial, designing and production team. Their endless efforts to recruit the best for this project, has resulted in the accomplishment of this book. They are a veteran in the field of academics and their pool of knowledge is as vast as their experience in printing. Their expertise and guidance has proved useful at every step. Their uncompromising quality standards have made this book an exceptional effort. Their encouragement from time to time has been an inspiration for everyone.

The publisher and the editorial board hope that this book will prove to be a valuable piece of knowledge for researchers, students, practitioners and scholars across the globe.

List of Contributors

Pradeep Kumar Biswal and Santosh Biswas
Department of Computer Science and Engineering, Indian Institute of Technology, Guwahati, India

Ping-Min Hsu
Automotive Research & Testing Center, Lukang, Changhua 50544, Taiwan

Chun-Liang Lin
Department of Electrical Engineering, National Chung Hsing University, Taichung 402, Taiwan

Yanhui Li, Chao Zhang and Xiujie Zhou
College of Electrical and Information Engineering, Northeast Petroleum University, Daqing, Heilongjiang Province, 163318, People's Republic of China

Thomas Stockley, Kary Thanapalan, Mark Bowkett and Jonathan Williams
Faculty of Computer, Engineering and Sciences, University of South Wales, Pontypridd, Wales

Rui Yan, Tao Liu, Fengwei Chen and Shijian Dong
Institute of Advanced Control Technology, Dalian University of Technology, Dalian 116024, People's Republic of China

G.-L. Osorio-Gordillo
Tecnológico Nacional de México, Centro Nacional de Investigación y Desarrollo Tecnológico, CENIDET, Interior Internado Palmira S/N, Col. Palmira, 62490 Cuernavaca, Mor. Mexico
CRAN-CNRS (UMR 7039), Université de Lorraine, IUT de Longwy, 186, Rue de Lorraine, 54400 Cosnes et Romain, France

M. Darouach and L. Boutat-Baddas
CRAN-CNRS (UMR 7039), Université de Lorraine, IUT de Longwy, 186, Rue de Lorraine, 54400 Cosnes et Romain, France

C.-M. Astorga-Zaragoza
Tecnológico Nacional de México, Centro Nacional de Investigación y Desarrollo Tecnológico, CENIDET, Interior Internado Palmira S/N, Col. Palmira, 62490 Cuernavaca, Mor. Mexico

Ricardo J. Mantz
Laboratorio de Electrónica Industrial, Control e Instrumentación (LEICI), FI. Universidad Nacional de La Plata UNLP., CC 91., La Plata 1900, Argentina
Comisión de Investigaciones Científicas de la Provincia de Buenos Aires (CIC), Argentina

Jigui Jian and Zhihua Zhao
College of Science, China Three Gorges University, Yichang, Hubei 443002, People's Republic of China

Abdo Abou Jaoude
Department of Mathematics and Statistics, Faculty of Natural and Applied Sciences, Notre Dame University - Louaize, Zouk Mosbeh, Lebanon

Mohammed Jamal Alden and Xin Wang
Department of Electrical and Computer Engineering, Southern Illinois University Edwardsville, Edwardsville, IL 62026, USA

Yoni Mandel
Core Photonics, Tel Aviv 6971035, Israel

George Weiss
School of EE, Tel Aviv University, Tel Aviv 6997801, Israel

Y.V. Venkatesh
Electrical Sciences Division, Indian Institute of Science, Bangalore, India
Department of ECE, National University of Singapore, Singapore

Shoulin Hao, Tao Liu, Ximing Sun and Chongquan Zhong
Institute of Advanced Control Technology, Dalian University of Technology, Dalian 116024, People's Republic of China

Jie Zhang
School of Chemical Engineering and Advanced Materials, Newcastle University, Newcastle upon Tyne NE1 7RU, UK

Afef Fekih and Shankar Seelem
Department of Electrical and Computer Engineering, University of Louisiana at Lafayette, Lafayette, LA, USA

Carsten Knoll and Klaus Röbenack
Faculty of Electrical and Computer Engineering, Institute of Control Theory, Technische Universität Dresden, Dresden, Germany

Abdelkarim M. Ertiame, Dingli Yu and J.B. Gomm
Control System Research Group, School of Engineering, Liverpool John Moores University, Byrom Street, Liverpool L3 3AF, UK

Feng Yu
School of Electronic Information, Changchun Architecture &Civil Engineering College, Changchun, Jilin Province, People's Republic of China

Maryam Sarboland
Department of Mathematics, Saveh Branch, Islamic Azad University, PO Box 39187-366, Saveh, Iran

Azim Aminataei
Faculty of Mathematics, Department of Applied Mathematics, K. N. Toosi University of Technology, PO Box 16315-1618, Tehran, Iran

Chaman Singh
Department of Mathematics, Acharya Narendra Dev College (University of Delhi), New Delhi - 19, India

S.R. Singh
Department of Mathematics, D.N. College, Meerut, Uttar Pradesh, India

Kairu Li
School of Control Science and Engineering, Dalian University of Technology, Dalian 116024, People's Republic of China
School of Engineering, The University of Warwick, Coventry CV4 7AL, UK

Cheng Shao
School of Control Science and Engineering, Dalian University of Technology, Dalian 116024, People's Republic of China

Index

www.ingramcontent.com/pod-product-compliance
Lightning Source LLC
Chambersburg PA
CBHW061950190326
41458CB00009B/2837
* 9 7 8 1 6 3 2 3 8 5 9 3 2 *